Preparative Chromatography for Separation of Proteins

Wiley Series in Biotechnology and Bioengineering

Significant advancements in the fields of biology, chemistry, and related disciplines have led to a barrage of major accomplishments in the field of biotechnology. **Wiley Series in Biotechnology and Bioengineering** focuses on showcasing these advances in the form of timely, cutting-edge textbooks and reference books that provide a thorough treatment of each respective topic.

Topics of interest to this series include, but are not limited to, protein expression and processing; nanotechnology; molecular engineering and computational biology; environmental sciences; food biotechnology, genomics, proteomics and metabolomics; large-scale manufacturing and commercialization of human therapeutics; biomaterials and biosensors; and regenerative medicine. We expect these publications to be of significant interest to the practitioners both in academia and industry. Authors and editors were carefully selected for their recognized expertise and their contributions to the various and far-reaching fields of biotechnology.

Preparative Chromatography for Separation of Proteins
by Arne Staby, Anurag S. Rathore, Satinder (Sut) Ahuja

Vaccine Development and Manufacturing
by Emily P. Wen (Editor), Ronald Ellis (Editor), Narahari S. Pujar (Editor)

Risk Management Applications in Pharmaceutical and Biopharmaceutical Manufacturing
by Hamid Mollah (Editor), Harold Baseman (Editor), Mike Long (Editor)

Emerging Cancer Therapy: Microbial Approaches and Biotechnological Tools
by Arsenio Fialho (Editor), Ananda Chakrabarty (Editor)

Quality by Design for Biopharmaceuticals: Principles and Case Studies
by Anurag S. Rathore (Editor), Rohin Mhatre (Editor)

Preparative Chromatography for Separation of Proteins

Edited by
Arne Staby, Anurag S. Rathore, and Satinder Ahuja

WILEY

Published by John Wiley & Sons, Inc., Hoboken, New Jersey

Published simultaneously in Canada

Library of Congress Cataloging-in-Publication Data:

Names: Staby, Arne, 1964– editor. | Rathore, Anurag S. (Anurag Singh), 1973– editor. | Ahuja, Satinder, 1933– editor.
Title: Preparative chromatography for separation of proteins / edited by Arne Staby, Anurag S. Rathore, Satinder Ahuja.
Description: Hoboken, New Jersey : John Wiley & Sons Inc., [2017] | Series: Wiley series in biotechnology and bioengineering | Includes bibliographical references and index. Identifiers: LCCN 2016042817 | ISBN 9781119031109 (hardback)
Subjects: LCSH: Chromatographic analysis. | Separation (Technology) | BISAC: TECHNOLOGY & ENGINEERING / Chemical & Biochemical. | SCIENCE / Life Sciences / Biochemistry. | SCIENCE / Chemistry / Analytic.
Classification: LCC QD272.C4 P75 2017 | DDC 572/.64028–dc23
LC record available at https://lccn.loc.gov/2016042817

Set in 10/12pt Warnock by SPi Global, Pondicherry, India

Cover Design: Wiley
Cover Image Credit: Courtesy of L. Sejergaard, H. Ahmadian, T.B. Hansen, A. Staby, and E.B. Hansen.

Printed in the United States of America

10 9 8 7 6 5 4 3 2 1

Contents

List of Contributors

Haleh Ahmadian
Protein Purification Technology,
Novo Nordisk A/S, Måløv, Denmark

Satinder Ahuja
Ahuja Consulting, Calabash, NC,
USA

Niklas Andersson
Chemical Engineering, Lund
University, Lund, Sweden

Chris Antoniou
Biogen, Cambridge, MA, USA

Suvrajit Banerjee
Department of Chemical and
Biological Engineering, Rensselaer
Polytechnic Institute, Troy,
NY, USA

Marc Bisschops
Pall Corporation, Medemblik, the
Netherlands

Martin P. Breil
DAPI Modelling & Optimization,
Novo Nordisk A/S, Bagsværd,
Denmark

Nina Brestrich
Institute of Engineering in Life
Sciences, Section IV: Biomolecular
Separation Engineering, Karlsruhe
Institute for Technology, Karlsruhe,
Germany

Mark Brower
Merck & Co., Inc., Kenilworth, NJ,
USA

Steven Cramer
Department of Chemical and
Biological Engineering, Rensselaer
Polytechnic Institute, Troy,
NY, USA

Patrick Diederich
Peptide & Protein Purification,
Novo Nordisk A/S, Bagsværd,
Denmark

Søren S. Frederiksen
Mathematical Modelling, Novo
Nordisk A/S, Bagsværd, Denmark

Ernst B. Hansen
Mathematical Modelling, Novo
Nordisk A/S, Bagsværd, Denmark

Thomas B. Hansen
Mathematical Modelling, Novo Nordisk A/S, Bagsværd, Denmark

Sigrid Hansen
Protein Purification Development, Novo Nordisk A/S, Gentofte, Denmark

Milton T. W. Hearn
Victorian Centre for Sustainable Chemical Manufacturing, Monash University, Melbourne, Australia

Jürgen Hubbuch
Institute of Engineering in Life Sciences, Section IV: Biomolecular Separation Engineering, Karlsruhe Institute for Technology, Karlsruhe, Germany

Stephen Hunt
KBI Biopharma, Boulder, CO, USA

Francis Insaidoo
Biologics and Vaccines, Process Development and Engineering, Merck Research Labs, Kenilworth, NJ, USA

Steffen Kidal
Mathematical Modelling, Novo Nordisk A/S, Bagsværd, Denmark

Trent Larsen
KBI Biopharma, Boulder, CO, USA

Hong Li
Process Development and Engineering, Merck, Kenilworth, NJ, USA

Justin McCue
Biogen, Cambridge, MA, USA

Jørgen M. Mollerup
PrepChrom, Klampenborg, Denmark

Venkatesh Natarajan
Biogen, Cambridge, MA, USA

Bernt Nilsson
Chemical Engineering, Lund University, Lund, Sweden

Marcel Ottens
Department of Biotechnology, Delft University of Technology, Delft, the Netherlands

Silvia M. Pirrung
Department of Biotechnology, Delft University of Technology, Delft, the Netherlands

Jennifer Pollard
Process Development and Engineering, Merck, Kenilworth, NJ, USA

Tom Ransohoff
BioProcess Technology Consultants, Woburn, MA, USA

Anurag S. Rathore
Department of Chemical Engineering, Indian Institute of Technology, New Delhi, India

Frank Riske
BioProcess Technology Consultants, Woburn, MA, USA

David Roush
Biologics and Vaccines, Process Development and Engineering, Merck Research Labs, Kenilworth, NJ, USA

Lars Sejergaard
Mathematical Modelling, Novo Nordisk A/S, Bagsværd, Denmark

Sumit K. Singh
Department of Chemical Engineering, Indian Institute of Technology, New Delhi, India

Arne Staby
CMC Project Planning & Management, Novo Nordisk A/S, Bagsværd, Denmark

Jörg Thömmes
Biogen, Cambridge, MA, USA

Robert J. Todd
KBI Biopharma, Boulder, CO, USA

Nihal Tugcu
Process Development and Engineering, Merck, Kenilworth, NJ, USA

Shuichi Yamamoto
Bio-Process Engineering Laboratory, Graduate School of Medicine and School of Engineering, Biomedical Engineering Center (YUBEC), Yamaguchi University, Ube, Japan

Noriko Yoshimoto
Bio-Process Engineering Laboratory, Graduate School of Medicine and School of Engineering, Biomedical Engineering Center (YUBEC), Yamaguchi University, Ube, Japan

Qing Sarah Yuan
Biogen, Cambridge, MA, USA

Series Preface

The upcoming volumes will attest to the importance and quality of books in this series. I would like to acknowledge the fellow coeditors and authors of these books for their agreement to participate in this endeavor. Lastly, I would like to thank Ms. Anita Lekhwani, Senior Acquisitions Editor at John Wiley and Sons, Inc., for approaching me to develop such a series. Together, we are confident that these books will be useful additions to the literature that will not only serve the biotechnology community with sound scientific knowledge but also inspire as they further chart the course of this exciting field.

Anurag S. Rathore
Department of Chemical Engineering
Indian Institute of Technology
New Delhi, India

Preface

This book covers various aspects of preparative chromatography, with a unique combination of academic research and industrial applications. We expect it to appeal to those in academia and industry who are involved in process development and the production of peptides and proteins, an area where the industry is typically reluctant to publicly share their knowledge because of trade secret considerations. Most of these major developments have either not been disclosed at all or exist only as oral conference contributions. This book aims to alleviate some of these gaps as we aim to supplement the academic contributions with industrial contributions. This aspect makes the treatment quite novel and unique when compared with other texts on the topic.

The book is divided into two parts: basic modeling and reviews and industrial separations/case studies. The basic modeling section aims to describe the recent developments in chromatographic theory and general approaches to research to provide increased understanding of the fundamentals behind chromatographic separation and behavior of proteins in these environments. The aim of this section is to provide a solid background in the theory of chromatography to the readers and to better prepare them for industrial case studies. Topics covered comprise the application of various approaches of modeling including computer simulations and mechanistic modeling. Chapter 1, by the editors, is designated to the general background for use of the various modeling tools and approaches.

The first section of the book contains fundamental contributions, general overviews, and reviews. Chapter 2, by Mollerup, provides a general and thorough overview of the thermodynamic tools and isotherm description necessary to model process chromatography in a double chapter. The author proposes approaches for acquiring accurate experimental data from which the model parameters in the adsorption isotherms can be estimated, in order to facilitate the use of simulation tools to the design and optimization of a chromatographic separation process.

Simulation of the performance of chromatographic separation of proteins is a powerful tool, and Chapter 3, by Nilsson and Andersson, presents a summary of the many methodologies applied to various chromatographic techniques including ion exchange, affinity, and multimodal chromatography. Predictions of chromatographic behavior have been presented for a set of different separation problems, illustrating that a large number of common protein separation problems can be simulated quite easily with today's technology.

Chapter 4, by Yoshimoto and Yamamoto, describes simplified methods for understanding and designing chromatography processes for proteins and other biological products, with a focus on modeling of gradient elution chromatography. Simplified models based on the mechanistic model for linear gradient elution chromatography of proteins and other large molecule biological products are presented, together with several applications of the models to process design and process understanding and for bio-recognition.

Continuous processing, including chromatography, has gained much attention the last decade, and Chapter 5, by Riske and Ransohoff, presents industrial application of such multicolumn chromatography (MCC) systems for general capture. The authors suggest that the appropriateness and use of MCC in capture steps and in other parts of the downstream process depend on a number of factors, including the molecular characteristics and stability of the target molecule, the feed titer and product amount required, and the facility design and intention (multipurpose or dedicated). As industry gains more experience with MCC and other forms of continuous processing, the authors foresee that MCC is likely to be more commonly used throughout industry.

Molecular dynamics (MD) is another area that is getting much attention in recent years, and this approach will undoubtedly be key to better understanding of interactions on the molecular level and will ultimately result in better mechanistic models. This topic is described with case studies in Chapter 6, by Insaidoo, Banerjee, Roush, and Cramer. The authors summarize the current state of computational biophysics for determination of individual contributions of key interactions at an atomistic level. They conclude that there remains a significant gap in the linkage of experimental techniques (typically macroscopic) to biophysical modeling and that it is essential that these gaps be closed in order to realize the potential for rational process design.

Chapter 7, by Hansen, teaches the upscaling technique based on volumetric flow rate, which is founded in well-known chromatographic theory and equations, and the approach provides high process design flexibility. The chapter presents an overview of the underlying theory and also provides several examples of successful scale-ups on ion exchange and reversed-phase chromatography. A couple of industrial case studies related to these scale-ups are also presented. Finally, a step-by-step guide for scale-up is presented together with recommendations and a discussion of the challenges that a practitioner is likely to face.

The industrial separations section presents new and existing chromatographic unit operations and discusses how mechanistic and empirical modeling approaches are used to optimize equipment and methodologies. Equipment includes column hardware, scale-down equipment, continuous operation mode, etc., as well as tools for monitoring and control; for example, on-, in-, and at-line equipment for improved process development and manufacturing methods. Improved methodologies comprise scaling approaches, the use of models for validation, uncertainty and robustness evaluations, and process design. A mix of industrial, equipment vendor, and academic authors contributed to this section. Chapter 8, by Antoniou, McCue, Natarajan, Thömmes, and Yuan, provides a number of examples where modeling may help in scale-up of chromatography in industry and how computational fluid dynamics (CFD) has been applied. The authors explore why column packing is such an important criterion that has to be consistent across scales, and they discuss how models can be utilized to predict column packing across scales and to perform packing consistently in an industrial environment.

Chapters 9, 10, and 11 (by Pirrung and Ottens; Diederich and Hubbuch; and Li, Pollard, and Tugcu, respectively) present industrial applications of process development, optimization, and small-scale practice. Chapter 9, among others, demonstrates the use of the high-throughput process development (HTPD) setup to generate mechanistic model parameters for process development, optimization, and design. The authors have discussed the pros and cons of the various experimental approaches, including the one-factor-at-a-time (OFAT), design of experiments (DOE), mechanistic modeling, and hybrid approaches. Chapter 10 provides guidance to process development using robot systems, including modeling/simulation of peak shapes for mechanistic modeling and validation. Factors that have been examined include the influence of pipetting precision, absorption measurements in microtiter plates, peak fractionation, flow patterns, and salt step heights in gradient elution experiments. Separate and combined effects have been qualitatively and quantitatively investigated using both experiments and simulations based on a mechanistic model. The authors demonstrate that with a sufficient number of fractions collected per peak, a significant improvement in precision can be obtained despite low analytical precision. Finally, Chapter 11, focuses on DOE and OFAT in an HTPD setup and presents the state-of-the-art experimental process development approach. A methodology for lab-scale chromatography process development utilizing high-throughput tools in conjunction with traditional column-based methodologies has been presented. The proposed experimental plan for process development relies heavily on a DOE approach supplemented with OFAT experiments. It fully utilizes HTPD and transitions into lab-scale column experiments where additional confirmation is required for defining parameter ranges and scale-up.

Chapters 12, 13, and 14 (by Breil, Frederiksen, Kidal, and Hansen; Hunt, Larsen, and Todd; and Sejergaard, Ahmadian, T.B. Hansen, Staby, and E.B. Hansen, respectively,) present three industrial case studies of mechanistic modeling for use in-process development, optimization, challenge, and identification of critical process parameters, troubleshooting, deviation handling, control strategy setup, and establishing a design space for chromatographic purification. Also included are equation systems and computer coding that may help new applicants in setting up models. Chapter 13 presents an example where the general rate model has been used to describe transport behavior in the column and in the beads and the steric mass action binding model to describe protein binding to the resin matrix. This approach has been used successfully to describe the primary mechanisms involved in cation exchange chromatography of proteins. An open-source chromatography solver was used to estimate model parameters and evaluate the impact of operating parameters on process performance. Model parameters were estimated by performing a set of specific model calibration experiments. Pulse injection experiments were used to estimate the general rate model transport parameters, while steric mass action binding parameters were estimated by backfitting the model to a set of fractionated gradient elution runs. Chapter 14 discusses a specific application involving the use of a size-exclusion chromatography step for reducing aggregated product forms for the commercial production of turoctocog alfa. It has been illustrated how the different quality by design (QbD) elements of risk assessment and process knowledge can be linked through identification of key critical quality attributes (CQAs), which may be affected by the step and the different process parameters responsible for such influence on the CQAs.

Continuous processing including chromatography has gained much attention in the last decade, and Chapter 15, by Bisschops and Brower, presents industrial applications of such MCC systems for dynamic simulations as predictive models for MMC separation. This chapter describes a numerical simulation approach for predicting the performance of continuous chromatographic separations of biopharmaceutical proteins. The numerical simulations are based on the linear driving force model for mass transfer kinetics and a Langmuir isotherm for equilibrium behavior. The numerical simulations have been compared with the experimental capture efficiency of monoclonal antibodies on Protein A media in a continuous MCC system for two different monoclonal antibodies and two different (agarose based) Protein A media. The authors demonstrate the possibility of using simulation models for process characterization, thereby enabling the knowledge space with limited experimentation significantly speeding up the development program.

Chapter 16, by Rathore and Singh, presents the general state of the art of multivariate data analysis and review of current process analytical technology (PAT) methods available to facilitate process chromatography. This chapter presents a review of chemometrics applications in process chromatography. The various data preprocessing methods and modeling approaches have been discussed along with two case studies illustrating the utility of chemometrics in analyzing process chromatographic data.

Process control and PAT are topics of great interest in the industry, and new tools that may move the analytical release test burden to the unit operation process control are highly desirable. A recent tool exploiting UV spectra for this application is shown in Chapter 17, by Hansen, Brestrich, Staby, and Hubbuch. The proposed tool has a response time of <1 s and allows real-time pooling decisions. Both the screening and the PAT tool have been based on partial least squares (PLS) regression models, correlating mid-UV protein absorption spectra with selective protein concentrations. The fundamentals of intrinsic protein absorption and PLS as well as their application for selective protein quantification have been also addressed.

Finally, Chapter 18, by Hearn, presents the more sustainable and green approach to chromatographic separation and to many practical considerations needed in future manufacturing. This chapter examines recent progress toward the incorporation of the concepts underlying sustainable manufacturing of protein-based products, with emphasis of the downstream aspects of the recovery and purification of value-added protein products derived from biotechnological procedures. Lessons gained from the use of similar approaches developed within the chemical, traditional pharmaceutical, and food ingredient industries have been examined in terms of their applicability to the downstream processing of protein products derived from genetic engineering, cell culture, and associated biotechnology strategies.

The book may be read for individual contributions; however, all of the book chapters complement each other with state-of-the-art implementation of modeling in the biopharmaceutical industry and academic research within the field. All chapters of the book have been peer reviewed. We would like to thank all authors for their valuable contributions and hope the academic, industrial, and regulatory scientists will benefit from this book.

Arne Staby
Anurag S. Rathore
Satinder Ahuja

20 December 2016

1

Model-Based Preparative Chromatography Process Development in the QbD Paradigm

Arne Staby[1], Satinder Ahuja[2], and Anurag S. Rathore[3]

[1] *CMC Project Planning & Management, Novo Nordisk A/S, Bagsværd, Denmark*
[2] *Ahuja Consulting, Calabash, NC, USA*
[3] *Department of Chemical Engineering, Indian Institute of Technology, New Delhi, India*

1.1 Motivation

Preparative chromatography for separation of proteins and peptides continues to be the primary workhorse in purification of biopharmaceuticals. Numerous papers and books exist describing theory and implementation of preparative chromatography; however, this is the first book that combines academic progress in modeling with industrial implementation. Although theory and models have been available for many years, industrial usage of these tools has been scarce due to labor- and material-intensive requirements. However, with the biotech industry moving to implement the expectations underlined in the recent regulatory initiative of quality by design (QbD), interesting and outspread applications of modeling tools for commercial process development and manufacture have emerged.

1.2 Regulatory Context of Preparative Chromatography and Process Understanding

QbD expectations to biopharmaceutical production including preparative chromatography are described in the ICH quality guidelines Q8, Q9, Q10, and Q11 [1–4]. Further, ICH Q8-R2 [1] provides the overall definition of QbD in a regulatory context.

A systematic approach to development that begins with predefined objectives and emphasizes product and process understanding and process control based on sound science and quality risk management.

The focus of this book is on the underlined parts of this definition, and the framework of QbD may be outlined as presented in Figure 1.1. In the top part of the figure, the primary focus of biopharmaceuticals is the patient, and the patient needs are defined through the quality target product profile (QTPP), which in turn is affected by chemistry, manufacturing, and controls (CMC) activities. Fulfilling patients' needs places some requirements on the product, and these elements are obtained through linkage of the QTPP to the list of critical quality attributes (CQAs). The CQAs will have acceptable ranges for the manufacturer to comply with, and to obtain product of the desired quality, the process needs to be run within acceptable ranges of process parameters. Proper knowledge of how process parameters affect the product quality may be obtained through process models that may end up in a regulatory, enhanced application for approval of a design space. To control process parameters within defined ranges, process models and/or even a design space will provide some requirements to the GMP facility and linkage to the control strategy, which will include various process monitors, process analytical technology (PAT) tools, process validation, and release tests and specifications. All elements are linked through risk assessment exercises to address the risk-based approach of QbD in a regulatory setting.

Figure 1.1 (bottom) displays an example of QbD elements contained in the QbD framework for a preparative chromatography step. A key patient need is of course to get efficient treatment, and one element affecting this is to get a proper dose of the biopharmaceutical. To obtain proper dosing, the purity and among others the bioactivity of the biopharmaceutical needs to be correct. Purity is significantly affected by the peak collection criteria used in preparative chromatography, and a well-known methodology for peak collection is by UV monitoring as part of the control strategy (e.g., see Chapters 12 and 17). A proper understanding and control of the preparative chromatography process may be obtained by a mechanistic or statistical model and their boundary conditions that may define an operational design space. Thus, the idea of this linkage exercise is to obtain a complete overview of the process in a way that will elucidate, for example, how a defect in or removal of a UV monitor in a preparative chromatographic purification step will affect the patient through cascading back in the figure through a series of risk assessments. The focus of this book is to obtain "process understanding and process control based on sound science" as described earlier, and it can be visualized by observing the elements within the red circle in Figure 1.1 (top).

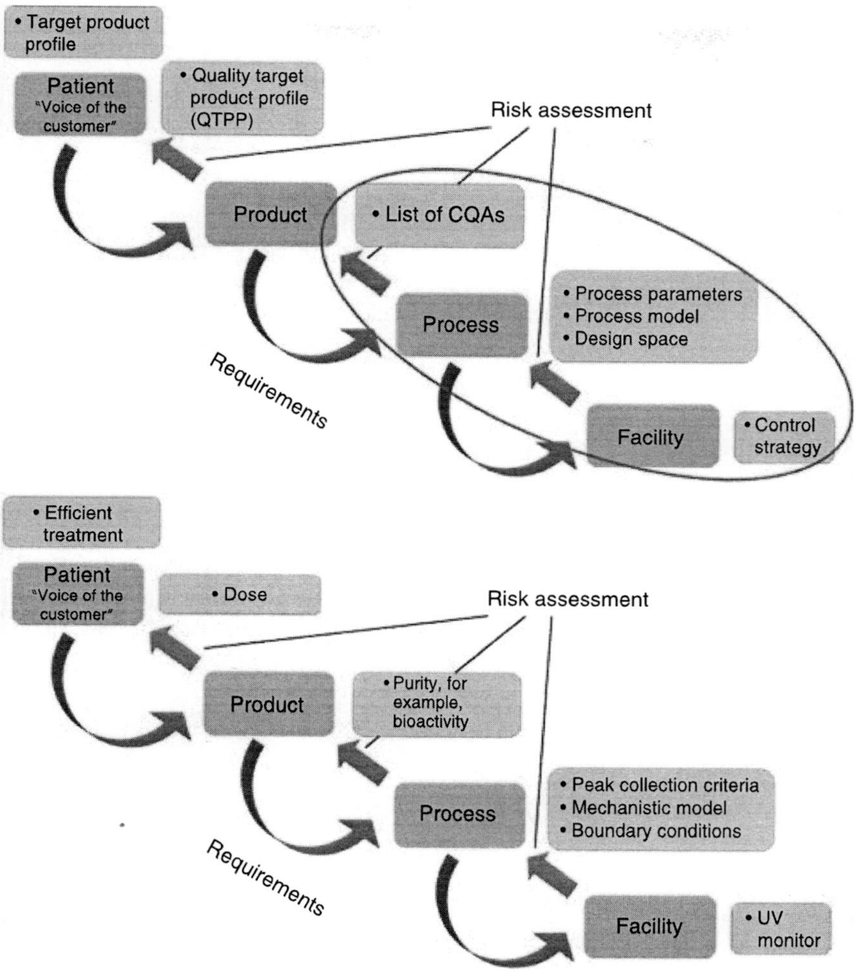

Figure 1.1 (Top) The framework of QbD. (Bottom) Example of QbD elements contained in the QbD framework for a preparative chromatography step. (*See insert for color representation of the figure.*)

A proper control strategy is achieved through sufficient process understanding. Traditionally, process understanding in the biopharmaceutical industry was obtained through a combination of theoretical knowledge based on the following: (i) education; (ii) experience from other projects and proteins optionally of similar nature, for example, mAbs; (iii) preliminary experimentation of less systematic nature; and (iv) "one parameter at a time" (OPAT) experimentation where all variables are kept constant while systematically altering one variable. This concept has worked well for many years, and most legacy products have been developed using this approach. Figure 1.2 presents the general level of knowledge obtained by the different methodologies including more recent concepts. Although some companies have also used multivariate methods for development and documentation of legacy products, the extensive use of more advanced methods for process understanding has been affected by implementation of QbD concepts. The general methodology used in the industry today is based on multivariate statistical analysis such as design of experiments (DoE) often combined with various high-throughput process development (HTPD) techniques (see e.g., Chapter 11). DoE is a very broad and important tool that does not require mechanistic understanding prior to implementation, and it works quite efficiently if the user has prior knowledge of which parameters are significant and if the number of parameters is limited. Today, the most comprehensive application of statistical methods to support QbD and a true enhanced approach filing has been accomplished by Genentech/Roche with its recent regulatory approval of Gazyva. Disadvantages of DoE include less optimal identification of assumptions and the general lack of opportunities for extrapolation outside the experimental area used to set up the statistical models. DoE is used extensively for validation of parameter ranges in preparative chromatography; however for other unit operations

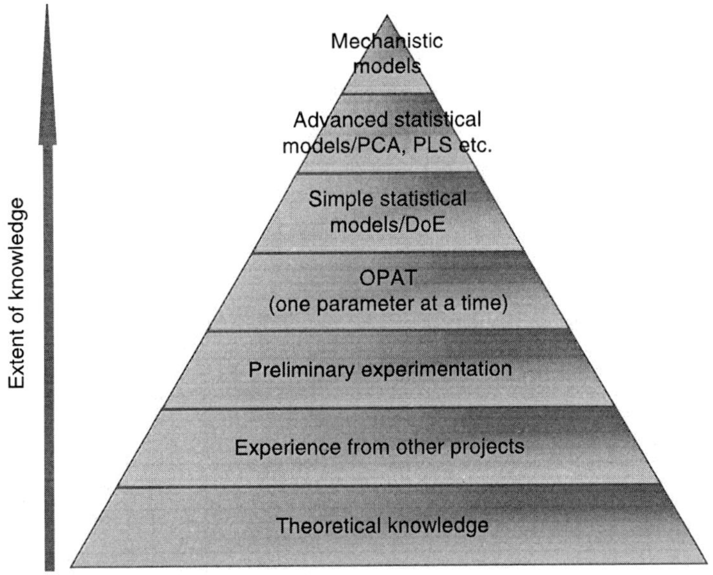

Figure 1.2 General extent of knowledge and process understanding obtained employing various methodologies and approaches.

such as fermentation, more advanced statistical methods like principal component analysis (PCA), partial least squares (PLS) methods, etc. are used due to their capability to handle very high number of variables (see also Chapter 16). At the top of the pyramid in Figure 1.2 and at the highest extent of knowledge obtainable are models based on mechanistic principles because full mechanistic process understanding is typically achieved. Depending on assumptions, these mechanistic models are also referred to as first-principle models, and they provide optimal evaluation of assumptions as well as opportunities for extrapolation outside the experimental area of parameter estimation.

An example of the difference in process understanding achieved from application of mechanistic modeling and a DoE approach for a preparative SEC step is presented in Figure 1.3 [5] (see also Chapter 14). The figure shows the effect of the feed concentration of a biopharmaceutical on the content of high molecular weight proteins (HMWP)—a typical CQA in the drug substance addressed by purification. The different experimental values for a given feed concentration (red diamonds) are due to controlled variation of other variables. Predictions based on a mechanistic model and on a statistical model by DoE are shown with full green and light blue colors, respectively. It is noticed that the model based on DoE cannot predict the worst-case conditions at a feed concentration of 0.75 g/L (indicated by the green, dashed circle) and instead the DoE-based model predicts the lowest concentration of 0.5 g/L as the worst-case conditions (indicated by the light blue, dashed circle). Further, the prediction error increases if extrapolation is performed outside the experimental area. The problem is partly caused by the general setup of

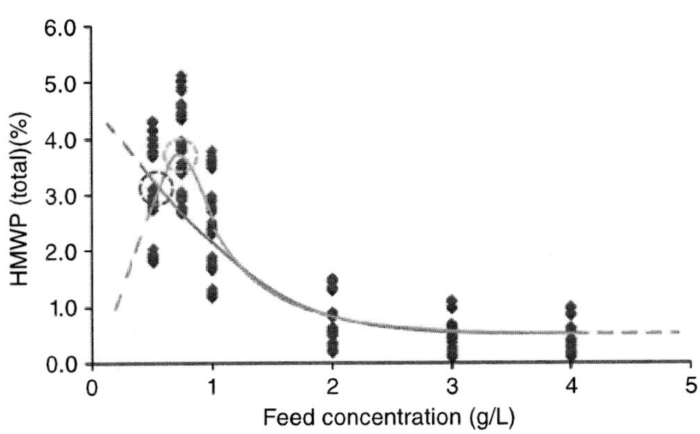

Figure 1.3 HMWP content after purification on SEC for a biopharmaceutical as a function of feed concentration. ◊, experimental results; —, model prediction by mechanistic model; and —, model prediction by statistical model based on DoE. (*See insert for color representation of the figure.*)

experiments supporting DoE where center points and parameter range limits are often applied (in the current case ~2 g/L and 0.5 and 4 g/L, respectively). DoE-based models are good in capturing monotonous functions, but they have problems capturing functions containing inflection points, and it would require a very comprehensive experimental setup for DoE-based models to capture functions with inflection points—far more than what is used in general in the industry. The experimental setup to obtain mechanistic models is typically not more comprehensive, but it is different. This example illustrates some of the pitfalls of applying DoE the way it is usually performed in the biopharmaceutical industry and how a mechanistic model may provide more process understanding.

1.3 Application of Mathematical Modeling to Preparative Chromatography

Mathematical models and modeling tools have been available for decades in academia, for example, Van Deemter [6], Giddings [7], Guiochon et al. [8], Melander and Horváth [9], Brooks and Cramer [10], Yamamoto et al. [11], Hearn et al. [12], Lenhoff [13], Carta and Jungbauer [14], Frech et al. [15], Łącki et al. [16], Hansen and Mollerup [17], Ottens et al. [18], Bracewell et al. [19] and many, many more, and the tools have been applied to academic problems such as separation of standard proteins like BSA, lysozyme, etc. and occasionally to more industry-relevant proteins. The experimental burden required and essential access to large amounts of pure experimental material made it very difficult and in fact too cumbersome for the biopharmaceutical industry to implement the methodology for many years. Motivation and requirements have, however, changed over the last years. The regulatory environment as described earlier [1–4] access to HTPD techniques [20, 21] facilitating fast experimentation and low demands of experimental material, and, in the specific case of polishing chromatography, proper assumptions and approaches to minimize the experimental task of generating preparative modeling parameters [22]. These aspects have aided the industry into initiating application of mechanistic modeling, and this book also presents numerous examples of such implementation for preparative chromatography.

Another aspect challenging the biopharmaceutical industry in implementation of mechanistic modeling tools is access to skilled personnel that can master modeling and computer coding at an expert level as well as to have comprehensive insight into preparative chromatography at manufacturing scales. Many implementation attempts in industry have failed due to lack of management support and critical mass of skilled personnel. In contrast, statistical modeling based on DoE or similar methods are much more easily implemented. An approach to initiation of implementation of mechanistic modeling is collaboration between academics or specialized consultants and the biopharmaceutical industry, and numerous examples of such collaboration exist, for example, Borg et al. [23], Ghosh et al. [24], Rathore et al. [25], Nfor et al. [26], and many more. Another approach may be to look at trends in the small molecule pharmaceutical area, which are typically several years ahead of the biopharmaceutical industry in implementation of new tools and approaches.

New trends and hot topics in the industry include the utilization of semi- and continuous techniques (see also Chapters 5 and 15), PAT method implementation (see also Chapter 17), production of antibody–drug conjugates (ADCs) and other conjugates, and manufacturing of biosimilars, and many of these applications will benefit from the use of mechanistic modeling. As examples, the insulin purification method using MCSGP presented in Figure 1.4 was modified extensively from the original batch process by a mathematical model (L. Aumann et al., Chromacon AG, internal report to Novo Nordisk), and conjugate products that require reactions may benefit from reaction models as presented elsewhere [27]. Finally, the manufacture of biosimilars could significantly benefit from access to mechanistic modeling of preparative chromatography and other unit operations to demonstrate optimal process understanding, identification of critical process parameters, PAT-based process control, and demonstration of consistently achieving product profile that is similar to that of originator products.

Once a mechanistic model for a preparative chromatography step has been developed, the applications of the model are numerous depending on the approach and assumptions made. Figure 1.5 lists some common applications of mechanistic and statistical modeling in industry. Topics presented in black text in the figure represent themes that are covered by the subsequent book chapters, and a more thorough guidance to the individual chapters is given in the preface.

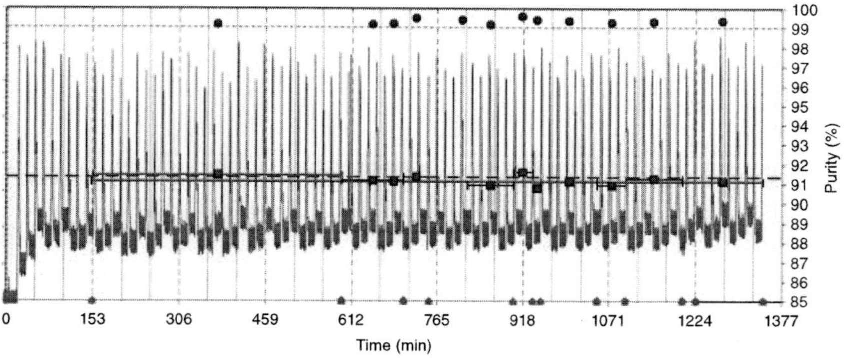

Figure 1.4 Chromatogram and purity of a three-column MCSGP unit as a function of time for a 23 h semi-continuous chromatographic purification of insulin (L. Aumann et al., Chromacon AG, internal report to Novo Nordisk).

- Process and analytical development
- Process optimization
- Process validation/challenge and critical parameters
- Plant design
- Process control (PAT)
- Troubleshooting and deviation handling
- Process understanding and design space
- Scaling of chromatography

Figure 1.5 Application options of mechanistic modeling.

Acknowledgements

The editors would like to thank all contributors of this book, as well as Thorbjørn Strøm-Hansen, Michael Schousboe, Lars Sejergaard, Marcus Degerman and Erik Skibsted, Novo Nordisk, and Lars Aumann, ChromaCon, for help with the figures of this chapter.

References

1. ICH Q8(R2) Pharmaceutical Development (2009), International Conference of Harmonization.
2. ICH Q9 Quality Risk Management (2005), International Conference of Harmonization.
3. ICH Q10 Pharmaceutical Quality System (2008), International Conference of Harmonization.
4. ICH Q11 Development and Manufacturing of Drug Substances (Chemical Entities and Biotechnological/Biological Entities) (2012), International Conference of Harmonization.
5. M. Degerman, E. Hansen, L. Sejergaard, T.B. Hansen, J. Krarup, S. Frederiksen, S. Kidal, A. Staby. Model based process development in biopharmaceutical industry—case studies. Oral keynote presentation at BPP2011, Puerto Vallarta, Mexico, September 18–22, 2011.
6. J.J. van Deemter, F.J. Zuiderweg, A. Klinkenberg. Longitudinal diffusion and resistance to mass transfer as causes of nonideality in chromatography. *Chemical Engineering Science* 5, 271–285 (1956).
7. J.C. Giddings. *Unified Separation Science*, John Wiley & Sons, Inc, New York (1991).
8. G. Guiochon, A. Felinger, D.G. Shirazi, A.M. Katti. *Fundamentals of Preparative and Nonlinear Chromatography*, 2nd edition, Academic Press, Amsterdam (2006).

9. W. Melander, C. Horváth. Salt effects on hydrophobic interactions in precipitation and chromatography of proteins: an interpretation of the lyotropic series. *Archives of Biochemistry and Biophysics* 183, 200–215 (1977).
10. C.A. Brooks, S.M. Cramer. Steric mass-action ion exchange: displacement profiles and induced salt gradients. *AIChE Journal* 38, 1969–1978 (1992).
11. S. Yamamoto, K. Nakanishi, R. Matsuno. *Ion-Exchange Chromatography of Proteins*, Marcel Dekker, New York (1988).
12. F. Lin, W. Chen, M.T.W. Hearn. Thermodynamic analysis of the interaction between proteins and solid surfaces: application to liquid chromatography. *Journal of Molecular Recognition* 15, 55–93 (2002).
13. A.M. Lenhoff. Significance and estimation of chromatographic parameters. *Journal of Chromatography A* 384, 2, 285–299 (1987).
14. G. Carta, A. Jungbauer. *Protein Chromatography: Process Development and Scale-up*, Wiley VCH Verlag GmbH & Co., Weinheim (2010).
15. M. Schmidt, M. Hafner, C. Frech. Modeling of salt and pH gradient elution in ion-exchange chromatography. *Journal of Separation Science* 37, 1–2, 5–13 (2014).
16. X. Zhang, B.A. Grimes, J.-C. Wang, K.M. Łącki, A.I. Liapis. Analysis and parametric sensitivity of the behavior of overshoots in the concentration of a charged adsorbate in the adsorbed phase of charged adsorbent particles: practical implications for separations of charged solutes. *Journal of Colloid and Interface Science* 273, 1, 22–38 (2003).
17. E. Hansen, J. Mollerup. The influence of retention on the plate height in ion exchange chromatography. *Separation Science and Technology* 39, 2011–2030 (2004).
18. B.K. Nfor, D.S. Zuluaga, P.J.T. Verheijen, P.D.E.M. Verhaert, L.A.M. van der Wielen, M. Ottens. Model-based rational strategy for chromatographic resin selection. *Biotechnology Progress* 27, 6, 1629–1643 (2011).
19. S. Gerontas, M.S. Shapiro, D.G. Bracewell. Chromatography modelling to describe protein adsorption at bead level. *Journal of Chromatography A* 1284, 44–52 (2013).
20. J.L. Coffman, J.F. Kramarczyk, B.D. Kelley. High-throughput screening of chromatographic separations: 1. Method development and column modeling. *Biotechnology and Bioengineering* 100, 605–618 (2008).
21. M. Bensch, P. Schulze Wierling, E. von Lieres, J. Hubbuch. High throughput screening of chromatographic phases for rapid process development. *Chemical Engineering & Technology* 28, 11, 1274–1284 (2005).
22. S. Kidal, T.B. Hansen, A. Staby. Methods of optimizing chromatographic separation of polypeptides. WO/2008/028974 (2008).
23. N. Borg, Y. Brodsky, J. Moscariello, S. Vunnum, G. Vedantham, K. Westerberg, B. Nilsson. Modeling and robust pooling design of a preparative cation-exchange chromatography step for purification of monoclonal antibody monomer from aggregates. *Journal of Chromatography A* 1359, 170–181 (2014).

24. P. Ghosh, K. Vahedipour, M. Lin, J.H. Vogel, C. Haynes, E. von Lieres. Computational fluid dynamic simulation of axial and radial flow membrane chromatography: mechanisms of non-ideality and validation of the zonal rate model. *Journal of Chromatography A* 1305, 30, 114–122 (2013).
25. A.S. Rathore, O. Kaltenbrunner, P. Engel, J.M. Mollerup, J. McCue. Modeling of biopharmaceutical processes. Part 2: process chromatography unit operation. *BioPharm International* 21, 8, 28–42 (2008).
26. B.K. Nfor, T. Ahamed, G.W.K. van Dedem, P.D.E.M. Verhaert, L.A.M. van der Wielen, M.H.M. Eppink, E.J.A.X. van de Sandt, M. Ottens. Model-based rational methodology for protein purification process synthesis. *Chemical Engineering Science* 89, 185–195 (2013).
27. L. Sejergaard, M.H. Hansen, A.A. Olsen, P. Østergaard, T. Strøm-Hansen, P. Østfeldt, S. Valentin, A. Staby. Implementation of unit operation design spaces: experiences with regulatory authorities. Oral presentation at 245th ACS National Meeting, New Orleans, LA, April 7–11, 2013.

2

Adsorption Isotherms

Fundamentals and Modeling Aspects

Jørgen M. Mollerup

PrepChrom, Klampenborg, Denmark

2.1 Introduction

The use of process simulation in computational fluid dynamics and in computer-aided design enables the chemical engineer to analyze, design, and optimize chemical processes. Chemical engineers use process simulators to scale up processes from bench scale to production scale in a single step. A prerequisite for doing this is the availability of models of the unit operations involved as well as models of the physical–chemical properties of the various fluids and solutions. The fulcrum point to the design of separation processes like distillation, extraction, absorption, and adsorption is the phase equilibrium properties.

To the design of a large distillation column, a chemical engineer will not use a small distillation column to investigate how to purify a product by trial and error. The engineer will use dedicated equipment to measure the vapor–liquid equilibrium data needed to determine the parameters in the phase equilibrium model and use a process simulator to design and optimize the distillation column.

The agenda in process chromatography is very similar. There is no need to collect thousands of data from a robotic system and look for some "optimal" separation conditions. The agenda is to measure some very accurate experimental data from which the model parameters in the adsorption isotherms can be estimated in order to facilitate the use of simulation tools to the design and optimization of a chromatographic separation process. The be-all and end-all are the isocratic retention volumes at analytical load from which most of the parameters in the adsorption isotherms can be estimated. These data can be performed using a rather crude feedstock and consume very little material.

The parameters that determine the nonlinear behavior of the adsorption isotherm can be estimated from chromatographic runs at high load and from batch adsorption data.

2.2 Definitions

This section gives the definitions of the nomenclature used in this chapter. A schematic representation of the column is shown in Figure 2.1. There are three phases in the column. The *mobile phase* is the volume of the column where the liquid is flowing through the column. The *adsorbent* is a particle with a very high porosity and a large surface area. The ligands are not bonded directly to the surface of the solid structure of the adsorbent, but they are immobilized by means of a chain of small molecules, a spacer, in order to make it easier for the large macromolecules to associate with the ligands. Some spacers are grafted and carry several ligands. There are no ligands in the third phase, the solid support of the particle.

The *interstitial bed porosity*, ε_b, is the volume of the cavities that are between and around the adsorbents. It is difficult to measure but it is of the order of 0.37–0.45. The *particle porosity*, ε_p, is the fraction of the particle volume that is occupied by the mesopores inside the particle. The *total porosity*, ε_t, of the bed is the volume fraction of liquid phase in the bed, that is, $\varepsilon_t = \varepsilon_b + (1-\varepsilon_b)\varepsilon_p$.

Mobile phase	Adsorbent
Volume, ε_b	Solid support
Solute concentrations, c_i^m	Solid volume $(1-\varepsilon_b)(1-\varepsilon_p)$
	Pore volume $(1-\varepsilon_b)\varepsilon_p$
	Available pore volume fractions, $k_{d,i}$
	Ligand concentration, Λ
	Solute concentrations, c_i
	Adsorbate concentrations, q_i

Figure 2.1 A schematic representation of a unit volume of a chromatographic column. Gray shading represents the solid-phase support which is unaffected by the interaction with mobile phase and solutes.

$k_{d,i}$ is an *exclusion factor* that represents the fraction of the stationary phase pore volume that is available for diffusion of a given solute species *i*; consequently $k_{d,i}$ has a value between unity and zero. By definition, $k_{d,i}$ is unity for ions of common salts like sodium chloride.

Concentrations are usually reported as molarity (moles per liter of solution), M, or millimoles per liter, mM. An *adsorbate* is a molecule that associates with a number of ligands in the adsorbent, q_i denotes the concentration of the adsorbate of species *i* in the pore volume, and c_i is the concentration of a solute molecule of the same species in the pore volume. q_i and c_i are equilibrium concentrations. The mobile-phase concentration of species *i*, c_i^{m}, is in general not in equilibrium with the corresponding pore volume concentration, c_i, due to the resistance to mass transfer. However, when isocratic retention data at analytical load are utilized to determine the parameters in the adsorption isotherm, it is assumed that the first moment of a peak represents an equilibrium property.

The *interstitial volume*, V_0, is the volume of the bed in the column, V_{c}, times the interstitial porosity, ε_{b}, that is, $V_0 = V_{\mathrm{c}}\varepsilon_{\mathrm{b}}$, and this volume is identical with the volume of the mobile phase in the column.

The *interstitial velocity*, υ, is the average linear velocity in the interstices between the particles, $\upsilon = Q/A_{\mathrm{c}}\varepsilon_{\mathrm{b}}$, where Q is the volumetric flow rate, A_{c} is the cross-sectional area of the column, and $A_{\mathrm{c}}\varepsilon_{\mathrm{b}}$ is the average cross-sectional flow area.

The holdup time or the *residence time* of the solvent, t_0, is the interstitial volume, V_0, divided by the volumetric flow rate, Q, that equals the bed height, L, divided by the interstitial velocity, that is, $t_0 = V_0/Q = L/\upsilon$.

The *phase ratio*, ϕ, is defined as

$$\phi = \frac{\text{pore volume}}{\text{interstitial volume}} = \frac{(1-\varepsilon_{\mathrm{b}})\varepsilon_{\mathrm{p}}}{\varepsilon_{\mathrm{b}}} \tag{2.1}$$

The amount of material of species *i* adsorbed in a unit bed volume is $q_i(1-\varepsilon_{\mathrm{b}})\varepsilon_{\mathrm{p}}$.

There is an extensive list of glossary of terms in Ref. [1]. In this chapter the following terms are utilized:

Convex and concave: A function is convex if the second derivative, $f''(x)$, is positive, and if the second derivative, $f''(x)$, is negative, the function is concave. The point where $f''(x)$ is zero is an inflection point. In the literature convex is also called convex downward or concave upward, and concave is also called concave downward or convex upward.

Curvature: The curvature is the second derivative of a function.

Eluate: The mobile phase at the column exit containing some feed components, the eluites.

Elution: Purification or separation of the feed components by an eluant flowing through the column.

Eluant: Synonymous with the pure mobile phase at the column inlet.
Eluite: A component in the feed, dissolved in the mobile phase, at the column exit.
Elution profile: The plot of the concentration of the eluite component versus the volume of the mobile phase passed. Usually, it is a wide, unsymmetrical peak, such as those obtained under nonlinear conditions. Guiochon et al. prefer the term band profile.
Fronting: A type of peak asymmetry in which the rear is steep and the front rises slowly from the baseline. It is often due to a convex isotherm.
Modulator: Additives, usually salts and solvents, added to the eluant to adjust the elution of components in the feed that can adsorb to the ligands.
Solute: A component in solution in a fluid phase. In this chapter, a distinction is made between solutes, salts, and solvents.
Solvent: A water-soluble liquid added to the eluant or to the feed solution.
Tailing: A type of peak asymmetry in which the front is steep and the rear returns slowly to the baseline. It is often due to a concave isotherm.

2.3 The Solute Velocity Model

The solute velocity model is fundamental to the understanding and interpretation of the elution profiles. The model will be utilized to calculate the shape of elution profiles in ideal chromatography. Ideal chromatography [1, page 958] is a model of chromatography assuming no axial dispersion and no mass transfer resistance. It permits an easy study of the influence of the thermodynamics of phase equilibrium (i.e., of the adsorption isotherm) on the elution profiles.

The average linear velocity of the mobile phase is the interstitial velocity, v, but a molecule that adsorbs will move at a lower velocity because the adsorption–desorption process slows down the solute velocity. When a molecule of species i associates with the ligands, the overall concentration, s_i, in the pore volume increases from $s_i = c_i$ to $s_i = c_i + q_i$. Guiochon et al. [1, Section 8.1] have shown that the *solute wave velocity*, u_i, of a single solute at isocratic conditions is

$$u_i = \frac{v}{1+\phi k_{d,i} \partial s_i / \partial c_i} = \frac{v}{1+\phi k_{d,i}\left(1+\partial q_i / \partial c_i\right)} \tag{2.2}$$

The solute wave velocity depends on the slope of the adsorption isotherm, $\left(\partial q_i / \partial c_i\right)$, and that explains why the shape of the elution profile depends on the slope of the adsorption isotherm.

The classical adsorption isotherm has a *negative curvature*, that is, $\left(\partial^2 q_i / \partial c_i^2\right) < 0$. An example is shown in Figure 2.3, which shows the adsorption isotherm of a GLP-1 derivative, the slope of the isotherm, and q_i/c_i. When the adsorption isotherm has a negative curvature, the slope of the adsorption isotherm decreases with increasing solute concentration; wherefore the solute velocity will increase with increasing solute concentration as shown in Figure 2.4. Theoretically this will lead to a situation where the lower part of a leading edge of an elution profile moves at a lower velocity than the top of the leading edge. This is physically impossible. The difference in solute velocities leads to a compression of the front of the elution profile resulting in a shock wave [1, Section 14.1.4] moving at the velocity

$$u_i = \frac{\upsilon}{1+\phi k_{d,i}\Delta s_i/\Delta c_i} = \frac{\upsilon}{1+\phi k_{d,i}\left(1+\Delta q_i/\Delta c_i\right)} \tag{2.3}$$

Here $(\Delta q_i/\Delta c_i)$ is the slope of the cord of the adsorption isotherm of species i, and Δc_i is the step-up in the concentration at the leading edge. $\Delta q_i/\Delta c_i$ is not necessarily identical with q_i/c_i unless the step-up is from a concentration equal to zero. The diffuse wave velocity and the shock wave velocity of a GLP-1 derivative are shown in Figure 2.4.

In the absence of dispersion forces, the shock wave will be vertical; wherefore the shape of a leading edge is determined by the competition between the dispersion forces and the shock wave that counteracts the dispersion forces. At the rear of the elution profile, where the concentration decreases, the dispersion forces and the decreasing solute velocity will act in the same direction and create an elution profile with a positive curvature, a *diffuse wave.* A series of simulated elution profiles of a GLP-1 derivative are shown in Figure 2.5.

Some adsorption isotherms have a *positive curvature,* that is, $\left(\partial^2 q_i/\partial c_i^2\right) > 0$ at low solute concentration due to self-association of the proteins. This means that proteins act as ligands themselves. An example is shown in Figure 2.17, which shows the adsorption isotherm, the slope of the adsorption isotherm, and the ratio q_i/c_i of a GLP-1 analogue. When the adsorption isotherm has a positive curvature, the slope of the adsorption isotherm increases with increasing solute concentration; wherefore the solute velocity will decrease with increasing solute concentration. The result is that the leading edge of the elution profile will be a diffuse wave and the rear of the elution profile will be a shock wave at low solute concentration. Figure 2.19 shows the diffuse wave velocity and the shock wave velocity of a GLP-1 analogue, and a series of simulated elution profiles are shown in Figure 2.20.

The elution time of a wave is

$$\frac{t_i}{t_0} = \frac{\upsilon}{u_i} \tag{2.4}$$

If volume units are preferred, the corresponding equation is

$$\frac{V_i}{V_c} = \frac{\upsilon \varepsilon_b}{u_i} \tag{2.5}$$

A Dirac's delta function is a function that has unit area, the width h, and the height $1/h$ where $h \to 0$. If the adsorption isotherm is linear and a sample that resembles a Dirac's delta function is injected into the column, the resulting response at the column outlet is an elution profile that has the shape of a Gaussian distribution. A linear adsorption isotherm has a constant slope where

$$\lim_{c_i \to 0}\left(\frac{\partial q_i}{\partial c_i}\right) = \lim_{c_i \to 0}\left(\frac{q_i}{c_i}\right) \equiv A_i \tag{2.6}$$

A_i is the *thermodynamic retention factor* of species i. When the solute velocity is independent of the solute concentration, the retention time, $t_{\mathrm{R},i}$, the first moment of the Gaussian peak in time units, is calculable from Equations 2.2, 2.4, and 2.6, that is,

$$t_{\mathrm{R},i} = t_0\left(1 + \phi k_{d,i}\left(1 + A_i\right)\right) \tag{2.7}$$

If volume units are preferred, the corresponding equation is

$$V_{\mathrm{R},i} = V_{\mathrm{c}}\left(\varepsilon_{\mathrm{b}} + \left(1 - \varepsilon_{\mathrm{b}}\right)\varepsilon_{\mathrm{p}} k_{d,i}\left(1 + A_i\right)\right) \tag{2.8}$$

The second central moment, the variance, σ^2, of the Gaussian peak is determined by the dispersion forces and the retention time alone. The model of the second central moment is known as the van Deemter equation [2], and its applications are analyzed in Ref. [1, Section 6.2.2] and in [3, 4].

In size exclusion chromatography, as well as in other chromatographic techniques, the porous structure retains the species simply because the molecules can diffuse into some of the openings. When there is no adsorption–desorption taking place, the thermodynamic retention factor, A_i, in Equation 2.8 is zero, and the retention volume due to the pore diffusion is

$$V_{NA,i} = V_{\mathrm{c}}\left(\varepsilon_{\mathrm{b}} + \left(1 - \varepsilon_{\mathrm{b}}\right)\varepsilon_{\mathrm{p}} k_{d,i}\right) \tag{2.9}$$

The retention volume of a common salt, where $k_{d,i}$ by definition is unity, is

$$V_{\mathrm{salt}} = V_{\mathrm{c}}\left(\varepsilon_{\mathrm{b}} + \left(1 - \varepsilon_{\mathrm{b}}\right)\varepsilon_{\mathrm{p}}\right) = V_{\mathrm{c}}\varepsilon_{\mathrm{t}} \tag{2.10}$$

The retention factor, k_i', in gas chromatography is

$$k_i' = \frac{t_{\mathrm{R},i} - t_0}{t_0} = \frac{V_{\mathrm{R},i} - V_0}{V_0} \tag{2.11}$$

k_i' is a convenient measure of the relative retention in an open tubular column where the stationary phase is coated to the interior surface of the tube. However, in liquid column chromatography where the stationary phase is porous beads, it is convenient to correct for the pore diffusion and use the quantity

$$R_i = \frac{V_{R,i} - V_{NA,i}}{V_c} = (1-\varepsilon_b)\varepsilon_p k_{d,i} A_i \tag{2.12}$$

as a measure of the *relative retention* due to the adsorption. This quantity is the retention due to the adsorption per unit column volume, and it can be calculated directly from experimental data and does not require any knowledge of the porosities.

2.4 Introduction to the Theory of Equilibrium

The reason to include this section is to analyze similarities and differences in the application of the equilibrium theory for three different systems. The first example is an analysis of the equilibrium between two fluid phases. The second example is a reversible reaction in a single-phase system, and the last example is an analysis of a reversible adsorption. The traditional-phase equilibrium system is included in the analysis to illustrate the difference between adsorption equilibria and traditional-phase equilibria.

2.4.1 Phase Equilibria

It is a consequence of the second law of thermodynamics that an isolated system will approach equilibrium by increasing its entropy and that the equilibrium state is a stationary point of maximum entropy [5].

In consequence of this, two fluid phases are in equilibrium when the temperature, T, and the pressure, p, of the two phases are identical and when an intensive variable called the chemical potential, μ_i, of any component, i, is identical in the two phases. If the two phases are denoted by α and β, respectively, the necessary conditions that characterize a system at equilibrium are

$$T^{(\alpha)} = T^{(\beta)} \tag{2.13}$$

$$p^{(\alpha)} = p^{(\beta)} \tag{2.14}$$

$$\mu_i^{(\alpha)} = \mu_i^{(\beta)}; \quad \text{at all species i in the system} \tag{2.15}$$

The chemical potentials are calculable as partial molar derivatives of the Gibbs energy:

$$G(T, p, \mathbf{n}) = \sum_1^N n_i \mu_i (T, p, \mathbf{n}) \Leftrightarrow \mu_i = \left(\frac{\partial G}{\partial n_i}\right)_{T,p} \tag{2.16}$$

where $\mathbf{n}$ is a composition vector, in mole numbers, of the system, $n_1, n_2, \ldots, n_i, i = 1, N$, and where N is the number of components in the system. If electrolytes are present, the specification of the charges of the ionic species and the dielectric properties of the solvent are needed to characterize the system, a subject analyzed in the Appendix.

The equilibrium conditions, Equations 2.13–2.15, are necessary but not sufficient conditions to ensure that the system is stable. The number of phases in a system at equilibrium is determined from the condition that the value of the Gibbs energy is at its global minimum. That is, the system is stable if, for any changes that can be made in the closed system [5, Chapter 8],

$$(\Delta G)_{T,p} > 0 \tag{2.17}$$

If a three-phase equilibrium system has a lower Gibbs energy than the corresponding two-phase system, the three-phase system will be the stable one. A trivial case is that a one-phase system has the lowest Gibbs energy of all.

The condition that the value of the Gibbs energy in a closed system at equilibrium is at its global minimum is a consequence of the fact that the equilibrium state is a state where the value of the entropy is at its global maximum, that is, a stationary point of maximum entropy.

2.4.2 Reversible Chemical Reaction

The second system is a single-phase system where a reversible chemical reaction takes place. The system is an isothermal and isobaric system where the components A and B react to form a new component D. The reaction scheme is

$$\mathrm{A} + \nu\mathrm{B} \rightleftharpoons \mathrm{D} \tag{2.18}$$

where ν is a stoichiometric coefficient. Equation 2.18 is the overall reaction scheme, and it does not provide any information of the reaction pathway and the kinetics.According to the condition given in Equation 2.17, the system is at equilibrium when the Gibbs energy of the system is at its global minimum; wherefore

$$\Delta G = \mu_\mathrm{D} - \mu_\mathrm{A} - \nu\mu_\mathrm{B} = 0 \tag{2.19}$$

that is, when any displacement in the reaction equilibrium will increase the Gibbs energy of the system.

2.4.3 Adsorption of a Single Component

Reversible adsorption equilibria are thermodynamically speaking similar to reversible reaction equilibria because the proteins associate with the immobilized ligands and this association adds a new "component" to the system.

The condition that determines the equilibrium is therefore the one given in Equation 2.19 and not the condition given in Equation 2.15.

When a protein molecule, P, associates with ν ligands, L, to form an association complex, PL_ν, the equilibrium scheme is

$$\mathrm{P} + \nu\mathrm{L} \rightleftharpoons \mathrm{PL}_\nu \tag{2.20}$$

and at equilibrium, it must hold that the value the Gibbs energy is at its global minimum, that is,

$$\Delta G = \mu_{\mathrm{PL}_\nu} - \mu_{\mathrm{P}} - \nu\mu_{\mathrm{L}} = 0 \tag{2.21}$$

It is convenient to analyze Equation 2.21 in detail and work out a simple model for the adsorption isotherm. In order to utilize the equilibrium conditions given in Equation 2.21 to calculate the equilibrium concentrations, a relationship between the chemical potentials and the mole fractions, x_i, must be set up. The general expression for the chemical potentials is analyzed in Section 2.A.1.2. The general expression is

$$\mu_i(T, p, \mathbf{n}) = \mu_i^0(T, p) + \mu_i^E(T, p, \mathbf{n}) + RT\ln x_i \tag{2.22}$$

where $\mu_i^0(T, p)$ are the reference state chemical potentials and $\mu_i^{\mathrm{E}}(T, p, \mathbf{n})$ are the excess potentials. For simplicity, it is in this analysis assumed that the excess potentials are zero because the scope of the analysis is to illustrate the fundamental principles alone. When the chemical potential model is inserted in Equation 2.21, the result is an equation that enables calculation of the equilibrium composition:

$$\mu_{\mathrm{PL}_\nu}^0(T, p) - \mu_{\mathrm{P}}^0(T, p) - \nu\mu_{\mathrm{L}}^0(T, p) + RT\left(\ln x_{\mathrm{PL}_\nu} - \ln x_{\mathrm{P}} - \nu\ln x_{\mathrm{L}}\right) = 0 \tag{2.23}$$

The reference state chemical potentials define the thermodynamic equilibrium constants, K, where

$$RT\ln K = -\left(\mu_{\mathrm{PL}_\nu}^0(T, p) - \mu_{\mathrm{P}}^0(T, p) - \nu\mu_{\mathrm{L}}^0(T, p)\right) \tag{2.24}$$

The equilibrium constant is a function of temperature and pressure alone. The relationship between mole fractions, x_i, and molar concentrations, c_i, is

$$x_i = \frac{c_i}{\sum_{j=1}^{N} c_j} = \frac{c_i}{c} \tag{2.25}$$

where c is the molar density of the solution. Inserting Equations 2.24 and 2.25 in Equation 2.23 gives a relation between the composition, in either mole fractions or molarities, and the thermodynamic equilibrium constant:

$$K = \frac{x_{\mathrm{PL}_\nu}}{x_{\mathrm{P}} x_{\mathrm{L}}^{\nu}} = \frac{c_{\mathrm{PL}_\nu} c^{\nu}}{c_{\mathrm{P}} c_{\mathrm{L}}^{\nu}} \tag{2.26}$$

In a chromatographic adsorbent there is a constraint on the system; the density of the immobilized ligand, Λ (mol/L of pore volume), is fixed; wherefore a material balance provides the constraint that

$$\Lambda = c_{\mathrm{L}} + \nu c_{\mathrm{PL}_\nu} \tag{2.27}$$

The theoretical maximum adsorption capacity is Λ/ν. In reality, the maximum available capacity is less, often only 40–50% or less of the theoretical capacity. When Equation 2.27 is used to eliminate the concentration of available ligands, c_{L}, the resulting equation provides a relationship between the solute concentration, c_{P}, and the concentration of the protein associated with ν ligands, c_{PL_ν}:

$$K = \frac{c_{\mathrm{PL}_\nu}}{c_{\mathrm{P}}} \left(\frac{c}{\Lambda - \nu c_{\mathrm{PL}_\nu}} \right)^{\nu} \Leftrightarrow \frac{c_{\mathrm{PL}_\nu}}{c_{\mathrm{P}}} = K \left(\frac{\Lambda}{c} \right)^{\nu} \left(1 - \frac{c_{\mathrm{PL}_\nu}}{\Lambda/\nu} \right)^{\nu} \tag{2.28}$$

Using the traditional notation, the protein concentration in the adsorbed state, the adsorbate concentration, c_{PL_ν}, is replaced by q and Λ/ν by q^{max}, and the eventual result is

$$\frac{q}{c_{\mathrm{P}}} = K \left(\frac{\Lambda}{c} \right)^{\nu} \left(1 - \frac{q}{q^{\mathrm{max}}} \right)^{\nu} \tag{2.29}$$

If there is only one association site between the protein and the ligand, that is, $\nu = 1$, the model becomes identical with the Langmuir adsorption isotherm:

$$\frac{q}{c_{\mathrm{P}}} = K \left(\frac{\Lambda}{c} \right) \left(1 - \frac{q}{q^{\mathrm{max}}} \right) = A \left(1 - \frac{q}{q^{\mathrm{max}}} \right) \Leftrightarrow q = \frac{A q^{\mathrm{max}} c_{\mathrm{P}}}{q^{\mathrm{max}} + A c_{\mathrm{P}}} = \frac{b q^{\mathrm{max}} c_{\mathrm{P}}}{1 + b c_{\mathrm{P}}} \tag{2.30}$$

where $A = bq^{\mathrm{max}}$. Note that it has been assumed that the solution behaves like an ideal solution because the excess potentials have been disregarded. The model applies neither to ion exchange chromatography, where the adsorption–desorption of the counterion must be included in the equilibrium scheme, nor to hydrophobic interaction chromatographic techniques because the excess potentials are not included, but nonetheless, it has found widespread application.

If competitive adsorption takes place, the multicomponent form of Equation 2.30 is

$$q_i = \frac{A_i c_i}{1 + \sum_j A_j c_j / q_j^{\max}} \tag{2.31}$$

2.5 Association Equilibria

It is a fundamental assumption [6] that the classical adsorption equilibrium entails a reversible association of some functional groups of the proteins with the functional groups of the immobilized ligands of the stationary phase and that adsorption–desorption of proteins on ion exchange adsorbents involve desorption–adsorption of small ions [7]. The interactions may also involve reversible associations of proteins to form various kinds of oligomers, displacement of water and small ions from the surface of the proteins, and interactions of solvents and additives with the proteins as well as the ligands.

Models of adsorptions isotherms can be derived from a general association scheme and the corresponding equilibrium conditions. The general association scheme is

$$\nu_{i,1} b_{i,1} + \nu_{i,2} b_{i,2} + \cdots \rightleftharpoons \nu_{i,j} b_{i,j} + \cdots + \nu_{i,m} b_{i,m}\,; \quad i = 1, 2, \ldots, \alpha \tag{2.32}$$

where $b_{i,j}$ designates the various species, $\nu_{i,j}$ designates the corresponding stoichiometric coefficients, subscripts, $i = 1, 2, \ldots, \alpha$, indicate the various association equilibria, and subscripts, $j = 1, 2, \ldots, m$, indicate the species in adsorption equilibrium number i.

When the association reactions have reached equilibrium, it must hold that the value of the Gibbs energy is at its global minimum, that is,

$$\Delta G_i = \sum_{j=1}^{m} \nu_{i,j} \mu_{i,j}(T, p, \mathbf{n}) = 0; \quad i = 1, 2, \ldots, \alpha \tag{2.33}$$

Here G is the Gibbs energy, and $\mu_{i,j}(T, p, \mathbf{n})$ is the chemical potential of species j in association reaction number i.

A general thermodynamic expression for the chemical potentials is

$$\mu_{i,j}(T, p, \mathbf{n}) = \mu_{i,j}^{0}(T, p) + \mu_{i,j}^{E}(T, p, \mathbf{n}) + RT \ln x_{i,j} \tag{2.34}$$

where $\mu_{i,j}^{0}(T, p)$ are the reference state potentials, $\mu_{i,j}^{E}(T, p, \mathbf{n})$ are the excess potentials, and $x_{i,j}$ are the mole fractions. The excess potentials include all potentials except the reference potential, that is, when electrolytes and other charged species are present in the solution, the electric potentials must be included in the excess potentials as analyzed in Section 2.A.1.5.

For the sake of brevity, the arguments will be omitted in the rest of this section and in all subsequent sections, but the Appendix provides the necessary details.

Combining Equations 2.33 and 2.34 shows that at equilibrium, for all $i = 1, 2, \ldots, \alpha$,

$$\sum_{j=1}^{m} \nu_{i,j}\mu_{i,j}^{0} + \sum_{j=1}^{m} \nu_{i,j}\mu_{i,j}^{E} + RT\sum_{j=1}^{m} \nu_{i,j} \ln x_{i,j} = 0 \tag{2.35}$$

It is convenient to define two auxiliary variables, namely, the equilibrium constants, K_i, and the equilibrium excess functions, Γ_i. The thermodynamic equilibrium constants, K_i, are defined as

$$RT \ln K_i \equiv -\sum_{j=1}^{m} \nu_{i,j}\mu_{i,j}^{0}; \quad i = 1,2,\ldots,\alpha \tag{2.36}$$

Similarly, the equilibrium excess functions, Γ_i, which account for all deviations from ideal solution behavior, are defined as

$$RT \ln \Gamma_i \equiv -\sum_{j=1}^{m} \nu_{i,j}\mu_{i,j}^{E}; \quad i = 1,2,\ldots,\alpha \tag{2.37}$$

The excess potentials are calculable from excess Gibbs energy models, electrostatic models, and equations of state or similar state function-based models, a subject analyzed in the Appendix.

The equilibrium relations, Equation 2.35, can be written in a neater form:

$$\ln K_i + \ln \Gamma_i = \sum_{j=1}^{m} \nu_{i,j} \ln x_{i,j}; \quad i = 1,2,\ldots,\alpha \tag{2.38}$$

or

$$K_i \Gamma_i = \prod_{j=1}^{m} x_{i,j}^{\nu_{i,j}}; \quad i = 1,2,\ldots,\alpha \tag{2.39}$$

The constraints are the material and the charge balances.

2.5.1 The Asymmetric Reference Potential

Excess potentials are zero for pure substances, but the pure state is not a convenient reference state for substances that are solids in the pure state like salts and proteins. The problem can be circumvented by changing the reference state potentials to include the excess potentials at infinite dilution in a solvent at a predefined constant solvent composition. The excess potentials at infinite dilution must be calculated at a predefined solvent composition in order to assure that the reference state potentials can be applied for solutions in pure and in mixed solvents. Due to its ubiquity, it is convenient to use pure water, w, as the reference solvent and use the symbol $\mu_{i,j}^{E_w^\infty}$ for the excess potentials at infinite dilution in pure water. The asymmetric excess potentials are defined as

$$\tilde{\mu}_{i,j}^{E} = \mu_{i,j}^{E} - \mu_{i,j}^{E_w^\infty} \tag{2.40}$$

and the asymmetric reference potentials are defined as

$$\tilde{\mu}_{i,j}^{0} = \mu_{i,j}^{0} + \mu_{i,j}^{E_w^\infty} \tag{2.41}$$

The introduction of the asymmetric excess potentials and the asymmetric reference potentials does of course not change the values of the potentials because

$$\tilde{\mu}_{i,j}^{0} + \tilde{\mu}_{i,j}^{E} = \mu_{i,j}^{0} + \mu_{i,j}^{E} \tag{2.42}$$

However, it allows for a greater flexibility in the development of models for excess potentials including salts and proteins. The details are discussed in Sections 2.A.1.3–2.A.1.5.

When the asymmetric reference potentials are used, the thermodynamic equilibrium constants, $\tilde{K}_{i,j}$, and the equilibrium excess functions, $\tilde{\Gamma}_{i,j}$, must be defined accordingly. The definitions are

$$RT\ln\tilde{K}_i = -\left(\sum_{j=1}^{m}\nu_{i,j}\mu_{i,j}^{0} + \sum_{j=1}^{m}\nu_{i,j}\mu_{i,j}^{E_w^\infty}\right) = -\sum_{j=1}^{m}\nu_{i,j}\tilde{\mu}_{i,j}^{0} \tag{2.43}$$

$$RT\ln\tilde{\Gamma}_i = -\left(\sum_{j=1}^{m}\nu_{i,j}\mu_{i,j}^{E} - \sum_{j=1}^{m}\nu_{i,j}\mu_{i,j}^{E_w^\infty}\right) = -\sum_{j=1}^{m}\nu_{i,j}\tilde{\mu}_{i,j}^{E} \tag{2.44}$$

Note that, although $K_i \neq \tilde{K}_i$ and $\Gamma_i \neq \tilde{\Gamma}_i$, it holds that $K_i\Gamma_i = \tilde{K}_i\tilde{\Gamma}_i$; wherefore

$$\tilde{K}_i\tilde{\Gamma}_i = K_i\Gamma_i = \prod_{j=1}^{m} x_{i,j}^{\nu_{i,j}} \tag{2.45}$$

Equation 2.45 will be used in the analysis of the various adsorption models.

2.6 The Classical Adsorption Isotherm

The classical adsorption isotherm originates when proteins in solution associate with the immobilized ligands of the stationary phase. The analysis presents a common form of the adsorption isotherm that includes the ion exchange adsorption isotherm and the hydrophobic adsorption isotherm. Conformational changes of the proteins do most likely take place when the proteins bind to the ligands, but in the classical model, the proteins do not associate, that is, act as ligands themselves. This subject will be analyzed in Section 2.9.

The classical ion exchange adsorption isotherm is analyzed in Section 2.7, and the adsorption isotherm of a GLP-1 derivative including simulations and an analysis of the simulated elution profiles are presented in Sections 2.7.1–2.7.1.4. The classical hydrophobic adsorption isotherm is analyzed in Section 2.8, and two examples are given. The first example in Section 2.8.1 is an analysis of the retention of lysozyme in hydrophobic interaction chromatography (HIC), the second example in Section 2.8.2 shows an analysis of the retention of three insulin variants on similar HIC adsorbents, and in Section 2.8.3 an analysis of differences and similarities between the two examples is presented.

2.6.1 Protein Association to Immobilized Ligands

Where proteins in solution associate with the immobilized ligands of the stationary phase, the general form of the association equilibria is derived from Equation 2.32:

$$\mathrm{P}_i^{z_i} + \nu_i \mathrm{L}^{z_\mathrm{L}}\mathrm{S}_\sigma^{z_\mathrm{S}} \rightleftharpoons \mathrm{P}_i^{z_i}\mathrm{L}_{\nu_i}^{z_\mathrm{L}}\mathrm{S}_{\sigma\nu_i(1-\beta_i)}^{z_\mathrm{S}} + \nu_i\beta_i\sigma \mathrm{S}^{z_\mathrm{S}}; \quad i = 1,2,\ldots,\alpha \tag{2.46}$$

The stoichiometric coefficient, ν_i, of species i is the number of immobilized ligands, L, associated with the protein, $\mathrm{P}_i^{z_i}$, carrying the charge z_i, and $\nu_i\beta_i$ is the fraction of ligands where the counterions, S, are displaced. $\beta_i = 0$ corresponds to pure hydrophobic interactions, and $\beta_i = 1$ corresponds to ionic interactions, that is, pure ion exchange. $\sigma = z_\mathrm{L}/z_\mathrm{S}$ where z_L is the charge number of the ligand and z_S is the charge number of the counterion. If the ligands do not carry any charged groups, $\sigma = 0$.

The equilibrium relation is given in Equation 2.45, and traditionally, mole fractions are replaced by ratios of molar concentrations. The mole fractions of the adsorbates, $\mathrm{P}_i^{z_i}\mathrm{L}_{\nu_i}^{z_\mathrm{L}}\mathrm{S}_{\sigma\nu_i(1-\beta_i)}^{z_\mathrm{S}}$, are q_i/c, the mole fraction of the free counterion, $\mathrm{S}^{z_\mathrm{S}}$, is c_S/c, the mole fractions of the solute proteins, $\mathrm{P}_i^{z_i}$, are c_i/c, and the mole fraction of the ligand associated with the counterion, $\mathrm{L}^{z_\mathrm{L}}\mathrm{S}_\sigma^{z_\mathrm{S}}$, is q_L/c, where c is the overall molar concentration in the pore volume. When the mole fractions are inserted in Equation 2.45, the eventual result is

$$\tilde{K}_i\tilde{\Gamma}_i = \frac{q_i}{c}\left(\frac{c_\mathrm{S}}{c}\right)^{\nu_i\beta_i\sigma}\frac{c}{c_i}\left(\frac{c}{q_\mathrm{L}}\right)^{\nu_i} \tag{2.47}$$

If $\tilde{\mu}^0$ denotes an asymmetric reference potential in the fluid phase and $\overline{\tilde{\mu}}^0$ denotes an asymmetric reference potential in the immobilized state and these potentials are inserted in Equation 2.43, the expression for thermodynamic equilibrium constants is

$$RT\ln\tilde{K}_i = -\left(\overline{\tilde{\mu}}_i^0 - \tilde{\mu}_i^0\right) - \nu_i\beta_i\sigma\left(\tilde{\mu}_S^0 - \overline{\tilde{\mu}}_S^0\right) = -\Delta\tilde{G}_i^0 + \nu_i\beta_i\sigma\Delta\tilde{G}_S^0 \tag{2.48}$$

where $\tilde{\mu}_i^0$ are the asymmetric reference potentials of the solute proteins, $\overline{\tilde{\mu}}_i^0$ are the asymmetric reference potentials of the immobilized proteins in the protein–ligand complexes, $P_i^{z_i}L_{\nu_i}^{z_L}S_{\sigma\nu_i(1-\beta_i)}^{z_S}$, $\tilde{\mu}_S^0$ is the asymmetric reference potential of the free counterion, and $\overline{\tilde{\mu}}_S^0$ is the asymmetric reference potential of the immobilized counterion. Finally, $\nu_i\beta_i\sigma$ is the number of counterions that changes state, that is, they are displaced by the proteins. The ligands do not undergo a phase change because they only exist in the immobilized state; wherefore they do not contribute. $\Delta\tilde{G}_i^0$ is the change in the reference potential of a protein when it goes from the immobilized state to the solute state, and $\Delta\tilde{G}_S^0$ is the change in the reference potential of the counterion when it goes from the immobilized state to the solute state.

The equilibrium excess functions, Γ_i, are calculable from the excess potentials. If $\tilde{\mu}^E$ denotes an asymmetric excess potential in the fluid phase and $\overline{\tilde{\mu}}^E$ denotes an asymmetric excess potential in the immobilized state, the expression for the equilibrium excess functions, defined in Equation 2.44, is

$$RT\ln\tilde{\Gamma}_i = -\left(\overline{\tilde{\mu}}_i^E - \tilde{\mu}_i^E\right) - \nu_i\beta_i\sigma\left(\tilde{\mu}_S^E - \overline{\tilde{\mu}}_S^E\right) \tag{2.49}$$

where $\tilde{\mu}_i^E$ are the asymmetric excess potentials of the solute proteins, $\overline{\tilde{\mu}}_i^E$ are the asymmetric excess potentials of the proteins in the association complexes, $P_i^{z_i}L_{\nu_i}^{z_L}S_{\sigma\nu_i(1-\beta_i)}^{z_S}$, $\tilde{\mu}_S^E$ is the asymmetric excess potential of the free counterion, and $\overline{\tilde{\mu}}_S^E$ is the asymmetric excess potential of the counterion bound to the ligand. The nature of the co-ion influences the excess properties of the proteins especially if the co-ion associates with the proteins as many divalent cations do.

The material balance including unavailable ligands, available ligands, and ligands bound to proteins is

$$\Lambda = q_L + \sum_{j=1}^{\alpha}(\nu_j + \lambda_j)q_j \tag{2.50}$$

Λ is the ligand density of the chromatographic adsorbent, and according to Brooks and Cramer [8], λ_j is the number of ligands unavailable for species j due to, for example, an unfavorable pore size distribution, shielding, or electrostatic exclusion, and q_L is the density of available ligands.

The maximum adsorption capacity is

$$q_j^{\max} = \frac{\Lambda}{(\lambda_j + \nu_j)} \tag{2.51}$$

and when Equation 2.50 is used to eliminate q_L from Equation 2.47, the model for the classical adsorption equilibria is

$$\frac{q_i}{c_i} = \tilde{K}_i \tilde{\Gamma}_i \left(\frac{c}{c_S}\right)^{\nu_i \beta_i \sigma} \left(\frac{\Lambda}{c}\right)^{\nu_i} \left(1 - \sum_{j=1}^{\alpha} \frac{q_j}{q_j^{\max}}\right)^{\nu_i} \tag{2.52}$$

The stoichiometric coefficients, especially in the ion exchange mode, depend on pH of the solution, and it is uncertain whether λ_i is independent of pH. Only experimental data can provide the correct information of the pH dependence of the maximum adsorption capacity.

2.7 The Classical Ion Exchange Adsorption Isotherm

In the classical ion exchange model, derived from Equation 2.52, β_i is unity:

$$\frac{q_i}{c_i} = \tilde{K}_i \tilde{\Gamma}_i \left(\frac{c}{c_S}\right)^{\nu_i \sigma} \left(\frac{\Lambda}{c}\right)^{\nu_i} \left(1 - \sum_{j=1}^{\alpha} \frac{q_j}{q_j^{\max}}\right)^{\nu_i} \tag{2.53}$$

The charge ratio $\sigma = z_L / z_S$ is a known quantity, usually unity. In this exposition it is assumed that the charge of the ion exchanger is impervious to pH, that is, it is a strong ion exchanger. The initial slope of the ion exchange adsorption isotherm is calculated from Equation 2.6 and is

$$A_i = \tilde{K}_i \tilde{\Gamma}_i \left(\frac{c}{c_S}\right)^{\nu_i \sigma} \left(\frac{\Lambda}{c}\right)^{\nu_i} \tag{2.54}$$

A straightforward way to calculate the thermodynamic retention factors, A_i, and the stoichiometric coefficients, ν_i, is to measure the isocratic retention volumes, $V_{R,i}$, at low to moderate salt concentrations and the isocratic retention volumes, $V_{NA,i}$, at high salt concentration and calculate the relative retentions, R_i, using Equation 2.12. The result is

$$\ln R_i = \ln(1 - \varepsilon_b)\varepsilon_P k_{d,i} + \ln \tilde{K}_i + \nu_i\left((\sigma - 1)\ln c + \ln \Lambda\right) - \nu_i \sigma \ln c_S + \ln \tilde{\Gamma}_i \tag{2.55}$$

In order to reduce the influence of the experimental uncertainty, it is recommended that $\ln R_i \geq 0.3$.

It is convenient to depict the experimental values of $\ln R_{A,i}$ versus $\ln c_S$. If the lines are straight lines, it shows that values of $\ln \tilde{\Gamma}_i$ are independent of the salt concentration. The equilibrium excess functions, $\tilde{\Gamma}_i$, account for the excess potentials, and if they are independent of the salt concentration, it is a strong indication of that the excess potentials neutralize. If the plot is a straight-line plot, as it very often is, the slopes of the straight lines give very accurate estimates of the stoichiometric coefficients. The stoichiometric coefficients, ν_i, depend on the pH of the eluant. The equilibrium constants are calculable from the values of the intersections of the straight lines with the ordinate at $\ln c_S = 0$ and numbers of the interstitial porosity, ε_b, the porosity of the adsorbent, ε_p, and the ligand density, Λ:

$$\ln \tilde{K}_i \tilde{\Gamma}_i = \ln R_i(\ln c_S = 0) - \ln(1-\varepsilon_b)\varepsilon_P k_{d,i} - \nu_i((\sigma - 1)\ln c + \ln \Lambda) \quad (2.56)$$

The equilibrium constant is independent of the salt concentration, and if the equilibrium excess functions do not display dependence of the salt concentration, one cannot distinguish $\tilde{K}_i$ and $\tilde{\Gamma}_i$ from one another. If a solvent, for example, ethanol or polyethylene glycol is added to the eluant, it will change the equilibrium excess functions, but it will not change the equilibrium constants if the excess potentials are normalized properly as analyzed in Section 2.5.1.

According to Equation 2.48, $\ln \tilde{K}_i$ is a linear function of the stoichiometric coefficient, which depends on pH. The change in the reference potential of a protein of species i, $\Delta\tilde{G}_i^0$, is specific to the protein–adsorbent pair. It is independent of the concentration of solvents, and it is reasonable to assume that it is independent of pH. The change in the reference potential of the salt, $\Delta\tilde{G}_S^0$, is specific to the salt–adsorbent pair, and it is independent of pH for strong electrolytes. If retention data are measured for several proteins using the same salt–adsorbent pair, $\Delta\tilde{G}_S^0$ is a common parameter.

If one saturates the column with a solution of a nitrate salt and injects a small pulse of pure water, the drop in the UV signal, when the pulse passes the detector, gives a good estimate of the retention volume of the salt. The retention volume of the salt provides the value of the total porosity, and the retention volumes of the proteins at a high salt concentration provide information of the parameters $k_{d,i}$. It is usually difficult to get information about the particle porosity; wherefore it is convenient to fix the interstitial porosity, ε_b, at a reasonable value of 0.37–0.45, and calculate ε_p and $k_{d,i}$, and check if the numbers make sense. $k_{d,i}$ is less than unity and decreases with increasing size of the molecule, and the particle porosity is usually larger than 0.5 and can be as high as 0.95 for soft gels.

The maximum adsorption capacities, $q_i^{\max}$, can be determined from a few static capacity measurements and breakthrough experiments or by fitting the model to experimental elution profiles obtained at high column load. An example is analyzed in Section 2.7.1.4.

2.7.1 The Adsorption Isotherm of a GLP-1 Derivative

In Section 2.7.1.1 it is explained how the parameters in the adsorption isotherm were estimated from measurements of the isocratic retention volumes supplemented with measurements of the adsorption capacity. The parameters are utilized to calculate the adsorption isotherm, the diffuse wave velocity, and the shock wave velocity of a GLP-1 derivative.

In Section 2.7.1.2, the isocratic elution profiles are simulated at loads from 0.1 mM·CV (column volume) up to 8.8 mM·CV.

In Section 2.7.1.3 it is analyzed how the interplay between the diffuse wave velocity and the shock wave velocity shapes the elution profile, and the result is utilized to analyze the simulated elution profiles.

In Section 2.7.1.4 it is shown how the slope of the trailing edge of an elution profile at a high load can be utilized to calculate approximate values of the second derivative of the classical adsorption isotherm.

2.7.1.1 The Adsorption Isotherm and the Wave Velocities

The parameters in the adsorption isotherms of a GLP-1 derivative and six contaminants were estimated from the isocratic retention measurements supplemented with some capacity measurements. The adsorbent is Source 30 Q and the modulator is sodium chloride. The interstitial porosity, $\varepsilon_b = 0.45$, the particle porosity, $\varepsilon_p = 0.57$, the ligand density, $\Lambda = 0.30$ mol Eq/L pore volume, and the reference Gibbs energy change of to counterion, Cl^-, and $\Delta G_s^0/RT = 0.320$ are from Ref. [9]. Four of the components have been used in a simulation study published in Ref. [10].

Figure 2.2 shows the measured isocratic elution data at analytical load for a GLP-1 derivative and six contaminants (T. B. Hansen and S. Kidal, Novo Nordisk A/S, private communication). It is important to observe that the injected amounts are so low that the adsorption isotherm is linear. This is tested by varying the injected amount in order to prove that the first moments of the recorded peaks are independent of the injected amount.

The adsorption capacity was measured for a GLP-1 derivative but not for the impurities. In order to calculate the maximum adsorption capacities for the impurities, it was assumed that for all components

$$q_{\max,i} = \frac{\Lambda}{v_i + \lambda} \tag{2.57}$$

A value of $\lambda = 4$ was estimated from the capacity data measured for a GLP-1 derivative, and this number was used for the contaminants as well. The parameters are summarized in Table 2.1. Figure 2.3 shows the calculated adsorption isotherm, the slope of the isotherm, and q/c of a GLP-1 derivative with 30 mM sodium chloride in the eluant.

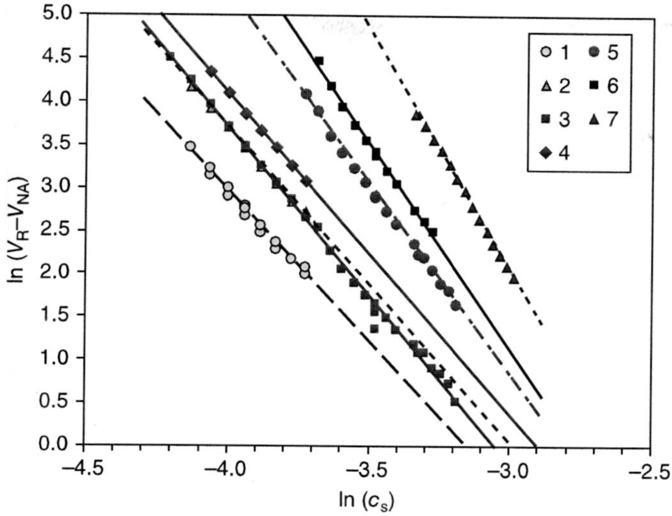

Figure 2.2 A log–log plot of the measured isocratic retention volumes of a GLP-1 derivative (■) and six contaminants on a Source 30 Q adsorbent at various sodium chloride concentrations in the eluant. (*See insert for color representation of the figure.*)

Table 2.1 Parameters for a GLP-1 derivative and six impurities estimated from the isocratic retention measurements on a Source 30 Q adsorbent.

Eluite	ν	K	$\frac{\Delta G_i^0}{RT}$	q_{max} (M)
1	3.47	0.00365	−6.72	0.0402
2	3.66	0.00459	−6.55	0.0392
3	3.90	0.00235	−7.30	0.0380
4	3.72	0.00591	−6.32	0.0389
5	4.43	0.00251	−7.40	0.0356
6	4.78	0.00185	−7.82	0.0342
7	5.53	0.00120	−8.49	0.0315

The shape of an elution profile is the result of the interplay between the diffuse wave velocity, the shock wave velocity, and the dispersion forces. Figure 2.4 depicts the calculated diffuse wave velocity and the shock wave velocity in the column with 30 mM sodium chloride in the eluant. The adsorption isotherm of the GLP-1 derivative has a negative curvature at all concentrations, and this means that the diffuse wave velocity will increase with increasing solute

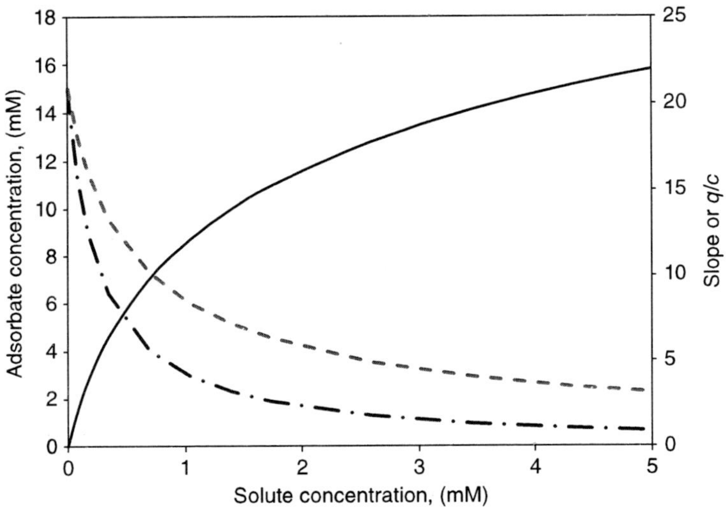

Figure 2.3 The adsorption isotherm (full line) of a GLP-1 derivative on a Source 30 Q adsorbent with 0.03 M sodium chloride in the eluant. The dashed–dotted line is the slope of the isotherm, and the dashed line is q/c.

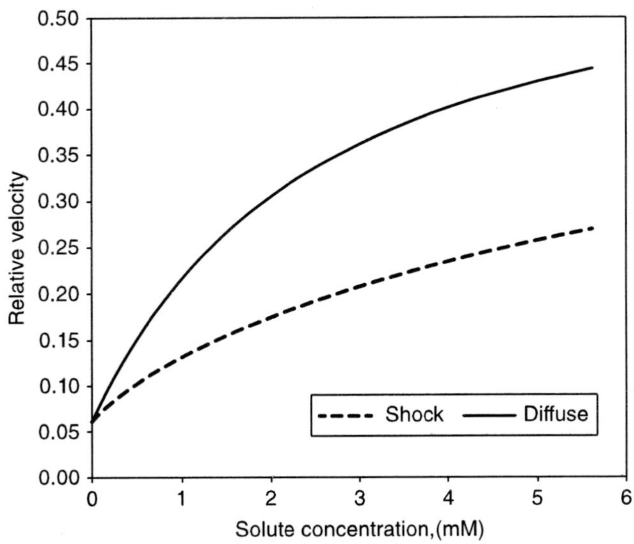

Figure 2.4 The calculated relative velocities of the diffuse wave and the shock wave of a GLP-1 derivative on a Source 30 Q adsorbent with 0.03 M sodium chloride in the eluant.

concentration. This leads to a compression at the leading edge of the elution profile; wherefore a shock wave determines the velocity at the leading edge of the elution profile. The trailing edge of the elution profile can be much curved because the diffuse wave velocity decreases with decreasing concentration and that will, in addition to the dispersion forces, increase the distance between the top and the base of the rear of the elution profile.

2.7.1.2 Simulations

Figure 2.5 depicts the result of a series of simulations. The load was varied from 0.1 to 8.8 mM·CV. The column was equilibrated with a 5 mM sodium chloride solution, and the load was a solution of 0.2 mM of a GLP-1 derivative in a 5 mM sodium chloride solution. The eluant was a 30 mM sodium chloride solution. The simulated elution profiles display the curved trailing edge that is a characteristic of the classical adsorption isotherm. The plateau at high load indicates that the elution profile reached a steady state at the leading edge. The maximum load was 8.8 mM·CV. When the load was increased to 9 mM·CV, there was a breakthrough during the load step, and part of the feed was lost. Except in a region close to the maximum of an elution profile, where the curvature is changing, the trailing edges of the simulated elution profiles are alike. In the absence of dispersion forces, all elution profiles will end at a retention volume of 7.34 CV.

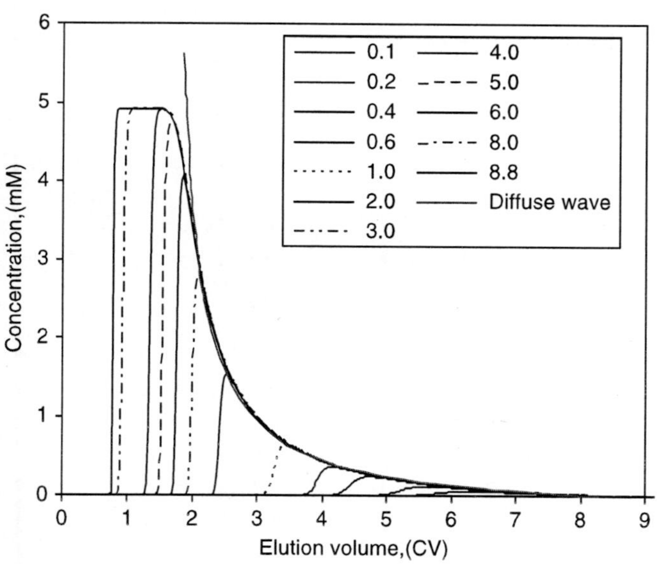

Figure 2.5 Simulated isocratic elution profiles of a GLP-1 derivative on a Source 30 Q adsorbent with 0.03 M sodium chloride in the eluant. The numbers in the legend indicate the load in mM·CV. The concentration of GLP-1 in the feed was 0.2 mM. The solid line that extends above the top of the simulated elution profiles shows the shape of the trailing edge of the ideal elution profile created by the diffuse wave.

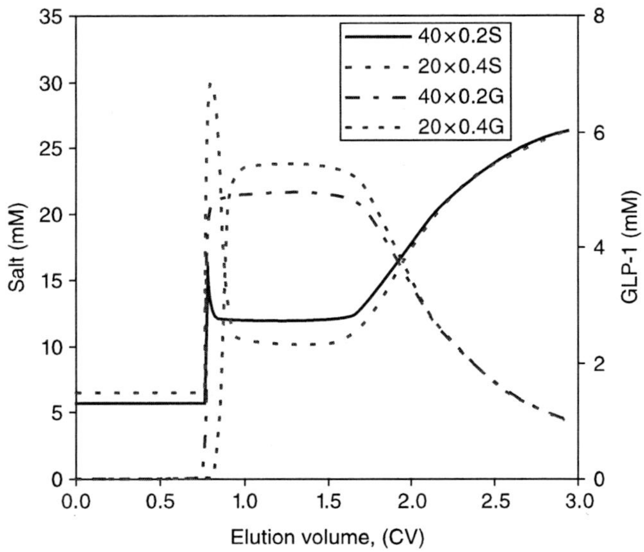

Figure 2.6 Simulated isocratic elution profiles of a GLP-1 derivative on a Source 30 Q adsorbent with 0.03 M sodium chloride in the eluant and a load of 8.8 mM·CV using two different feed concentrations. The first number in the legend is the number of column volumes loaded, and the second number is the mM concentration of GLP-1 in the feed. G means the GLP-1 elution profile, and S means the salt elution profile.

To analyze the influence of the feed concentration on the shape of the simulated elution profiles, simulations were performed where the feed volumes were decreased to one half of the values used in the first series of simulations and the concentration in the feed doubled in order to keep identical loads. At loads up to 4 mM·CV, there was no difference between the simulated elution profiles. However, when the load was increased to 5 mM·CV, a minor difference was observed, a difference that increased with increasing load. Figure 2.6 shows the elution profiles at a load of 8.8 mM·CV and the corresponding salt profiles at two different feed concentrations. The spike in the salt concentration profile started at a volume corresponding to the total porosity. The higher feed concentration gave a higher adsorbate concentration; wherefore the response from the load with the higher feed concentration gave a higher value of the concentration at the plateau. Consequently, the leading edge of this profile must elute a little later because the area under the two elution profiles must be identical as they have identical loads.

The simulations were carried out using the Craig model because it is much simpler to use especially when the adsorption equilibria become complex due to oligomer formation. In the Craig model, the dispersion due to the longitudinal dispersion and the mass transfer resistance in the system is characterized

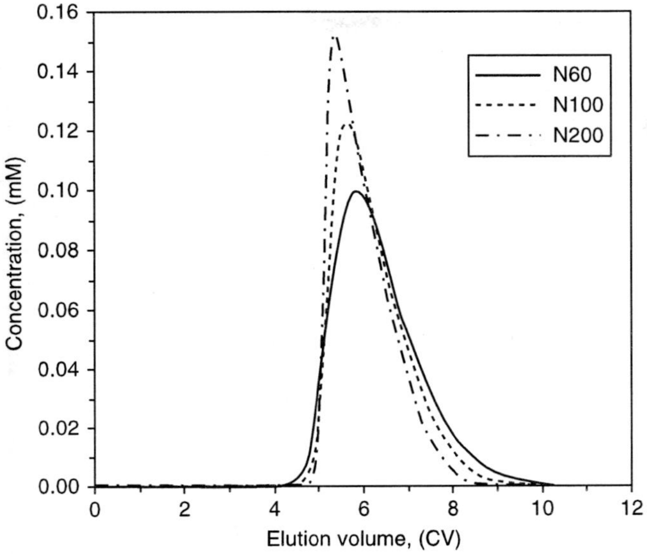

Figure 2.7 Simulated isocratic elution profiles of a GLP-1 derivative on a Source 30 Q adsorbent using 60, 100, and 200 cells in the Craig model. The eluant was a 0.03 M sodium chloride solution, and the load was 0.2 mM·CV.

by the number of cells; a high number of cells correspond to a low dispersion [1]. One hundred cells correspond to a moderate dispersion. To simulate a symmetric peak required more than a thousand cells and an extremely tiny load. The simulations shown in Figures 2.5 and 2.6 were performed using 100 cells. Figure 2.7 shows how the number of cells influenced the dispersion at a load of 0.2 mM·CV. It is easy to observe the influence of the dispersion at this low load where the adsorption isotherm is almost linear and the dispersion forces have a great influence on the second central moment of the elution profile. The influence of the number of cells on the simulated elution profiles decreases strongly with increasing load.

2.7.1.3 How the Wave Velocities Shape the Elution Profiles

The scope of this section is to analyze the simulated elution profiles shown in Figure 2.5. Chromatography is a rate-controlled separation where the shape and the position of an elution profile is the result of the interplay between the diffuse wave velocity, the shock wave velocity, and the dispersion forces. However, at preparative load, the main features of an elution profile are explained by the interplay between the two wave velocities of the solute alone. Disregarding the dispersion forces, the relation between the concentration of the eluite and the elution volume is calculable from the wave velocities using Equation 2.5 alone.

The first example is depicted in Figure 2.8. The salt concentration in the feed pulse and in the isocratic elution was 30 mM. The corresponding diffuse wave and shock wave velocities are depicted in Figure 2.4, and it is observed that the diffuse wave moves faster than the shock wave. The feed concentration was 5 mM and the feed started at $V_i = 0$ and it lasted for one column volume, that is, the load was 5 mM·CV. The feed pulse is shown as the dotted square function from 0 to 1 CV. The front of the feed pulse at $V_i = 0$ represented a step-up in the concentration from 0 to 5 mM, and this step-up created a shock wave that moved at a velocity of 0.256 times the interstitial velocity corresponding to an elution volume of 1.75 CV.

The step-down in the concentration at the end of the feed pulse at $V_i = 1$ created a diffuse wave. The diffuse wave velocity decreased with decreasing concentration, which means that the top of the trailing edge moved faster than the base of the trailing edge and that resulted in a strong tailing. At a concentration of 5 mM, the diffuse wave velocity was 0.429 times the interstitial velocity corresponding to an elution volume of 1.05 CV, and at the baseline the relative velocity was 0.061 corresponding to an elution volume of 7.34 CV.

At a concentration of 5 mM, the diffuse wave created at the rear of the feed pulse moved faster, $V_i = 1.05$, than the shock wave created at the front of the feed pulse, $V_i = 1.75$, and thus if the feed volume had been <0.7 CV, the diffuse wave at the top of the trailing edge had run down the shock wave at the leading edge and eliminated the plateau of the feed pulse completely. In this example, the feed lasted 1 CV, and this means the shock wave left the column 0.3 CV before the top of the diffuse wave reached the column exit.

The elution volume of the shock wave at the leading edge of the elution profile is shown in Figure 2.8 as the vertical dashed–dotted line at 1.75 CV, and the elution profile of the trailing edge is depicted as the curved line starting at $V_i = 1.05 + 1$ CV and ending at $V_i = 7.34 + 1$ CV. The original plateau of 1 CV of the feed pulse was reduced to a plateau of 0.3 CV. The area encircled by the elution profile and the baseline is 5 mM·CV. The arrows connected with a dashed–dotted line indicate the distance between the front of the feed pulse and the resulting shock wave, and the arrows connected with a solid line indicate the rear of the feed pulse and the resulting trailing edge of the elution profile.

The second example is depicted in Figure 2.9. The feed concentration and the salt concentration are identical with the concentrations in the previous example. The feed pulse started at $V_i = 0$, and it lasted for half a column volume, that is, the load was 2.5 mM·CV. The feed pulse is shown as the dotted square function from 0 to 0.5 CV. In the first example, the shock wave generated at the front of the feed pulse made it to the column exit because the duration of the feed pulse was larger than 0.7 CV, but in the second example, the duration of the feed pulse was reduced to 0.5 CV. This means that the top of the diffuse wave ran down the shock wave before the column exit and that diminished the

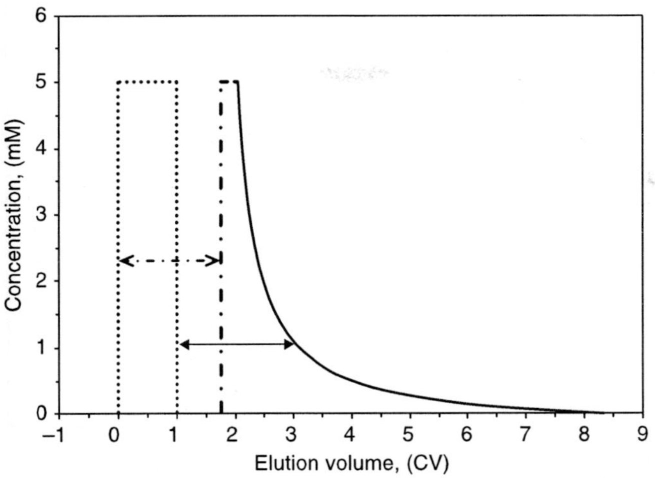

Figure 2.8 The ideal elution profile of a GLP-1 derivative on a Source 30 Q adsorbent with 0.03 M sodium chloride in the feed and in the eluant. The dotted square from 0 to 1 CV represents a load of 5 mM·CV. The front of the feed pulse enters at $V_i = 0$. The curved line is the trailing edge of the elution profile, and the vertical dashed–dotted line at $V_i = 1.75$ CV is the leading edge of the elution profile. The arrows connected with a dashed–dotted line indicate the distance between the front of the feed pulse and the resulting shock wave. The arrows connected with a solid line indicate the rear of the feed pulse and the resulting trailing edge of the elution profile.

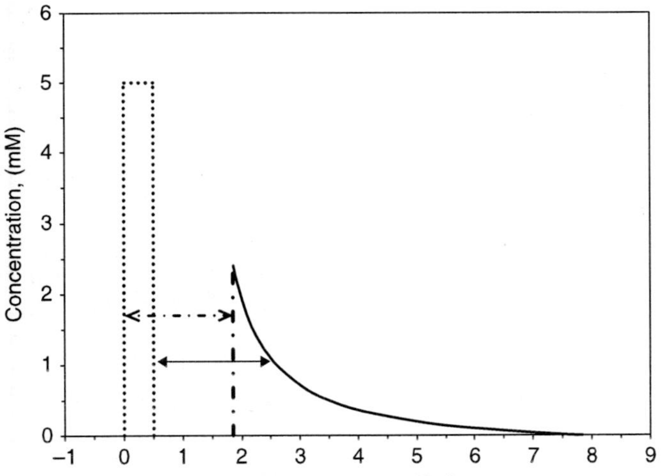

Figure 2.9 The ideal elution profile of a GLP-1 derivative on a Source 30 Q adsorbent with 0.03 M sodium chloride in the feed and in the eluant. The dotted square from 0 to 0.5 CV represents a load of 2.5 mM·CV. The front of the feed pulse enters at $V_i = 0$. The curved line is the trailing edge of the elution profile, and the vertical dashed–dotted line at $V_i = 1.86$ CV is the leading edge of the elution profile. The arrows connected with a dashed–dotted line indicate the distance between the front of the feed pulse and the resulting shock wave. The arrows connected with a solid line indicate the rear of the feed pulse and the resulting trailing edge of the elution profile.

concentration at the leading edge of the concentration profile in the column and slowed down the speed of the shock wave. The elution volume at the leading edge must therefore be determined from the material balance.

The elution profile at the trailing edge, resulting from a diffuse wave, is the full line ending at $V_i = 7.34 + 0.5$ CV. The position of the shock wave at the leading edge of the elution profile was determined by the fact that the area under the trailing edge of the elution profile from $V_i = 7.84$ CV to the leading edge of the elution profile must be equal to the area of the feed pulse, that is, 2.5 mM·CV. The material balance showed that the shock wave eluted at $V_i = 1.86$ CV. The leading edge is shown as the vertical dashed–dotted line at $V_i = 1.86$ CV on the figure. The arrows connected with a dashed–dotted line indicate the distance between the front of the feed pulse and the resulting shock wave. The arrows connected with a solid line indicate the rear of the feed pulse and the resulting trailing edge of the elution profile.

When comparing Figures 2.5, 2.8, and 2.9, it is easy to identify some very important similarities. The result of the analysis, the elution profiles shown in Figures 2.8 and 2.9, explains the curvature of the trailing edges observed in the simulated elution profiles and the steep leading edges as well. The simulations also account for the dispersion forces, and that explains why the leading edges in the simulated elution profiles are not vertical. In the calculations shown in Figures 2.8 and 2.9, the development of the leading edge of the elution profile, resulting from the step-up in the feed concentration, started at $V_i = 0$ CV, and the development of the trailing edge of the elution profile, resulting from the step-down in the feed concentration, started when the feed pulse ended at $V_i = 1$ and 0.5 CV, respectively. In a simulation, the load was focused in the column at low salt concentration at the column entrance before the elution started. This means that the development of the trailing edge of an elution profile started at $V_i = 0$ at all loads; wherefore all trailing edges of the elution profiles form a common envelope up to the inflection point on the trailing edge.

The trailing edge of the ideal elution profile is indicated in Figure 2.5 as the thin full line that extends above the plateau of the elution profiles. It is depicted with a minor horizontal displacement in order to make it coincide with the envelope of the trailing edges of the simulated elution profiles. There is a good agreement between the curvatures of the two profiles at concentrations up to 3 mM. The agreement will probably depend on the dispersion in the column.

2.7.1.4 Modeling the Trailing Edge of a Peak at High Load

The slope of the trailing edge of the ideal elution profile, $\partial V_i / \partial c$, is proportional to the second derivative of the adsorption isotherm. The analysis presented in Section 2.7.1.3 showed the similarity in the slope of the trailing edge of the ideal elution profile and the slope of the envelope of the trailing edges of the simulated elution profiles at high load. The comparison is depicted in Figure 2.5. This observation was utilized to calculate the second derivative of the adsorption isotherm from the simulated data, and finally, the calculated values of the second derivative were used to estimate the maximum adsorption capacity. If the retrieved value of the maximum adsorption capacity is close to the model parameter, it shows that this approach can be utilized with some confidence when experimental data replace the simulated data.

In order to model the trailing edge of an elution profile at high load, the following approximations are used:

$$V_i(c) \simeq d + V_{\text{diffuse}} = d + V_{NA,i} + (1-\varepsilon_b)\varepsilon_p k_{d,i}\frac{\partial q}{\partial c}$$
$$= d + V_{NA,i} + (1-\varepsilon_b)\varepsilon_p k_{d,i}\frac{\partial F}{\partial c} \tag{2.58}$$

where d is a horizontal displacement that accounts for the dispersion and $F(c)$ is a function utilized to fit the curvature of the trailing edge. It must have the property that $F'(0) = A_i$, and it must be applicable up to the inflection point where $\partial^2 V/\partial c^2 \le 0$. A quadratic isotherm [1] is a suitable choice for $F(c)$:

$$F(c) = \frac{A_i c + ac^2}{1 + Bc + bc^2} \tag{2.59}$$

That is,

$$\frac{\partial F}{\partial c} = \frac{A_i + 2ac}{1 + Bc + bc^2} - \frac{(A_i c + ac^2)(B + 2bc)}{(1 + Bc + bc^2)^2} \tag{2.60}$$

and according to Equation 2.58

$$\frac{\partial^2 F}{\partial c^2} = \frac{\partial^2 q}{\partial c^2} \tag{2.61}$$

To test the hypothesis, the model, Equation 2.58, was fitted to the trailing edge of the elution profile at a load of 5 mM·CV in the interval $7.34 \ge V_i \ge 2$ CV, corresponding to a concentration interval of $0.03 \le c_i \le 3.6$ mM. The number 7.34 CV is the value of the ideal elution volume at $c_i = 0$, and at elution volumes <2 CV, the curvature of the elution profile changes. The fitted parameters, a, B, b, and d, are given in Table 2.2.

To test the validity of the approximation the residual sum of squares,

$$\text{RSSQ} = \Sigma\left(\frac{\partial^2 F}{\partial c^2} - \frac{\partial^2 q}{\partial c^2}\right)^2 \tag{2.62}$$

Table 2.2 Estimated parameters in $F(c)$, Equation 2.59.

A_i (mM^{-1})	a (mM^{-2})	B (mM^{-1})	b (mM^{-2})	d (CV)
20.98	25.5	3.21	0.946	0.64

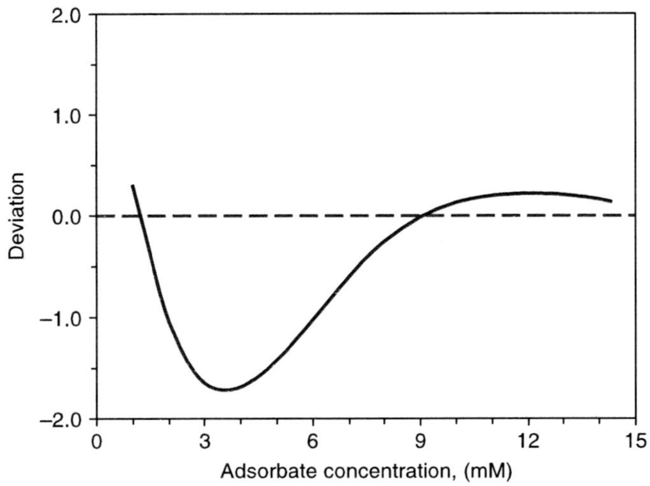

Figure 2.10 The solid line shows the deviation between the second derivative of the isotherm and the second derivative of the $F(c)$ function. Details are given in Section 2.7.1.4.

was minimized using 14 specified adsorption capacities from 1 to 14 mM at intervals of 1 mM by adjusting the maximum adsorption capacity. The corresponding solute concentrations ranged from 0.05 to 3.3 mM. The second derivative $\partial^2 q/\partial c^2$ was calculated from Equation 2.53. The estimated value of the maximum adsorption capacity was 39.5 mM, which is almost identical with the value used in the simulations, which was 39.16 mM. The deviation plot is shown in Figure 2.10.

As for any least square parameter estimation method, this method will be sensitive to the selected data interval. The influence of the load, the salt, and the solute concentrations in the feed has not been explored. A high load is most likely preferable. The influence of the dispersion on the reliability of the results was not investigated either.

2.8 Hydrophobic Adsorbents, HIC and RPC

Adsorbents in HIC are less hydrophobic than adsorbents in reversed-phase chromatography (RPC). The modulator in HIC is salt, and the adsorbate is eluted by reducing the salt concentration in the eluant. In RPC the modulator is an organic water-soluble solvent, and the adsorbate is eluted by increasing the solvent concentration in the eluant.

The classical adsorption isotherm model is derived from Equation 2.52 when $\sigma = 0$:

$$\frac{q_i}{c_i} = \tilde{K}_i \tilde{\Gamma}_i \left(\frac{\Lambda}{c}\right)^{\nu_i} \left(1 - \sum_{j=1}^{\alpha} \frac{q_j}{q_j^{\max}}\right)^{\nu_i} \tag{2.63}$$

This adsorption isotherm has, like the corresponding ion exchange adsorption isotherm, a negative curvature, that is, the solute velocity increases with increasing concentration; wherefore a shock wave is formed at the leading edge of the elution profile, and a diffuse wave shapes the trailing edge of the elution profile. The excess potentials play a crucial role in the modeling in contrast to what is commonly observed in ion exchange chromatography. The excess potentials of macro-ions are functions of the nature and the concentrations of salt and solvents and of the temperature and pH. In Section 2.A.4, the Kirkwood theory of dipolar ions is analyzed, a theory that accounts for salting-in and salting-out potentials of macro-ions.

The initial slope of a classical hydrophobic adsorption isotherm is

$$A_i = \left(\frac{\Lambda}{c}\right)^{\nu_i} \tilde{K}_i \tilde{\Gamma}_i = \left(\frac{\Lambda}{c}\right)^{\nu_i} \exp\left(-\frac{\Delta G_i^0}{RT}\right) \tilde{\Gamma}_i = A_{0,i} \tilde{\Gamma}_i \tag{2.64}$$

Isocratic retention data measured on HIC adsorbents can be utilized to develop models of the equilibrium excess functions of the retained components. $A_{0,i}$ varies with the solvent concentration, and $\tilde{\Gamma}_i$ depends on the salt and solvent concentrations. When measuring reversed-phase retention data, it is thus essential to measure the retention data at constant solvent composition and vary the salt concentration because otherwise one cannot distinguish the variation of $A_{0,i}$ with the solvent composition from the variation of $\tilde{\Gamma}_i$ with the solvent composition.

The fundamental equation for the relative retention in HIC and RPC is

$$\ln R_i = \ln(1 - \varepsilon_b)\varepsilon_P k_{d,i} + \ln \tilde{K}_i + \nu_i \ln\left(\frac{\Lambda}{c}\right) + \ln \tilde{\Gamma}_i \tag{2.65}$$

where

$$RT \ln \tilde{\Gamma}_i = -\left(\overline{\tilde{\mu}}_i^E - \tilde{\mu}_i^E\right) \tag{2.66}$$

Since $\ln \tilde{\Gamma}_i$ is a function that accounts for the contributions of the excess potentials, it is common practice to depict the experimental values of $\ln R_i$

versus the salt concentration. The graph will show how $\ln \tilde{\Gamma}_i$ varies with the salt concentration at constant solvent composition, and in many cases, a linear dependence of the salt concentration can be observed. In HIC, it is common to observe salting out, and in RPC one often observes salting in.

In Section 2.8.1, an analysis of the modeling of the retention of lysozyme on the Toyopearl Butyl-650S, Phenyl-650S, and Toyopearl Ether-650S adsorbents is presented. In Section 2.8.2, the retention data of three insulin components measured on the Toyopearl Butyl-650S and Toyopearl Phenyl-650S adsorbents are analyzed and modeled. Finally, in Section 2.8.3 similarities and differences in the retention behavior of lysozyme and the insulin components are analyzed, and some preliminary conclusions are drawn.

2.8.1 The Adsorption of Lysozyme

The isocratic retention volumes of lysozyme were determined on the Toyopearl Butyl-650S, Phenyl-650S, and Ether-650S adsorbents, from Tosoh Bioscience at pH 7 [11]. These data were utilized to develop a model of the equilibrium excess function of the lysozyme component at infinite dilution. In order to develop a model that also accounts for the influence of the lysozyme concentration, the model was extended to include correlation of the solubility of lysozyme.

Figure 2.11 depicts the experimental and correlated isocratic retention volumes of lysozyme on Toyopearl Butyl-650S at pH 7 and the predicted retention

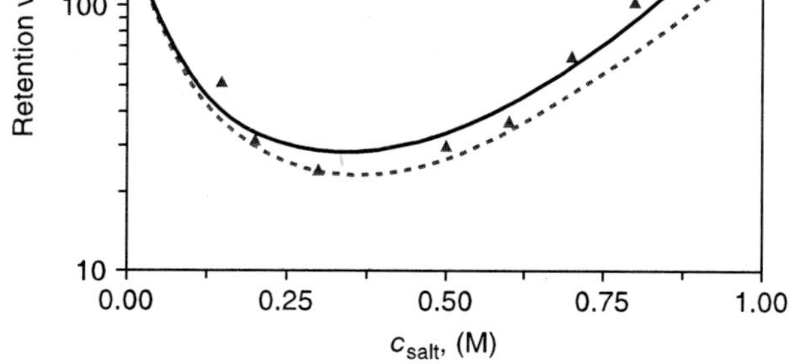

Figure 2.11 Experimental and modeled isocratic retention volumes of lysozyme on a Tosoh Butyl-650S adsorbent at pH 7 with ammonium sulfate in the eluant. The triangles show the experimental data, and the solid line shows the correlation. The dashed line is the predicted retention volume at pH 4.

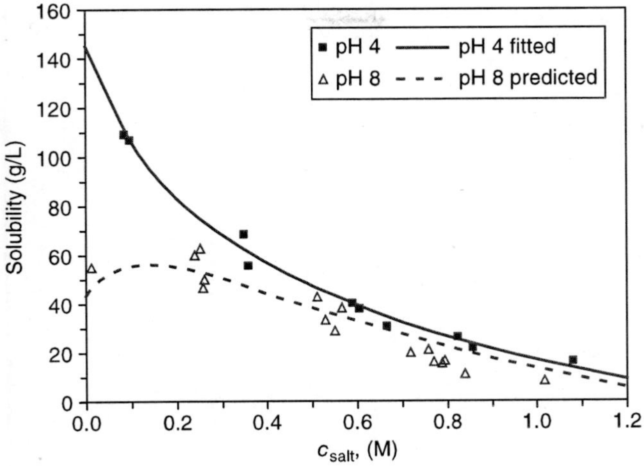

Figure 2.12 Experimental and calculated solubility curves of lysozyme in an aqueous solution of ammonium sulfate. The squares and the triangles are experimental data, and the full line is the correlated solubility curve at pH 4. The dotted line is the predicted solubility curve at pH 8.

volumes at pH 4. In Ref. [12] it was shown that $A_{0,i}$ for lysozyme did not display any observable pH dependence. Figure 2.12 depicts the correlated solubility of lysozyme at pH 4 and the predicted solubility at pH 8. The excess potential model is summarized in Equation 2.67. The $\ln\tilde{\Gamma}_i^{K_w}$ term was developed from Kirkwood's theory of dipolar ions in an electrolyte solution, and the additional terms are derived from an excess Helmholtz energy model analyzed in Section 2.A.1.4.

$$\ln\tilde{\Gamma}_\mathrm{L} = \ln\tilde{\Gamma}_\mathrm{L}^{K_w} + a_\mathrm{S}c_\mathrm{S} + a_\mathrm{L}c_\mathrm{L} + a_\mathrm{SL}c_\mathrm{S}c_\mathrm{L} \tag{2.67}$$

Here a_L, a_S, and $\mathrm{a_{SL}}$ are adjustable pH-independent parameters, c are molar concentrations, subscript L denotes lysozyme, and subscript S denotes salt. The parameters were determined from the lysozyme solubility data at pH 4. What is interesting in this context is that the solubility data at pH 4 do not display a maximum in the solubility, whereas the predicted solubility at pH 8 does in agreement with the experimental data. Figure 2.13 shows the values of the equilibrium excess function at pH 4 and 8 as function of the lysozyme concentration at four different salt concentrations.

The measured retention data of lysozyme on the Toyopearl Butyl-650S and Ether-650S adsorbents also included retention measurements with 6.1 weight % ethanol in the eluant. Figure 2.14 depicts the experimental and the calculated thermodynamic retention factors with and without the admixture of ethanol. The solid lines are the model using a pure aqueous solvent, and the dashed

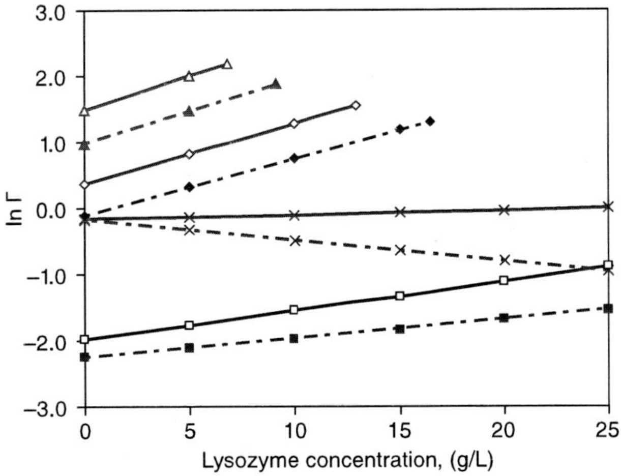

Figure 2.13 Calculated equilibrium excess function of lysozyme in four ammonium sulfate solutions at pH 4 (dashed–dotted lines) and at pH 8 (full lines). The ammonium sulfate concentrations are × 0.01 M, squares 0.3 M, diamonds 1 M, and triangles 1.2 M. The lines at 1 and 1.2 M salt terminate at the solubility limit of lysozyme.

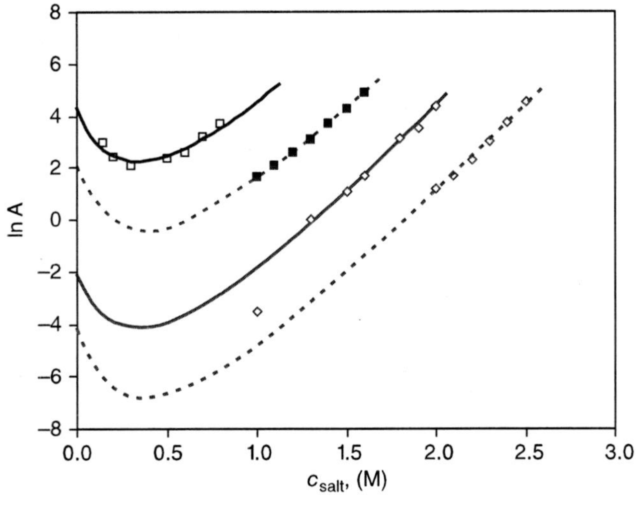

Figure 2.14 Experimental and correlated thermodynamic retention factors of lysozyme at pH 7 on the Toyopearl Butyl-650S (squares) and the Toyopearl Ether-650S (diamonds) adsorbents. The symbols show the experimental values, and the lines are the correlations. The eluant is an aqueous solution of ammonium sulfate. Solvents: solid lines, pure water; dashed lines, water with an admixture of 6.1 weight % ethanol.

lines are the model using an eluant with an admixture of ethanol. There are two important observations to make. Firstly, the two solid lines are parallel lines, and the two dashed lines are also parallel lines, but the slope has changed a little due to the admixture of ethanol. Secondly, the vertical distance between the two solid lines is identical with the vertical distance between the two dashed lines. That is, the decrease in $\ln A_{0,i}$ due to the admixture of ethanol is independent of the adsorbent. These observations will be included in the analysis of the retention data of three insulin components on Toyopearl Butyl-650S and Toyopearl Phenyl-650S in Section 2.8.2.

2.8.2 The Retention of Three Insulin Components on Two HIC Adsorbents

The data utilized in this section were part of a study published by Johansson et al. [13]. The project included retention studies of three insulin variants on the Toyopearl Butyl-650S and Toyopearl Phenyl-650S adsorbents and two reversed-phase adsorbents, a C4 and a C18 adsorbent. The measured retention data on the RPC adsorbents were measured at constant salt concentration. The data showed that salting in is predominating in RPC as predicted by the model developed from the Kirkwood theory [14].

The insulin variants were insulin aspart, insulin desB30, and an insulin ester. The modulator in the eluant was potassium chloride, and three different eluants were used. The first eluant was pure water, the second eluant was water with an admixture of 5 weight % ethanol, and the third eluant had an admixture of 10 weight % ethanol. Perfect straight lines could represent the data of ln A_i versus the salt concentration in the eluant, but the parallel behavior observed for the retention of lysozyme, analyzed in the previous section, did not materialize [13].

The model that was used to correlate the thermodynamic retention factor of the three insulin variants was

$$\ln A_i = \ln \tilde{K}_i + \nu_i \ln\left(\frac{\Lambda}{c}\right) + \ln \tilde{\Gamma}_i = \ln A_{0,i} + b_i c_{\mathrm{KCl}} \tag{2.68}$$

With a minor adjustment of the original parameters, a systematic change in $\ln A_{0,i}$ was observed for the three insulins. A comparison of the old and the new model for the insulin ester is shown in Figure 2.15. Figure 2.16 depicts the parameters for the insulin ester versus the ethanol concentration in the eluant. The dashed lines are trend lines to indicate that b_i is not a linear function of the ethanol concentration. Table 2.3 gives the values of the new parameters $\ln A_{0,i}$ and b_i. It is observed that the difference between the $\ln A_{0,i}$ parameters on the two adsorbents is independent of the solvent concentration. For the insulin ester, the difference is 1.35, for insulin desB30 the difference is 1.04, and for

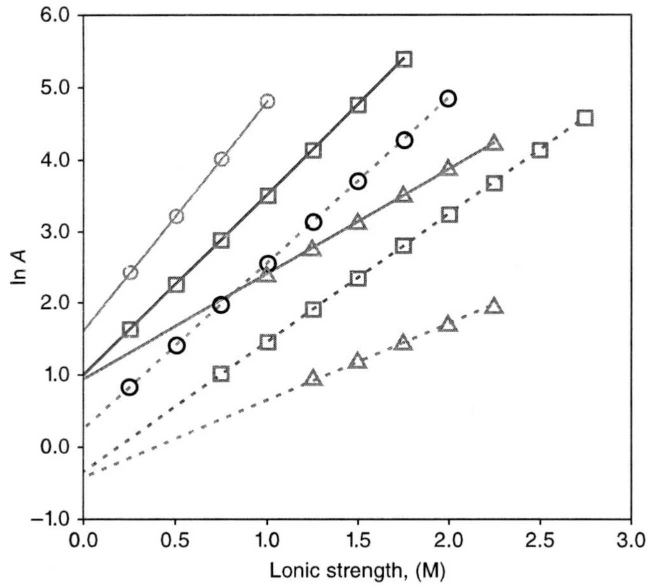

Figure 2.15 A comparison of the old and the new correlation of the thermodynamic retention factor of the insulin ester. The symbols represent the old correlation, and the lines represent the new correlation: circles 0% ethanol, squares 5% ethanol, and triangles 10% ethanol. Adsorbents, full lines, Toyopearl Butyl-650S; dashed lines, Toyopearl Phenyl-650S.

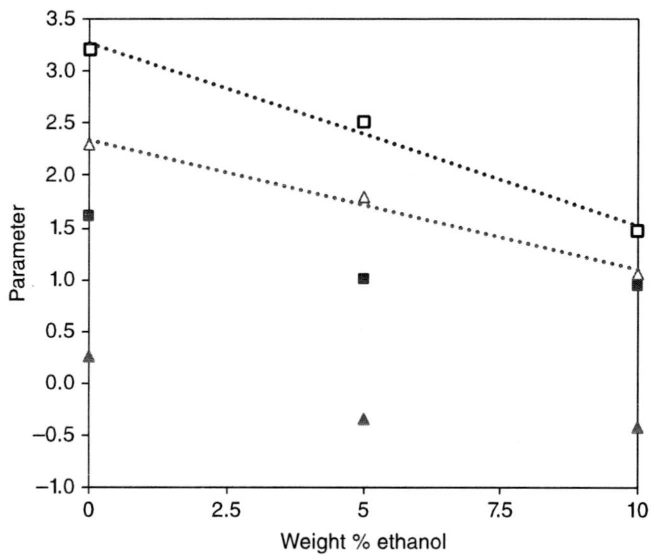

Figure 2.16 The parameters in the retention model of an insulin ester on the Toyopearl Butyl-650S (squares) and the Toyopearl Phenyl-650S (triangles) adsorbents with different concentrations of ethanol in the eluant. The open symbols are the values of the linear slopes of the retention model, b_i, and the solid symbols are the values of $\ln A_{0,i}$. The dashed lines are trend lines to indicate that b_i is not a linear function of the ethanol concentration.

Table 2.3 Isocratic retention parameters for three insulin variants on Toyopearl Butyl-650S and Toyopearl Phenyl-650S adsorbents.

Solvent		**Water**		**5% Ethanol in water**		**10% Ethanol in water**	
Insulin	**Adsorbent**	**ln A_0**	***b***	**ln A_0**	***b***	**ln A_0**	***b***
Insulin aspart	Butyl	0.260	3.67	0.000	2.55	0.171	1.35
	Phenyl	−0.644	2.24	−0.904	1.54	−0.733	0.75
Insulin desB30	Butyl	0.854	3.45	0.430	2.47	0.489	1.36
	Phenyl	−0.189	2.23	−0.613	1.60	−0.554	0.84
Insulin ester	Butyl	1.618	3.21	1.014	2.51	0.939	1.47
	Phenyl	0.269	2.29	−0.335	1.79	−0.411	1.06

Table 2.4 The calculated additional loss of the wetted surface area parameter, $\eta_r \Delta \mathbb{A}$, of the insulin variants during adsorption on Toyopearl Butyl-650S.

Insulin	**$\eta_r \Delta(\Delta \mathbb{A})$ (Å^2)**
Insulin aspart	514
Insulin desB30	441
Insulin ester	330

insulin aspart it is 0.90. This observation is in agreement with the observation made for the lysozyme retention data where the difference in $\ln A_{0,i}$ was 2.07.

The calculated thermodynamic retention factors for lysozyme, depicted in Figure 2.14, show that the slopes, $(\partial \ln A / \partial c_{\text{salt}})$, are independent of the adsorbent but depend on the admixture of a solvent. This means that the estimated interaction between lysozyme and the adsorbent did not depend on the individual adsorbent. However, the measured retention data for the insulin variants on the Toyopearl adsorbents show that the slopes, b_i, depend on the nature of the adsorbent and the solvent concentration. The ratios of b_i for the butyl to the phenyl adsorbents are almost independent of the solvent concentration. For the insulin ester, this ratio is 1.4. The chemical potential of the insulin ester in the mobile phase is independent of the adsorbent; wherefore the 40% increase in the parameter b_i must be due to a stronger hydrophobic interaction between the insulin ester and the butyl adsorbent when compared with the phenyl adsorbent. The hydrophobic interaction between the insulin and the adsorbent changes the conformation of the adsorbate; the insulin partly unfolds and that reduces the surface area exposed to the solvent. The change in the surface potential is proportional to the loss of the wetted surface area exposed to the solvent as explained in Section 2.A.1.6. Equation A28 shows that

$$\bar{\mu}_i = \bar{\mu}_i^0 - N_A \eta_i \Delta\mathbb{A} \tag{2.69}$$

N_A is Avogadro's number, η_i is the interfacial tension between the surface of the molecule and the solvent, and $\Delta\mathbb{A}$ is the loss of the wetted surface area of a molecule. The more a molecule unfolds and exposes its surface to the adsorbent, the greater is the loss of the wetted surface area, and thus the decrease in the potential in the adsorbed state, and the stronger is the retention. The interfacial tension between the insulin molecule and the solution is not known, but it is assumed that the change in the interfacial tension works in parallel with the change in the surface tension of the solution. If the ratio of the interfacial tension between the solvent and the protein and the surface tension of the solution is η_r, the asymmetric excess potential is

$$\overline{\tilde{\mu}}_i^E = \bar{\mu}_i^E - \bar{\mu}_i^{E_w^\infty} = -N_A \eta_r \left(\eta_{\text{solution}} - \eta_{\text{water}}\right)\Delta\mathbb{A} \tag{2.70}$$

where η is the surface tension of the solution, and according to Equation 2.66 the equilibrium excess function is

$$RT \ln \tilde{\Gamma}_i = \tilde{\mu}_i^E + N_A \eta_r \left(\eta_{\text{solution}} - \eta_{\text{water}}\right)\Delta\mathbb{A} \tag{2.71}$$

The increment in the interfacial tension is most often proportional to the salt concentration, and if $\ln \tilde{\Gamma}_i$ increases linearly with the salt concentration, one cannot distinguish an increase in $\tilde{\mu}_i^E$ from a decrease in $\overline{\tilde{\mu}}_i^E$. However, if the retentions of an insulin component on the butyl and the phenyl adsorbents show different salt dependence, it can be explained as a difference in the contact area between the insulin molecule and the adsorbent.

The molar surface tension increment, δ, of an aqueous solution of potassium chloride was calculated from the surface tension data from Ref. [15]. The difference in the loss of the wetted surface area was calculated from the equation

$$\eta_r \Delta\left(\Delta\mathbb{A}\right) = RT \frac{\left((b)_{\text{butyl}} - (b)_{\text{phenyl}}\right)}{N_A \delta} \tag{2.72}$$

The molar surface tension increment, δ, of an aqueous solution of potassium chloride was estimated to $1.132 \cdot 10^{-3}$ J/m^2/M^1 using the data from 0 to 2.5 M salt, and the calculated numbers for $\eta_r \Delta\left(\Delta\mathbb{A}\right)$ in Å^2 are given in Table 2.4.

2.8.3 Concluding Remarks

The thermodynamic retention factor, A_i, calculated from the isocratic retention volumes on different adsorbents displayed some characteristic features. The difference in the logarithm of the thermodynamic retention factors of the same component, $\Delta \ln A_i$, measured on the adsorbent α and adsorbent β is

$$\Delta_{\alpha\beta} \ln A_i = (\ln A_i)_{\text{adsorbent } \alpha} - (\ln A_i)_{\text{adsorbent } \beta} = \Delta_{\alpha\beta} \ln A_{i,0} + \Delta_{\alpha\beta} \ln \tilde{\Gamma}_i \tag{2.73}$$

From the results of the investigation of the retention data of lysozyme and the insulin variants, it is possible to draw some preliminary conclusions.

$\Delta_{\alpha\beta} \ln A_{i,0}$ was independent of the ethanol content of the eluant, but the value of $\Delta_{\alpha\beta} \ln A_{i,0}$ depended on the retained component i.

The size of $\Delta_{\alpha\beta} \ln \tilde{\Gamma}_i = \Delta_{\alpha\beta} \overline{\tilde{\mu}}_i^E$ depends on the interaction between the adsorbate and the adsorbent. If the interactions are alike, $\Delta_{\alpha\beta} \overline{\tilde{\mu}}_i^E$ is zero. If a difference is observed, it can be explained as a difference in the size of the contact area between the adsorbate and the adsorbent and thus in the degree of unfolding. The contact area represents a loss in the wetted surface area of the protein exposed to the solvent and thus a loss in the surface energy.

In order to discriminate between changes in $\tilde{\mu}_i^E$ and changes in $\overline{\tilde{\mu}}_i^E$, molecular light scattering can be a very useful tool to trace variations in the excess potentials, $\tilde{\mu}_i^E$, at low and at high protein concentrations. Section 2.A.5 provides a brief note on the thermodynamics of molecular light scattering.

2.9 Protein–Protein Association and Adsorption Isotherms

The formation of oligomers, like dimers, tetramers, or hexamers, modifies the adsorption isotherm. The classical adsorption isotherm has a negative curvature. When oligomer formation takes place due to protein–protein interactions, the adsorption isotherm will display a positive curvature at low protein concentration, and as the protein concentration increases, the adsorption isotherm will display an inflection point and the curvature will change from positive to negative. The change of the curvature will have a remarkable influence on the elution profile. The difference between the two types of isotherms is easy to observe if one compares Figures 2.3 and 2.17. The corresponding wave velocities are shown in Figures 2.4 and 2.19, respectively.

When the adsorption isotherm has the classical form, the wave velocity increases with increasing solute concentration; wherefore a shock wave is

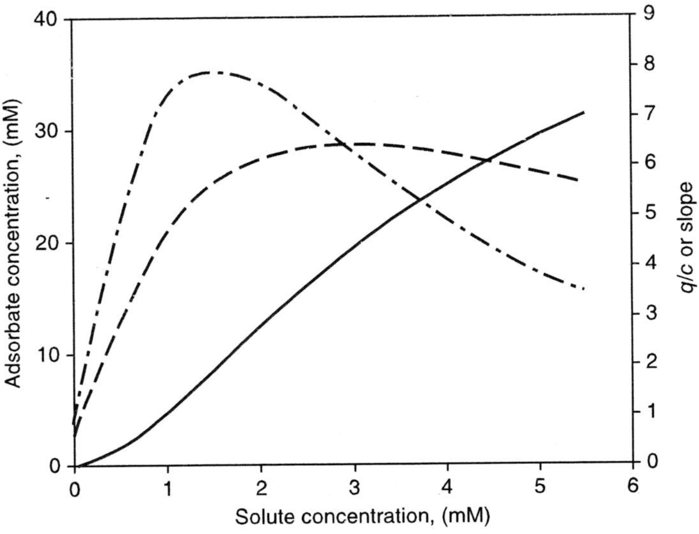

Figure 2.17 The calculated adsorption isotherm (full line) of a GLP-1 analogue on a Source 30 Q adsorbent with 0.065 M sodium chloride in the eluant. The dashed–dotted line is the slope of the isotherm, and the dashed line is q/c.

formed at the leading edge of the elution profile, and a diffuse wave shapes the trailing edge of the profile, as shown in Figures 2.5, 2.8, and 2.9.

At low protein concentration, where the curvature of the adsorption isotherm is positive, the diffuse wave velocity decreases with increasing concentration; wherefore the leading edge of the elution profile starts as a diffuse wave, and a shock wave shapes the rear of the elution profile. An example is shown in Figure 2.20.

When the concentration increases, the curvature changes from positive to negative, and the diffuse wave velocity starts to increase with increasing concentration; consequently, the wave changes from a diffuse wave to a shock wave at the upper part of the leading edge of the elution profile, whereas the upper part of the rear will be shaped by a diffuse wave, and the lower part by a shock wave as shown in Figure 2.23.

The oligomers can be formed in the fluid phase as well as in the immobilized state. The models are derived in Sections 2.9.1 and 2.9.2, and in Section 2.9.3 the equivalence between the two approaches is analyzed.

2.9.1 Protein–Protein Association in the Fluid Phase

Proteins in solution often form association species. The association may be due solely to protein–protein interactions or enhanced by a soluble ligand, M. The association equilibria are

$$\tau_i P_i^{z_i} + \tau_j P_j^{z_j} + \psi M \rightleftharpoons M_\psi P_{\tau_i}^{z_i} P_{\tau_j}^{z_j}; \quad i = 1, 2, \ldots, h \text{ and } j = 1, \ldots, i \tag{2.74}$$

where τ_i, τ_j, and ψ are stoichiometric coefficients and h is the number of protein species forming oligomers. If the protein–protein association takes place in the fluid phase without a soluble ligand, ψ is zero. The equilibrium relations are

$$\frac{c_{ij}}{c} = \tilde{K}_{ij}^F \tilde{\Gamma}_{ij}^F \left(\frac{c_i}{c}\right)^{\tau_i} \left(\frac{c_j}{c}\right)^{\tau_j} \left(\frac{c_M}{c}\right)^{\psi} \tag{2.75}$$

where c_{ij} are the concentrations of the solute association complexes, $M_\psi P_{\tau_i}^{z_i} P_{\tau_j}^{z_j}$, c_i and c_j are the concentrations of the solute proteins, $P_i^{z_i}$ and $P_j^{z_i}$, respectively, and c_M is the concentration of the soluble ligand, M. $\tilde{K}_{ij}^F$ are the equilibrium constants of the solute association complexes, and $\tilde{\Gamma}_{ij}^F$ are the corresponding equilibrium excess functions.

When the association complexes, formed in the fluid phase, associate with the ligands of the stationary phase, the equilibria between the solute protein–protein complexes and the immobilized ligands are

$$M_\psi P_{\tau_i}^{z_i} P_{\tau_j}^{z_j} + \nu_{ij} L^{z_L} S_\sigma^{z_S} \rightleftharpoons M_\psi P_{\tau_i}^{z_i} P_{\tau_j}^{z_j} L_{\nu_{ij}}^{z_L} S_{\sigma\nu_{ij}(1-\beta_{ij})}^{z_S} + \nu_{ij}\beta_{ij}\sigma S^{z_S} \tag{2.76}$$

The corresponding equilibrium relations are

$$\tilde{K}_{ij}^I \tilde{\Gamma}_{ij}^I = \frac{q_{ij}}{c}\left(\frac{c_S}{c}\right)^{\nu_{ij}\beta_{ij}\sigma} \frac{c}{c_{ij}}\left(\frac{c}{q_L}\right)^{\nu_{ij}} \tag{2.77}$$

where q_{ij} are the concentrations of the oligomer association complexes, $M_\psi P_{\tau_i}^{z_i} P_{\tau_j}^{z_j} L_{\nu_{ij}}^{z_L} S_{\sigma\nu_{ij}(1-\beta_{ij})}^{z_S}$; c_S is the concentration of the counterion, S; c_{ij} are the concentrations of the solute protein–protein complexes, $M_\psi P_{\tau_i}^{z_i} P_{\tau_j}^{z_j}$; q_L is the concentration of available ligands, $L^{z_L} S_\sigma^{z_S}$; $\tilde{K}_{ij}^I$ are the equilibrium constants of the association equilibria between the solute protein–protein complexes and ν_{ij} immobilized ligands; and $\tilde{\Gamma}_{ij}^I$ are the corresponding equilibrium excess functions.

The material balance is

$$\Lambda = q_L + \sum_{k=1}^{\alpha}(\nu_k + \lambda_k)q_k + \sum_{i=1}^{h}\sum_{j=1}^{i}(\nu_{ij} + \lambda_{ij})q_{ij} \tag{2.78}$$

where λ_k and λ_{ij} are steric hindrance factors as explained in Section 2.6.1.

The adsorption isotherms for monomer species result from combining Equation 2.47 with the material balance, Equation 2.78, to eliminate q_L, and noting that $q_k^{\max} = \Lambda/(\nu_k + \lambda_k)$ and $q_{ij}^{\max} = \Lambda/(\nu_{ij} + \lambda_{ij})$, the eventual result is

$$\frac{q_k}{c_k} = \tilde{K}_k \tilde{\Gamma}_k \left(\frac{c}{c_S}\right)^{\nu_k\beta_k\sigma} \left(\frac{\Lambda}{c}\right)^{\nu_k} \left(1 - \sum_{k=1}^{\alpha}\frac{q_k}{q_k^{\max}} - \sum_{i=1}^{h}\sum_{j=1}^{i}\frac{q_{ij}}{q_{ij}^{\max}}\right)^{\nu_k} \tag{2.79}$$

The adsorption isotherms for the protein–protein complexes are obtained by combining Equations 2.77 and 2.78 to eliminate q_L. The result is

$$\frac{q_{ij}}{c_{ij}} = \tilde{K}_{ij}^{I}\tilde{\Gamma}_{ij}^{I}\left(\frac{c}{c_S}\right)^{\nu_{ij}\beta_{ij}\sigma}\left(\frac{\Lambda}{c}\right)^{\nu_{ij}}\left(1-\sum_{k=1}^{\alpha}\frac{q_k}{q_k^{\max}}-\sum_{i=1}^{h}\sum_{j=1}^{i}\frac{q_{ij}}{q_{ij}^{\max}}\right)^{\nu_{ij}} \tag{2.80}$$

The concentrations of the solute protein–protein complexes, c_{ij}, are calculable from Equation 2.75.

2.9.2 Protein Association to Immobilized Protein

Oligomer formation may also take place when proteins associate with adsorbed proteins. The associations may be due solely to protein–protein–ligand interactions or enhanced by a soluble ligand, M. When τ_j proteins in solution, $P_j^{z_j}$, and ψ soluble ligands associate with θ_j available ligands and ω_i immobilized proteins, $P_i^{z_i}L_{\nu_i}^{z_L}S_{\sigma\nu_i(1-\beta_i)}^{z_S}$, the association equilibria are

$$\begin{aligned}&\tau_j P_j^{z_j} + \psi M + \theta_j L^{z_L}S_\sigma^{z_S} + \omega_i P_i^{z_i}L_{\nu_i}^{z_L}S_{\sigma\nu_i(1-\beta_i)}^{z_S} \rightleftharpoons \\ &M_\psi P_{\omega_i}^{z_i}P_{\tau_j}^{z_j}L_{\omega_i\nu_i+\theta_j}^{z_L}S_{\sigma(\omega_i\nu_i(1-\beta_i)+\theta_j(1-\beta_j))}^{z_S} + \theta_j\beta_j\sigma S^{z_S}\end{aligned} \tag{2.81}$$

for $i = 1, 2,\ldots, h$ and $j = 1,\ldots, i$.

If the associations take place without a soluble ligand, ψ is zero. The equilibrium relations are

$$\tilde{K}_{ij}^{II}\tilde{\Gamma}_{ij}^{II} = \frac{q_{ij}}{c}\left(\frac{c_S}{c}\right)^{\theta_j\beta_j\sigma}\left(\frac{c}{c_j}\right)^{\tau_j}\left(\frac{c}{c_M}\right)^{\psi}\left(\frac{c}{q_L}\right)^{\theta_j}\left(\frac{c}{q_i}\right)^{\omega_i} \tag{2.82}$$

where q_{ij} are the concentrations of the oligomer association complexes, $M_\psi P_{\omega_i}^{z_i}P_{\tau_j}^{z_j}L_{\omega_i\nu_i+\theta_j}^{z_L}S_{\sigma(\omega_i\nu_i(1-\beta_i)+\theta_j(1-\beta_j))}^{z_S}$; c_S is the concentration of the counterion, S; c_j are the concentrations of the solute proteins, $P_j^{z_j}$; c_M is the concentration of the soluble ligand, M; q_i are the concentrations of the association complexes of monomer proteins with the ligands, $P_i^{z_i}L_{\nu_i}^{z_L}S_{\sigma\nu_i(1-\beta_i)}^{z_S}$; q_L is the concentration of the available ligands, $L^{z_L}S_\sigma^{z_S}$; $\tilde{K}_{ij}^{II}$ are the equilibrium constants of the association equilibria between τ_j solute protein species, $P_j^{z_j}$, ψ soluble ligands, θ_j available ligands, and ω_i immobilized proteins, $P_i^{z_i}L_{\nu_i}^{z_L}S_{\sigma\nu_i(1-\beta_i)}^{z_S}$; and $\tilde{\Gamma}_{ij}^{II}$ are the corresponding equilibrium excess functions.

The material balance is

$$\Lambda = q_L + \sum_{k=1}^{\alpha}(\nu_k + \lambda_k)q_k + \sum_{i=1}^{m}\sum_{j=1}^{i}(\omega_i\nu_i + \theta_j + \lambda_{ij})q_{ij} \tag{2.83}$$

where λ_k and λ_{ij} are steric hindrance factors as explained in Section 2.6.1.

The adsorption isotherm for monomer species is similar to Equation 2.79, and the equilibrium relationship for the oligomer association complexes is

$$\frac{q_{ij}}{c}\left(\frac{c}{c_j}\right)^{\tau_j}=\tilde{K}_{ij}^{\mathrm{II}}\tilde{\Gamma}_{ij}^{\mathrm{II}}\left(\frac{q_i}{c}\right)^{\omega_i}\left(\frac{c_{\mathrm{M}}}{c}\right)^{\psi}\left(\frac{c}{c_{\mathrm{S}}}\right)^{\theta_j\beta_j\sigma}\left(\frac{\Lambda}{c}\right)^{\theta_j}\left(1-\sum_{k=1}^{\alpha}\frac{q_k}{q_k^{\max}}-\sum_{i=1}^{h}\sum_{j=1}^{i}\frac{q_{ij}}{q_{ij}^{\max}}\right)^{\theta_j} \quad (2.84)$$

where $q_{ij}^{\max}=\Lambda/(\omega_i\nu_i+\theta_j+\lambda_{ij})$ and h is the number of protein species forming oligomers.

2.9.3 The Equivalence Between the Models in 2.9.1 and 2.9.2

Unless some supplementary information about the mechanism of oligomer formation is available, it is not possible to discriminate the models presented in Sections 2.9.1 and 2.9.2 from one another if the number of protein units and the number of ligands in an association complex are identical, that is, if $\tau_i=\omega_i$, $\nu_{ij}=\omega_i\nu_i+\theta_j$, and $\beta_i=\beta_j=\beta_{ij}$ because it is straightforward to prove that

$$\tilde{K}_{ij}^{\mathrm{I}}\tilde{\Gamma}_{ij}^{\mathrm{I}}\tilde{K}_{ij}^{F}\tilde{\Gamma}_{ij}^{F}=\tilde{K}_{ij}^{\mathrm{II}}\tilde{\Gamma}_{ij}^{\mathrm{II}}\left(\tilde{K}_i\Gamma_i\right)^{\omega_i} \quad (2.85)$$

where I is the immobilized phase, F is the fluid phase, and II is the protein–protein association in immobilized phase.

This shows that either models will give identical results. This means that it is immaterial whether the association takes place in the fluid phase or in the associated state. The adsorption equilibria will be alike.

2.10 The Adsorption Isotherm of a GLP-1 Analogue

2.10.1 The Adsorption Isotherm and the Wave Velocities

One of the first publications on the application of thermodynamic modeling of dimer formation in process chromatography is the paper by Mollerup et al. [16]. The model from this publication is utilized to calculate the adsorption isotherm, the first and the second derivatives of the adsorption isotherm, the diffuse wave velocity, and the shock wave velocity. The elution profiles are simulated at loads from 0.001 up to 15 mM·CV, and the interplay between the diffuse wave velocity and the shock wave velocity is utilized to analyze the shape of the simulated elution profiles.

The adsorption equilibrium model of the monomer is from Equation 2.47 where $\sigma=1$, $\beta_i=1$, and $\tilde{\Gamma}_i=1$; wherefore

$$\tilde{K}_i=\frac{q_i}{c_i}\left(\frac{c_{\mathrm{S}}}{q_{\mathrm{L}}}\right)^{\nu_i} \quad (2.86)$$

The association equilibrium between the monomer protein in the fluid phase and the adsorbed monomer protein is derived from the adsorption scheme in Equation 2.81:

$$\tau_j P_j^{z_j} + \psi M + \theta_j L^{z_L} S_\sigma^{z_S} + \omega_i P_i^{z_i} L_{\nu_i}^{z_L} S_{\sigma\nu_i(1-\beta_i)}^{z_S} \rightleftharpoons$$
$$M_\psi P_{\omega_i}^{z_i} P_{\tau_j}^{z_j} L_{\omega_i\nu_i+\theta_j}^{z_L} S_{\sigma(\omega_i\nu_i(1-\beta_i)+\theta_j(1-\beta_j))}^{z_S} + \theta_j \beta_j \sigma S^{z_S}$$

The oligomer is modeled as a dimer formation of two identical species; wherefore $i=j$, $\tau_j=1$, and $\omega_i=1$. No available ligands are involved in the formation of the association complex, that is, $\theta_j=0$, and finally, $\psi=0$ because the dimer formation is formed by a self-association without a soluble ligand, that is, the association scheme is

$$P_i^{z_i} + P_i^{z_i} L_{\nu_i}^{z_L} S^{z_S} \rightleftharpoons P_i^{z_i} P_i^{z_i} L_{\nu_i}^{z_L} S^{z_S} \tag{2.87}$$

The equation for the equilibrium model, Equation 2.82, is simplified accordingly, and when it is assumed that $\tilde{\Gamma}_{ij}^{II} = 1$, the equilibrium model is

$$\tilde{K}_{ii}^{II} = \frac{q_{ii}}{c}\left(\frac{c}{c_i}\right)\left(\frac{c}{q_i}\right) = \frac{q_{ii}}{c_i}\left(\frac{c}{q_i}\right) \tag{2.88}$$

The material balance is from Equation 2.83, and it simplifies to

$$\Lambda = q_L + (\nu_i + \lambda_i) q_i + (\nu_i + \lambda_{ii}) q_{ii} \tag{2.89}$$

Furthermore, it was assumed that

$$\lambda_i = 2.1 \tag{2.90}$$

and

$$(\nu_i + \lambda_{ii}) = 2(\nu_i + \lambda_i) \tag{2.91}$$

wherefore

$$q_{ii}^{\max} = \frac{q_i^{\max}}{2} \tag{2.92}$$

The interstitial porosity of the column, ε_b, is 0.45; the adsorbent, Source 30 Q, has a ligand density, $\Lambda = 0.30$ mol Eq/L pore volume, particle porosity, $\varepsilon_p = 0.57$, and stoichiometric coefficient, $\nu_i = 2.44$. The calculated maximum adsorption capacity is $q_i^{\max} = \Lambda/(\nu_i + 2.1) = 0.0661$ M, the thermodynamic

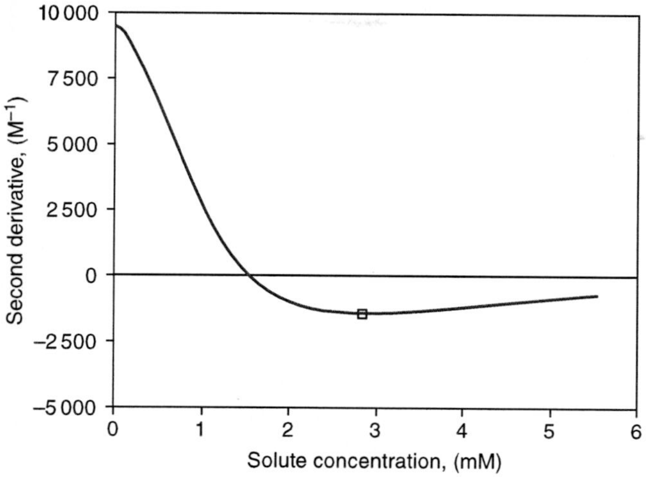

Figure 2.18 The calculated second derivative of the adsorption isotherm of a GLP-1 analogue on a Source 30 Q adsorbent with 0.065 M sodium chloride in the eluant. The open square indicates the location of minimum of the second derivative.

equilibrium constant for the monomer is $\tilde{K}_i = 0.01763$, and the equilibrium constant for the dimer is $\tilde{K}_{ii}^{II} = 1.8 \cdot 10^5$.

Figure 2.17 depicts the calculated adsorption isotherm of the GLP-1 analogue, the slope of the adsorption isotherm, and the slope of the cord from the origin to (c, q) at a sodium chloride concentration of 0.065 M in the eluant, and Figure 2.18 shows the second derivative of the adsorption isotherm. The adsorption isotherm has an inflection point corresponding to a maximum of the first derivative. The second derivative is zero at the inflection point and reaches a minimum at the inflection point of the first derivative. This point is indicated by a square on Figure 2.18. Furthermore, at the maximum of the q/c-curve, the slope of the adsorption isotherm equals q/c because

$$\frac{\partial (q/c)}{\partial c} = \frac{1}{c}\frac{\partial q}{\partial c} - \frac{q}{c^2} = 0 \Rightarrow \frac{\partial q}{\partial c} = \frac{q}{c} \tag{2.93}$$

Figure 2.19 depicts the calculated diffuse wave velocity and the shock wave velocity at a salt concentration of 0.065 M in the eluant. The adsorption isotherm has a positive curvature at concentrations below the concentration at the inflection point and a negative curvature at concentrations above this point. This means that the diffuse wave velocity decreases with increasing solute concentration at concentrations below the inflection point, and at concentrations above the inflection point, the velocity of the diffuse wave starts to increase with increasing solute concentration.

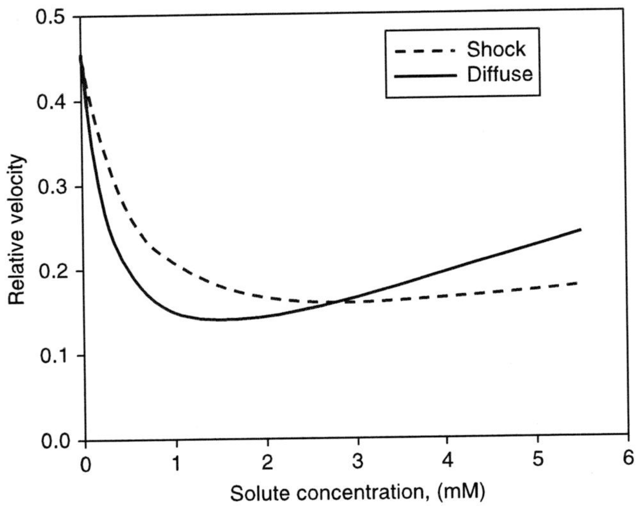

Figure 2.19 The relative velocity of the diffuse wave and the shock wave of a GLP-1 analogue on a Source 30 Q adsorbent with 0.065 M sodium chloride in the eluant.

2.10.2 Simulations

The simulations were performed with the Craig model using 100 cells corresponding to a moderate dispersion. The column was equilibrated with a 0.005 M sodium chloride solution, the salt concentration in the feed was 0.005 M sodium chloride, and the eluant was a 0.065 M sodium chloride solution. At low loads, from 0.001 to 0.05 mM·CV, the feed concentration was 0.1 mM, and at the loads from 0.2 to 15 mM·CV, the concentration in the feed was 1 mM. The reason for using a lower feed concentration at low load is that the load must be at least one cell volume, that is, 0.01 CV. Figure 2.20 depicts the result of a series of simulations at low loads, and Figure 2.21 shows the development of the elution profiles at high loads. The loads in Figure 2.21 range from 0.2 to 15 mM·CV.

It is a characteristic of oligomer formation that the simulated elution profiles, at low load, develop almost identical slopes at the leading edge and that the end of the elution profiles displays the characteristic shape formed by a shock wave. The shapes of the elution profiles are fairly identical up to a load of 1 mM·CV. When the load increases further and the concentration in the mobile phase increases to a concentration above at the inflection point of the adsorption isotherm, a shock wave is formed at the leading edge, and the elution profile rises sharply. When the maximum concentration is reached and the concentration starts to drop, the rear of the elution profile is shaped by the diffuse wave until the inflection point is reached once more; wherefore the end

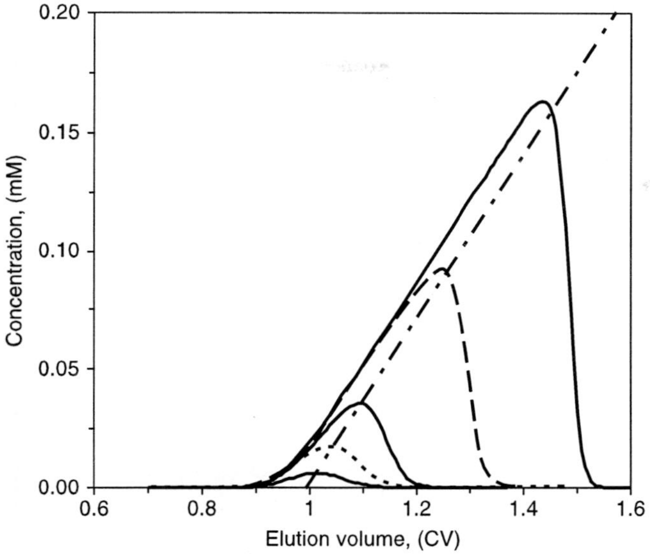

Figure 2.20 Simulated isocratic elution profiles of a GLP-1 analogue on a Source 30 Q adsorbent with 0.065 M sodium chloride in the eluant. The loads were 0.001 (full line), 0.002 (dotted line), 0.005 (full line), 0.02 (dashed line), and 0.05 (full line) mM·CV. The concentration of GLP-1 in the feed was 0.1 mM. The dashed–dotted line shows the calculated ideal elution profile created by the diffuse wave.

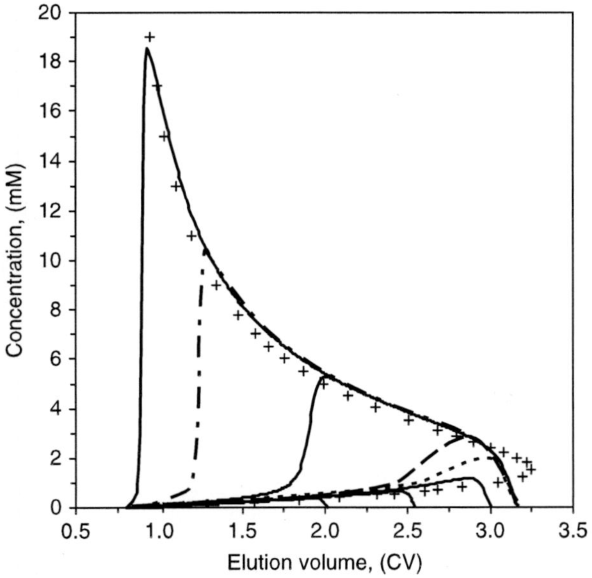

Figure 2.21 Simulated isocratic elution profiles of a GLP-1 analogue on a Source 30 Q adsorbent with 0.065 M sodium chloride in the eluant. The loads were 0.2, 0.5, 1 (full lines), 1.5 (dotted line), 2 (dashed line), 5 (full line), 10 (dashed–dotted line), and 15 (full line) mM·CV. The concentration of GLP-1 in the feed was 1 mM. The crosses show the ideal elution profile created by the diffuse wave.

of the elution profiles is determined by a common shock wave that develops when the concentration drops below the concentration at the inflection point of the adsorption isotherm.

2.10.3 How the Wave Velocities Shape the Elution Profiles

The scope of this section is to analyze the shape of the simulated elution profiles shown in Figures 2.20 and 2.21. Chromatography is a rate-controlled separation where the shape and the position of an elution profile result as a perturbation of the ideal elusion profile by the dispersion forces. The main features of an elution profile are therefore explained by the interplay between the two wave velocities of the solute alone; wherefore the relation between the concentration of the eluite and the elution volume is calculable from the wave velocities using Equation 2.5. The elution profile is the mirror image of the concentration profile formed by the solute movement in the column.

The diffuse wave and the shock wave velocities are depicted in Figure 2.19. The depicted shock wave velocity is only applicable if the step in concentration starts or ends at the baseline; wherefore it only applies to the shock wave that terminates the elution profile.

The elution profile created by the diffuse wave is depicted in Figure 2.22 as the curved solid line. It starts at a retention volume of 0.99 CV, increases with increasing solute concentration, reaches a maximum value at 3.25 CV, and

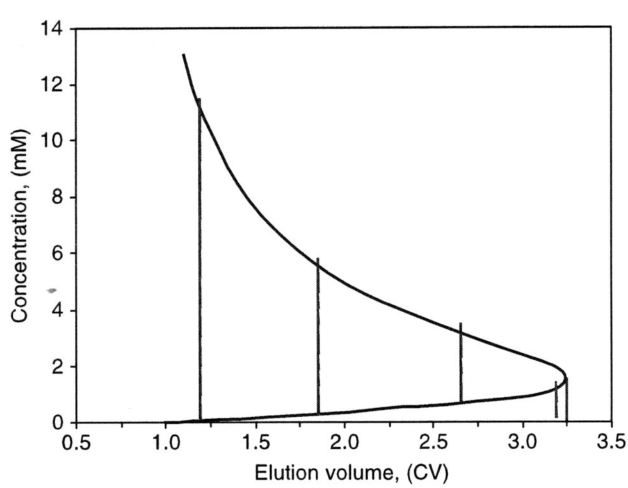

Figure 2.22 The ideal elution profile of a GLP-1 analogue on a Source 30 Q adsorbent with 0.065 M sodium chloride in the feed and in the eluant. The vertical lines are shock waves at different loads (mM·CV): short dashed, 1; dashed–dotted, 2; dashed–double-dotted, 5; and long dashed, 10. The full vertical line at 3.25 CV is a shock wave common to all loads >1.1 mM·CV.

decreases. The lower branch of the diffuse wave shapes the leading edge of the elution profile, whereas the upper branch shapes the upper part of the rear of the elution profile.

The leading edge of any elution profile starts at 0.99 CV, and if the maximum concentration in the mobile phase stays below the concentration at the inflection point of the adsorption isotherm, that is, below 1.5 mM, the elution profile is terminated by a shock wave, a vertical line. The position of the shock wave depends on the load, which equals the area under the elution profile. The vertical short dashed line in Figure 2.22 at 3.2 CV corresponds to a load of 1 mM·CV.

When the load increases above 1.1 mM·CV, there will be two fixed points in the elution profile because the concentration in the mobile phase rises above the concentration at the inflection point of the adsorption isotherm. The first fixed point is, as mentioned previously, a retention volume of 0.99 CV where the leading edge starts, and the second fixed point is a shock wave at the rear of the elution profile at 3.25 CV. The shock wave starts at the baseline and ends at the concentration where the curvature of the isotherm changes, that is, at 1.5 mM. The relative shock wave velocity is 0.176. At concentrations above 1.5 mM, where the curvature of the adsorption isotherm is negative, the diffuse wave controls the curvature of the upper part of the rear of the elution profile. It starts at $(V, c) = (3.25, 1.5)$ on Figure 2.22, and it terminates where the shock wave formed at the upper part of the leading edge intersects with the diffuse wave. The position of the shock wave at the upper part of the leading edge of the elution profile must be determined from the material balance.

If one follows the lower part of the elution profile from where it starts at 0.99 CV until it meets the long dashed vertical line at 1.2 CV, follows the vertical line up to where it meets the upper part of the diffuse elution profile, moves down the curved line until it reaches its maximum value at 3.25 CV, and then follows the vertical solid line to the baseline, this path shows the shape of an ideal elution profile corresponding to a load of 10 mM·CV. The vertical dashed–double-dotted line at 1.9 CV is the shock wave formed at the upper part of the leading edge corresponding to a load of 5 mM·CV, and the vertical dashed–dotted line at 2.7 CV is the shock wave formed at the upper part of the leading edge corresponding to a load of 2 mM·CV.

In summary, if the load is high, the leading edge of the elution profile will start with a diffuse wave, develop into a shock wave, and reach a maximum value, where a diffuse wave will shape the upper part of the rear of the elution profile, and finally, a shock wave will terminate the elution profile. That is, the profile is shaped by two diffuse waves and two shock waves. The elution profile corresponding to a load of 5 mM·CV is shown in Figure 2.23. The full line is the ideal elution profile, and the dashed line is that part of the diffuse wave elution profile that has been erased by the shock wave that rose at the leading edge. The corresponding simulated profile is shown as a dashed–dotted line. The figure shows that the ideal elution profile explains the main features of the

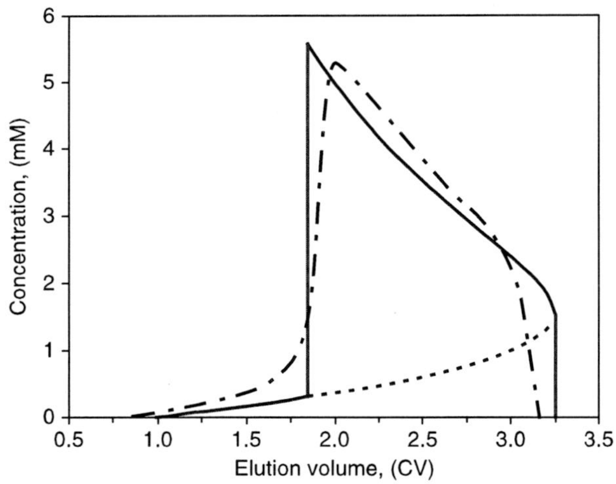

Figure 2.23 The ideal elution profile (full line) and the simulated elution profile (dashed–dotted line) of a GLP-1 analogue on a Source 30 Q adsorbent with 0.065 M sodium chloride in the eluant at a load of 5 mM·CV. The curved full lines are the part of the elution profile calculated from the diffuse wave velocity, and the dashed line in between is the part that disappeared due to the shock wave formation at the upper part of the leading edge at 1.9 CV. The vertical line at 3.25 CV is a shock wave that terminates the rear of the ideal elution profile.

elution profiles. The calculated elution profile created by the diffuse wave is also shown in Figure 2.20 as a dashed–dotted line, and in Figure 2.21 it is represented by a series of the small crosses.

2.10.4 Calculation of Second Derivatives from Simulated Elution Profiles

The dashed–dotted line in Figure 2.20 shows the leading edge of the ideal elution profile and a series of simulated elution profiles. The leading edges of the simulated and the ideal profile are not identical, but the slopes are similar except at low load. This observation can be utilized to calculate the second derivative of the adsorption isotherm from experimental elution profiles as explained in Section 2.7.1.4. To test the hypothesis, the simulated profiles are used as experimental data because that enables one to check the values calculated from the elution profiles against the values of the second derivatives of the adsorption isotherm. The slopes of the simulated elution profiles were calculated from a linear fit to the elution profiles at the loads 0.05, 0.2, and 0.5 mM·CV, respectively, in a concentration interval from 0.075 to 0.10 mM, and for the loads 0.2 and 0.5 mM·CV, a concentration interval from 0.28 to 0.30 mM was used. The calculated second derivatives were extrapolated to zero load to correct for the influence of the dispersion. The results are given in

Table 2.5 The second derivative of the isotherm calculated from the slope of the leading edges of simulated elution profiles compared with the value calculated from the adsorption isotherm.

Simulation			Second derivative (mM^{-1})	
Load (mM)·CV	**Slope (mM/CV)**	**Concentration (mM)**	**From slope**	**Isotherm**
0.053	0.349	0.075–0.10, average 0.088	9.15	9.29 at 0.088 mM
0.2	0.360		8.87	
0.5	0.370		8.61	
0	Extrapolation to zero load		9.28	
0.2	0.396	0.28–0.30, average 0.29	8.06	8.30 at 0.29 mM
0.5	0.414		7.70	
0	Extrapolation to zero load		8.30	

Table 2.5. The second derivative of the adsorption isotherm was calculated at the average concentration of the interval given in the table. The agreement is good, but further investigations are necessary to assess the accuracy of the proposed method.

2.11 Concluding Remarks

Thermodynamics is applicable to all kinds of phase equilibrium calculations including the adsorption equilibria in protein chromatography. The classical adsorption equilibrium entails a reversible association of some functional groups of the proteins with the functional groups of the immobilized ligands. This has been recognized long ago and used in model development. What is new in this exposition is the rigorous thermodynamic approach to the modeling using well-known and well-defined chemical engineering concepts. All quantities used in the modeling must have a proper definition that fits into the framework of the classical thermodynamics. A proper definition of a thermodynamic quantity is a definition that relates the quantity to one or more of the derivatives of a thermodynamic state function.

The excess potentials play a crucial role in the modeling of the solution behavior in any aspects of chemical engineering and naturally in chromatography as well. HIC and RPC cannot be understood and interpreted correctly without a proper knowledge of the theory of excess potentials. No solution not even in chromatography is an ideal solution. Modulation of the retention in ion exchange chromatography by admixture of an organic solvent to the eluant is due to modulation of the excess potentials and can be modeled accordingly.

The modeling of oligomer formation is a fairly new subject. This phenomenon can also be modeled using simple chemical engineering concepts. An indication of the presence of oligomer formation in chromatographic investigations is the observation of fronting at low load. If oligomer formation can be controlled actively, it has a great separation potential as demonstrated in Ref. [17].

Last but not the least, the fulcrum point in any investigation as regards process development and modeling of chromatographic separations is the isocratic retention data. A subject that is most often ignored and only recognized occasionally.

Investigation of the thermodynamic properties of proteins in electrolyte solutions in aqueous and mixed solvents is one of the most important subjects in chemical engineering today. The isocratic retention measurements are a key to the understanding of the properties of proteins in solution at infinite dilution of the proteins. These data supplement the traditional measurements like solubility measurements, osmotic measurements, light scattering data, and calorimetric measurements.

Appendix 2.A Classical Thermodynamics

2.A.1 Chemical, Electrical, and Surface Potentials

2.A.1.1 Ideal Mixtures

The ideal mixture is a hypothetical state, which no real solution strictly follows, but it is a convenient reference of normal fluid behavior. There is no unique definition of an ideal mixture, but it is most convenient that the definition of the chemical potentials of an ideal mixture conforms to the chemical potentials of a perfect gas mixture. That is, an ideal mixture is a mixture whose chemical potentials at any composition x_i conform to the equation:

$$\mu_i^{\mathrm{id}}(T,p,\mathbf{n}) = \left(\frac{\partial G^{\mathrm{id}}(T,p,\mathbf{n})}{\partial n_i}\right)_{T,p} = \mu_i^0(T,p) + RT\ln x_i \tag{A1}$$

$G^{\mathrm{id}}(T,p,\mathbf{n})$ is the Gibbs energy of the ideal system, $\mu_i^0(T,p)$ is the reference chemical potential taken as a state of pure species i at the temperature, T, and the pressure, p, of the solution, and x_i is the mole fraction of species i. A mole fraction is the relative number of moles in a given phase, that is, $x_i = n_i / \sum_j^N n_j$, where n_i is the number of moles of species i in a solution of N components and $\mathbf{n}$ is the corresponding vector of mole numbers $(n_1, n_2, \ldots, n_i, \ldots, n_N)$.

The reference chemical potential, $\mu_i^0(T,p)$, is an arbitrary reference because an absolute value of the chemical potentials cannot be calculated unless a state where the entropy $S = 0$ is defined. This academic problem is immaterial because what matters are changes in the chemical potentials and not the absolute values.

2.A.1.2 Chemical Potentials

The classical chemical potentials, $\mu_i(T, p, \mathbf{n})$, which account for the short-range attractive and repulsive intermolecular forces, are the compositional derivatives of the Gibbs energy of the system:

$$\mu_i\left(T,p,\mathbf{n}\right) \equiv \left(\frac{\partial G\left(T,p,\mathbf{n}\right)}{\partial n_i}\right)_{T,p} = \mu_i^0\left(T,p\right) + \mu_i^E\left(T,p,\mathbf{n}\right) + RT\ln x_i \\ = \mu_i^{id}\left(T,p,\mathbf{n}\right) + \mu_i^E\left(T,p,\mathbf{n}\right) \tag{A2}$$

$G(T, p, \mathbf{n})$ is the Gibbs energy of the system, $\mu_i^0(T, p)$ are the reference chemical potentials, and $\mu_i^E(T, p, \mathbf{n})$ are the excess potentials accounting for the nonideal behavior of the solution. Applying Euler's theorem of homogeneous that functions to the Gibbs energy shows that [5]

$$G\left(T,p,\mathbf{n}\right) = \sum_{i=1}^{N} n_i \mu_i\left(T,p,\mathbf{n}\right) \tag{A3}$$

The Helmholtz energy, $A(T, V, \mathbf{n})$, has volume rather than pressure as an independent variable, but the other independent variables are identical. The relationship between the two state functions is

$$A\left(T,V,\mathbf{n}\right) = G\left(T,p,\mathbf{n}\right) + \left(\frac{\partial A\left(T,V,\mathbf{n}\right)}{\partial V}\right)_{T,\mathbf{n}} V = G\left(T,p,\mathbf{n}\right) - pV \tag{A4}$$

An equivalent way to calculate the chemical potentials is

$$\mu_i(T,V,\mathbf{n}) = \left(\frac{\partial A(T,V,\mathbf{n})}{\partial n_i}\right)_{T,V} = \mu_i^0(T,V) + \mu_i^E(T,V,\mathbf{n}) + RT\ln x_i \tag{A5}$$

Note that while

$$\mu_i(T,p,\mathbf{n}) = \mu_i(T,V,\mathbf{n}) \tag{A6}$$

this does not hold for the excess potentials

$$\mu_i^E(T,p,\mathbf{n}) \neq \mu_i^E(T,V,\mathbf{n}) \tag{A7}$$

because a conversion from pressure to volume as an independent variable in the excess potentials requires a correction term because the reference potentials $\mu_i^0(T,p)$ and $\mu_i^0(T,V)$ are not identical except for perfect gases. The conversion from pressure to volume and vice versa is analyzed in Section 2.A.1.3.

2.A.1.3 Excess Potentials and Residual Properties

The excess Gibbs energy is the Gibbs energy of the real system minus that of the equivalent ideal system:

$$\begin{aligned} G^E(T,p,\mathbf{n}) &= G(T,p,\mathbf{n}) - G^{\text{id}}(T,p,\mathbf{n}) = \sum_{i=1}^{N} n_i \mu_i^E(T,p,\mathbf{n}) \\ &= RT\sum_{i=1}^{N} n_i \ln \Gamma_i(T,p,\mathbf{n}) \end{aligned} \tag{A8}$$

The second line of this equation shows an equivalent expression for the excess potentials that defines the activity coefficients, γ_i. There is a variety of empirical excess Gibbs energy models available in the literature. They are functions of temperature and composition but have no explicit pressure dependence.

However, most of the thermodynamic models, including models of electrolyte solutions, are contributions to the Helmholtz energy function. It is therefore important to derive equations that enable derivation of models for the excess potentials at state (T, p, $\mathbf{n}$) from models that have volume rather than pressure as an independent variable. The residual properties provide the link between the two types of state functions.

The residual Helmholtz energy is the energy of a system at state (T, V, $\mathbf{n}$) minus that of the equivalent system in a hypothetical perfect gas state (pg). A hypothetical perfect gas state is hypothetical because it cannot be realized experimentally, but its thermodynamic properties are well defined and can be calculated. The residual Helmholtz energy is by definition

$$A^r(T,V,\mathbf{n}) = A(T,V,\mathbf{n}) - A^{\text{pg}}(T,V,\mathbf{n}) \tag{A9}$$

Note that the last term, A^{pg}, is the Helmholtz energy of a hypothetical perfect gas at state (T, V, $\mathbf{n}$) where V is the volume of the mixture, and this volume is not identical with the volume of the corresponding perfect gas mixture. The residual Helmholtz energy is calculable from the p–V–T properties of the fluid using the equation

$$A^r(T,V,\mathbf{n}) = -\int_{\infty}^{V}\left(p - p^{\text{pg}}\right)dV = -\int_{\infty}^{V}\left(p - \frac{nRT}{V}\right)dV \tag{A10}$$

In order to calculate the excess Gibbs energies from the residual Helmholtz energy, it is necessary to derive a relation between $G^E(T,p,\mathbf{n})$ and $A^r(T,V,\mathbf{n})$. The eventual result is

$$\begin{aligned} G^E(T,p,\mathbf{n}) &= G^r(T,p,\mathbf{n}) - \sum_{i=1}^{N} n_i g_i^r(T,p) \\ &= A^r(T,V,\mathbf{n}) - \sum_{i=1}^{N} n_i a_i^r(T,v_i) \\ &\quad + p\left(V - \sum_{i=1}^{N} n_i v_i\right) + RT\sum_{i=1}^{N} n_i \ln\left(\frac{n v_i}{V}\right) \end{aligned} \tag{A11}$$

where g_i^r are the residual molar Gibbs energies of the pure species, a_i^r are the residual Helmholtz energies of the pure species, v_i are the molar volumes of the pure species, and n is the total number of moles in the system. Inspection of Equation A4 explains the first term in the last line term of Equation A11. The last term is due to the conversion from volume to pressure as an independent variable because the hypothetical perfect gas properties are calculated from the real state of the system. The hypothetical perfect gas properties of a liquid are calculated from the actual p–V–T properties where V is a liquid volume.

The application of Equation A11, to calculate the excess potentials from the residual Helmholtz energies, is straightforward. The derivative with respect to the mole number n_i at constant temperature and pressure is

$$\mu_i^E(T,p,\mathbf{n}) = \left(\frac{\partial A^r(T,V,\mathbf{n})}{\partial n_i}\right)_{T,V} - a_i^r(T,v_i) + RT - p v_i + RT\ln\left(\frac{v_i}{v}\right) \tag{A12}$$

where $v = V/n$ is the molar volume of the mixture.

This equation is easy to apply for gases and liquids where the volumetric properties of the pure components can be measured and modeled. However, it is inconvenient to apply to solids like salts and proteins. This problem can be circumvented by changing the reference state potentials to include the excess potentials at infinite dilution in a solvent at a predefined constant solvent composition. The solvent composition must be a predefined and constant solvent composition in order to assure that the reference state potentials can be applied for solutions in pure and in mixed solvents. Due to its ubiquity, it is convenient to use pure water, w, as the reference solvent. We use the symbol $\mu_i^{E_w^\infty}$ for this potential. The asymmetric potentials are

$$\begin{aligned} \tilde{\mu}_i^E(T,p,\mathbf{n}) &= \mu_i^E(T,p,\mathbf{n}) - \mu_i^{E_w^\infty}(T,p) \\ &= \left(\frac{\partial A^r(T,V,\mathbf{n})}{\partial n_i}\right)_{T,V} - \lim_{n\to n_w}\left(\frac{\partial A^r(T,V,\mathbf{n})}{\partial n_i}\right)_{T,V} + RT\ln\left(\frac{v_w}{v}\right) \\ &= \mu_i^r(T,V,\mathbf{n}) - \lim_{n\to n_w}\mu_i^r(T,V,\mathbf{n}) + RT\ln\left(\frac{v_w}{v}\right) \end{aligned} \tag{A13}$$

where v is the molar volume of the mixture and v_w is the molar volume of pure water.

The reference state potentials, $\mu_i^0(T, p)$, must be changed accordingly, that is,

$$\tilde{\mu}_i^0(T, p) = \mu_i^0(T, p) + \mu_i^{E_w^\infty}(T, p) \tag{A14}$$

Note that $\mu_i^{E_w^\infty}$ is not a pure component property like $\mu_i^0(T, p)$, but it is independent of the configuration of the mixture.

This paraphrase of the reference potentials and of the excess potentials does not change the chemical potential itself because

$$\begin{aligned}\mu_i(T, p, \mathbf{n}) &= \mu_i^0(T, p) + \mu_i^E(T, p, \mathbf{n}) + RT\ln x_i \\ &= \tilde{\mu}_i^0(T, p) + \tilde{\mu}_i^E(T, p, \mathbf{n}) + RT\ln x_i\end{aligned} \tag{A15}$$

but it circumvents the need of knowing the volumetric properties of the pure solid components.

2.A.1.3.1 Worked Example: Excess Model from van der Waals Equation of State

The first empirical equation of state was proposed in 1873 by van der Waals. It is a very simple equation of state, which has a repulsive and a single attractive term, but it can predict the qualitative behavior of fluids correctly. For one mole of a pure substance, the pressure equation of state is

$$p = \frac{RT}{v - b} - \frac{a}{v^2} \tag{I}$$

where a and b are two parameters that depend on the properties of the fluid.

The residual Helmholtz energy is calculated by integration of the pressure equation of state, Equation I, as shown in Equation A10. The residual Helmholtz energy of a pure component is

$$A_i^r(T, v_i) = -RT\ln\left(1 - \frac{b_i}{v_i}\right) - \frac{a_{ii}}{v_i} \tag{II}$$

where v_i is the molar volume and b_i and a_{ii} are the pure component parameters. The corresponding expression for n moles of a mixture is

$$A^r(T, V, \mathbf{n}) = -nRT\ln\left(1 - \frac{B}{V}\right) - \frac{D}{V} \tag{III}$$

where B and D are two mixture parameters calculable from the molar parameters b_i and a_{ii}. In this example it is convenient to apply the conventional mixing rules, but more complex mixing rules, see Equation XII, are often advantageous. The conventional mixing rules are

$$B = nb = \sum_{i=1}^{N} n_i b_i \tag{IV}$$

$$D = \sum_{i=1}^{N} n_i \sum_{j=1}^{N} n_j a_{ij} \tag{V}$$

The parameters for the unlike interactions a_{ij} are calculable from the combining rule

$$a_{ij} = \sqrt{a_{ii} a_{jj}} \left(1 - k_{ij}\right) \tag{VI}$$

k_{ij} is a binary interaction parameter where $k_{ij} = k_{ji}$ and $k_{ii} = k_{jj} = 0$

The compositional derivatives of the Helmholtz model are

$$\left(\frac{\partial A^r(T, V, \mathbf{n})}{\partial n_i}\right)_{T,V} = -RT \ln\left(1 - \frac{B}{V}\right) + \frac{nRTB_i}{V - B} - \frac{D_i}{V} \tag{VII}$$

where B_i and D_i are the compositional derivatives of B and D, respectively,

$$B_i = \left(\frac{\partial B}{\partial n_i}\right) = b_i \tag{VIII}$$

$$D_i = \left(\frac{\partial D}{\partial n_i}\right) = 2\sum_{j=1}^{N} n_j a_{ij} \tag{IX}$$

The limiting value of the compositional derivative of the residual Helmholtz energy of species i in pure water, w, is calculated from Equation VII:

$$\lim_{n \to n_w} \left(\frac{\partial A^r(T,V,\mathbf{n})}{\partial n_i}\right)_{T,V} = -RT \ln\left(1 - \frac{b_w}{v_w}\right) + \frac{RTb_i}{v_w - b_w} - \frac{2a_{iw}}{v_w} \tag{X}$$

at all $n_j = 0$ and $j \neq$ w.

The asymmetric excess potentials are calculated from Equation A13 using Equations X and VII:

$$\tilde{\mu}_i^E(T, p, \mathbf{n}) = \left(\frac{\partial A^r(T, V, \mathbf{n})}{\partial n_i}\right)_{T,V} - \lim_{n \to n_w} \left(\frac{\partial A^r(T, V, \mathbf{n})}{\partial n_i}\right)_{T,V} + RT \ln\frac{v_w}{v}$$

$$= RT \ln\left(\frac{v_w - b_w}{v - b}\right) + \frac{RTb_i}{v - b} - \frac{RTb_i}{v_w - b_w} - \frac{2\sum_j x_j a_{ij}}{v} + \frac{2a_{iw}}{v_w} \tag{XI}$$

Each component is characterized by two parameters a_{ii} and b_i; v is the molar volume of the mixture, and v_w is the molar volume of pure water, and a_{ij} is calculated using the combining rule in Equation VI. The "pure component" parameters for electrolytes and soluble solids are obtained by fitting the model to measured properties of solutions of these molecules.

Sometimes more sophisticated mixing rules are advantageous to use especially for mixtures where the excess properties are large. Equation XII combines the van der Waals equation of state with the local composition concept from Wilson's activity coefficient model. The last term in Equation XII is the density-dependent local composition term derived from Wilson's local composition theory [18]. The model of the residual Helmholtz energy is

$$A^r(T, V, \mathbf{n}) = -nRT\ln\left(1-\frac{B}{V}\right)-\frac{D_0}{V}-RT\sum_{j=}^{N} n_i \ln\sum_{j=1}^{N}\frac{n_k}{n}E_{ji} \tag{XII}$$

where

$$D_0 = \sum_{i=1}^{N} n_i \sum_{j=1}^{N} n_j \sqrt{a_{ii}a_{jj}} \tag{XIII}$$

The exponential function, E_{ij}, accounts for the deviation between the real value of the a_{ij} parameter and the geometric mean value $\sqrt{a_{ii}a_{jj}}$:

$$E_{ji} = \exp\left(\frac{n\left(a_{ij}-\sqrt{a_{jj}a_{kk}}\right)}{RTV}\right) = \exp\left(-\frac{n\sqrt{a_{jj}a_{kk}}\,k_{jk}}{RTV}\right) \tag{XIV}$$

When the parameter for the adjustment of the unlike interaction, k_{ij}, is small, it can be shown that D_0/V plus a series expansion of the local composition term equals the original D/V term.

2.A.1.4 Excess Helmholtz Energy

An excess Helmholtz energy function of a mixture of N components can be defined as

$$A^E(T, V, \mathbf{n}) = A^r(T, V, \mathbf{n}) - \sum_{i=1}^{N} n_i a_i^r(T, v_i) \tag{A16}$$

The excess chemical potentials are the compositional derivatives of the excess Helmholtz energy at constant temperature and volume, that is,

$$\mu_k^E(T, V, \mathbf{n}) = \left(\frac{\partial A^E}{\partial n_k}\right)_{T,V} \tag{A17}$$

The asymmetric excess potentials at state (T, p, $\mathbf{n}$) are calculable from the excess Helmholtz potentials using Equation A18, an equation that is the equivalent of Equation A13:

$$\tilde{\mu}_i^E(T,p,\mathbf{n}) = \mu_i^E(T,V,\mathbf{n}) - \lim_{n \to n_w} \mu_i^E(T,V,\mathbf{n}) + RT\ln\left(\frac{v_w}{v}\right) \tag{A18}$$

An old and simple empirical excess Gibbs energy model is

$$\frac{G^E}{nRT} = \frac{1}{2}\sum_{i=1}^{N} x_i \sum_{j=1}^{N} x_j a_{ij} = \sum_{i=1}^{N} x_i \sum_{j=i+1}^{N} x_j a_{ij} \tag{A19}$$

where N is the number of components, $a_{ij} = a_{ji}$, and $a_{ii} = a_{jj} = 0$ because the excess energy of a pure substance is zero.

A modest empirical excess Helmholtz energy model, similar to Equation A19, is developed from a series expansion of the excess Helmholtz function

$$\begin{aligned}\frac{A^E}{nRT} &= a^E = a_0^E + \sum_{j=1}^{N} c_j \left(\frac{\partial a^E}{\partial c_j}\right)_{T,V} + \frac{1}{2}\sum_{i=1}^{N} c_i \sum_{j=1}^{N} c_j \left(\frac{\partial^2 a^E}{\partial c_i \partial c_j}\right)_{T,V} \\ &= a_0^E + \sum_{j=1}^{N} c_j a_j + \frac{1}{2}\sum_{i=1}^{N} c_i \sum_{j=1}^{N} c_j a_{ij} = \frac{1}{2}\sum_{i=1}^{N} c_i \sum_{j=1}^{N} c_j a_{ij} = \sum_{i=1}^{N} c_i \sum_{j=i+1}^{N} c_j a_{ij}\end{aligned} \tag{A20}$$

where $c_i = n_i/V$ and $a_{ij} = a_{ji}$. The excess energy of a pure substance is zero; wherefore $a_0^E = 0$, $a_j = 0$, and $a_{ii} = a_{jj} = 0$.

The excess potential of species k in a mixture of N components is the compositional derivative of A^E

$$\frac{\mu_k^E}{RT} = \frac{n}{V^2}\sum_{i=1}^{N} n_j a_{ik} + \frac{1}{V^2}\sum_{i=1}^{N} n_i \sum_{j=i+1}^{N} n_j a_{ij} \tag{A21}$$

and the asymmetric potential of species 3 in a ternary mixture is

$$\begin{aligned}\frac{\tilde{\mu}_3^E}{RT} &= cc_1a_{13} + cc_2a_{23} + c_1c_2a_{12} + c_1c_3a_{13} + c_2c_3a_{23} - c_w^2 a_{13} \\ &= cc_1a_{13} - c_w^2 a_{13} + (ca_{23} + c_1a_{12})c_2 + c_1c_3a_{13} + c_2c_3a_{23}\end{aligned} \tag{A22}$$

The concentration of water is in general not known; wherefore it is convenient to make some approximations. If subscript 1 denotes the solvent water, the following approximations are made: $cc_1a_{13} - c_w^2 a_{13} = 0$, $ca_{23} + c_1a_{12} = A_2$, and $c_1a_{13} = A_3$ where A_2 and A_3 are two adjustable parameters, that is,

$$\frac{\tilde{\mu}_3^E}{RT} = A_2c_2 + A_3c_3 + a_{23}c_2c_3 \tag{A23}$$

A more complex excess Helmholtz energy model, similar in form to Wilson's excess Gibbs energy model, is a local composition excess Helmholtz energy model derived from the density-dependent local composition concept, that is, a density-dependent Boltzmann distribution of the excess energy. The local composition excess Helmholtz energy model is obtained by integration of the local composition excess energy with respect to $1/T$ at constant V. The eventual result is

$$\frac{A^E}{RT} = -\sum_{i=1}^{N} n_i \ln \sum_j \frac{\psi_j}{\sum_k \psi_k} \exp\left(\frac{n\Delta a_{ji}}{RTV}\right) + \sum_{k=1}^{N} n_k \ln \frac{nr_k}{\sum_j n_j r_j} \tag{A24}$$

where ψ_j are the statistical weights in the Boltzmann distribution and r_j are appropriate size parameters.

2.A.1.5 Electric Potentials

Electrolytes are substances that dissociate into ions in an aqueous solution. If it dissociates completely it is called a strong electrolyte, and substances that dissociate partially are named weak electrolytes. The substances are those called acids, bases, ampholytes, and salts. The solvent is considered as a continuum characterized by the volume, V, and the temperature, T, of the system and the static permittivity, ε, of the solvent. The static permittivity is the permittivity at zero frequency. The permittivity has the unit $C^2/(J{\cdot}m)$. The relative permittivity is the permittivity divided by the permittivity of free space, $\varepsilon_0 = (4\pi c_0^2)^{-1} 10^7\, F \cdot m^{-1}$, where c_0 is the speed of light in free space. The number of ions, the nature of the ions, and the nature of the medium define the assembly completely. The permittivity of the solvent varies with the nature of the solvent and the solvent composition. The addition of ions changes the dielectric properties of the solution due to dielectric saturation. However, this exposition does not account for the dielectric saturation of the solvent, and the permittivity, ε, is, in this context, the permittivity of the pure solvent or the solvent mixture.

There are three major contributions to the chemical potentials of species carrying charges, μ_i^{ch}:

$$\mu_i^{\mathrm{ch}} = \left(\frac{\partial A^{\mathrm{ch}}}{\partial n_i}\right)_{T,V} = \mu_i^{\mathrm{self}}(T, \varepsilon, z_i) + \mu_i^{\mathrm{el}}(T, V, \varepsilon, \mathbf{q}, \boldsymbol{\delta}) + z_i F \psi \tag{A25}$$

where z_i is the charge number of the ion of species i, q_i is the charge of n_i moles of ions of species i, $\mathbf{q}$ is the corresponding charge vector for all the ions, and $\boldsymbol{\delta}$ is the vector of the electric dipole moments of the macromolecules.

The self potentials, μ_i^{self}, are the potentials due to the self-energy of charging each ion in the solution in the effective absence of the other ions, and these potentials are independent of the configuration of the charged molecules, that is, they do not depend on the concentration of the ions. The self potentials are also known as the Born potentials [19].

The electric potentials, μ_i^{el}, originate from the energy due to the mutual electric interactions of the various ions and the interactions between ions and dipolar ions.

The last term in Equation A25 is due to the work done when moving one mole of ions of charge $z_i F$ in an electric field of potential ψ. $F = eN_A$ is Faraday's number, 96 485.34 C/mol, e is the unit charge, and N_A is Avogadro's number.

The last term in Equation A25 shows that equilibrium calculations involving ions require special treatment because the electric field potential contributes to the chemical potential of a single ion by an amount $z_i e\psi$. The first and the second terms are insufficient to characterize the ion–ion equilibria across a membrane or in a multiphase system where the electric field potentials of the various phases are not identical due to a different electrolyte concentration in the phases.

Equation A13 gives the formula for the conversion from volume to pressure as an independent variable,

$$\tilde{\mu}_i^{\mathrm{el}}(T, p, \mathbf{q}, \delta) = \mu_i^{\mathrm{el}}(T, V, \mathbf{q}, \delta) + RT\ln\left(\frac{v_{\mathrm{w}}}{v}\right) \tag{A26}$$

Note that all electric potentials are zero at infinite dilution of all charged species, that is, $\lim_{n_i \to 0} \mu_i^{\mathrm{el}} = 0$ at all $n_i \neq n_{\mathrm{w}}$.

The presence of charged species in a phase introduces a new constraint on each of the phases in the system, the electroneutrality condition:

$$\Sigma v_j z_j = 0 \tag{A27}$$

It also introduces a new variable: the electric field potential, ψ, or more precisely the difference in the electric field potential between a picked phase and each of the other phases. The subject is addressed in Section 2.A.2.

2.A.1.6 Surface Potentials

In heterogeneous macroscopic systems, the chemical potentials include a contribution due to the surface energy. When η is the interfacial tension and $\mathbb{A}$ is the surface area, the contribution is

$$\mu_i^{\sigma} = \eta\left(\frac{\partial \mathbb{A}}{\partial n_i}\right)_{T,p} \tag{A28}$$

The interfacial tension is the surface energy per unit surface area. The surface tension of a solution is the interfacial tension between the solution and the air, and this is not identical with the interfacial tension between the solution and a macromolecule or a crystal in the solution.

When a macromolecule adsorbs to a hydrophobic adsorbent, it partly unfolds, and due to the contact with the hydrophobic ligands, the wetted surface area of the macromolecule is diminished, and that decreases the chemical potential of the molecule in the adsorbed state.

The surface potential is also very important for the calculation of the chemical potentials of clusters formed in mother liquor. The surface area per unit volume of a body decreases when the volume of the body increases. For a cubic crystal, this ratio is $6/L$, and in general, this ratio is β/L where β is a shape factor and L is a characteristic length. Including Equation A28 in the classical chemical potential shows that the chemical potential of a crystal having a characteristic length L is

$$\mu_i(L) = \mu(\infty) + \eta\left(\frac{\partial \mathbb{A}(L)}{\partial L}\right)\left(\frac{\partial L}{\partial n_i}\right) = \mu_i(\infty) + \frac{2}{3}\frac{\eta v_c \beta}{L} \tag{A29}$$

where v_c is the molar volume of the crystal and $\mu_i(\infty)$ is the chemical potential at equilibrium. Equation A29 explains why small crystals have a higher chemical potential than large crystals, and that is the reason why small crystals go into solution while large crystals grow, because the surface to volume ratio decreases when the volume of the crystal increases, and this lowers the surface energy per unit volume of the crystals. During this process, the Gibbs energy of the system decreases and it becomes more stable, but it is a very slow process because it is diffusion controlled.

2.A.2 Membrane and Ion–Ion Equilibria

The usual equilibrium conditions in an isothermal system imply that the pressure is identical in all the phases of the system. However, if a semipermeable membrane that retains some macromolecules or charged species separates two phases, the pressure in the two phases will not necessarily be exactly alike. A pressure difference may also arise without inserting a semipermeable membrane if one of the phases contains immobilized charges as in ion exchange adsorbents.

Whether a pressure difference is established depends on whether the equilibrium conditions of non-charged species can be fulfilled under the appropriate experimental conditions. The system is isothermal, but the pressures in the two phases are not identical as indicated in the argument to the potentials in Equation A30. It is assumed that the electrolyte species are fully dissociated. The pressure in phase α is $p\alpha$, and the pressure in phase β is $p\beta$. The equilibrium condition for a molecule, j, that can distribute between the two phases and that has a net charge of z_j is

$$\begin{aligned}&\tilde{\mu}_j^{\alpha}\left(p^{\alpha}\right)+\mu_j^{\mathrm{self},\alpha}\left(p^{\alpha}\right)+\tilde{\mu}_j^{\mathrm{el},\alpha}\left(p^{\alpha}\right)+z_j^{\alpha}\psi^{\alpha}\boldsymbol{F}=\\&\tilde{\mu}_j^{\beta}\left(p^{\beta}\right)+\mu_j^{\mathrm{self},\beta}\left(p^{\beta}\right)+\tilde{\mu}_j^{\mathrm{el},\beta}\left(p^{\beta}\right)+z_j^{\beta}\psi^{\beta}\boldsymbol{F}\end{aligned}\tag{A30}$$

If all potentials in phase β are adjusted to the pressure $p\alpha$, the equilibrium condition is

$$\begin{aligned}&\tilde{\mu}_j^{\alpha}\left(p^{\alpha}\right)+\mu_j^{\mathrm{self},\alpha}\left(p^{\alpha}\right)+\tilde{\mu}_j^{\mathrm{el},\alpha}\left(p^{\alpha}\right)+z_j^{\alpha}\psi^{\alpha}F=\\&\tilde{\mu}_j^{\beta}\left(p^{\alpha}\right)+\mu_j^{\mathrm{self},\beta}\left(p^{\alpha}\right)+\tilde{\mu}_j^{\mathrm{el},\beta}\left(p^{\alpha}\right)+z_j^{\beta}\psi^{\beta}F+\bar{V}_j^{\beta}\left(p^{\beta}-p^{\alpha}\right)\end{aligned}\tag{A31}$$

where $\bar{V}_j^{\beta}$ are the partial molar volumes of the charged species in phase β. The equation of course also applies to the proton, which means that pH in the two phases will not be identical. This also holds for pH in the pore volume and in the mobile phase of an ion exchange adsorbent.

When the electrolyte is a strong electrolyte, the charge numbers z_j of the ions are independent of pH, that is, $z_j^{\alpha}=z_j^{\beta}=z_j$, the equilibrium condition is

$$\begin{aligned}&\tilde{\mu}_j^{\alpha}\left(p^{\alpha}\right)-\tilde{\mu}_j^{\beta}\left(p^{\alpha}\right)+\mu_j^{\mathrm{self},\alpha}\left(p^{\alpha}\right)-\mu_j^{\mathrm{self},\beta}\left(p^{\alpha}\right)+\tilde{\mu}_j^{\mathrm{el},\alpha}\left(p^{\alpha}\right)-\tilde{\mu}_j^{\mathrm{el},\beta}\left(p^{\alpha}\right)\\&=z_j\boldsymbol{F}\left(\psi^{\beta}-\psi^{\alpha}\right)+\bar{V}_j^{\beta}\left(p^{\beta}-p^{\alpha}\right)\end{aligned}\tag{A32}$$

Even if the pressure in the two phases were quite the same, the right-hand side will not be zero because the two electric field potentials ψ^{α} and ψ^{β} will not be exactly alike unless all the charges are distributed evenly between the two phases. The difference in the field potentials of the two phases is a dependent variable. A worked example is given in Chapter 6 in Ref. [5].

The equilibria for nonionic species, w, like water and other non-dissociating species are calculable from the equation

$$\tilde{\mu}_{\mathrm{w}}^{\alpha}\left(p^{\alpha}\right)-\tilde{\mu}_{\mathrm{w}}^{\beta}\left(p^{\alpha}\right)+\mu_{\mathrm{w}}^{\mathrm{el},\alpha}\left(p^{\alpha}\right)-\mu_{\mathrm{w}}^{\mathrm{el},\beta}\left(p^{\alpha}\right)=\bar{V}_{\mathrm{w}}^{\beta}\left(p^{\beta}-p^{\alpha}\right)\tag{A33}$$

Whether a pressure difference $(p^{\beta}-p^{\alpha})$ will be established depends on whether a pressure difference is necessary in order to satisfy Equation A33 for all nonelectrolyte species w. The electric contributions to the chemical potentials of nonelectrolyte species, w, are minor, but the electric Helmholtz energy will contribute to the pressure of the system, and in mixed solvents, the permittivity of the solvent depends on the solvent composition and that will contribute to the chemical potentials of the solvent components. Equation A33 explains the origin of the osmotic pressure as well as the swelling pressure in ion exchange adsorbents.

2.A.3 The Debye–Hückel Model

The model derived by Debye and Hückel in 1923 [20] is a model of the electric contributions to the Helmholtz energy of a solution of small ions. The model includes two contributions that give rise to the Born [19] potential and to the Debye–Hückel potential. See also Chapter 6 in Ref. [5].

The model derived by Debye and Hückel includes the self-energy often denoted the Born model

$$A^{\text{self}} = \frac{F^2}{8\pi N_A \varepsilon} \sum_i \frac{n_i z_i^2}{b_i} \tag{A34}$$

and the energy due to the ion–ion interactions

$$A^{DH} = -\frac{k_B T V}{4\pi \sum_j n_j z_j^2} \sum_i \frac{n_i z_i^2}{a_i^3} \left[\ln\left(1+\kappa a_i\right) - \kappa a_i + \frac{1}{2}\left(\kappa a_i\right)^2 \right] \tag{A35}$$

For each ion, b_i is the diameter of the hydrated ion, a_i is the distance of closest approach, and k_B is Boltzmann's constant. The Debye length or the shielding length, $1/k$, is an important parameter in the Poisson–Boltzmann equation and thus in many electrolyte models, and it includes the effect of the ions as well as the permittivity of the solvent, ε. The definition is

$$\kappa^2 = \frac{F^2}{\varepsilon RTV} \sum_j n_j z_j^2 \tag{A36}$$

The chemical potentials of the ions are the compositional derivatives of the Helmholtz functions at constant temperature and volume:

$$\mu_k^{\text{self}} = \left(\frac{\partial A^{\text{self}}}{\partial n_k} \right)_{T,V} = \frac{F^2 z_k^2}{8\pi N_A \varepsilon b_k} \tag{A37}$$

$$\mu_k^{DH} = \left(\frac{\partial A^{DH}}{\partial n_k} \right)_{T,V} = -\frac{A^{DH} z_k^2}{\sum_j n_j z_j^2} - \frac{k_B T V z_k^2}{4\pi \sum_j n_j z_j^2} \left(\chi_k + \frac{\kappa}{2} \frac{\sum_i n_i z_i^2 \sigma_i}{\sum_j n_j z_j^2} \right) \tag{A38}$$

where

$$\sigma_i = \frac{1}{a_i^2}\left[\frac{1}{1+\kappa a_i} + \kappa a_i - 1\right] = \frac{1}{a_i^2}\frac{(\kappa a_i)^2}{1+\kappa a_i} \tag{A39}$$

The pressure of the system changes due to the charging process by an amount p^e because the charging process takes place at constant temperature and constant volume. The electric pressure is calculable as the volume derivative of the electric contribution to the Helmholtz function:

$$\begin{aligned} -p^e = \left(\frac{\partial A^e}{\partial V}\right)_{T,\mathbf{n}} &= -\frac{A^{\text{self}}}{\varepsilon}\left(\frac{\partial \varepsilon}{\partial V}\right)_{T,\mathbf{n}} \\ &+ \frac{A^{DH}}{V} + \frac{k_B T V}{8\pi}\left(\frac{\sum_i n_i z_i^2 \sigma_i}{\sum_j n_j z_j^2}\right)\left(\frac{\kappa}{V} + \frac{\kappa}{\varepsilon}\left(\frac{\partial \varepsilon}{\partial V}\right)_{T,\mathbf{n}}\right) \end{aligned} \tag{A40}$$

The composition dependence of the permittivity of a mixed solvent contributes to the chemical potentials of the solvent components, h, and this contribution amounts to

$$\mu_h^e = \left(\frac{\partial A^e}{\partial n_h}\right)_{T,V} = \left[-\frac{A^{\text{self}}}{\varepsilon} + \frac{k_B T V}{8\pi}\frac{\sum_i n_i z_i^2 \sigma_i}{\sum_j n_j z_j^2}\frac{\kappa}{\varepsilon}\right]\left(\frac{\partial \varepsilon}{\partial n_h}\right)_{T,V} \tag{A41}$$

The derivatives $(\partial\varepsilon / \partial n_h)_{T,V}$ and $(\partial\varepsilon / \partial V)_{T,\mathbf{n}}$ must be estimated from an appropriate correlation. If no correlation is available, the composition dependence of the permittivity of a mixed solvent can be calculated as the volume fraction average of the permittivities of the pure solvents using the equation

$$\varepsilon = \frac{\sum_{\text{solvents}} n_i v_i \varepsilon_i}{\sum_{\text{solvents}} n_i v_i} \tag{A42}$$

where v_i are the molar volumes of the pure solvents and ε_i are the permittivities of the pure solvents.

The Debye–Hückel potential applies to classical electrolytes, but it is unclear how to apply the Debye–Hückel potential to large charged macromolecules. Large macro-ions are soft ions, which mean that small mobile ions and solvent molecules can penetrate into the domain of the macro-ion. When small mobile ions penetrate into the domain of the macro-ion, they will interact with the various ionic sites of the macro-ion. Furthermore, the acidic and basic groups of the amino acids are weak electrolytes and charges within the macro-ion fluctuate. Thus, one cannot consider a soft macro-ion to behave like an ion having a fixed point charge that is identical with the net charge of the macro-ion.

2.A.4 Kirkwood's Theory of Dipolar Ions

The content in this section is an application of the theory of dipolar ions in an electrolyte solution developed by Scatchard and Kirkwood [21] and Kirkwood [22, 23]. An ion having both positive and negative charges at the same time is called a dipolar ion or a zwitterion. Amino acids, peptides, and proteins are zwitterions, and the number of charges that these molecules carry differs from the net charge of the molecule. The fundamental equation is analyzed in Chapter 6 in Ref. [5], and the model has been applied to correlate experimental retention volumes of lysozyme in HIC using ammonium sulfate as the modulator in the eluant [11].

The Helmholtz energy of zwitterions in an electrolyte solution is derived from Scatchard and Kirkwood's theory of solutions of molecules containing widely separated charges. The electrostatic interactions between the charges of the small ions of the electrolyte and the ion dipole of the macro-ion give rise to a "salting-in" potential of the macro-ion as well as a "salting-out" potential.

2.A.4.1 Salting in

The "salting-in" potential of a macro-ion or a zwitterion component represents an attraction due to the forces between the electric dipole of the macro-ion and the surrounding small ions. It gives rise to a chemical potential that decreases with increasing salt concentration. The charges of the ion dipole create a cavity of low permittivity inside the macro-ion due to dielectric saturation. The cavity inside the macro-ion has the permittivity ε_{in}, and $\varepsilon_R = \varepsilon_{in}/\varepsilon$ is the ratio of the permittivity inside the cavity to the permittivity of the solvent, ε. One must expect ε_R to depend on the permittivity of the solvent.

The Helmholtz function for the attractive forces is

$$A^{KW}(T,V,\kappa) = -\frac{3N_A}{32\pi\varepsilon}\frac{1}{\kappa^2}\sum_{i=1}^{MC} n_i \eta_i \Theta_i \tag{A43}$$

where MC is the number of macro-ions, k is the inverse Debye length, and the parameter η_i is

$$\eta_i = \frac{\delta_i^2}{a_i \tau_{0,i}^2}\frac{1}{(1-\varepsilon_R)^2} = \frac{\eta_{0,i}}{(1-\varepsilon_R)^2} \tag{A44}$$

where δ_i is the dipole moment of the macro-ion, a_i is the mean distance of closest approach of the macro-ion to the small ions in the solution, and a sphere of radius b_i represents the cavity of low permittivity created inside the macro-ion. The parameter $\tau_{0,i}$ is

$$\tau_{0,i} = \frac{1}{6}\frac{b_i^3}{a_i} \tag{A45}$$

and

$$\Theta_i = \tau_i\kappa^2 - \ln\left(1 + \tau_i\kappa^2\right) \tag{A46}$$

The parameter τ_i is proportional to $\tau_{0,i}$, and in addition it depends on the permittivity ratio, ε_R, in the following manner:

$$\tau_i = \tau_{0,i}\frac{2 - 2\varepsilon_R}{2 + \varepsilon_R} \tag{A47}$$

The electrostatic potentials of the macro-ion components are calculable as the compositional derivatives of A^{KW} with respect to mole number n_k, and the eventual result is

$$\frac{\mu_k^{KW}}{RT} = -\frac{A^{KW} z_k^2}{RT\sum_j n_j z_j^2} - \frac{3N_A}{32\pi F^2}\frac{V}{\sum_j n_j z_j^2}\left(\eta_k\Theta_k + \sum_{i=1}^{MC} n_i\eta_i\theta_i\frac{\kappa}{2}\frac{z_k^2}{\sum_j n_j z_j^2}\right) \tag{A48}$$

where θ_i is the derivative of Θ_i with respect to the k, that is,

$$\theta_i = \frac{\partial\Theta_i}{\partial\kappa} = \frac{2\tau_i^2\kappa^3}{1 + \tau_i\kappa^2} \tag{A49}$$

2.A.4.2 Salting out

Kirkwood [23] also developed a limiting law for the "salting-out" potential of a zwitterion component of an electrolyte solution. The forces between the small ions and a zwitterion are modeled as a repulsion between the small ions and an image distribution in the cavity of low permittivity created by the zwitterion in the solvent. For simplicity, Kirkwood supposed that the mutual interaction of the zwitterions is negligible and thus limited himself to solutions in which the concentration of dipolar ions is itself small. The cavity of low permittivity with an image distribution is viewed as a sphere of radius b, and the excluded region is a sphere of radius a. The eventual model of the salting-out potential of a zwitterion component, k, at infinite dilution is

$$\frac{\mu_k^{kw}}{RT} = \frac{3}{2}\underline{\kappa}^2 \tau_k \phi\left(\frac{b_k}{a_k}\right) \tag{A50}$$

where $\underline{\kappa}$ is the inverse Debye length including small ions only, and when $x = b_k / a_k$

$$\phi(x) = \frac{2}{3x^4}\left[\frac{x^3}{2}\ln\frac{1+x}{1-x} - \ln\left(1-x^2\right) - x^2\right] \tag{A51}$$

The adjustable parameters in the model are η_k, $\tau_{0,k}$, and x. If a strong correlation between the parameters is observed, a default value $x = 2/3$ can be used, or if available, one can use the experimental value of the dipole moment and adjust a_k and b_k.

2.A.5 Light Scattering

Light scattering is not an equilibrium property but caused by fluctuations in the permittivity that give rise to a turbidity of the fluid. Equilibrium properties are independent of the choice of independent variables used in the calculations. That is, equilibrium properties derived from the Helmholtz, the Gibbs, or the McMillan–Mayer state functions are exactly alike. This is not true for fluctuations. For each environment and therefore for each set of independent variables, the problem is different.

If the light scattering experiments are conducted at constant temperature and pressure, the model for the turbidity arising from local fluctuations in the number density is

$$\tau = \frac{8\pi^3 V}{3\lambda^4}\frac{RT}{N_A}\sum_{j=1}\sum_{k=1}\left(\frac{\partial \varepsilon_r}{\partial n_j}\right)_{T,p}\left(\frac{\partial \varepsilon_r}{\partial n_k}\right)_{T,p}\left\{\mathbf{G}^{-1}\right\}_{jk} \tag{A52}$$

where λ is the wavelength in vacuum, ε_r is the relative permittivity, and the elements of the **G** matrix are the second derivatives of the Gibbs energy function:

$$\left\{\mathbf{G}\right\}_{jk} = \left(\frac{\partial \mu_j}{\partial n_k}\right)_{T,p} = \left(\frac{\partial \mu_k}{\partial n_j}\right)_{T,p} = \left(\frac{\partial^2 G}{\partial n_j \partial n_k}\right)_{T,p} \tag{A53}$$

and

$$\left(\frac{\partial \varepsilon_{\mathrm{r}}}{\partial n_i}\right)_{T,p} = \left(\frac{\partial \varepsilon_{\mathrm{r}}}{\partial n_i}\right)_{T,V} + \bar{V}_i \left(\frac{\partial \varepsilon_{\mathrm{r}}}{\partial V}\right)_{T,n} \tag{A54}$$

At the frequency of visible light, the relative permittivity, ε_r, is equal to the refractive index squared. The refractive index, n, of a solution is a function of the polarizability of the solvents and solute components. The Clausius–Mossotti model, Equation A55, provides an accurate correlation of these properties:.

$$\frac{\varepsilon_{\mathrm{r}} - 1}{\varepsilon_{\mathrm{r}} + 2} = \frac{n^2 - 1}{n^2 + 2} = \frac{\mathrm{N_A}}{3\varepsilon_0 V} \sum_i n_i \alpha_i (T) \tag{A55}$$

where ε_0 is the permittivity in free space and α_{i} is the polarizability of species i. Further details are given in Ref. [24].

The isocratic retention measurements at analytical load are a source of information of the chemical potentials of the retained species at infinite dilution, and the solubility data provide information of the chemical potentials at saturation. Light scattering data can be used to close the gap between the infinite dilution limit and the saturated state.

The second osmotic virial coefficient, B_{22}, is often used to correlate light scattering data, but it is less advantageous to use than the exact equations given in this section, because the second osmotic virial coefficient lacks a proper definition in a classical thermodynamic context. A proper definition of a thermodynamic quantity is a definition that relates the quantity to one or more of the derivatives of a state function.

References

1. G. Guiochon, A. Felinger, D.G. Shirazi, A.M. Katti, *Fundamentals of Preparative and Nonlinear Chromatography*, 2nd Edition, Academic Press, Amsterdam, 2006.
2. J.J. van Deemter, F.J. Zuiderweg, A. Klinkenberg, Longitudinal diffusion and resistance to mass transfer as causes of nonideality in chromatography. *Chemical Engineering Science* 5 (1956) 271–285.
3. G. Guiochon, Review—Preparative chromatography. *Journal of Chromatography A* 965 (2002) 129–161.
4. E. Hansen, J. Mollerup, The influence of retention on the plate height in ion exchange chromatography. *Separation Science and Technology* 39 (2004) 2011–2030.

5. M.L. Michelsen, J.M. Mollerup, *Thermodynamic Modelling: Fundamentals and Computational Aspects*, 2nd Edition, Tie-Line Publications, Holte, 2007. (Chapters 1–6 were converted to a LaTeX document by the publisher. A list of misprints is available from J.M. Mollerup at jm@prepchrom.com).
6. W. Melander, C. Horwárth, Salt effects on hydrophobic interactions in precipitation and chromatography of proteins: an interpretation of the lyotropic series. *Archives of Biochemistry and Biophysics* 183 (1977) 200–215.
7. F. Helfferich, *Ion Exchange*, Dover Publications, New York, 1995. A translation of *Ionenaustauscher*, Verlag Chemie, Germany 1959.
8. C.A. Brooks, S.M. Cramer, Steric mass–action ion exchange: displacement profiles and induced salt gradients. *AIChE Journal* 38 (1992) 1969–1978.
9. L. Pedersen, J.M. Mollerup, E. Hansen, A. Jungbauer, Whey proteins as a model system for chromatographic separation of proteins. *Journal of Chromatography B* 790 (2003) 161–173.
10. J.M. Mollerup, T.B. Hansen, S. Kidal, A. Staby, Quality by the design—thermodynamic modelling of chromatographic separation of proteins. *Journal of Chromatography A* 1177 (2008) 200–206.
11. J.M. Mollerup, M.P. Breil, C. Vogelpohl, G. Sadowski, Simultaneous correlation of hydrophobic interactions in HIC and protein solubility in aqueous salt solutions and mixed solvents. *Fluid Phase Equilibria* 301 (2011) 163–170.
12. J.M. Mollerup, Applied thermodynamics: a new frontier for biotechnology. *Fluid Phase Equilibria* 241 (2006) 205–215.
13. K. Johansson, S.S. Frederiksen, M.P. Breil, J.M. Mollerup, B. Nilsson, Combined effects of potassium chloride and ethanol as mobile phase modulators on hydrophobic interaction and reversed-phase chromatography of three insulin variants. *Journal of Chromatography A* 1381 (2015), 64–73.
14. J.M. Mollerup, Chemical engineering thermodynamics and protein adsorption chromatography. *Journal of Chemical & Engineering Data* 59 (2014) 991–998.
15. O. Ozdemir, S.I. Karakashev, A.V. Nguyen, J.D. Miller, Adsorption and surface tension analysis of concentrated alkali halide brine solutions. *Minerals Engineering* 22 (2009) 263–271.
16. J.M. Mollerup, T.B. Hansen, S. Kidal, L. Sejergaard, E. Hansen, A. Staby, Use of quality by the design for the modelling of chromatographic separations. *Journal of Liquid Chromatography & Related Technologies* 32 (2009) 1577–1597.
17. J.M. Mollerup, Modelling oligomer formation in chromatographic separations. *Journal of Chromatography A* 1218 (2011) 8869–8873.
18. J.M. Mollerup, Correlation of thermodynamic properties of mixtures using a random-mixture reference state. *Fluid Phase Equilibria* 15 (1983) 189–206.
19. M. Born, Volumen und Hydratationswärme der Ionen. *Zeitschrift für Physik* 1 (1920) 45–48.

20. P. Debye, E. Hückel, Zur Theorie der Elektrolyte. *Physikalische Zeitschrift* 24 (1923) 179–207. An English translation is available in the monograph: *The Collected Papers* by Peter J. W. Debye. A reprint published by Ox Bow Press, Woodbridge, CT, 1988.

21. G. Scatchard, J.G. Kirkwood, Das Verhalten von Zwitterionen und von mehrwertigen Ionen mit weit entfernten Ladungen in Elektrolytlösungen. *Physikalische Zeitschrift* 33 (1932) 297–300.

22. J.G. Kirkwood, Theory of solutions of molecules containing widely separated charges with special application to zwitterions. *Journal of Chemical Physics* 2 (1934) 351–361.

23. J.G. Kirkwood. The theoretical interpretation of the properties of solutions of dipolar ions. In (E.J. Cohn, J.T. Edsall Eds.) *Proteins, Amino Acids and Peptides as Ions and Dipolar Ions.* American Chemical Society Monograph Series, Reinhold Publishing Corporation, New York, 1943.

24. M.P. Breil, J.M. Mollerup, Modelling of molecular light scattering. *Fluid Phase Equilibria* 310 (2011) 120–128.

3

Simulation of Process Chromatography

Bernt Nilsson and Niklas Andersson

Chemical Engineering, Lund University, Lund, Sweden

3.1 Introduction

Simulation is a powerful tool for study and analysis of the performance in chromatographic separation of protein. Simulation means to mimic the real system in a virtual system, and today the virtual system is almost always a computer simulator. The benefits to do these virtual studies through simulations are many, but in the following text, three main purposes are discussed, namely, replace expensive experiments, document performance, and gain insight for design.

Simulations can replace experiments when experiments on the real system are expensive, time consuming, or hazardous. In industries where safety is important, simulations have been a major tool for a long time. Examples are flight industries and atomic power industries. In pharmaceutical industry experiments can be very expensive based on the lack of material or time consuming based on a large number of analysis that have to be done. In these cases simulations can complement experimental study. Simulations can also be used to do studies that are almost impossible to do experimentally or to do virtual studies of advanced experiments to point out critical aspects and to maximize the output of the experiments.

Models are formal documentation of our knowledge, and simulation studies can therefore serve as a way to illustrate the behavior of our knowledge and as a documentation of its performance. In pharmaceutical industry there are needs for extensive documentation and validation of system performance, particularly from a quality and risk perspective. Examples are large and rigorous parametric studies of critical process parameters and their impact on the critical quality attributes. The combination of extensive simulation studies with a set of well-chosen validation experiments has been proven to be a fruitful method.

Simulation studies are used to gain insight and to increase understanding of complex processes with simultaneous phenomena. These studies can be used to point out critical aspects of the design and operation and finally optimize process performance. In pharmaceutical industry there is by tradition a focus on robust operation with large quality margins. Simulations of the separation behavior and its impact on product quality can be used to quantify quality margins during production and integrate robust quality in to the design.

Simulations of chromatographic separations have been performed for a long time [1]. Gu presented a general computational platform for multicomponent separation already 20 years ago [2]. Today this technology is widely spread and easy to use, and a number of simulation tools for chromatography simulations are available [3–5]. The simulations presented in this chapter were performed using a MATLAB-based toolbox presented in Borg et al. [3]. The outline of the chapter is as follows. In the first section predictions of chromatographic behavior are presented for a set of different separation problems, which shows that a large number of common protein separation problems can be simulated quite easily with today's technology. To perform simulations a numerical method has to be selected, and the model has to be calibrated where its parameters are estimated based on experiments. This is discussed in the two following sections. In the fifth section parametric analysis for documentation of performance is discussed. In the last section optimization of separation performance is discussed and exemplified.

3.2 Simulation-Based Prediction of Chromatographic Processes

Replacement of experiment with simulation requires both formal definition of the experimental procedure and definition of the model of the chromatographic system, including its physical and adsorption parameters. The first requirement is just the mimic of the experimental run. The second one is based on the need to have a mathematical model that describes the separation good enough. This includes the selection of model structure and the performance of calibration experiment with the associated parameter estimation.

Protein chromatography is dominated by separations enhanced by a gradient of the mobile-phase conditions. This is often referred to as bind-and-elute operation compared to the classic isocratic operation. In isocratic operation the mobile-phase conditions are the same throughout the whole procedure, and the adsorption behavior can be assumed to be constant. Simulation of gradient chromatography adds to the complexity of dynamic mobile-phase conditions and changing adsorption behavior. Adsorption models for gradient protein chromatography require model structures that can handle changing mobile-phase conditions, that is, desorption induced by gradients [6]. This section shortly presents models and simulations of common chromatographic separation problems and exemplifies the technique on a set of separation cases.

3.2.1 Size Exclusion Chromatography

Size exclusion chromatography (SEC) uses only the physical size and shape of the molecule to make a separation. Large molecules cannot penetrate small pores in the stationary phase, and therefore these molecules will have a lower residence time in the column. The pore structure of the stationary phase is important to create the separation between molecules of different size. One extreme case is buffer exchange where all pores are too small for the protein, forcing it to stay in the void volume, while the mobile phase is assumed to penetrate the whole pore structure, traveling with much lower speed.

A differential material balance of the column behavior gives a partial differential equation (PDE) in time and in the axial dimension of the column, called the column equation. It is based on a lot of assumptions and is often referred to as a lumped rate model [7]. The superficial velocity is the flow rate per column cross section. The SEC specific parameters are hence the specific porosity for the component in question and the apparent dispersion coefficient:

$$\frac{\partial c_i}{\partial t} = D_{\mathrm{app},i} \frac{\partial^2 c_i}{\partial z^2} - \frac{v_{\mathrm{sup}}}{\varepsilon_i} \frac{\partial c_i}{\partial z} \tag{3.1}$$

c_i is the concentration of component i in the mobile phase, t is time, z is the axial dimension, $D_{\mathrm{app},i}$ is the specific apparent dispersion coefficient, v_{sup} is the superficial velocity, and ε_i is the specific total porosity:

$$\varepsilon_i = \varepsilon_\mathrm{c} + \left(1 - \varepsilon_\mathrm{c}\right)\varepsilon_\mathrm{p} K_{\mathrm{D},i} \tag{3.2}$$

The specific total porosity, ε_i, is composed of the void, ε_c, the porosity of the stationary phase, ε_p, and the specific exclusion factor, $K_{\mathrm{D},i}$, for each component, which is one for ions and less than one for proteins and zero for totally excluded molecules.

The column model mentioned earlier needs two boundary conditions to be complete together with the initial conditions. At the inlet of the column, $z = 0$, a flux boundary condition, BC, is the most suitable one. At the outlet, $z = L$, a no diffusive flux boundary condition is a common choice. Changing inflow conditions are introduced through the inlet value parameter in the inlet flux BC. In the rest of the chapter, these BC are assumed for the column equation.

This first SEC model mentioned earlier with these boundary conditions is the basis for all transport-dispersive and reaction-dispersive models, which will be presented and discussed in the following text. The initial conditions are set to zero except for the modulator, which have suitable value for loading.

For more detailed studies the behavior in the stationary phase has to be included in the model, introducing an extra domain in the particle radius with an additional differential material balance, the particle equation. It is often called the general rate model [7]. This model introduces an additional term for the mass transfer between the void and the particle surface. The particle equation exists in the axial dimension making the pore concentration dependent both in r and z:

$$\frac{\partial c_i}{\partial t} = D_{\text{ax}} \frac{\partial^2 c_i}{\partial z^2} - \frac{v_{\text{sup}}}{\varepsilon_{\text{c}}} \frac{\partial c_i}{\partial z} - \frac{1-\varepsilon_{\text{c}}}{\varepsilon_{\text{c}}} \frac{3}{R} k_{\text{f}} \left(c_i - c_{\text{p},i,R} \right) \tag{3.3a}$$

$$\frac{\partial c_{\text{p},i}}{\partial t} = D_{\text{e},i} \left(\frac{\partial^2 c_{\text{p},i}}{\partial r^2} + \frac{2}{r} \frac{\partial c_{\text{p},i}}{\partial r} \right) \tag{3.3b}$$

c_i is the concentration of component i in the mobile phase, t is time, z is the axial dimension, D_{ax} is the dispersion coefficient, v_{sup} is the superficial velocity, ε_{c} is the void, k_{f} is the mass transfer coefficient, $c_{\text{p},i}$ is the pore concentration of component i in the stationary phase, $c_{\text{p,i},R}$ is the pore concentration of component i at the surface of the particle, r is the radius dimension, and $D_{\text{e},i}$ is the specific efficient diffusion coefficient for component i.

The column equation will have corresponding boundary conditions as the first model mentioned earlier. The second equation, the particle equation, has a no-flux condition in the center of the particle, $r = 0$. On the particle surface, $r = R$, there is a flux condition connecting the void concentration to the pore concentration. This second SEC model mentioned earlier with these boundary conditions is the basis for all general rate models with different adsorption descriptions.

Example 3.1 Size Exclusion Chromatography

A size exclusion column is used for buffer exchange after an affinity-based capture step in a downstream process for antibody purification. The pool from the affinity step has high salt concentration and a low pH [8]. The antibodies are assumed to have a low exclusion factor, around 0.1, and their retention times are much shorter than the buffer. Simulation of the SEC separation is seen in Figure 3.1, and the buffer exchange property is well illustrated.

3.2.2 Ion Exchange Chromatography

In ion exchange chromatography (IEX) charged molecules in the mobile phase interacts with ligands with opposite charge on the stationary-phase surface.

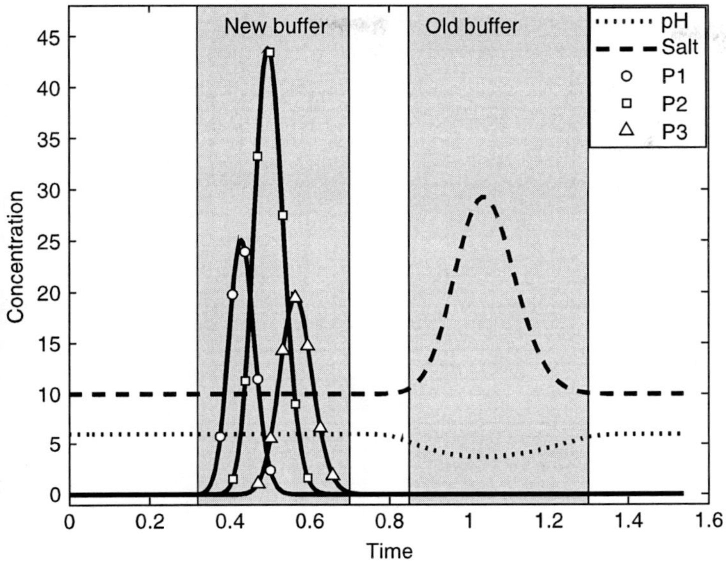

Figure 3.1 Simulation of buffer change using a SEC model in Example 3.1. Three proteins enter in a buffer with low pH and high salt concentration and exit with a desired buffer.

The charged molecules penetrate the porous material and interact with the ligands that are coated on the pore surface. Therefore they travel slower than the mobile phase through the column. The interaction is often strong for many proteins, and the interaction have to be weakened by the mobile-phase conditions to create suitable elution volumes, that is, gradient chromatography.

The column equation from the first SEC model mentioned earlier together with an adsorption term and an additional stationary-phase equation is defined, describing the amount of the component i that is adsorbed on the stationary phase. This model structure is a lumped rate model:

$$\frac{\partial c_i}{\partial t} = D_{\text{app},i} \frac{\partial^2 c_i}{\partial z^2} - \frac{v_{\text{sup}}}{\varepsilon_i} \frac{\partial c}{\partial z} - \frac{1-\varepsilon_c}{\varepsilon_i} r_{\text{ads},i} \tag{3.4a}$$

$$\frac{\partial q_i}{\partial t} = r_{\text{ads},i} \tag{3.4b}$$

c_i is the concentration of component i in the mobile phase and q_i in the stationary phase, t is time, z is the axial dimension, $D_{\text{app},i}$ is the specific apparent dispersion coefficient, v_{sup} is the superficial velocity, ε_i is the specific total porosity, ε_c is the void, and $r_{\text{ads},i}$ is the adsorption rate. The adsorption term can be expressed in many ways but in principle there are three main ways to express the adsorption rate. The first is as an internal film mass transfer, k_f, where the driving force is the difference between the equilibrium value that can be adsorbed, q^*, and amount of actually adsorbed, q. The second is an external film mass transfer, k_o, where the driving force instead is between the mobile-phase concentration, c, and the equilibrium concentration at the stationary phase expressed in mobile-phase concentration, c^*. A third version is to assume kinetics between adsorption, k_{ads}, and desorption, k_{des}. The three versions are exemplified with a Langmuir adsorption description as follows, H is the equilibrium Henry coefficient and q_{max} is the maximum capacity:

$$\begin{aligned} r_{ads,f,i} &= k_{f,i}\left(q_i^* - q_i\right) = k_f\left(H_i c_i\left(1+\sum_{j=1}^{n} K_j c_j\right)^{-1} - q_i\right) \\ r_{ads,o,i} &= k_{o,i}\left(c_i - c_i^*\right) = \frac{k_{o,i}/H_i}{1-\sum_{j=1}^{n} q_j / q_{max,j}}\left(H_i c_i\left(1-\sum_{j=1}^{n} q_j / q_{max,j}\right) - q_i\right) \\ r_{ads,kin,i} &= k_{ads,i} c_i\left(q_{max,i} - \sum_{j=1}^{n} q_j\right) - k_{des,i} q_i = k_{kin,i}\left(H_i c_i\left(1-\sum_{j=1}^{n} q_j / q_{max}\right) - q_i\right) \end{aligned} \tag{3.5}$$

The first two versions are called transport-dispersive and the third reaction-dispersive. The first version is the dominating one and is easy to use when an explicit adsorption description is available. The drawback is when the adsorption description is described implicit, creating a nonlinear equation system. This drawback is taken care of in the third version where the adsorption is expressed on reaction kinetic form only generating explicit expressions. The second version is rarely used. Note that for low concentrations all three expressions are identical and $k_f = k_o/H = k_{kin}$.

The dominating ion exchange adsorption description is the implicit steric mass action (SMA) description, which is based on multivalent ion exchange with steric hindrance of ligands due to the size of large proteins [9]. Here it is written on kinetic form:

$$\begin{aligned} r_{SMA,i} &= k_{kinS,i}\left(K_i c_i\left(\Lambda - \sum_{j=1}^{n}\left(\nu_j + \sigma_j\right) q_j\right)^{\nu_i} - s^{\nu_i} q_i\right) \\ 0 &= \sum_{i=1}^{n} r_{SMA,i} \end{aligned} \tag{3.6}$$

c_i is the concentration of component i in the mobile phase and q_i in the stationary phase, s is the salt concentration in the mobile phase, $r_{SMA,i}$ is the adsorption rate, K_i is the equilibrium coefficient, k_{kinS} is the SMA kinetic coefficient,

Λ is the ligand density, ν_i is the characteristic charge, and σ is the shielding factor. SMA describes the counterion in ion exchange also as a component, and the sum of all adsorption has to be equal to zero to keep electroneutrality. When SMA is used in a reaction-dispersive model, the equation system becomes a partial differential algebraic equation (PDAE) system, which means that a PDAE-solver is needed to solve the equation systems. It is also very non-linear, which can make it hard to calibrate and to estimate its parameters. A set of simplifications have been introduced to overcome these problems.

The first is to decouple the counterion from the proteins and calculate the counterion adsorption as the sum of all the other:

$$r_{\text{SMA},n} = -\sum_{i=1}^{n-1} r_{\text{SMA},i} \tag{3.7}$$

A second simplification is that the amount of desorbed counterion is neglected compared to its concentration in the mobile phase. This is valid at moderate protein concentrations. This simplified version of SMA can be interpreted as a Langmuir description with a mobile-phase modulator on salt [10], and it is called the MPM model (Karlsson et al. [11]):

$$r_{\text{MPM},i} = k_{\text{kinM},i}\left(K_{0,i} q_{\max,i} c_i \left(1 - \frac{q_i}{q_{\max,i}}\right) - s^{\nu_i} q_i\right) \tag{3.8}$$

c_i is the concentration of component i in the mobile phase and q_i in the stationary phase, s is the salt concentration in the mobile phase, $r_{\text{MPM},i}$ is the adsorption rate, $K_{0,i}$ is the modified equilibrium coefficient, ν_i is the characteristic charge, k_{kinM} is the MPM kinetic coefficient, and $q_{\max}$ is the maximum capacity.

Example 3.2 Ion Exchange Chromatography

A complex protein separation problem is illustrated in Figure 3.2 where model and parameters are taken from Auman et al. [12]. The product, P, is separated from 1 weak impurity, KW1, with additional two strong impurities, KS1 and KS2. The model is a transport-dispersive model with an internal mass transfer kinetic and a bi-Langmuir adsorption description with salt-based mobile-phase modulator on both equilibrium and maximum capacity coefficients. Note that the main separation is between the product and the weak impurity, which is very hard to perform based on a linear gradient, well illustrated in Figure 3.2.

To increase the validity range of the simulations of IEX models, the model parameters need to be changed in order for the models to predict behavior at new operating conditions. As mentioned earlier the models can handle changing salt concentrations. They can also handle temperature and pH changes

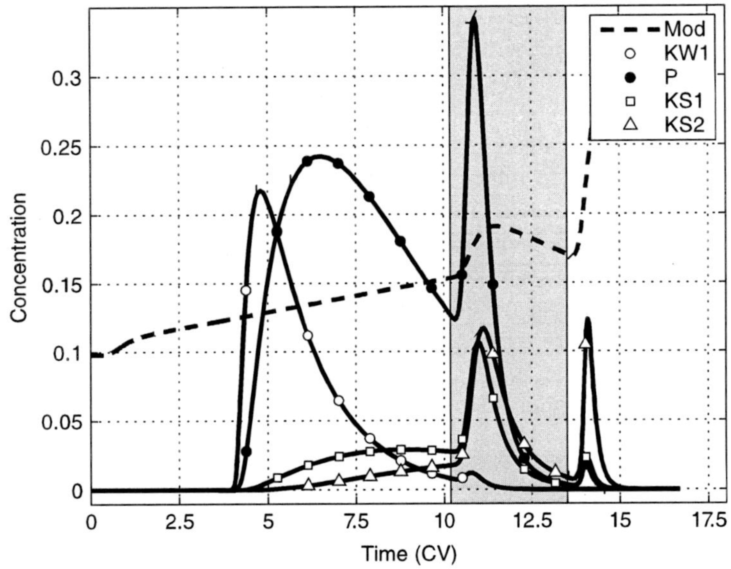

Figure 3.2 Simulation of an IEX separation in Example 3.2.

within certain margins. In a certain range the characteristic charge, ν_i, can be assumed to follow a modification of the Henderson–Hasselbalch equation, which makes it possible to estimate the charge as a function of pH:

$$\nu_i = A_i + \frac{B_i}{1 + 10^{\mathrm{p}K_a - \mathrm{pH}}} \tag{3.9}$$

The equilibrium constant K_i also changes due to the characteristic charge, ν_i, but also as a function of temperature, T. Here is ΔG_i the standard Gibbs energy for adsorption of the protein and ΔG_s for the counterion:

$$K_i = e^{\frac{-\Delta G_i^0}{RT} + \nu_i \frac{\Delta G_s^0}{RT}} \tag{3.10}$$

With these modifications of the adsorption parameters, ion exchange models can predict changes in protein load, protein composition, buffer compositions, salt gradients, pH, and temperature.

Example 3.3 Ion Exchange Chromatography

An advanced study of the purification of a therapeutic protein using ion exchange is presented in Westerberg et al. [13]. The study is based on a reaction-dispersive model with a kinetic adsorption description with a self-association term and with mobile-phase modulator based on temperature and pH on

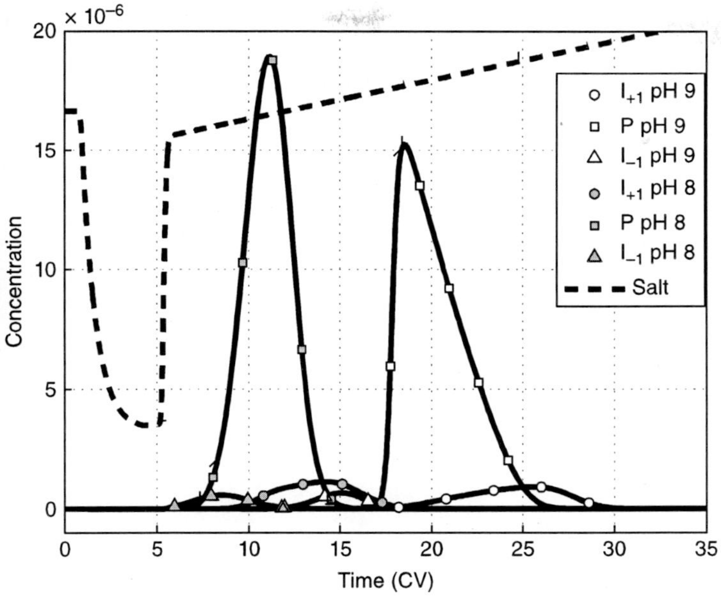

Figure 3.3 Simulation of the IEX separations of three proteins at two different pH.

the equilibrium coefficients [14]. The self-association description, called SAS, is a generalization of SMA that capture anti-Langmuir behavior:

$$r_{\mathrm{SAS},i} = k_{\mathrm{kin},i}\left(\gamma_i K_i c_i\left(1+2\gamma_i K_{2,i}c_i\right)\left(\Lambda-\sum_{j=1}^{n}\left(\nu_j+\sigma_j\right)q_j\right)^{\nu_i} - s^{\nu_i}q_i\right) \quad (3.11)$$

Additional parameters compared to the SMA description in Equation 3.6 are $K_{2,i}$, the self-association equilibrium coefficient, and γ_i, the activity coefficient for component i. Simulation of the separation at two different pH is seen in Figure 3.3, clearly indicating how pH changes the position, shape, and separation.

3.2.3 Hydrophobicity-Based Chromatography

Separations based on hydrophobicity are reversed-phase chromatography (RPC) and hydrophobic interaction chromatography (HIC). In RPC the proteins are adsorbed on the nonpolar surface, and they are desorbed when an organic modulator in the mobile phase is introduced. The hydrophobic interaction in RPC is often very strong making it a powerful tool for separation. This property also means that it is hard to simulate with often steep concentration fronts. In HIC the proteins are adsorbed on the stationary phase due to very high salt concentration in the mobile phase forcing the proteins to be adsorbed on the nonpolar surface. The hydrophobicity interaction in HIC is much weaker compared with RPC. By decreasing the mobile-phase ionic strength, that is, salt concentration, the proteins are desorbed into a less polar mobile phase.

Langmuir-type adsorption models with mobile-phase modulators are the most common type of chromatography models used for simulations of RPC and HIC. A reaction-dispersive model with a kinetic version of Langmuir adsorption with a mobile-phase modulator is seen as follows. The equilibrium constant is modified by an exponential term that is dependent on the modulator concentration, s. The modulator dependency is described by the parameter γ, and for positive values the adsorption strength increases with modulator, also called salting-out effect, and for negative values it decreases, called salting-in effect [10].

$$r_{\mathrm{MPM},i} = k_{\mathrm{kinM},i}\left(K_{0,i}e^{\gamma_i s}q_{\max,i}c_i\left(1-\sum_{j=1}^{n}\frac{q_j}{q_{\max,j}}\right)-q_i\right) \quad (3.12)$$

c_i is the concentration of component i in the mobile phase and q_i in the stationary phase, s is the modulator concentration in the mobile phase, $r_{\mathrm{MPM},i}$ is the adsorption rate, $K_{0,i}$ is the modified equilibrium coefficient, γ_i is the characteristic hydrophobicity coefficient, k_{kinM} is the MPM kinetic coefficient, and $q_{\max}$ is the maximum capacity.

The reaction-dispersive model, Equations 3.4 and 3.12, is used to simulate both RPC and HIC with the main difference that the modulator is different and the sign of the characteristic hydrophobicity coefficient is opposite. The modulator in RPC is an organic solvent, and its concentration in the mobile phase, s, is used to change the adsorption strength, and it decreases with the modulator concentration, and therefore the parameter γ becomes negative. In HIC the modulator is the salt concentration in the buffer, and increased salt concentration increases the adsorption strength, that is, the parameter γ becomes positive.

Example 3.4 RPC and HIC Simulations
An RPC separation of an insulin analog from a product-related impurity is seen in Figure 3.4a. It is a reaction-dispersive model with Langmuir reaction kinetics with a mobile-phase modulator using the organic solvent. Model and parameters are taken from Westerberg et al. [15]. A HIC separation of one insulin analog from two product-related impurities is seen in Figure 3.4b from Johansson et al. [16]. The model is the same as mentioned earlier but using the salt concentration as modulator. Model and parameter are taken from Holmqvist et al. [17].

3.2.4 Affinity-Based Chromatography

Affinity chromatography (AC) is often specific and unique for a particular separation. It is not a common separation step, and therefore the number of simulations studies is limited. One exception is of course protein A capture of

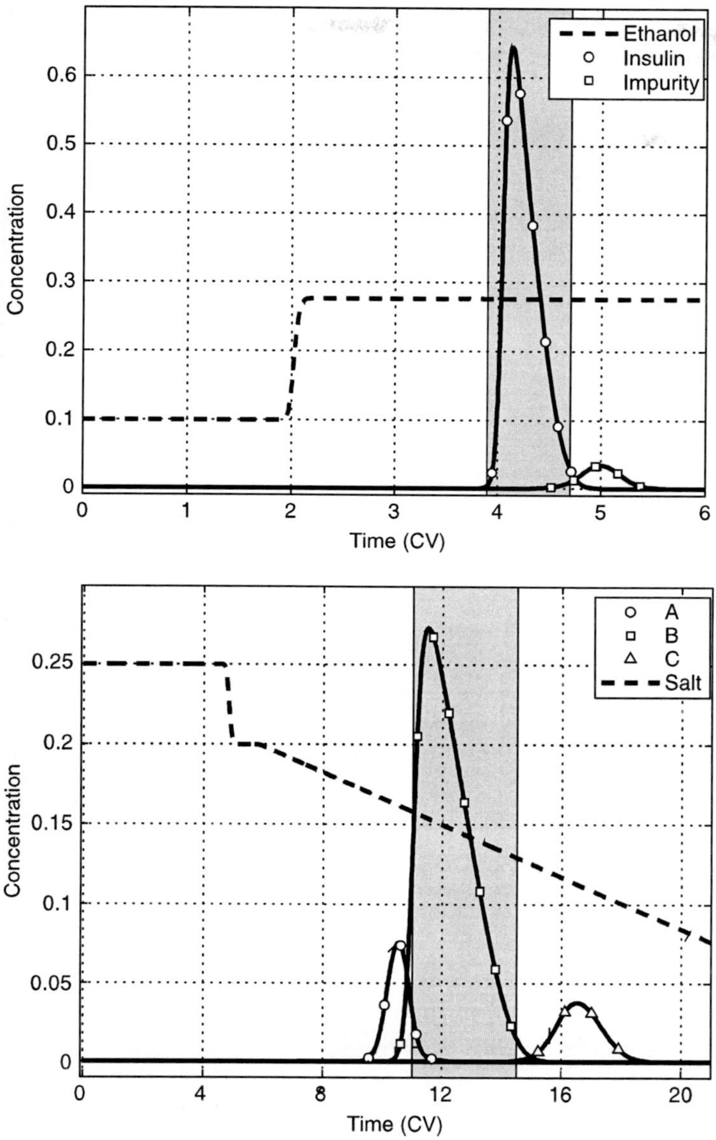

Figure 3.4 Simulation of insulin separation from product-related impurities. (Top) Ethanol step elusion in RPC. (Bottom) Decreasing salt gradient in HIC.

antibodies, which is a very important step in antibody downstream processing. The most common way to describe AC is to use the modeling framework presented earlier, such as the transport-dispersive model structure for the mobile phase, Equation 3.4, with the internal mass transfer kinetic description, from Equation 3.5, together with a modified Langmuir adsorption. In IgG–protein A affinity separation, the adsorbed IgG is eluted by low pH mobile phase:

$$r_{\mathrm{IgG,f}} = k_{\mathrm{f}} \left(\frac{K_0 \left(\mathrm{pH}/\mathrm{pH}_{\mathrm{ref}}\right)^n q_{\max} c_{\mathrm{IgG}}}{\left(1 + K_0 \left(\mathrm{pH}/\mathrm{pH}_{\mathrm{ref}}\right)^n c_{\mathrm{IgG}}\right)} - q_{\mathrm{IgG}} \right) \tag{3.13}$$

c_{IgG} is the concentration of IgG in the mobile phase and q_{IgG} in the stationary phase, $r_{\mathrm{IgG,f}}$ is the adsorption rate, K_0 is the modified equilibrium coefficient, pH is the changing pH in the mobile phase, $\mathrm{pH}_{\mathrm{ref}}$ is a reference pH coefficient, n is an exponent parameter for the pH dependency, k_{f} is the internal mass transfer coefficient, and $q_{\max}$ is the maximum capacity. The property of the description mentioned earlier is strong adsorption at pH higher than the reference pH ($n > 1$) and almost no adsorption below the reference pH.

Example 3.5 Simulation of an Affinity-Based Capture Step
A transport-dispersive model with internal mass transfer kinetics and Langmuir adsorption description with a pH-based mobile-phase modulator is simulated where model and parameters are taken from Ng et al. [18]. The chromatogram is seen in Figure 3.5 with a long loading phase, a short wash phase, and a fast and efficient elution phase.

Closely related to AC is the mixed-mode chromatography (MMC), which exploits the possibility to use more than one interaction form to create the separation. One such example is presented in the following text. The MMC is based on a ligand with both an ion exchange part and a hydrophobic part. This enhances the adsorption of molecules that have this property. The following is an example where a reaction-dispersive model with SMA adsorption description is used to predict a MMC separation.

Example 3.6 MMC Simulations
Separation of monomers and dimers of modified human hormones is modeled by a transport-dispersive model using SMA adsorption description on kinetic form and simulated using parameters from Sejergaard et al. [19].

Simulation of process chromatography is well established, and a set of applications are presented and exemplified in almost all types of chromatography modes. As presented earlier a wide range of separation problems can be studied using simulations, from size exclusion to mixed mode using a limited number of modeling concepts and reliable numerical methods (Figure 3.6). There are examples mentioned earlier showing salt, ethanol and pH gradients, overloading conditions, competitiveness effects, displacement and tag-along effects, self-association effects, pH, and temperature dependency. Still there are many challenges in chromatography simulations. Some of them will be highlighted in the following sections.

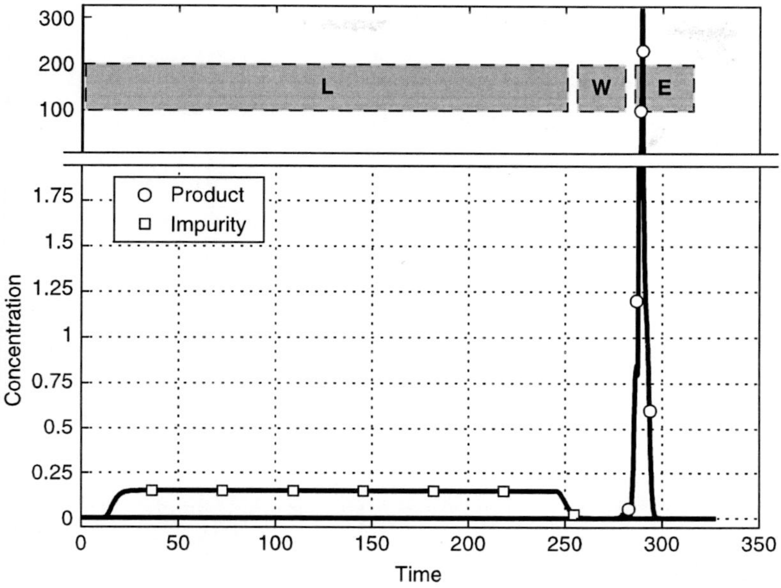

Figure 3.5 Simulation of affinity-based capture of an antibody.

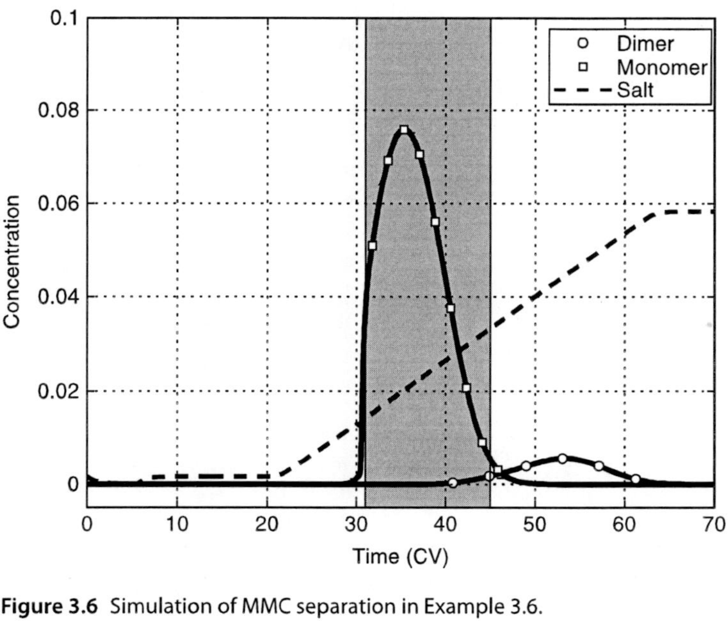

Figure 3.6 Simulation of MMC separation in Example 3.6.

3.3 Numerical Methods for Chromatography Simulation

The simulation problems presented earlier are all based on mathematical models consisting of PDEs in the axial column domain together with kinetic adsorption description. In the case of general rate models, there are also PDEs in the particle radius domain. In the majority of the examples presented in the previous section, the adsorption description is explicitly formulated. In some cases the adsorption is implicitly described by a set of nonlinear algebraic equations, and the chromatography models become a PDAE system.

The dominating numerical techniques are all based on the concept method of lines (MOL), in which the space dimensions are discretized and all space derivatives are approximated using the discrete space values. The result is a large set of ordinary differential equations (ODEs) in time. The solution to the problem is found by solving the ODE system in time by an ODE-solver. MOL is a successful approach to decompose the simulation problem into three subproblems: (i) space approximation, (ii) selection of ODE-solver, and (iii) handling of large equations systems.

The dominating techniques for space approximations are finite element methods (FEM) and finite volume methods (FVM). There are two other methods available, the finite difference method that can be interpreted as a simplified version of FVM and orthogonal collocation method that can be seen as a special version of FEM.

The FEM is based on the idea of approximating the unknown solution with a set of finite elements, composed of basic functions. The integral of finite element approximation multiplied by a weight is forced to zero. This is called the method of weighted residuals. The most common technique in FEM is to use a weight function that is the same as the basis function, which is called the Galerkin method. FEM is a well-balanced approximation with high accuracy for many process chromatography problems [4, 20].

The FVM is based on the idea of discretization of the space domains into a set of finite volumes and based on these volumes approximate all space-dependent expressions. In general the FVM is not as well balanced as FEM, but the technique has developed a set of methods to handle discontinuities, such as upwind approximations or flux limiters [3, 5, 21]. Discontinuities are shock waves that propagate through the column and are created during saturation of Langmuir systems. In systems with fast mass transfer and small amount of dispersion, these shock waves can be very steep; see the RPC in Example 3.4.

The discretized PDE system, generating a large set of ODEs, is solved by an ODE-solver. The ODE system is almost always stiff resulting in the need of an implicit ODE-solver. Many implicit ODE-solvers can also handle DAE systems, which makes it possible to solve implicit adsorption descriptions directly.

The implicit ODE-solver solves the time-stepping problem by formulating an implicit nonlinear equation system that is solved by a Newton-type solver in each time step. It is therefore natural for many of these solvers to solve algebraic expressions.

The ODE system for complex separations and with steep profiles results in large or very large equation systems due to the need of fine grid approximations for accurate solutions. There is a need for good computational resources. To decrease the computation time and memory allocation, the equation solver has to handle sparse matrix computations and efficient program coding to minimize memory allocation, particularly to help the ODE-solver with the sparse Jacobian information, and in some cases iterative linear solvers, in the Newton method, are needed [5].

Example 3.7 Front-Capturing in Finite Volume Methods

In Figure 3.7 the RPC chromatogram from Example 3.4 is shown using 16 different numerical methods for front-capturing on a course grid [3]. In FVM there are a large set of different front-capturing techniques available. The simplest method is the so-called low order upwind, which is robust and reliable. As can be seen in Figure 3.6, this method does not perform as good as more advanced flux-limiter methods, which come with a price of computation time. The WENO method has the best performance [5], but Swedby15 [21] and van Albaba 1 are good choices based on lower computation time compared to WENO.

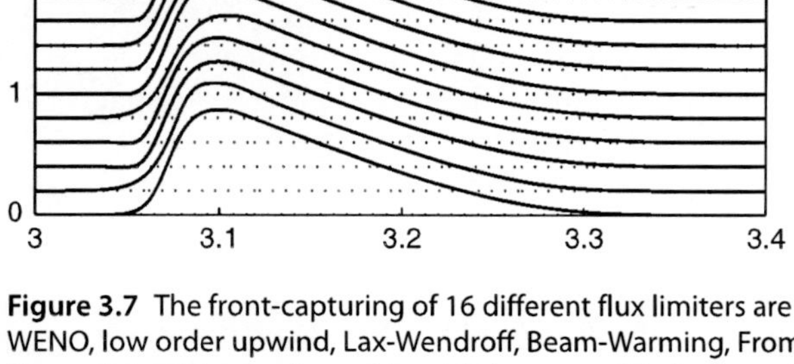

Figure 3.7 The front-capturing of 16 different flux limiters are seen above, from the top: WENO, low order upwind, Lax-Wendroff, Beam-Warming, Fromm, van Leer, umist, Sweby15, Ospre, Osher15, mc, minmod, hquick, hcus, van Albaba2, van Albaba1.

3.4 Simulation-Based Model Calibration and Parameter Estimation

The usefulness and success of simulation-based studies rely heavily on the accuracy and validity range of the mathematical model and its potential to predict the real separation system. Model calibration procedure to adjust the model response to predict experimental response is perhaps the most important part in model-based engineering of chromatographic separation. The procedure can also be called parameter estimation when the focus is on estimation of the parameter value rather than to minimize the prediction error, as is the purpose in model calibration. Model calibration consists of three parts: (i) parameterized mathematical model, (ii) set of experimental data, and (iii) minimization objective.

Simulation-based model calibration is also called the inverse method. Simulation is used to generate a model response, which is compared with the experimental response, and the result is a residual. The residual is then minimized by a minimization method. If the comparison between the model and experimental responses is measured as the square of the difference, see Equation 3.14, the minimization method is called least square method:

$$Q = \sum_{i=1}^{n} \left(y_{\text{exp},i} - y_{\text{mod},i} \right)^2 = R^T R \tag{3.14}$$

Model calibration and parameter estimation based on least square methods also generate interesting statistical information about the experimental data, estimated parameters, and model predictions. Least square calibration based on single response is a classical and commonly used method in all type of engineering, but as in all minimization problems, there are problems when the model contain phenomenon that is not seen in the experiment, that is, the experimental conditions do not excite the behavior. In these situations the model is over-parameterized and the problem becomes ill conditioned. To calibrate a model over a wide range of experimental conditions, which is important in design and optimization, the model calibration have to be performed using a set of experiments that span the whole parameter space. In these situations there are experiments that are more important than others, for example, they can contain more information or are performed close to interesting operational conditions. In all these cases the experimental data are weighted, w, compared to each other. A weighted multiple experiment and normalized data objective function is shown in Equation 3.15:

$$Q = \sum_{j=1}^{m} \sum_{i=1}^{n} w_{i,j} \left(\frac{y_{\text{exp},i,j} - y_{\text{mod},i,j}}{y_{\text{exp},i,j}} \right)^2 \tag{3.15}$$

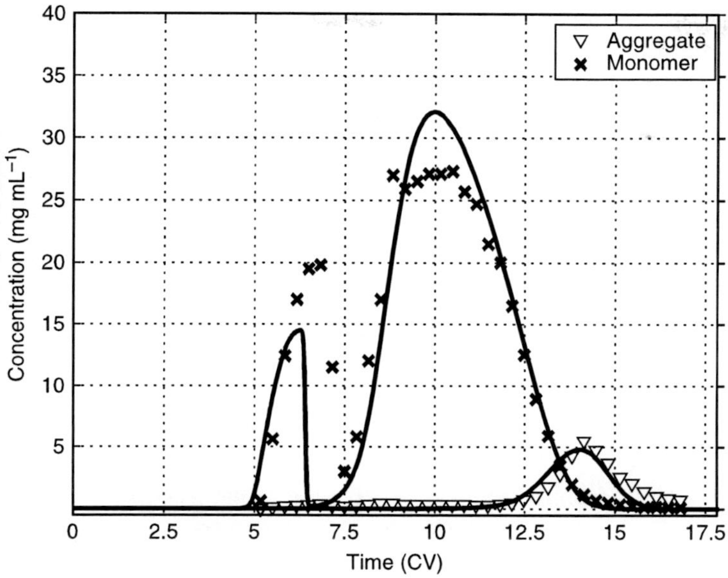

Figure 3.8 One simulation of 113 model responses done during calibration. *Source*: Borg et al. [22]. Reproduced with permission of Elsevier.

Model calibration of a complex separation problem involves many experiments, and the model is parameterized with many parameters. In these situations the effort to produce a sufficient accurate model cannot be exaggerated. Even with efficient numerical methods and a good mathematical model, the model calibration is time consuming and implies a set of engineering consideration.

Example 3.8 Model Calibration of an IEX Step Based on Multiple Experiments
In Borg et al. [22] monomers of a monoclonal antibody was separated from its dimers and higher aggregates using a cation exchanger. The model was a general rate model with an SMA adsorption description on kinetic form. The model was also capable to handle small deviations in pH and in temperature. The model calibration estimated 18 model parameters using 113 experiments in total, containing linear retention experiments, overloaded experiments, and fractionation analysis data. To make this possible all model responses was run on a computer cluster to speed up the procedure (Figure 3.8).

3.5 Simulation-Based Parametric Analysis of Chromatography

Well-calibrated models that can predict chromatographic behavior with a desired accuracy makes simulation-based parametric analysis an important tool in performance and quality assessment. Model-based simulations can be used to quantify the impact of a parameter on an objective or sensitivity analysis or quantify the uncertainty of an objective, uncertainty, or robustness analysis. Examples of sensitivity analysis are studies of the impact of process parameters on quality attribute, that is, indicate critical process parameters and critical quality attributes. An example of robustness analysis is a study of the distribution of uncertainty of a quality attribute.

A parametric analysis is based on two parts, namely, (i) the objective and (ii) the set of analysis parameters. The objective is a quantitative measure of the performance of interest. It can be the productivity or the yield of the separation step or it can be a quality attribute such as purity. The set of analysis parameters are the process parameters selected from all process parameters in the model to analyze their impact. The analysis can be performed by a deterministic methodology generating extreme scenarios, like worst case or best case, or by a stochastic methodology generating distributions, like the probability of failure.

A quality attribute is an integral measure over the target fraction out form the separation step. The target fraction, also called the product pool, is the output collected between two cut points. This means that the two cut points have to be selected before the objective can be calculated. This is often referred to as pooling, and in this context it can be a part of the analysis or done by an optimization problem. The most common one is to maximize the yield for a given chromatogram and a purity requirement [23]. The quality can be quantified as the purity, the amount of target product in relation to all products, or as a maximal allowable amount of a specific impurity.

Example 3.9 Pooling of the Target Component in an IEX Step

In Example 3.3 an IEX separation of a therapeutic protein is presented. A pooling strategy based on finding the best cut points for purity of 96% generates an overall yield of 94.7% (Figure 3.9).

Minimizing the parameter space, that is, the number of analysis parameters, is valuable when the number of process parameters is large, forcing the computation time to be high and increase the complexity of the analysis. Many of the process parameters will not have a significant impact of the objective at the particular case of interest. It is therefore a good choice to remove these from the study as early as possible. The process parameters that are significant can be interpreted as the critical process parameters at the specific operational point [24].

A sensitivity analysis is done by calculating the impact of each parameter, p_i, on the objective, Q. It is called one variable at a time or just OVAT [13, 25]. The sensitivity is based on the deviations from a normal operating point (NOP), which has to be defined a priori. To find a normalized sensitivity the objective deviation is scaled by the deviation between a NOP and a minimal requirement on the objective. For a purity objective the deviation is based on the nominal purity and the minimal purity allowed:

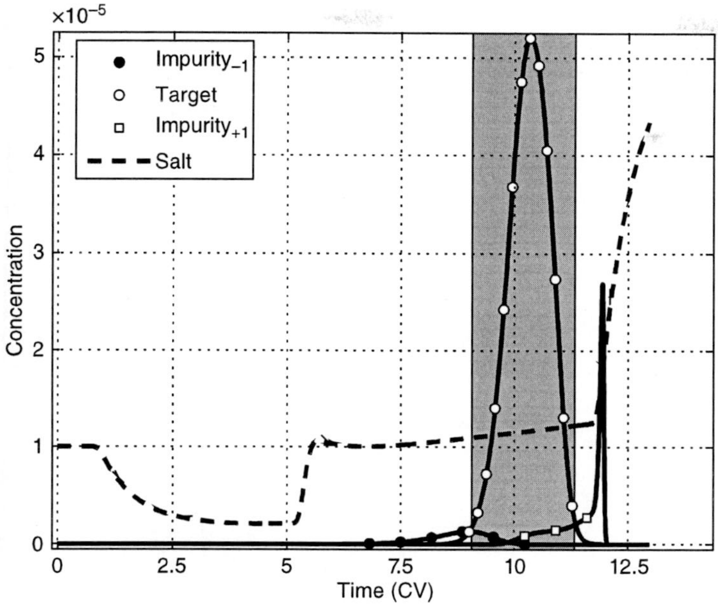

Figure 3.9 Maximal pooling of the target protein with a purity requirement of 96%.

$$S_{Q,i} = \frac{Q_{\text{NOP}} - Q_{\pm \Delta p_i}}{Q_{\text{NOP}} - Q_{\text{req}_i}} \tag{3.16}$$

Based on the sensitivity value, all parameters are ranked, and the parameters with high sensitivities are selected as critical process parameters, and the rest are removed from further analysis. A deterministic worst-case scenario is easy to point out where all critical process parameters have their largest negative impact on the objective. In a deterministic robustness analysis, this is the case that has to be taken care of in order to handle worst-case situations.

Example 3.10 Ranking of the Critical Process Parameter in an IEX Step
The case presented in Examples 3.3 and 3.9 has a set of parameters with a nominal operation point with a disturbance range, as seen in Table 3.1. A sensitivity analysis of the purity demand is performed where the purity is selected as the objective with the minimum requirement of 95%. The parameter ranking is listed in the following, and four parameters are selected to be critical and used to define a deterministic worst-case scenario. The chromatogram for this worst case is seen in Figure 3.10, and it has a purity of 92.8%. The probability that four critical process parameters are at their worst-case boundary at the same time is unlikely, making this prediction less interesting.

Table 3.1 Process parameter set point, relative standard deviation, and sensitivity with ranking.

Process parameter	Set point	RSD (%)	$S_{Q,I}$	Ranking
Buffer ($mol\,L^{-1}$)	0.020	5	0.28	6
Acid ($mol\,L^{-1}$)	0.011	5	0.31	5
Salt ($mol\,L^{-1}$)	**0.25**	**5**	**0.60**	**2**
Salt in feed ($mol\,L^{-1}$)	0.010	5	0.00	8
Protein load ($g\,L^{-1}$)	3.5	15	0.00	7
Weak impurity ($g\,g^{-1}$)	**0.027**	**25**	**0.60**	**3**
Strong impurity ($g\,g^{-1}$)	**0.079**	**25**	**0.57**	**4**
Protein measured ($g\,g^{-1}$)	**1.0**	**10**	**0.89**	**1**

The parameter ranking is listed and four parameters (bold face) are selected to be critical and used to define a deterministic worst-case scenario.

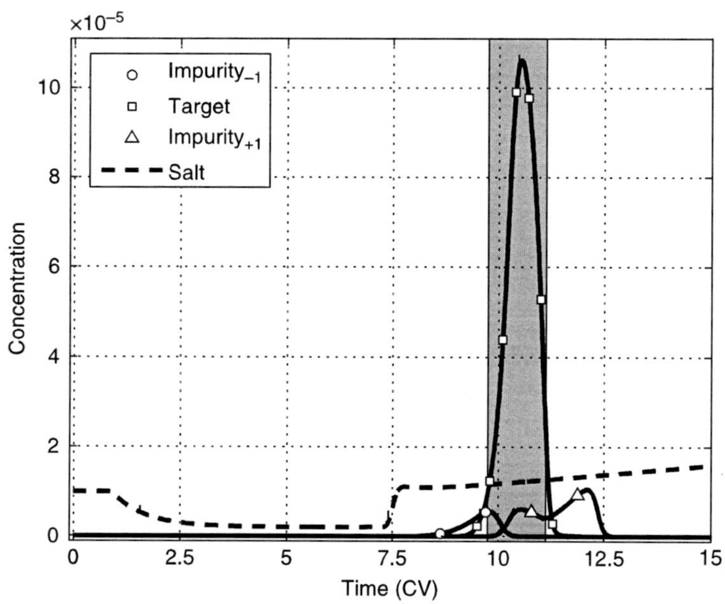

Figure 3.10 The chromatogram for the worst-case scenario.

Deterministic worst-case scenarios are often very conservative and point out scenarios that have low probability. It forces the analysis to be unnecessarily negative, exaggerating quality deviations. A better, but more complex, way is to include stochastic descriptions into the analysis. Stochastic analysis requires additional information, namely, models of the process parameter disturbance distributions and a sampling procedure of the stochastic parameter distributions. In general there are no limitations on the disturbance distribution descriptions, but a common and natural choice is to assume independent and normal distributed disturbances. In some cases the disturbances are correlated and it can easily be accounted for.

Stochastic analysis of a nonlinear system requires the simulations of a large set of parameter scenarios sampled from the disturbance distributions. One way to sample the multidimensional distribution is by Monte Carlo sampling based on randomly, or pseudo-randomly, selection of parameter scenarios. For small numbers of samples, a better method is Latin hypercube sampling, which ensures full coverage of the distribution [24]. The choice of number of samples is case dependent. If the intension is to measure the probability of an unlikely event, the number of samples has to be high enough, so the number of unlikely events becomes more than just a few individuals. For instance, to measure 5% probability a sampling volume of 100–200 samples is sufficient, but for a requirement on 0.1% the number of samples need to be up to 10 000 [26].

Example 3.11 Stochastic Sensitivity Analysis of an IEX Step
The disturbance range defined in Example 3.10 is used here as the width of a normal distribution added to the nominal value of the eight selected process parameters. This study is inspired by Westerberg et al. [13]. A Latin hypercube sampling is done generating 10 000 well-distributed parameter scenarios, which are all simulated. In Figure 3.11 on the left, the value of the objective, in this case the purity, is plotted in scatterplots as a function of each individual parameter. On the right a histogram of the purity uncertainty is shown. The 5% probability of failure is indicated in all plots and is found to be 95.5%. In this case the probability for purity below 95% is 0.02%.

3.6 Simulation-Based Optimization of Process Chromatography

To find the best operating conditions for a separation problem is a complex task, which involves all aspects of engineering. One tool that can be used in this task is model-based optimization. It can be used directly by calculating the optimal conditions based on a set of requirements. It can also be used in reversed engineering as a tool to find the objective and requirements needed to point out given conditions as optimal.

An optimization problem is defined by three parts: (i) the objective to be minimized, Q; (ii) a set of decision variables, u; and (iii) a set of constraints. In optimization community it is common to define the problem as a minimization of

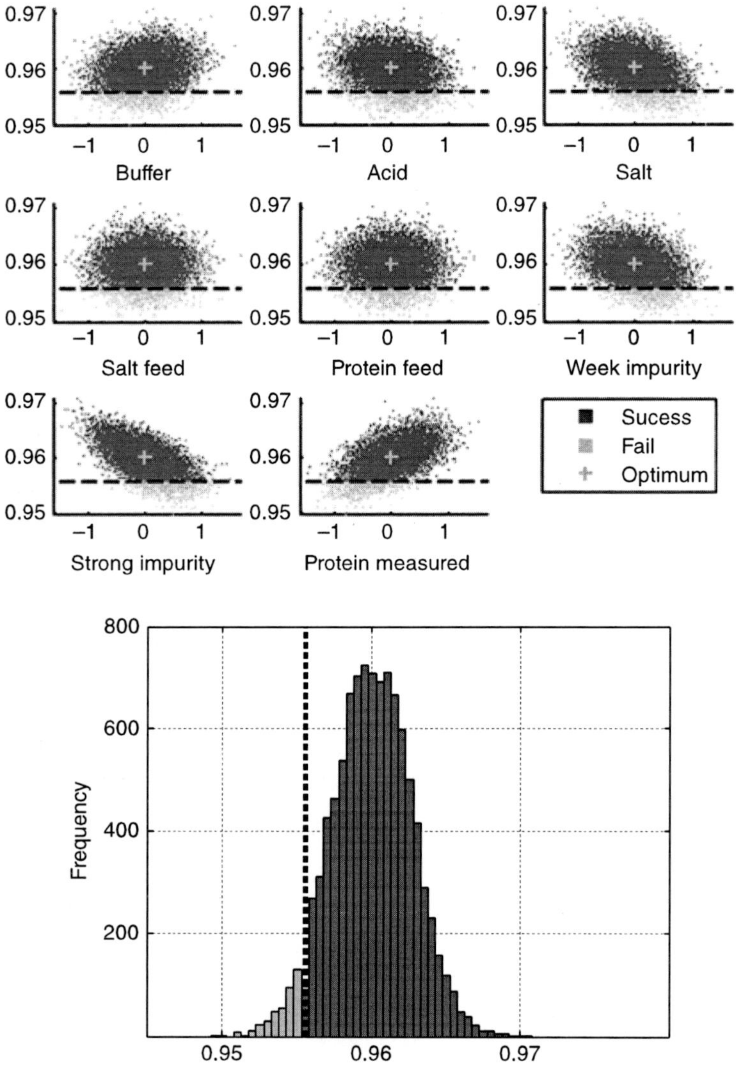

Figure 3.11 (Top) Scatterplot visualization of the distribution of the purity as a function of a set of process parameters, (bottom) the resulting purity distribution with average of 96%. There is 5% probability of purity below 95.5%.

the objective. It means that natural objectives in chromatographic separation, such as yield and productivity, have to be rewritten into a minimization objective. It is often done by just changing sign on the objective function. It can be formulated as follows:

$$\begin{aligned} &\min_{u} -Q(u) \\ &Pu \geq Pu_{\text{req}} \\ &u_{lb} \leq u \leq u_{ub} \end{aligned} \tag{3.17}$$

The selection of the objective is the most critical part because it will indirectly force the solution to be at its minima. A yield objective will optimize the recovery of the target, while a productivity objective will optimize the amount recovery per time unit, punishing processing time. An economic-based objective that takes both these aspects into account is a weighted, w, sum of the productivity, Pr, and yield, Y (Karlsson et al. [27]). To make it possible to sum productivity and yield, the productivity is normalized based on the maximal productivity, that is, the objective value at the optima for the weight equal to zero.

$$Q = (1-w)\frac{\text{Pr}}{\text{Pr}_{\text{max}}} + wY \tag{3.18}$$

Maximized productivity, w is zero, is the objective if the operational and feed costs are small compared with the investment costs, and maximized yield, w is one, is the best objective when the feed cost is dominating. A pure yield-based objective is not recommended because it is not a well-defined optimization problem without additional constraints on loading or productivity. An often used objective is the productivity times the yield, which is convenient because it is parameter-free and gives a balance result between productivity and yield [28].

The set of decision variables is recommended to be as small as possible because the computational effort increases rapidly with the number of variable. It is important to avoid decision variables that do not have an impact on the objective because they make the numerical problem ill conditioned. A sensitivity analysis and ranking of possible decision variables can be done in the same way as in the previous section. Dominating decision variables in optimization of process chromatography are variables that affect the loading factor; examples are loading time, column length, feed buffer conditions, and variables that define the elution conditions and pooling, such as gradient parameters and cut points [29].

Almost all optimization problems are performed under a set of constraints forcing the solution to fulfill a set of equality and inequality conditions. A set of basic constraints are upper and lower bounds on the decision variables. Other important constraints can be a purity requirement or a yield requirement. If the flow rate is a decision variable, then maximal allowable pressure drop is an important constraint. In some cases also constraints on protein concentrations or buffer conditions in the pool are defined to avoid precipitation. When the number of decision variables is high in a complex separation problem, it is hard to foresee the behavior. Some of the decision variable dimensions will not be convex, and their optimal value will be on the inequality constraint. It is recommended that these variables are set to the constraint value and removed from the set of decision variables.

Example 3.12 Optimization of an IEX Step
The IEX case from Example 3.3, which is analyzed in Examples 3.9–3.11, is optimized with the objective of productivity multiplied by yield, and the chromatogram is presented in Example 3.9:

$$Q = \text{Pr} \cdot Y \tag{3.19}$$

The decision variables are the load volume and the salt gradient parameters. A purity requirement is defined to be 96%. The optimal operation point is seen in Figure 3.9. In Figure 3.12 the impact of the decision variables on the normalized minimization objective, $-Q$, is seen, and the objective has a minimum at the optimal operational point for all decision variables.

The weight between productivity and yield, Equation 3.16, is perhaps not well defined, or it can be unknown at the design phase. This forces the designer to make multiple optimizations for different weights generating a set of optimal points for different weights. The optimal points form the so-called Pareto front of optimality. There are no operating points that fulfill all criteria beyond

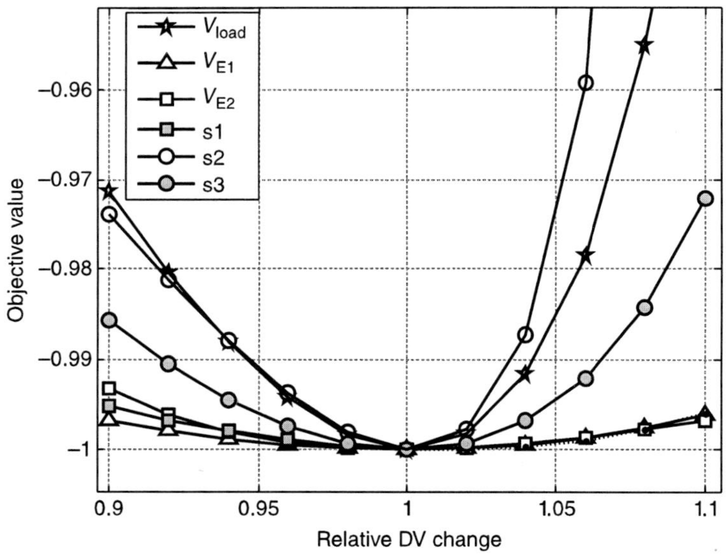

Figure 3.12 Decision variable impact on the objective minima in Example 3.12.

the Pareto front. Optimization based on multiple objective of the productivity and the yield is a common Pareto study in process chromatography [15, 26, 30]. To make this a well-defined optimization problem, a purity constraint is needed.

Example 3.13 Pareto Optimality of an IEX Step
The IEX case from Example 3.12 is optimized with a set of different weights generating multiple optimization problems, which are solved. The results are then illustrated in an objective plot, here the productivity as a function of yield, indicating the Pareto front. Seven different Pareto fronts are seen based on the purity requirements ranging from 96 to 99%. The objective used in Example 3.12 is indicated by solid triangles (Figure 3.13).

The numerical technique for optimization of process chromatography is dominated by the shooting methods. It means that the optimization method makes a complete simulation of the chromatographic step every time it needs to find a value of the objective. The optimization methods can be divided into three classes, namely, direct, indirect, and evolutionary methods. The direct methods use the objective calculations directly, while indirect methods make additional calculations, that is, find parameter Jacobians to generate better solutions. The direct method is therefore easy to use, but a well-designed

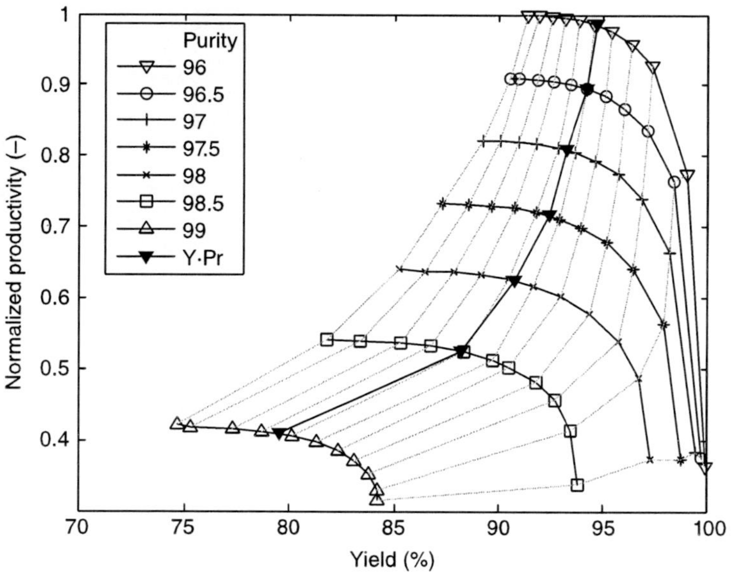

Figure 3.13 Seven Pareto fronts for the multi-objective optimization of separation problem in Example 3.3 based on increasing purity requirements from 96 to 99%. The triangles indicate the optimal point for the selected objective in Example 3.12.

indirect method for constraint problems is a very powerful tool. These methods are single-objective methods with local convergence properties. The third class of method is the evolutionary methods, which are based on the sequential evolution of a population, that is, a large set of simulations. Evolutionary methods are general, for example, can handle multi-objective problems, and are robust but very computationally heavy. None of the three classes of optimization methods dominate and therefore the selection is case dependent. The optimization in Examples 3.12 and 3.13 are done by an indirect Newton-type method using a numerical finite difference method for Jacobian calculations. A large set of parallel optimizations are performed to generate the Pareto plots.

An alternative to the numerical technique based on shooting methods, discussed earlier, is the simultaneous technique, where the chromatographic model is not simulated in time but instead discretized in both space and time, generating a large or very large nonlinear programming (NLP) problem. This large equation system is directly integrated into the optimization method as a constraint. The model and the optimization are solved simultaneously. This technique is illustrated in Example 3.14 showing how to find optimal general gradients.

Example 3.14 Optimization of a General Salt Gradient in HIC

The HIC case from Example 3.4 is optimized with a salt gradient based on 50 sequential small steps, uS, generating a changing salt gradient, cS (see Figure 3.14). Model and parameters are taken from Holmqvist et al. [17]. The objective is weighted productivity and yield, Equation 3.16, with a punishment on gradient variations and with a purity requirement on 99%. There is a weak impurity eluting before the target and a strong impurity eluting after. The optimal general gradient solution has ~30% higher performance compared with an optimal linear gradient solution. There are surprisingly many advantages like higher productivity, higher yield, higher pool concentration, and shorter batch cycles compared to optimal linear gradients case.

3.7 Summary

This chapter presents the use of simulations to predict, analyze, and optimize chromatographic separations. Many of the most common chromatographic separations can be simulated quite easy, fast, and robustly with available computer tools. The chapter illustrate how many of the most common protein separation problems can be simulated, from size exclusion to AC. Simulation-based parametric analysis and optimization are shown to give deep insight to complex separation problems. The quantification of quality margins and key performance is the key to engineering solutions to separation problems. Using model-based analysis to point out critical process parameters and critical quality attributes has been shown to be successful. Stochastic-based analysis

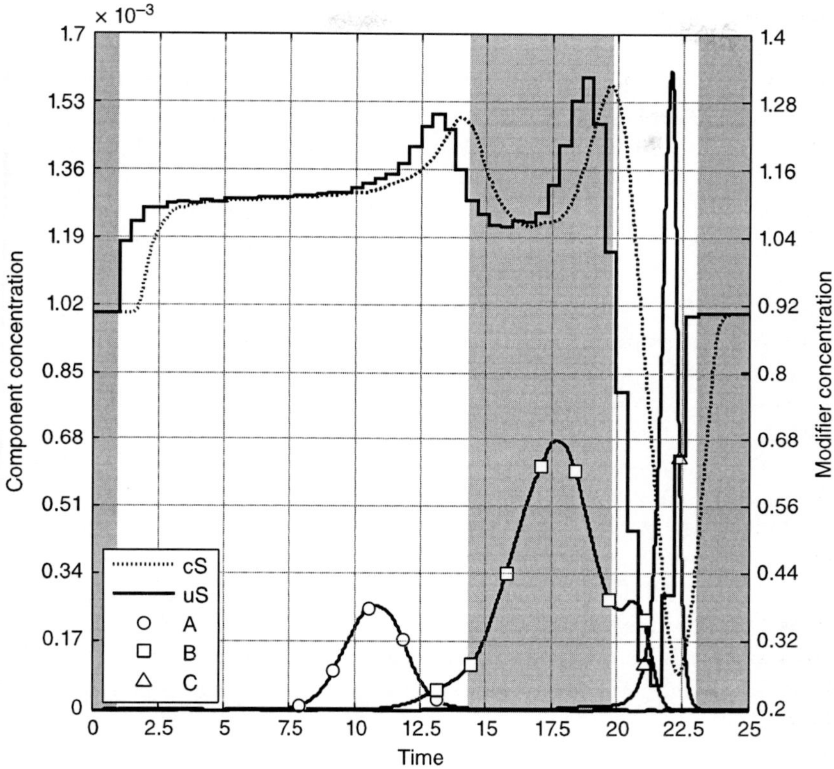

Figure 3.14 The optimal general gradient for the HIC case from Example 3.4.

makes it possible to calculate the probability to fulfill purity requirements. Simulation-based optimization can be used to find the landscape of possible optimal operation points, the Pareto surface, which can be used to move the operation point based on future demands. All these possibilities and features come with a prize. The use of model-based methodology, presented in this chapter, requires an accurate model that can predict the separation in question. The engineering effort needed for model calibration and a good experimental design cannot be exaggerated. Even though the methodology presented here is very powerful, it needs to be further developed with the incorporation of robustness and model uncertainty margins into analysis and optimization methods.

Acknowledgement

The authors thank Process Industry Centre at Lund University for financial support and special thanks to Anders Holmqvist and Anton Sellberg for rapid problem solving and help for generating simulations for some of the examples.

References

1. Guiochon, G., Shirazi, D.G., Katti, A.M. (1994). *Fundamentals of preparative and nonlinear chromatography*, Academic Press, Boston, MA.
2. Gu, T. (1995). *Mathematical Modeling and Scale-up of Liquid Chromatography*, Springer, Berlin.
3. Borg, N., Andersson, N., Nilsson, B. (2013). Preparative chromatography simulator—PCS. Lund University, Lund. Available at: www.chemeng.lth.se/pcs (accessed on September 8, 2016).
4. Hahn, T. (2015). ChromX, Karlsruhe Institute of Technology. Available at: https://gosilico.com/products/chromx/ (accessed on September 8, 2016).
5. Lieres, E.v., Andersson, J. (2010). A fast and accurate solver for the general rate model of column liquid chromatography, *Chemical Engineering & Technology*, 34, 1180–1191.
6. Carta, G., Jungbauer, A. (2010). *Protein Chromatography: Process Development and Scale-up*, John Wiley & Sons, Weinheim.
7. Smith-Traub, H., Schulte, M., Seidel-Morgenstern, A. (2012). *Preparative Chromatography*, Wiley-VCH, Weinheim.
8. Ojala, F. (2015). Modeling antibody aggregation in downstream processing. PhD thesis, Department of Chemical Engineering, Lund University, Lund, Sweden.
9. Brooks, C.A., Cramer, S.M. (1992). Steric mass-action ion exchange: displacement profiles and induced salt gradients, *AIChE Journal*, 38 (12), 1969–1978.
10. Melander, W.R., El Rassi, Z., Horváth, C. (1989). Interplay of hydrophobic and electrostatic interactions in biopolymer chromatography: effect of salts on the retention of proteins, *Journal of Chromatography A*, 469, 3–27.
11. Karlsson, D., Jakobsson, N., Axelsson, A., Nilsson, B. (2004). Model-based optimization of a preparative ion-exchange step for antibody purification, *Journal of Chromatography A*, 1055, 29–39.
12. Auman, L., Stroehlein, G., Morbidelli, M. (2007). Parametric study of a 6-column countercurrent solvent gradient purification (MCSGP) unit, *Biotechnology & Bioengineering*, 98 (5), 1029–1042.
13. Westerberg, K., Hansen, E.B., Degerman, M., Hansen, T.B., Nilsson, B. (2012). Model-based process challenge of an industrial ion-exchange chromatography step, *Chemical Engineering & Technology*, 35 (1), 183–190.
14. Mollerup, J. (2008). A review of the thermodynamics of protein association to ligands, protein adsorption, and adsorption isotherms, *Chemical Engineering & Technology*, 31 (6), 864–874.
15. Westerberg, K., Borg, N., Andersson, N., Nilsson, B. (2012). Supporting design and control of a reversed-phase chromatography step by mechanistic modeling, *Chemical Engineering & Technology*, 35 (1), 169–175.
16. Johansson, K., Frederiksen, S., Degerman, M., Breil, M., Mollerup, J., Nilsson, B. (2015). Combined effects of potassium chloride and ethanol as mobile

phase modulators on hydrophobic interaction and reversed-phase chromatography of three insulin variants, *Journal of Chromatography A*, 1381, 64–73.

17. Holmqvist, A., Magnusson, F., Nilsson, B. (2015). Dynamic multi-objective optimization of batch chromatographic separation processes. In Proceedings from 12th International Symposium on Process Systems Engineering and 25th European Symposium on Computer Aided Process Engineering, May 31,–June 4, 2015, Copenhagen, Denmark.
18. Ng, C.K., Osuna-Sanchez, H., Valéry, E., Sorensen, E., Bracewell, D.G. (2012). Design of high productivity antibody capture by protein A chromatography using an integrated experimental and modeling approach, *Journal of Chromatography B*, 899, 116–126.
19. Sejergaard, L., Karkov, H.S., Krarup, J.K., Hagel, A.B.B., Cramer, S.M. (2014). Model-based process development for the purification of a modified human growth hormone using multimodal chromatography, *Biotechnology Progress*, 30 (5), 1057–1064.
20. Javeed, S., Qamar, S., Seidel-Morgenstern, A., Warnecke, G. (2011). A discontinuous Galerklin method to solve chromatographic models, *Journal of Chromatography A*, 1218, 7137–7146.
21. Javeed, S., Qamar, S., Seidel-Morgenstern, A., Warnecke, G. (2011). Efficient and accurate numerical simulation of nonlinear chromatographic processes, *Computers & Chemical Engineering*, 35, 2294–2305.
22. Borg, N., Brodsky, Y., Moscariello, J., Vunnum, S., Vedantham, G., Nilsson, B. (2014). Modeling and robust pooling design of a preparative cation-exchange chromatography step for purification of monoclonal antibody monomer from aggregates, *Journal of Chromatography A*, 1359, 170–181.
23. Andersson, N., Knutson, H.-K., Max-Hansen, M., Borg, N., Nilsson, B. (2014). Model-based comparison of batch and continuous preparative chromatography in the separation of rare earth elements, *Industrial & Engineering Chemistry Research*, 53, 16485–16493.
24. Degerman, M., Westerberg K., Nilsson, B. (2009). Determining critical process parameters and process robustness in preparative chromatography—model-based approach, *Chemical Engineering & Technology*, 32 (6), 903–911.
25. Jakobsson, N., Karlsson, D., Axelsson, J.P., Zacchi, G., Nilsson, B. (2005). Using computer simulation to assist in the robustness analysis of an ion-exchange chromatography step, *Journal of Chromatography A*, 1063, 99–109.
26. Degerman, M., Westerberg K., Nilsson, B. (2009). A model-based approach to determine the design space of preparative chromatography, *Chemical Engineering & Technology*, 32 (8), 1195–1202.
27. Karlsson, D., Jakobsson, N., Brink, K.-J., Axelsson, A., Nilsson, B. (2004). Methodologies for model calibration to assist the design of a preparative ion-exchange step for antibody purification, *Journal of Chromatography A*, 1033, 71–82.

28. Felinger, A., Guiochon, G. (1996). Optimizing preparative separation at high recovery yield, *Journal of Chromatography A*, 752, 31–40.
29. Degerman, M., Jakobsson, N., Nilsson, B. (2006). Constraint optimization of a preparative ion-exchange step for antibody purification, *Journal of Chromatography A*, 1113, 92–100.
30. Nagrath, D., Bequette, B.W., Cramer, S.M., Messac, A. (2005). Multiobjective optimization strategies for linear gradient chromatography, *AIChE Journal*, 5 (2), 511–525.

4

Simplified Methods Based on Mechanistic Models for Understanding and Designing Chromatography Processes for Proteins and Other Biological Products-Yamamoto Models and Yamamoto Approach

Noriko Yoshimoto and Shuichi Yamamoto

Bio-Process Engineering Laboratory, Graduate School of Medicine and School of Engineering, Biomedical Engineering Center (YUBEC), Yamaguchi University, Ube, Japan

4.1 Introduction

Process-scale chromatography is essential as a unit operation in biopharmaceuticals (biologics) production. As the speed to the clinic is crucial, it is needed to establish a rapid and reliable procedure (method) for design, operation, and validation of the purification process.

For this purpose, various high-throughput process development (HTPD) systems have been developed. Most systems use a standard 96-well microplate format or its modified version. HTPD may reduce the time and the volume of the sample and the buffer solutions with the aid of design of experiments (DOE).

Another attractive method is to use mechanistic (phenomenological) chromatography models. Although this approach is not very new, a number of researchers have been investigating this approach recently because of the aforementioned reasons. Compared with DOE, which is commonly employed especially with HTS, the mechanistic model approach has several advantages:

- It provides better understanding of the factors affecting the separation.
- It allows for extrapolation if governing theories/models are valid.
- It enables you to reduce the number of experiments.

It is also useful for characterizing the new separation media/format and understanding the separation mechanism (i.e., biorecognition).

The problems or difficulties of this approach are rather complicated and laborious procedures to determine the parameter values needed for the simulation. Simplified versions of the model are useful not only for prediction of the separation but also for obtaining experimental data quickly, which are needed for the model simulations.

In this chapter simplified models based on the mechanistic model for linear gradient elution (LGE) chromatography of proteins and other large biological products are explained. Then, several applications of the models to process design and process understanding and for biorecognition study are presented.

4.1.1 Operation Mode of Chromatography and Zone Movement in the Column

Various different interactions are available for such chromatography steps as electrostatic interaction (charge-based separations, ion exchange chromatography (IEC)), hydrophobic interaction chromatography (HIC), and biospecific affinity (affinity chromatography (AFC)). Size-based separation is also possible (size-exclusion chromatography (SEC) or gel filtration chromatography (GFC)). There are also several elution modes based on the interaction as shown in Figure 4.1. Isocratic elution (IE) is a method where the mobile phase is the same during elution, whereas a modifier (modulator) concentration in the mobile phase is changed during non-IE such as LGE or stepwise elution (SE). For IEC and HIC, a salt is usually used as a modifier.

In order to characterize the zone movement (band migration) in the column, highly simplified models (equilibrium models), which neglect the zone-spreading effects, were developed and applied to isocratic and non-IE modes. Chromatography models including the equilibrium models are explained in various books and book chapters [1–6].

The simplified equilibrium models are still important to know how the distribution coefficient (or isotherm) affects the retention and the maximum resolution. The movement of a solute applied to the chromatography column is described by the following equation:

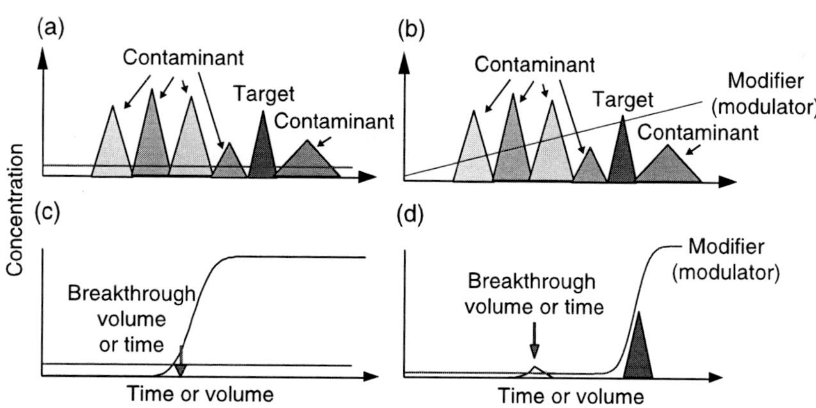

Figure 4.1 Elution methods (operation modes) for chromatography. (a) Isocratic elution. (b) Linear gradient elution. (c) Breakthrough curve (frontal analysis). (d) Stepwise elution.

$$\frac{dz}{dt} = \frac{u}{(1+HK)} = \frac{u}{(1+k')} \tag{4.1}$$

Here, z = distance from the top of the column, t = time, u = linear mobile-phase velocity, $H=(1-\varepsilon)/\varepsilon$ phase ratio, and $K=C_s/C$ = distribution coefficient of the solute between the stationary and the mobile phases. $k'=HK$ is the capacity factor. The linear mobile-phase velocity u is defined as

$$u = \frac{F_v}{(A_c\varepsilon)} = \frac{u_0}{\varepsilon} \tag{4.2}$$

F_v = volumetric flow rate, $A_c = \pi dc^2/4$ column cross-sectional area, and d_c = column diameter, and u_0 = superficial velocity.

The void fraction ε is the interparticle space in the column packed bed (void volume), which can be determined by the retention volume of a large molecule completely excluded from the pore of the stationary phase (packing material). There are several definitions of the distribution (partition) coefficient and the void fraction (see section Appendix for the details).

While the zone is traveling, it is spreading due to several mass transfer mechanisms as shown in Figure 4.2 such as pore diffusion, film diffusion, and axial dispersion. Various models considering the zone-spreading effects have been developed. First, a well-known Martin–Synge plate theory was proposed. The plate model regards a chromatography column (packed bed) as a series of plates. In each plate the equilibrium is instantaneously attained. The zone spreading is described simply by the value of the number of plates or the plate number N. For packed bed reactors or fixed-bed adsorbers, the rate models were developed, which consist of partial differentials equations for the mobile and the stationary phases. Table 4.1 summarizes most chromatography models. Although

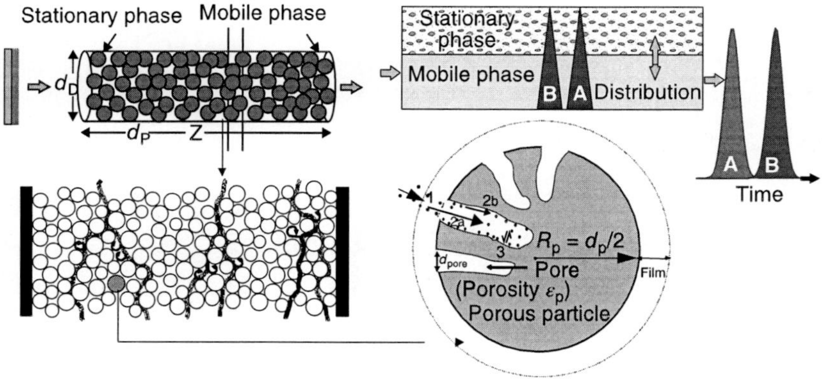

Figure 4.2 Schematic representation of separation of two components in chromatography (upper) and transport phenomena in the chromatography column (bottom).

Table 4.1 Basic equations and plate number equations for mechanistic chromatography models.

Model	Basic equations and plate number (N) equation	
1	$\frac{\partial C}{\partial t}+H\frac{\partial \overline{C_S}}{\partial t}=D_L\frac{\partial^2 C}{\partial z^2}-u\frac{\partial C}{\partial z}$	mobile phase
	$\frac{\partial C_S}{\partial t}=D_S\left[\frac{\partial^2 C_S}{\partial r^2}+\left(\frac{2}{r}\right)\frac{\partial C_S}{\partial r}\right]$	stationary phase
	$\frac{\partial \overline{C_S}}{\partial t}=\left(\frac{6k_F}{d_p}\right)\left(C-\frac{C_{S,i}}{K}\right)$	film mass transfer
	$\frac{1}{N}=2\frac{(D_L/u)}{Z}+\frac{d_p^2 R_a(1-R_a)u}{30D_S Z}+\frac{d_p R_a(1-R_a)Ku}{(3k_F Z)}$	
2	$\frac{\partial C}{\partial t}+H\frac{\partial \overline{C_S}}{\partial t}=D_{L,e}\frac{\partial^2 C}{\partial z^2}-u\frac{\partial C}{\partial z}$	mobile phase
	$\frac{1}{N}=\frac{2(D_{L,e}/u)}{Z}$	
3	$\frac{\partial C}{\partial t}+H\frac{\partial \overline{C_S}}{\partial t}=-u\frac{\partial C}{\partial z}$	mobile phase
	$\frac{\partial \overline{C_S}}{\partial t}=K_S\left[KC-\overline{C_S}\right]$	stationary phase (linear driving force)
	$\frac{1}{N}=\frac{2(1-R_a)^2}{\left[HKK_S(Z/u)\right]}=\frac{2(1-R_a)^2}{\text{NTU}}$	
Plate model	$\frac{dC_{(i)}}{dt}+H\frac{dC_{S(i)}}{dt}=N\left(\frac{u}{Z}\right)\left[C_{(i-1)}-C_{(i)}\right]$	
	$\frac{1}{N}=\frac{1}{N}$	

Models 1–3 are sometimes called rate-theory or rate-model; $R_a=1/(1+HK)$. N equations for linear isotherm (K = constant) NTU (the number of transfer units) $\text{NTU}=2N(1-R_a)^2$.

numerical solutions (and analytical solutions in special cases) of such models predict elution curves, it is needed to have simplified methods.

4.2 HETP and Related Variables in Isocratic Elution

For linear IE (K = constant), height equivalent to a theoretical plate (HETP) or plate height is a commonly employed measure for the column efficiency based on the peak width and the retention time (Figure 4.3):

$$\text{HETP}=\frac{Z}{N}=Z\left(\frac{\sigma}{t_R}\right)^2 \tag{4.3}$$

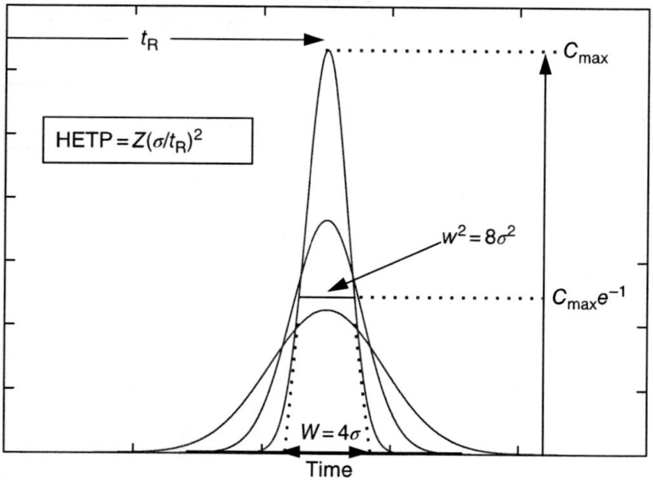

Figure 4.3 Chromatography elution curves (chromatogram, elution profile). The peak width at $C = 0.5C_{max}$ (half width) becomes 2.35σ.

Here Z = column bed length, N = plate number (the number of plates), σ = standard deviation of the elution curve, and t_R = peak retention time.

Although the peak retention time t_R and the variance of the curve σ^2 are most precisely determined by the first and the second moment defined by the following equations, the second moment is markedly influenced by the peak tailing. Therefore, HETP is often better to be directly calculated simply by measuring the peak width at $C = e^{-1}C_{max}$ and the peak maximum retention time [6–9]. Namely, $\mu_1' \approx t_R$, $\mu_2 \approx \sigma^2$.

$$\mu_1' = \frac{\int_0^\infty Ctdt}{\int_0^\infty Cdt} = \int_0^\infty \frac{Ctdt}{\mu_0} \tag{4.4}$$

Or by fitting the data to the modified Gaussian curve, $C = C_{max}(t_R/t)^{0.5} \exp\left[(t_R/t)(t-t_R)^2/(2\sigma^2)\right]$

$$\mu_2 = \int_0^\infty \frac{C\left(t-\mu_1'\right)^2 dt}{\mu_0} \tag{4.5}$$

where μ_0 = 0th moment (area of the curve). Note that for low plate numbers ($N < 30$) curves are skewed with tailing. Since the assumptions $\mu_1' \approx t_R$ and $\mu_2 \approx \sigma^2$ are no more valid for such curves, numerical integrations based on Equations 4.4 and 4.5 are needed. However, errors are always included due to the tailing part. Therefore, it is better to measure symmetrical curves by decreasing the flow velocity as shown later.

The retention time for linear IE (K = constant) is given by the following equation, which can be derived by integrating Equation 4.1 from $z=0$ to $z=Z(t=0$ to $t=t_R)$:

$$t_R=\left(\frac{Z}{u_0}\right)\left[\varepsilon+(1-\varepsilon)K\right]=\left(\frac{Z}{u}\right)(1+HK) \tag{4.6}$$

The retention volume is

$$V_R=F_v t_R=V_o+(V_t-V_o)K=V_o(1+HK) \tag{4.7}$$

When the HETP is plotted against the flow velocity (linear mobile-phase velocity) u, a so-called Van Deemter curve is obtained.

$$\begin{aligned}\text{HETP}&=\frac{Z}{N}=A^o+\frac{B^o}{u}+C^o u+D^o u\\&=\frac{2D_L}{u}+\frac{2\gamma_m D_m}{u}+\frac{HKd_p{}^2u}{\left[30D_s(1+HK)^2\right]}+\frac{HK^2d_pu}{\left[(3k_F)(1+HK)^2\right]}\end{aligned} \tag{4.8}$$

Here d_p = particle diameter, D_m = molecular diffusion coefficient, D_s = stationary-phase diffusion coefficient (in most cases, pore diffusion coefficient), D_L = axial dispersion coefficient (mixing in the mobile phase due to the channeling and the unequal path length of the interstitial spaces), k_F = film mass transfer coefficient around the particle (packing material), and γ_m = obstruction factor in the mobile phase. The first two terms (A^o+B^o/u) represent the dispersion (zone spreading) in the mobile phase. The third term $C^o u$ is for the dispersion due to the stationary-phase diffusion, and the fourth term is for dispersion due to the stagnant film mass transfer around the particle.

The previous form of the HETP equation was first derived by Van Deemter et al. [10]. Later a more exact form was proposed by Kubin [11] and Kucera [12] with the aid of the moment equations from the solution in the Laplace domain of the set of partial differential equations (see Model 1 in Table 4.1). Their final form is in our notation given by Equation 4.8.

In fixed-bed adsorption process design and operation, different parameters such as the number of transfer units (NTU), height of transfer units (HTU), and $K_f a_v$ (overall volumetric mass transfer coefficient) are employed. For linear isocratic elution chromatography, these parameters are easily related to HETP as HETP = 2HTU or NTU = $2N$ or $K_f a_v = K_S$.

It is convenient and useful to have a dimensionless form of the HETP equation as the data over a wide range of chromatography conditions such as d_p and

u can be compared in the same scale. With the dimensionless HETP $h = \text{HETP}/d_p$ and the dimensionless velocity $\nu = u \cdot d_p / D_m$, the HETP equation, Equation 4.8, can be rewritten as

$$h = A^* + \frac{B^*}{\nu} + C^* \nu + D^* \nu \tag{4.9}$$

$$h = \frac{\text{HETP}}{d_p} \tag{4.10}$$

$$\nu = u \frac{d_p}{D_m} = \text{ReSc} = \left(u \frac{d_p \rho}{\eta} \right) \left(\frac{\eta}{\rho D_m} \right) \tag{4.11}$$

$$A^* = \frac{2(D_L / u)}{d_p} = 2\lambda = \frac{2}{\text{Pe}} \tag{4.12}$$

$$B^* = 2\gamma_m \tag{4.13}$$

$$C^* = \frac{HK}{\left[30\gamma_s (1 + HK)^2 \right]} \tag{4.14}$$

$$D^* = \frac{HK^2}{\left[3\text{Sh}(1 + HK)^2 \right]} \tag{4.15}$$

$$\gamma_s = \frac{D_s}{D_m} \tag{4.16}$$

$\text{Sh}(= k_f d_P / D_m)$ is the Sherwood number. $\text{Re}(= \rho u d_P / \eta)$ is the Reynolds number based on the particle diameter and the linear mobile-phase velocity. $\text{Sc}(= \eta / \rho D_m)$ is the Schmidt number, and $\text{Pe}(= u d_p / D_L)$ is the Peclet number based on the particle diameter and the axial dispersion coefficient.

A^* is constant having a value between 2 and 5 for usual velocity ranges. Namely, Pe is 0.2–0.5. γ_s is an obstruction factor for stationary-phase diffusion, which implies the lowering of the molecular diffusion coefficient in the stationary phase (in most cases pore diffusion). A typical h–ν curve is shown in Figure 4.4 along with the contribution of each term. This type of plot is usually called "Van Deemter plot" based on their original equation [10]. Equation 4.8 has the same form as the original Van Deemter equation although their C and D terms are different from Equation 4.8.

Another empirical expression for h is given by

$$h = E^* \nu^n \tag{4.17}$$

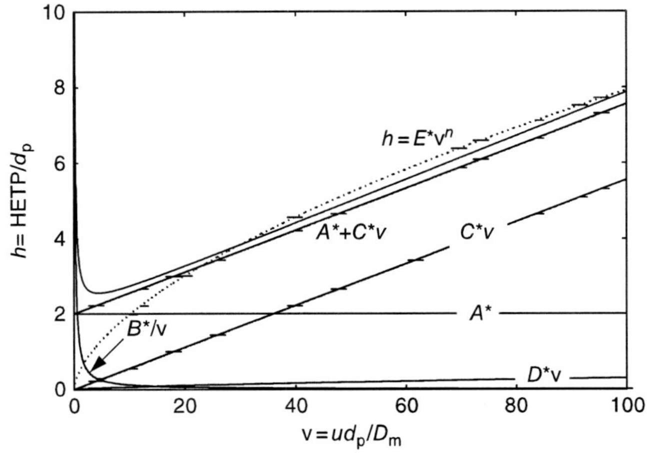

Figure 4.4 Relationship between *h* (dimensionless HETP) and ν (dimensionless velocity).

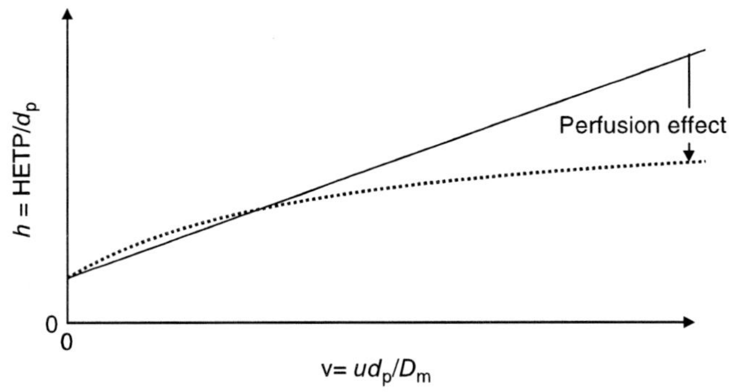

Figure 4.5 Perfusion effect.

This equation can describe the experimental data well over a certain range of velocity. In fixed-bed adsorption processes, the stationary-phase diffusion is dominant because of very large adsorbents (packing materials). So the assumption of $h = C^*\nu$ is commonly employed. This approximation is also valid for chromatography at very high dimensionless velocities ($\nu > 200$).

At very high dimensionless velocities and high pore-diameter/particle-diameter ratio values, convection through pores may enhance the mass transfer. As a consequence, *h* does not increase with ν significantly. This is called a "perfusion" mechanism (Figure 4.5). A modified, empirical expression for HETP considering the perfusion mechanism is given by [1]:

$$h = A^* + F^* \frac{\nu}{(1 + a\nu)} \tag{4.18}$$

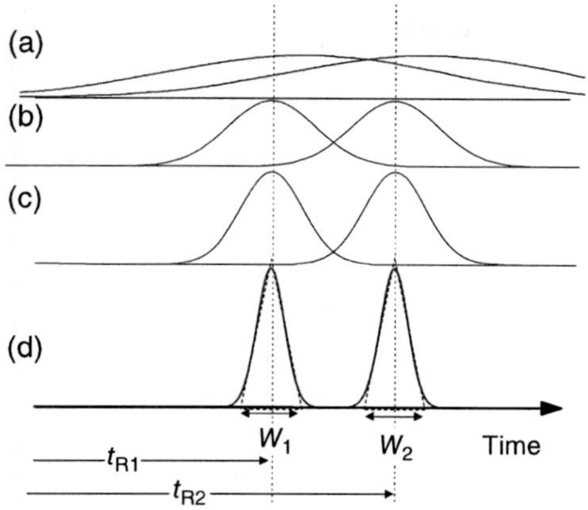

Figure 4.6 Resolution, R_s. (a, b) $R_s < 1$. (c) $R_s \approx 1$. (d) $R_s \approx 2$.

4.2.1 Resolution R_s in Isocratic Elution

In addition to the HETP (a measure of the column efficiency), another measure for the separation efficiency is needed, which considers the distance between the two adjacent peaks. The resolution R_s defined by the following equation is most commonly employed like HETP (Figure 4.6):

$$R_s = \frac{(t_{R2} - t_{R1})}{[(1/2)(W_1 + W_2)]} \tag{4.19}$$

When the two peaks are similar in terms of the peak height and the peak area, the complete resolution or the baseline separation is achieved when $R_s > 1.3$.

By assuming that $W_1 = W_2 = 4\sigma_1$ and inserting Equations 4.3 and 4.6 into the previous R_s equation, we obtain

$$R_s = \frac{\{H(K_2 - K_1)\}}{\{4(1 + HK_1)\}} N^{1/2} = \frac{\{H(K_2 - K_1)\}}{\{4(1 + HK_1)\}} Z^{1/2}\mathrm{HETP}^{-1/2} \tag{4.20}$$

This equation shows that R_s is proportional to the square root of the column length, Z.

Although the HETP equation, Equation 4.8, can be inserted into the previous equation in order to relate the R_s to various chromatography variables, it is more convenient to use Equation 4.9 instead:

$$R_s = \frac{H(K_2 - K_1)}{4(1 + HK_1)} Z^{1/2} u^{-n/2} d_p^{(-1-n)/2} D_m^{n/2} \tag{4.21}$$

By using this equation, we can predict how the R_s varies with such parameters as u and d_p. As D_m decreases with increasing molecular weight, the separation of larger molecules becomes difficult. In order to incorporate the effect of sample loading volume, the following equation is applicable until the peak shape does not deviate from the Gaussian curve:

$$\sigma^2 = \sigma_0{}^2 + \frac{t_F{}^2}{12} = t_R{}^2 \frac{\mathrm{HETP}}{Z} + \frac{t_F{}^2}{12} \tag{4.22}$$

Based on the previous discussion, it is easily understood why smaller particles result in high resolution at high velocities. However, at the same time the pressure drop Δp increases with decreasing particle diameter according to the Kozeny–Carman equation [1, 3, 13, 14]:

$$\Delta p = \frac{\kappa \mu Z}{d_p^2} \frac{(1-\varepsilon)^2}{\varepsilon^3} u_0 \tag{4.23}$$

Here κ is the constant (150–180). Δp is proportional to $u_0 (= u/\varepsilon)$, $1/d_p^2$, Z, and μ.

For a packed bed with compressible or semirigid gels, the linear range Δp–u becomes smaller when the column size (diameter and/or height) is increased [1, 3, 14–17]. This is an important issue when the process is scaled up. Because of the upper limit of the flow velocity due to the bed compression, the optimum condition must be carefully sought by changing the column geometry (height to diameter ratio) even for the same bed volume.

4.3 Linear Gradient Elution (LGE)

When a certain substance in the mobile phase (modulator) is changed continuously with time, good resolution is achieved, which is not easily obtained with IE. Major advantages of this method (LGE) are as follows [6, 18–21]:

1) Self-sharpening effect due to slower migrating velocities of the front part and faster velocities of the rear part exist. This can reestablish zone shape automatically even when the zone is distorted due to poor sample injection device and/or poorly packed bed (see Fig. 4.10).
2) Resolution can be changed very easily simply by changing the gradient slope as well as the flow velocity.
3) Multiple solutes can be eluted in a single run.
4) False peaks due to a discontinuous change of the mobile phase do not exist.

4.3.1 Retention in Linear Gradient Elution (LGE)

Compared with IE, LGE is complicated. At the beginning, the sample is tightly bound to the column stationary phase. In the case of electrostatic interaction-based chromatography (IEC), the salt concentration is set to be low. Then, linear increase of salt concentration at a fixed pH is introduced to the column. The sample zone does not move as it still binds to the column tightly. As the salt concentration increases, the isotherm shape changes as shown in Figure 4.7. Eventually, the sample zone starts moving in the column. The zone trajectory is shown in Figure 4.8. The zone movement can be described by Equation 4.1 when the distribution coefficient K as a function of I is known. Since K depends on I strongly, which will be shown later, the sample zone moves down the column for a very narrow range of I.

The linear increase of the modulator concentration (salt or counterion) for IEC is described by Equation 4.24. In the ideal case, the concentration at the column outlet is given by Equation 4.25. Based on the equilibrium model, it is

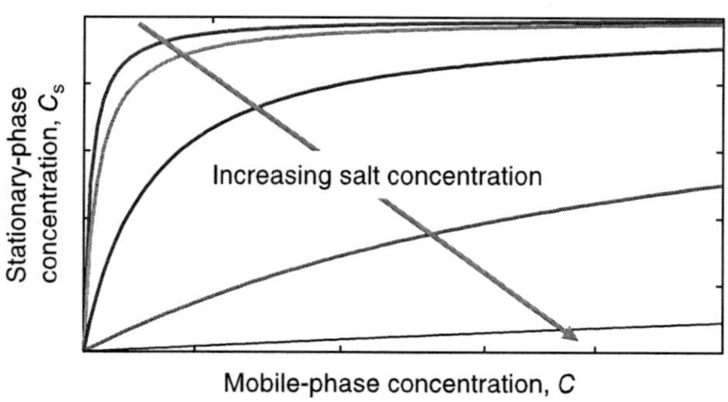

Figure 4.7 Adsorption isotherms as a function of salt concentration.

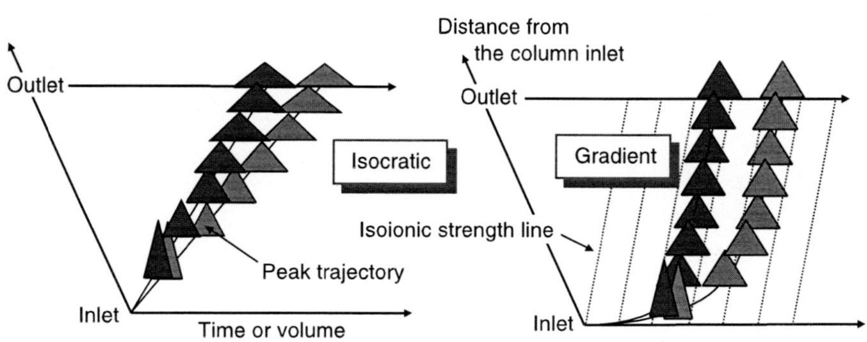

Figure 4.8 Zone movement during isocratic and gradient elution.

possible to calculate the zone movement numerically with Equation 4.1, $K(I)$, and Equation 4.26, which describes I as a function of time t and distance from the column inlet z:

$$\text{Column inlet} \quad I = I_0 + gF_v t = I_0 + gV \tag{4.24}$$

$$\text{Column outlet} \quad I = I_0 + g(F_v t - V') = I_0 + g(V - V') \tag{4.25}$$

$$\text{In the column} \quad I = I_0 + g\left[V - V'\left(\frac{z}{Z}\right)\right] = I_0 + g\left[F_v t - V_0(1 + HK')\left(\frac{z}{Z}\right)\right] \tag{4.26}$$

V' is the volume needed to elute the salt for gradient. K' is the distribution coefficient (constant) for the salt. After mathematical manipulations, the following equation is obtained with Eq. (4.1):

$$\begin{aligned}\frac{dI}{dt} &= G\left[\left(\frac{u}{Z}\right) - \frac{(1+HK')}{Z}\left(\frac{dz}{dt}\right)\right] \\ &= G\left(\frac{u}{Z}\right)\left[1 - \left(\frac{1+HK'}{1+HK}\right)\right] = \frac{G(u/Z)H(K-K')}{1+HK}\end{aligned} \tag{4.27}$$

$$\frac{dI}{dt} = \frac{dI}{dt}\frac{dt}{dz} = \frac{GH(K-K')}{Z} \tag{4.28}$$

This equation is integrated to give

$$GH = \int_{I_0}^{I_R} \frac{dI}{K - K'} \tag{4.29}$$

where

$$GH = (gV_o)H = g(V - V_o) \tag{4.30}$$

Differential form of Equation 4.29 is as follows:

$$\frac{d(GH)}{dI} = \frac{1}{[K(I) - K']} \tag{4.31}$$

When K is approximated to

$$K = AI^{-B} + K' \tag{4.32}$$

the following simple equation is obtained:

$$GH = \frac{{I_R}^{B+1}}{[A(B+1)]} \qquad (4.33)$$

Equation 4.33 tells that the peak shifts to larger elution volumes with decreasing gradient slope in LGE (Figure 4.9). The salt concentration at the peak position (peak retention volume, V_R), I_R, also decreases with decreasing gradient slope. The I_R values for different gradient slopes (at least four data points) are plotted against the normalized gradient slope *GH*. Once $GH–I_R$ plots are prepared, the data are usually fitted well by Equation 4.33 as shown in Figure 4.9. The *B* value is the number of binding sites (effective charges) involved in electrostatic interaction, whereas *A* includes the equilibrium coefficient, the binding site, and the ion exchange capacity (see Appendices 4.B.1 and 4.B.2).

The aforementioned equations do not include the effect of the solute concentration *C*. This is because up to a certain sample loading, the elution curve shows a geometrically similar profile. Then, in LGE, *K* varies with time during elution as the linear increase of *I* is introduced to the column. The t_R therefore changes with the initial salt concentration I_0 even when the same slope of the gradient *g* is used as seen in Table 4.2. However, we found that the peak salt concentration I_R values do not change with I_0. Similarly, the I_R value remains constant when the sample volume is increased although the V_R changes. $GH–I_R$

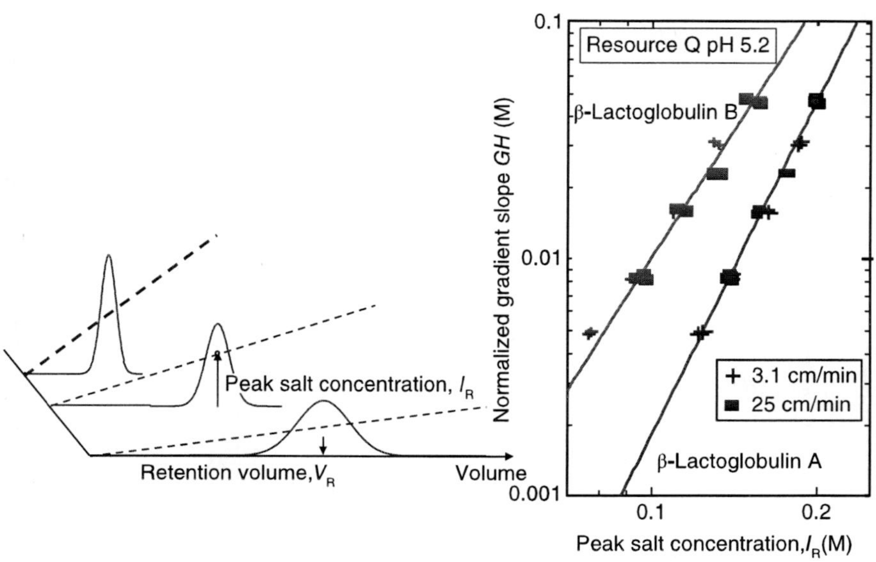

Figure 4.9 GH–I_R plots and LGE experiments. *Source*: Ishihara and Yamamoto [19] and Yamamoto [20]. Reproduced with permission of Elsevier.

Table 4.2 Effect of initial salt concentration (I_0) and sample loading on peak position and width in linear gradient elution (LGE).

C_0	V_F	I_0	W/W_r	$I_R/I_{R,r}$	$V_R/V_{R,r}$	CF
[mg/mL]	[mL]	[M]	[–]	[–]	[–]	[–]
0.01	66.2	0.03	0.97	1.01	1.04	23.64
0.05	12.4	0.03	0.98	1.00	1.05	4.36
1.00	1.76	0.03	1.13	1.00	1.03	0.54
1.00	**0.67**	**0.03**	**1.00**	**1.00**	**1.00**	**0.23**
1.00	0.67	0.05	1.04	1.01	1.01	0.23
1.00	0.67	0.10	1.03	1.00	0.91	0.23
1.00	0.67	0.15	1.05	1.00	0.86	0.23
1.00	0.67	0.20	0.99	0.98	0.68	0.23

Note: C_0 = initial lysozyme (sample) concentration, V_F = sample volume, GH = 0.096 M, u = 2.0 cm/min, W/Wr = relative peak width, $I_R/I_{R,r}$ = relative peak NaCl concentration, $V_R/V_{R,r}$ = relative retention volume. For these relative values, the **reference** values are those in the fourth row in the table (in boldface). C_0V_F, values are set to be equal except for the third row. CF = concentration factor. Column: SP Sepharose high performance, d_c = 0.9 cm, Z = 20 cm.
Source: Yamamoto [21]. Reproduced with permission of Wiley.

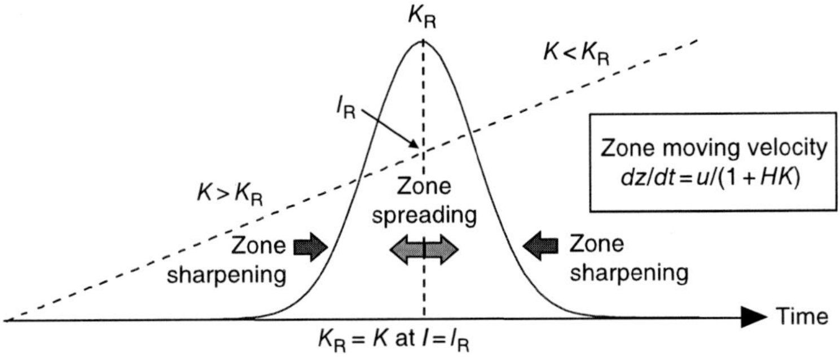

Figure 4.10 Linear gradient elution chromatography (zone-spreading and zone-sharpening effects).

plots can be employed for other chromatography such as HIC or reversed-phase chromatography [18, 22–24].

4.3.2 Peak Width, HETP, and R_s in Linear Gradient Elution

By the procedure shown in Section 4.3.1, we can also determine the distribution coefficient at I_R (K_R) in LGE. We define the apparent peak elution volume based on K_R as

$$V_R^{\text{app}} = V_o\left(1 + HK_R\right) \tag{4.34}$$

The peak width is also not dependent on I_0 or V_F as shown in Table 4.2. However, we also have to consider the zone-sharpening effect due to the linear gradient (Figure 4.10).

We define the zone-sharpening factor L as

$$L = \frac{W_G}{W_I} = \frac{\sigma_G}{\sigma_I} \tag{4.35}$$

where W is the peak width at the baseline and the subscript I is for the IE and G for the LGE. L is given as a function of M as follows. These correlations were obtained based on the experimental and numerical LGE curves. $L = M^{0.5}$ was derived from an asymptotic solution [6, 21]:

$$\begin{aligned} &L = \sqrt{M} && \text{for } M \le 0.25 \\ &L = \frac{3.22M}{1 + 3.13M} && \text{for } 0.25 < M < 12 \\ &L = 1 && \text{for } M \ge 12 \end{aligned} \tag{4.36}$$

$$M = \frac{1 + HK_R}{2GH \cdot J \cdot (1 + HK')} \tag{4.37}$$

where $J = |dK/dI|$ and M is a dimensionless group. L approaches 1.0 with increasing M. This means that the zone-sharpening effect is very weak and W is almost equal to that for the constant modulator concentration $I = I_E$ (IE). When M is <0.25, the relation $L^2 = M$ holds. Based on this relation, the resolution is proportional to a single dimensionless parameter [18]

$$R_s \propto Y_m = \left[\frac{(ZD_a I_a)}{(GHud_p^2)}\right]^{0.5} \tag{4.38}$$

I_a representing a dummy variable having a numerical value of 1 so that Y_m becomes dimensionless (Ym: Yamamoto number [14]).

A similar R_s equation is also derived from a completely different approach by Snyder et al. [25].

Equation 4.38 was verified experimentally for both IEC and HIC of proteins [18] and was also successfully employed for scaling up IEC columns from 15 L bench scale to 370 L process scale [26]. Calculation examples are shown in the book written by Harrison et al. [14].

However, it is still desirable to have more accurate methods, which can cover a wide range of flow rates. Because $GH \cdot J = B/(B+1)$,

$$M \approx \frac{(1 + HK_R)}{(1 + HK')} \frac{(B+1)}{2B} \tag{4.39}$$

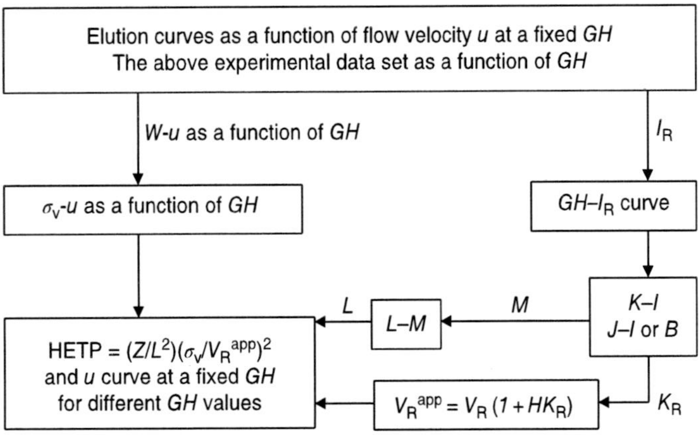

Figure 4.11 Flow sheet for determining HETP from linear gradient elution (LGE) curves. *Source*: Yamamoto [21]. Reproduced with permission of Wiley.

Finally, the HETP from the LGE curve $(\text{HETP})_{\text{LGE}}$ is determined as

$$\text{HETP}_{\text{LGE}} = \frac{Z}{L^2}\left(\frac{\sigma_V}{V_R^{\text{app}}}\right)^2 \tag{4.40}$$

A flow sheet for the determination of $(\text{HETP})_{\text{LGE}}$-$u$ is given in Figure 4.11.

Based on the method shown in Figure 4.11, HETP values can be determined from LGE curves as a function of the linear mobile-phase velocity u. It was confirmed that The HETP-u curve is not dependent on the column length Z. This is one of the fundamental assumptions in the HETP concept. The curve is also not affected by the initial salt concentration I_0 or by the sample loading. These results support that the present method is applicable to the determination of the column performance of the LGE–IEC of proteins.

This method was verified by several researchers [27–30].

4.3.3 Iso-Resolution Curve in Linear Gradient Elution (LGE)

Because we have one more tunable operating parameter for LGE, the same resolution can be obtained with different combinations of flow velocity and gradient slope even when the same column is used.

For predicting the resolution in LGE, the following parameter was proposed where $(\text{HETP})_{\text{LGE}}$ is HETP for LGE [31]. Note that by assuming that HETP is proportional to u, Y_m was derived from Equation 4.41:

$$O = \frac{(ZI_a)}{\left[G(\text{HETP})_{\text{LGE}}\right]} \tag{4.41}$$

The elution curves obtained for various combinations of experimental conditions such as Z, G, and u are very similar when the O values are set to be equal. Although the same resolution can be obtained with various combinations of operating–column variables (Figure 4.12), there may be some optimum condition because retention time t_R and retention volume V_R is a function of G and u.

For conventional porous bead chromatography, the peak broadens with increasing flow velocity due to the diffusion mass transfer in the pores. Therefore, when the flow velocity u is increased, the gradient slope g must become shallower in order to obtain the same O value. Similarly when u is decreased, g must be increased. With the aid of the $(\mathrm{HETP})_{\mathrm{LGE}}$-$u$ curve, this calculation can be done. The method for determining $(\mathrm{HETP})_{\mathrm{LGE}}$-$u$ relationships was already described.

Once the u and the g values are determined, the separation time t_S and the buffer consumption BC are calculated as follows:

$$t_S \approx t_R \approx \frac{(I_R - I_o)}{g} + V' \tag{4.42}$$

$$BC \approx \frac{V_R}{V_t} = \frac{F_v t_R}{V_t} = \frac{t_R u \varepsilon}{Z} \tag{4.43}$$

Here, V' is the elution volume for the salt.

We call the $t_R - V_R / V_t$ curve at a constant O value the "iso-resolution curve" as the same resolution with different separation time and buffer consumption can be obtained on this curve.

The same bed volume is assumed. The relative elution volume is often referred to as column volume (CV) and implies the relative buffer consumption. For the column length $Z = 5\,\mathrm{cm}$, the upper limit (symbol U) and lower limit (symbol L) are shown. The symbol A and B are the data from the elution curves shown in the inset (A) and (B), respectively.

In Figure 4.12, the calculated iso-resolution curves are shown. The open circles A and B are the data from the inset A and B, respectively. These data are in good agreement with the calculated iso-resolution curve. As shown in the figure, the separation time becomes longer as the elution volume decreases. On the contrary, large elution volume is needed for rapid separation. It is especially important to know where your separation conditions are located. For example, if your separation is carried out at point A in Figure 4.12, it is not wise to decrease the flow velocity for reducing the buffer consumption. Similarly, if your separation is performed at point B, it is not advantageous to reduce the separation time as very large buffer consumption is needed.

For the calculation of the iso-resolution curves, there are the upper and lower limits shown in the symbol U and L in the figures. The upper limit is

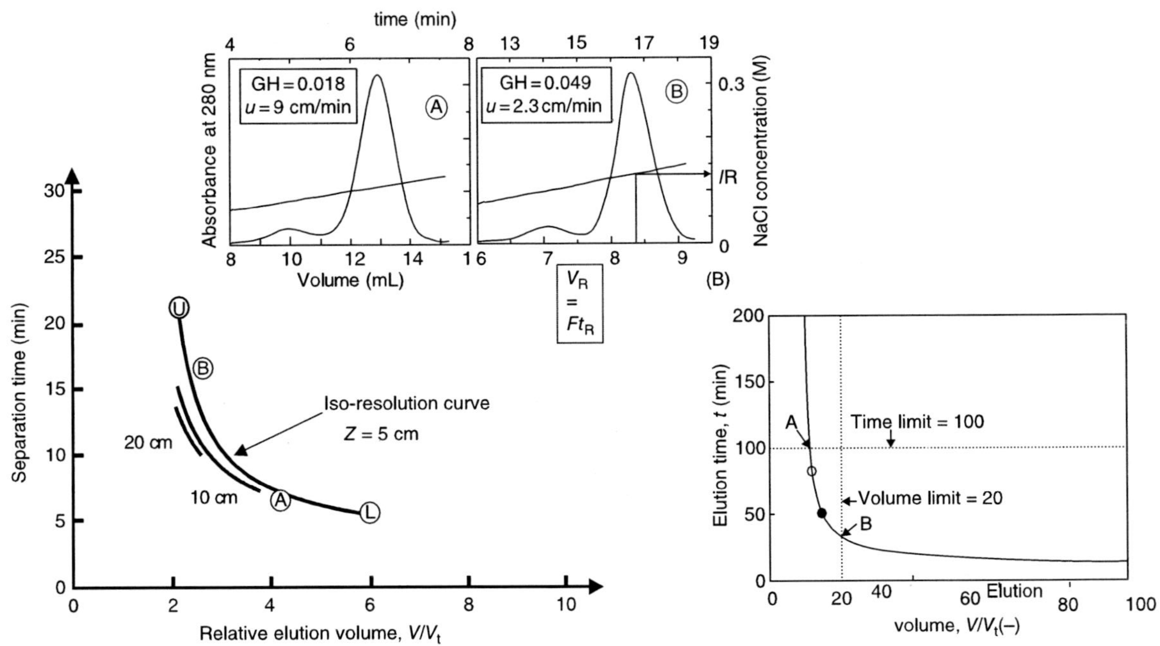

Figure 4.12 Separation time and relative elution volume relationships as a function of column length. *Source*: Yamamoto and Kita [31] and Ishihara et al. [29]. Reproduced with permission of Elsevier.

determined by the gradient slope. The gradient slope at the point U may be the limit for a very steep gradient. The lower limit L is due to the flow velocity or the pressure drop. When the column length is increased, the curve shifts to the left and the region between the limit U and L becomes smaller. This means that the buffer consumption decreases with increasing column length. But at the same time, short-time separation becomes difficult. This is especially true for large-scale or process-scale separation where the maximum operating flow velocity due to the pressure drop limit is significantly affected by the column bed height.

Figure 4.13 shows how to adjust the operating conditions according to the iso-resolution curve.

Although the effect of the sample loading is not included, the resolution is constant up to a certain sample loading. A further increase of the sample loading results in a lowering of the resolution. Nevertheless, it is still important to determine the optimum conditions at low sample loadings, from which we can examine the effect of the sample overloading.

Another important subject is whether an HETP-u curve can be predicted without any experimental data or with a very small amount of experimental data. The axial dispersion term (A°) that can be determined from the intercept of the $(\mathrm{HETP})_{\mathrm{LGE}}$-$u$ curve is quite similar to that obtained with the IE experiments.

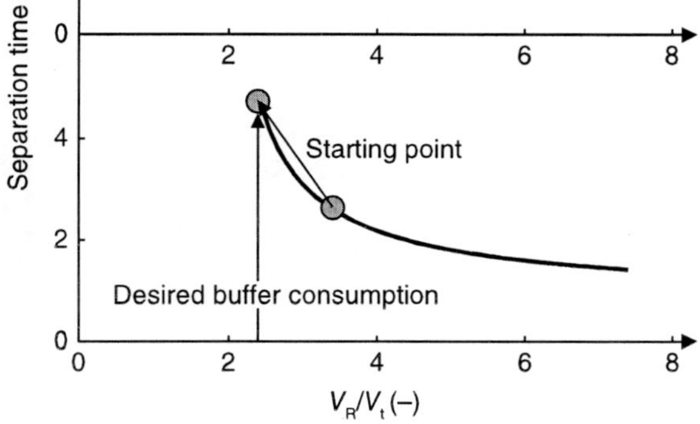

Figure 4.13 Illustration on how to adjust the flow velocity and the gradient slope for reducing separation time (upper figure) and buffer consumption (lower). *Source*: Yamamoto and Kita [31] and Ishihara et al. [29]. Reproduced with permission of Elsevier.

And the slope of $(\mathrm{HETP})_{\mathrm{LGE}}$-$u$ curve is also similar to that from the IE ones. This suggests that the detailed examination of the stationary-phase diffusion coefficient and the axial dispersion coefficient from $(\mathrm{HETP})_{\mathrm{LGE}}$-$u$ curve for different GH values may enable us to develop a simple method for predicting the column performance of the LGE–IEC.

The gradient slope is one of the most important variables in the LGE–IEC of proteins because it can be varied by a factor of 10–100. Other variables such as the column length and the linear velocity cannot be changed as much. Therefore, in IE such a drastic increase in the resolution cannot be achieved even when the flow rate is decreased and/or the column length is increased. However, the aforementioned three variables in LGE must be carefully tuned to obtain the highest productivity and the desired resolution together with other constraints and/or requirements for a certain protein purification process. The iso-resolution concept can be applied to other types of linear gradient chromatography. It is also useful for IE (or SE) chromatography in order to know the effect of modulator concentration [32].

4.4 Applications of the Model

So far several models, which we have developed, have been explained. It is important to understand how to use such simplified mechanistic models for different applications. Several examples on the application of our models are explained in the succeeding text.

4.4.1 Stepwise Elution (SE) Process Design Based on Linear Gradient Elution (LGE) Data

Although LGE is a very efficient elution method as already mentioned, for the industrial process SE is preferred because of the simple operation procedure. Basically, an SE process can be designed by using the mathematical model when $K(I)$ and mass transfer properties are known (examples are shown in Figure 4.14)

As shown in Figure 4.15, SE is grouped into two different mechanisms. In type I elution, a target protein is completely desorbed with an elution buffer. Namely, the distribution coefficient of the protein in the elution buffer K_{E} is less than that of the salt K'. The moving rate of the protein zone is higher than that of the salt. Therefore, it is regarded as gradient elution with a very steep salt gradient. Consequently, a very sharp peak along with a high concentration factor is expected as shown in Figure 4.14. Type I elution therefore is best suited for a capture step (initial step) where both purification and concentration (volume reduction) are important.

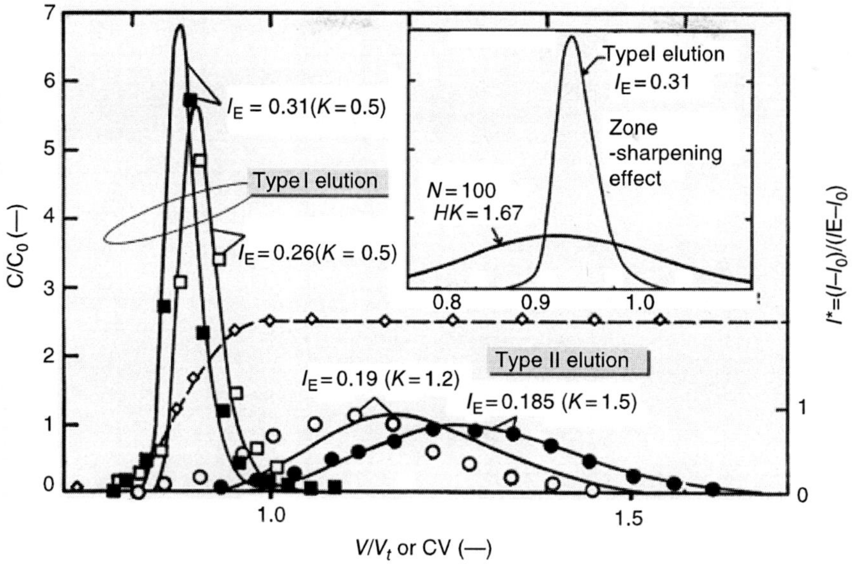

Figure 4.14 Typical types I and II elution behavior. Column: DEAE Sepharose CL6B. Sample: ovalbumin (0.2 kg/m^3) sample volume = 6 cm^3, column agarose-based beads, d_p = ca. 110 μm, d_c = 1.5 cm, Z = 10 cm, u_0 = 0.2 cm/min, pH 7.9. The inset shows the effect of zone-sharpening effect. Compared with an isocratic elution peak (the plate number, N = 100), type I elution peak (N = 100) is very sharp. *Source*: Yamamoto et al. [33]. Reproduced with permission of Wiley.

On the other hand, the protein zone moves after the column is saturated with the elution buffer. So it is similar to IE although in SE the sample can be concentrated on the top of the column before elution. Very difficult separation is achieved by type II elution provided that the conditions such as the elution buffer, the mobile-phase velocity, the column length, and so on are properly chosen. However, the peak becomes much broader with increasing K_E, and a small change of I_E affects the elution volume significantly as shown in Figure 4.14. This is one of the reasons why type II elution is unstable compared with LGE [35, 36]. Consequently, the I_E and pH of the elution buffer must be carefully prepared in order to maintain the elution volume [35, 36].

It is critical to determine the proper elution buffer salt concentration I_E for both types I and II elution. Empirically, the I_E value is determined from the I_R value in an LGE experiment. However, as shown previously I_E changes with the gradient slope, g. (This is the basis for our method for determining K as a function of I from the LGE experiment.) So, the determination of I_E from the I_R in LGE is not recommended as it is not certain whether it results in type I elution or type II elution of

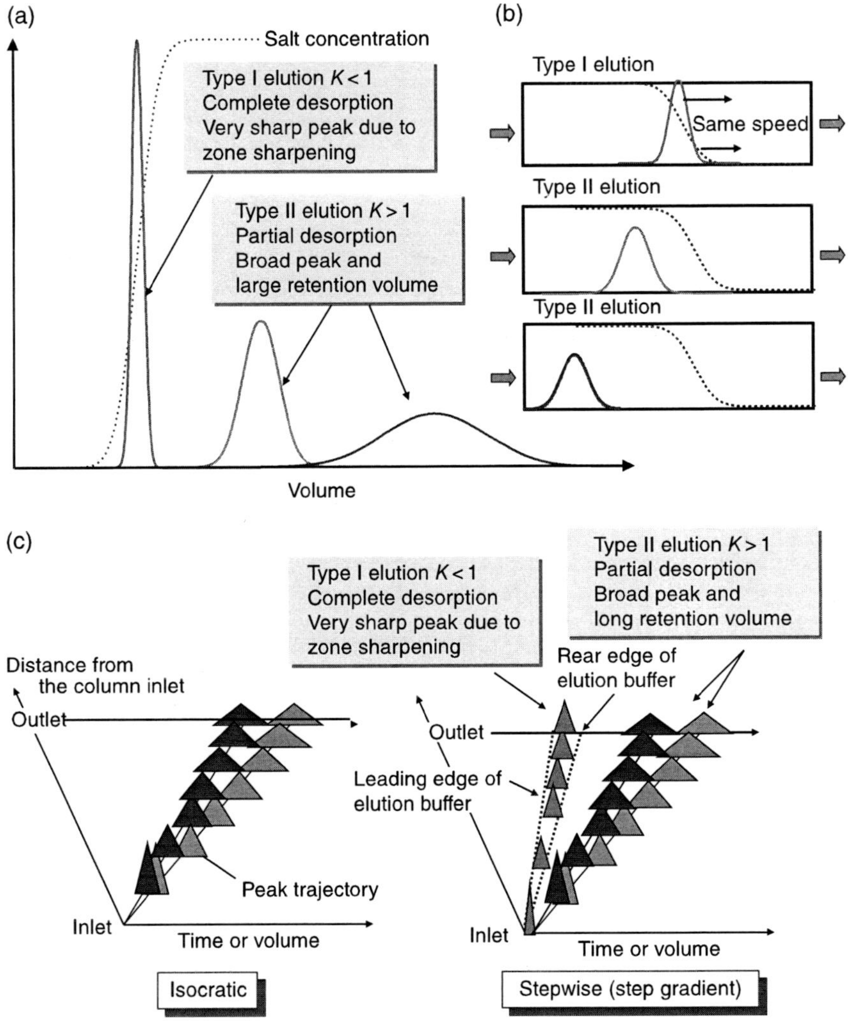

Figure 4.15 Stepwise elution. In the column the protein zone moves in the front-spreading boundary of the elution buffer in type I elution, and consequently at the column outlet the protein peak appears in the very steep salt gradient. For type II elution the peak retention volume V_R increases, and the peak becomes wider with decreasing buffer salt concentration I_E (increasing K_E). V_R is quite sensitive to a small change in I_E (see Figure 4.14). (a) Elution curves, (b) zone movement in the column, and (c) peak trajectory. *Source*: Yamamoto and Kita [34]. Reproduced with permission of Wiley.

very large elution volumes. Instead, K–I curves from GH–I_R curves can be used for designing type I SE as shown in Figures 4.16 and 4.17 [28].

It should be mentioned that although type I elution is best suited for obtaining a highly concentrated peak fraction in a capture step, attention must be paid

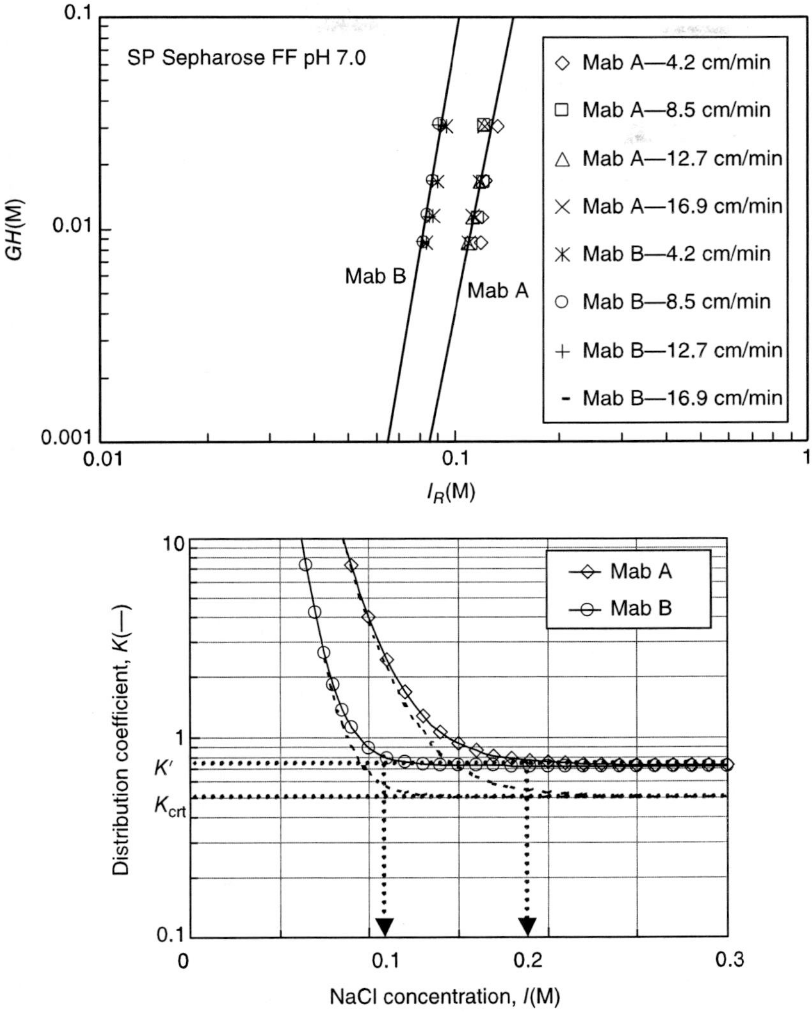

Figure 4.16 GH–I_R curves and *K*–*I* curves for monoclonal antibodies. *Source*: Ishihara and Yamamoto [28]. Reproduced with permission of Elsevier.

during scale-up and scale-down procedures. This is because the extra-column broadening in a small-scale separation affects the shape of the spreading boundary of the elution buffer [37]. Consequently, both the peak position and the peak width are likely to change significantly as shown schematically in Figure 4.18. If the peak concentration is high, the protein solubility may cause another problem as pointed out by Gallant et al. [38]. Another problem associated with scale-up is bed compression, which might affect the SE performance [15].

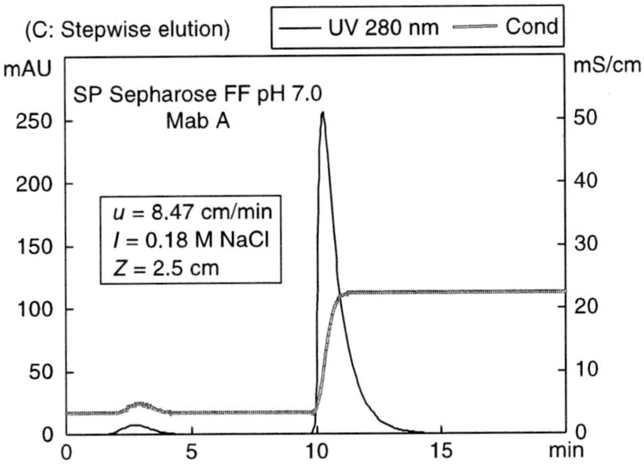

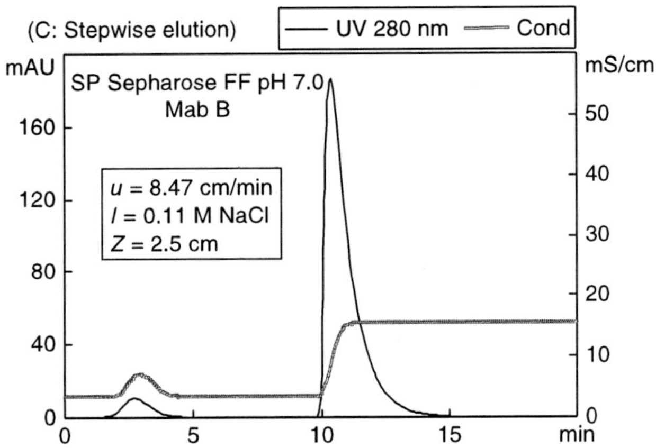

Figure 4.17 Type I stepwise elution-based on *K–I* curves from GH–I_R curves. *Source*: Ishihara and Yamamoto [28]. Reproduced with permission of Elsevier.

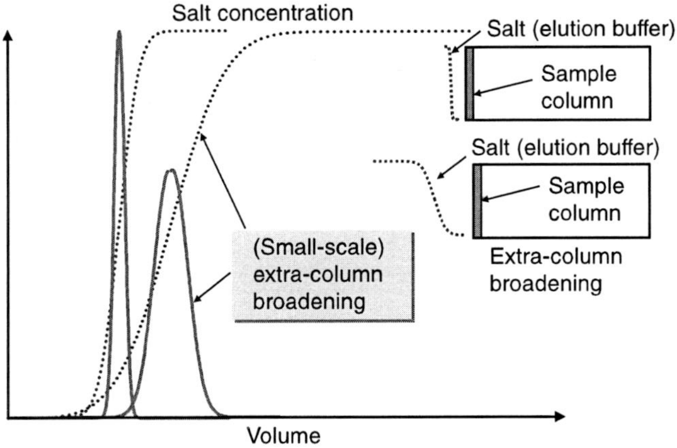

Figure 4.18 Extra-column broadening in type I stepwise elution. *Source*: Yamamoto and Kita [31] and Ishihara et al. [29]. Reproduced with permission of Elsevier.

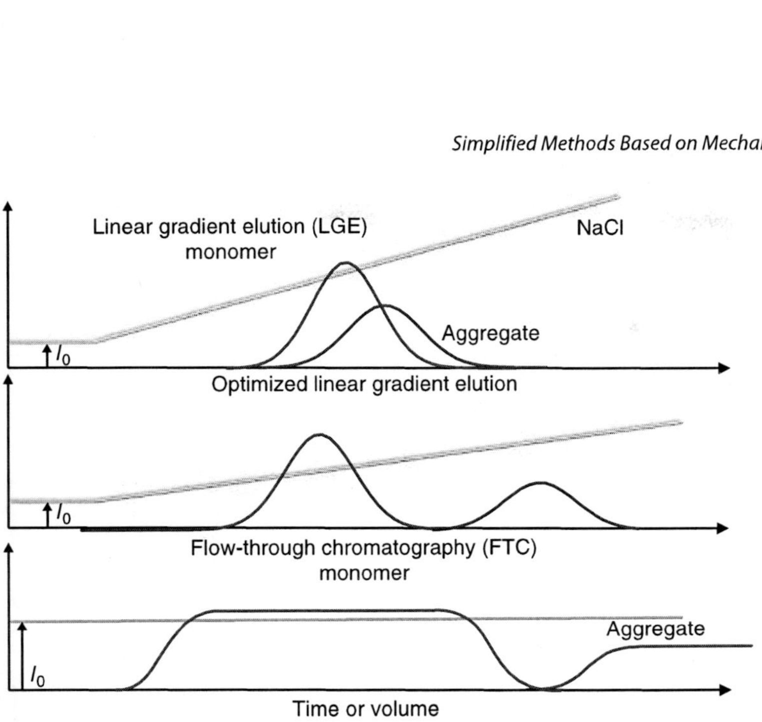

Figure 4.19 Flow-through chromatography (FTC) of monomer separation in comparison with linear gradient elution.

A method described in this chapter is applicable to other chromatography processes such as HIC or reversed-phase chromatography [18, 23, 32].

4.4.2 Flow-Through Chromatography

Flow-through chromatography (FTC) is considered to be best suited as a final polishing step. It is a type of pseudo-continuous (and disposable) chromatography process. A target material is not retained or weakly adsorbed on the column, whereas other contaminants are strongly adsorbed to the column (Figure 4.19).

FTC conditions in terms of K values are illustrated based on K–I curves in Figure 4.20. There is no clear definition in terms of weak adsorption for FTC although K of the target material may be below 5. The process design for FTC is somewhat similar to that for SE, which was already described.

Usually, the operating window is small as K is a strong function of I. FTC processes for aggregate separation by IEC were designed and carried out based on Yamamoto model (GH–I_R curves) [40]. They carried out LGE experiments of Mab monomer containing aggregates with two different IEC gels (SP Sepharose FF and SP Sepharose XL), and prepared the GH–I_R curves, from which the K–I curves for monomer and aggregates were calculated. They chose two different NaCl concentrations, which corresponded to $K = 1$

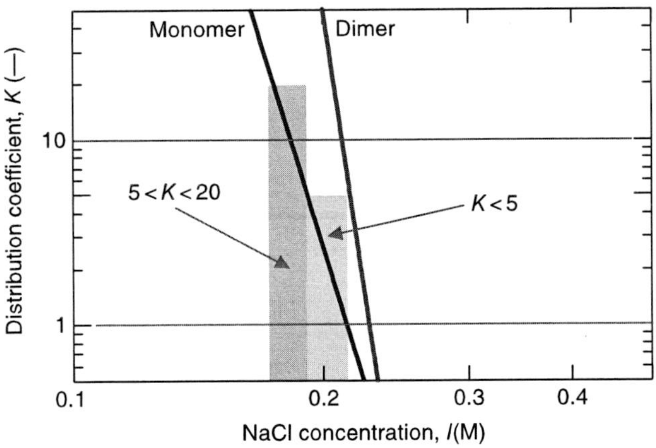

Figure 4.20 Operation windows for flow-through chromatography (FTC) on a distribution coefficient *K* versus salt concentration *I* graph. A similar conceptual graph was presented by Kelley et al. [39].

and $K = 2$ for the monomer, as their operating conditions. The performance was better for $K = 2$ in terms of the aggregate removal. It was pointed out that the operating window is narrower for the higher adsorption capacity resin. The mass transfer effect and the loading capacity were not examined.

Compared with LGE or SE, FTC operation is difficult as the UV signal does not indicate the separation performance. Although it is a pseudo-continuous (or continuous-flow) process, the sample feed has to be stopped before the breakthrough of the contaminant. This can be also calculated by the K–I curve and the mass transfer parameters.

4.4.3 Process Understanding and Analysis

To maintain process consistency, validated large-scale processes typically have a constraint to maintain elution volume within say ± 5%. This is often needed as tank volumes are fixed and elution must be completed within planned production times. While column size, flow rate, and protein loading are fixed by the manufacturing procedure, ion exchange resin ionic capacity, buffer pH, and salt concentration (ionic strength) will vary due to lot-to-lot variations. To assess the consistency of the separation, the model can be used to characterize and troubleshoot the effect of these fluctuations.

In this application, the model was used to elucidate the binding characteristics and to quantify changes in elution volume with variations media ion exchange capacity, buffer ionic strength, and pH [36, 41]. As a model system the separation of β-lactoglobulin variants (A and B forms) near its isoelectric point (p*I*) on a weak cation exchange gel, CM Sepharose at pH 5.2, was selected [41]. Although the

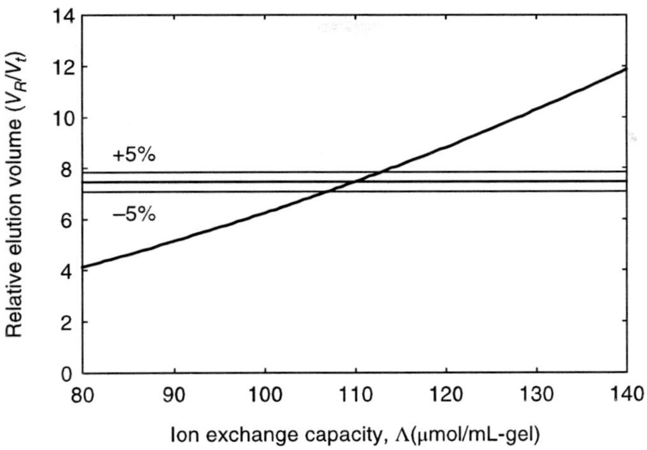

Figure 4.21 Relative elution volume versus ion exchange capacity. The reference value = V_R/V_t = 7.5, +5% value = 7.9, –5% value = 7.1. *Source*: Yamamoto et. al. [36]. Reproduced by permission of Elsevier.

isoelectric point of this protein is 5.1–5.2, it is retained on both anion and cation exchange chromatography columns at pH ~5.2 [19, 20, 36, 41, 42]. The values of $B = 2$ and $K_e\Lambda^B = 0.1179$ were obtained from LGE experiment data [19, 20, 36, 41].

For this system, SE is performed with an elution buffer of $I_E = 0.1$M, at which $K = 11.8$. According to the manufacturer, the ion exchange capacity Λ of CM Sepharose FF ranges from 90 to 130 μmol/mL-gel. Such variations can greatly affect the retention time since the distribution coefficient is related to Λ. The impact of such lot-to-lot variations on the relative elution volume V_R/V_t can be calculated as a function of Λ as $K = K_e \Lambda^B I^{-B} + K_{SEC}$ (B13).

$$\frac{V_R}{V_t} = \varepsilon + (1-\varepsilon)K \tag{4.44}$$

The void fraction ε is assumed to be 0.4. V_R/V_t at $\Lambda = 110$ was set to be a reference value where the protein eluted at 7.47 CVs. The target retention volume and the +5 and –5% retention volumes are shown in Figure 4.21. Even a relatively modest 5% lot-to-lot variation in Λ (105 or 116 μmol/mL-gel) significantly affects the elution volume well beyond the desired ± 5%.

In a production environment it is impractical to consistently obtain resin of a specific ionic capacity. In order to control the retention volume, it is necessary to adjust the salt concentration, I_E, of the elution buffer. Figure 4.22 shows the relationship between the relative elution volume and I_E. When the reference value curve **r** ($\Lambda = 110$) is compared with the low Λ value curve (1: $\Lambda = 90$) and with the high Λ value curve (2: $\Lambda = 130$), it is seen that the NaCl concentration

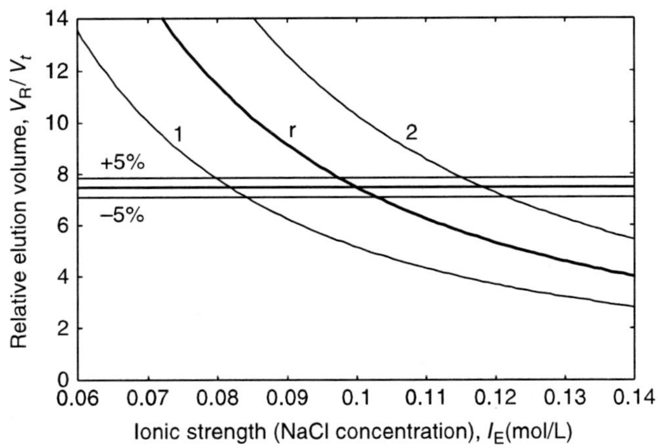

Figure 4.22 Relative elution volume versus ionic strength. Curve r = reference value $\Lambda = 110$, curve 1: $\Lambda = 90$, curve 2: $\Lambda = 130$. *Source*: Yamamoto et. al. [36]. Reproduced by permission of Elsevier.

must be adjusted by ± 0.015 mol/L to elute the protein within ± 5% of the reference elution volume. This is typically done by trial and error. However, as already explained, the distribution coefficient as a function of ionic strength can be obtained from LGE data. Once this K–I information is obtained for a given Λ, the elution volume can be predicted, and the elution buffer ionic strength can be adjusted. Hence, the chromatography model can serve as a convenient tool for tuning and troubleshooting very sensitive isocratic or type II SE chromatography processes. In addition, there is usually a variation in the salt concentration of elution buffers prepared at production scale. Figure 4.22 indicates that the salt concentration of the buffer must be within ± 0.002 mol/L in order to meet the ± 5% elution volume criteria.

Due to inherent variability during preparation, buffer pH will also vary at production scale. Buffer pH affects the charged state of the ion exchanger, which affects elution volume. The ion exchange capacity of CM Sepharose decreases with pH below 6 [43, 44]. The relative change in elution volume, resulting from changes in the ion exchange capacity, Λ, as a function of operating pH is shown in Figure 4.23. Although variations in the relative elution volume are smaller compared with those shown in Figures 4.21 and 4.22, it is still important that the buffer pH be within ± 0.1 pH unit. It is additionally important to control buffer pH, since as discussed previously, the interaction between the protein and the ion exchanger changes with pH especially near the isoelectric point. In particular, the values of B and $K_e\,\Lambda^B$ as well as Λ vary with pH in this region. However, controlling the effect of pH is more complex than controlling salt concentration or the ion exchange capacity.

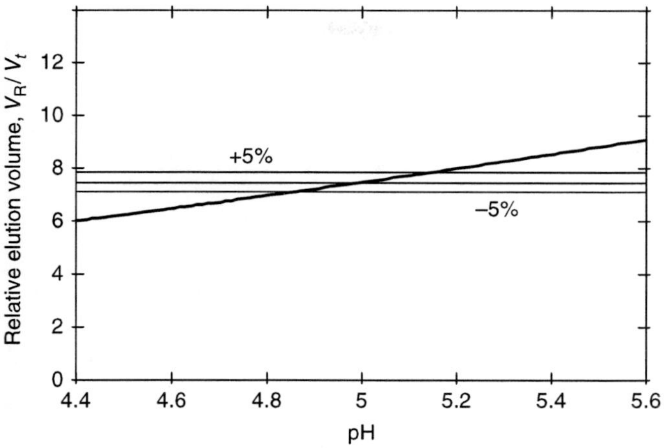

Figure 4.23 Relative elution volume versus pH. Reference value = pH 5. *Source*: Yamamoto et. al. [36]. Reproduced by permission of Elsevier.

4.4.4 High-Throughput Data Acquisition Method

Currently, fully automated robotic liquid-handling systems have been employed as an HTPD method (see several chapters of this book, e.g., Chapter 11). There are two types of experiments. One is a method using a 96-well filter plate with a gel slurry. This is a miniature version of batch adsorption experiments. Therefore, the equilibrium concentration in the gel (stationary phase), C_s, is determined as a function of the concentration in the solution (C).

When adsorbents (chromatography packing particles) are contacted with a solution containing a protein of interest in a mixing chamber, the governing equation is

$$V_L \frac{dC}{dt} + V_s \frac{d\overline{C_s}}{dt} = 0 \tag{4.45}$$

where V_L = liquid volume in the chamber, C = protein concentration in the chamber, V_s = total volume of the particles, $\overline{C_s}$ = average concentration of the protein in the particle, and t = time.

The previous equation is integrated from the start of the adsorption process till the equilibrium stage to give

$$V_L(C_0 - C_e) = V_s\left(\overline{C_{se}} - C_{s0}\right) = V_s\overline{C_{se}} \tag{4.46}$$

C_0 is the initial protein concentration in the chamber, C_e is the equilibrium protein concentration in the chamber, $\overline{C_{se}}$ is the equilibrium average protein concentration in the particle, and C_{s0} is the initial protein concentration in the particle (which is usually $C_{s0} = 0$). This is a usual working equation for determining $\overline{C_{se}}$ based on the equilibrium concentration of the chamber C_e.

The mass transfer properties such as the pore diffusion coefficient D_p can be determined by analyzing the uptake curve C versus t based on an appropriate diffusion model equation such as the pore diffusion with a rectangular isotherm.

Although the advantages of using a 96-well format for HTPD are claimed by many researchers, there are several precautions, constraints, requirements, or pitfalls. One of the most important requirements is a pure protein sample solution. This is quite serious for batch adsorption (absorbance at 280 nm) and protein precipitation (absorbance at 400 nm) experiments. Breakthrough curve measurements also need pure protein solutions.

Another constraint is the working volume (100–200 μL). When the absorbance at 280 or 400 nm is used, reducing the working volume results in low sensitivities due to a shorter path length for the microplate spectrophotometer. If additional separate analytical measurements are needed, this working volume is not big enough. For batch adsorption experiments additional two factors must be carefully examined: (i) how to transfer gels or resins into the well precisely and (ii) how to mix the slurry solution in the well [45].

Relatively small distribution coefficients, which are important for the SE process design as described in Section 4.4.1, are claimed to be determined by using the 96-well microplate batch adsorption experiments [39, 46].

The distribution coefficient values obtained by HTS batch binding, Yamamoto model with LGE data and isocratic elution (IE) were compared [46]. Their data are shown in Figure 4.24, where the distribution coefficient

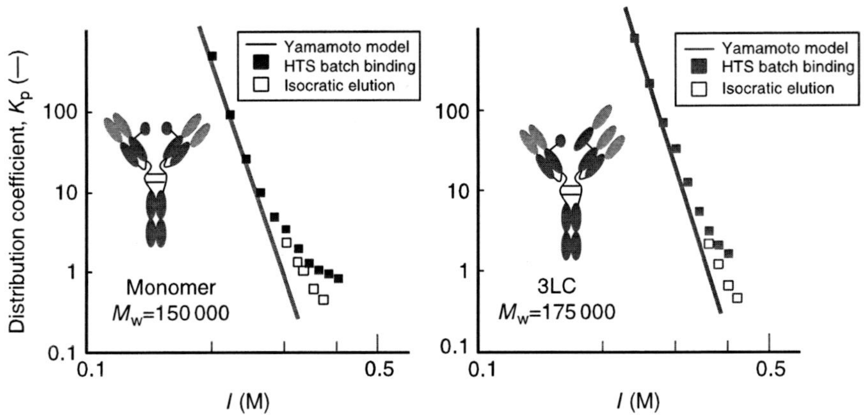

Figure 4.24 Distribution coefficient K as a function of salt concentration I determined by different methods for Mab and "triple light chain" (3LC) variant. The pI values for Mab monomer and 3LC are almost the same. HTS batch binding: robotic system with a 96-well filter plate batch adsorption method. Yamamoto model: K–I curves calculated from GH–I_R curves $K_P = (K - K_{SEC})/(1 - \varepsilon_p)$. *Source*: Fogle et al. [46].

values obtained by the different methods are plotted against the salt concentration. For small K_p values some deviations are observed among the three methods. For the HTS, the accuracy becomes lower for small K values as mentioned in the text. Similarly, for Yamamoto model the assumption $K' = K_{crt}$ may cause some errors for small K values ($K < 1$). The data were comparable with the small column data. However, for this case LGE column experiments with a fully automated LC station such as AKTA are quicker, easier, and reliable.

Ninety-six-well format packed columns, which can be operated by a fully automated robotic liquid-handling system, were also tested by several researchers (see e.g., Chapter 11). It is claimed that these columns can be used for the process design. However, the working volume constraint and the packing quality of tiny columns as well as the dead volume effect (see Figure 4.18) are still to be considered. The control of flow rate is rather difficult. A staircase increase of salt is needed for LGE experiments with this system. $GH–I_R$ curves obtained with 96-well monolith disk plates (staircase increase of salt) were similar to those with the column [47].

4.4.5 Characterization of Chromatography Stationary Properties and Binding of (Modified) Proteins or DNAs onto the Stationary Phase

So far we have explained how to use simplified mechanistic models for process design, analysis, and understanding. Such models can be also applied to understanding the binding mechanism of proteins or other biomolecules to the chromatography stationary phase. This is important not only for developing new and efficient chromatography stationary phases but also for optimizing the difficult separation such as similar molecules like modified proteins.

As shown in Appendices 4.B.1 and 4.B.2, the B value corresponds to the number of binding sites. The B values from the $GH–I_R$ curves can be used for examining the binding mechanisms. This was done by several researchers [48–52].

It is not clear how to choose the mobile-phase pH. Figure 4.25 shows that some proteins are bound to both anion and cation exchange chromatography even at the pI due to the asymmetrical charge distribution. The B value is smaller than the net charge as a part of the protein surface is involved for electrostatic interaction. It is also important to note that even when the B value is not large ($B = 2–5$), proteins may be retained strongly and result in better separations [19]. It is also important to note that the B values are similar for different IEC media, which can be understood based on the concept shown in Figure 4.26. Similar data were obtained for separation of protein monomer and dimer by LGE with different IEC columns [53]. Several researchers applied the same model for the understanding of the binding mechanism or the salt concentration dependence of the distribution coefficient [40, 48–52].

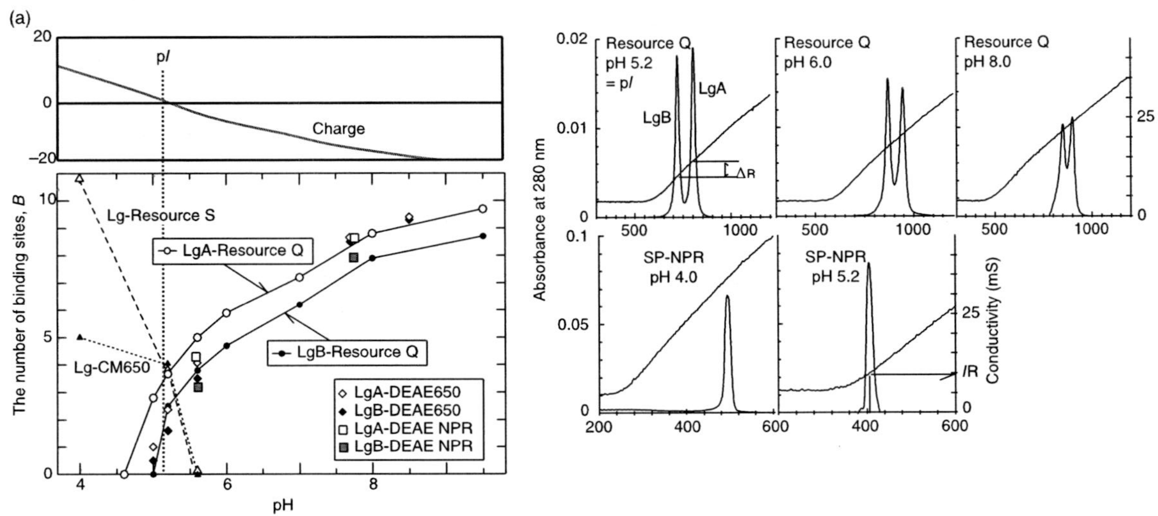

Figure 4.25 (a) B as a function of pH and related separation behavior. (b) A schematic drawing of B or Z_p versus pH. *Source*: Adapted from Ishihara and Yamamoto [19], Yamamoto [20] and Yamamoto and Ishihara [40]. Reproduced with permission of Taylor & Francis Group and Elsevier.

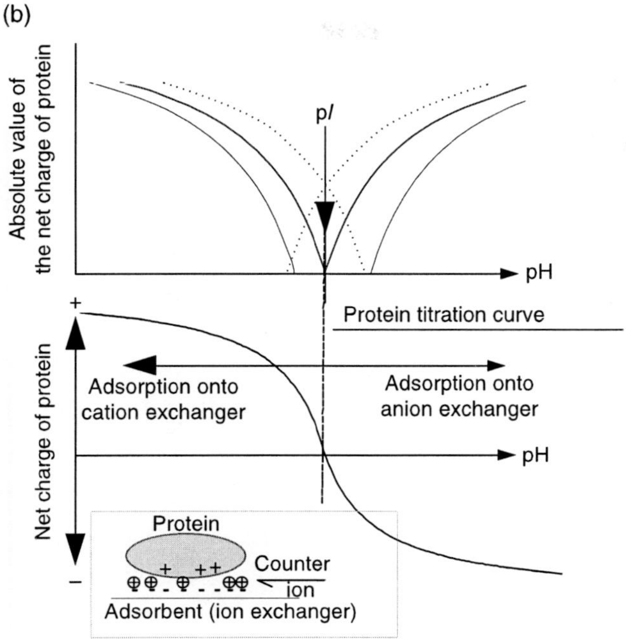

Figure 4.25 (Continued)

PEGylation is an attractive method for making smart protein drugs. It was already known that the retention of PEGylated proteins in IEC becomes weaker compared with that of native non-PEGylated protein. It was found that the binding site does not change when proteins are PEGylated, and the steric hindrance may be the reason for the weak retention (Figures 4.27 and 4.28). It was also shown that positional isomers can be separated by LGE–IEC efficiently although the binding site B does not change [56].

Aggregates are usually retained more strongly compared with their monomer protein because the aggregates have more charges (larger B values) [40, 53].

Refolding of proteins is important for the recombinant protein production from inclusion bodies. While on-column refolding is expected to be a promising method for protein renaturation, the binding mechanism of unfolded and refolded (native) proteins is not clearly understood. When lysozyme was unfolded in the presence of urea, the retention became weak, whereas the binding site did not change appreciably [57].

DNAs are known to bind to anion exchange chromatography stationary phase quite strongly (Figure 4.29). There is a good correlation between the number of charges of DNA molecule and the binding site value, B when $B < 10$

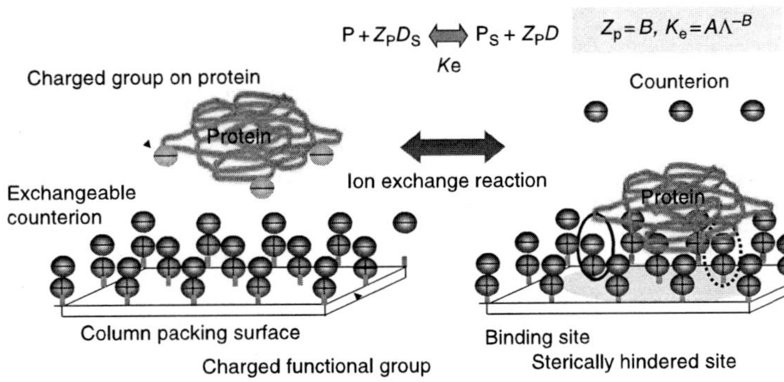

Figure 4.26 Stoichiometric displacement model or steric mass action (SMD) model. *Source*: Yamamoto [20]. Reproduced with permission of Wiley. (*See insert for color representation of the figure.*)

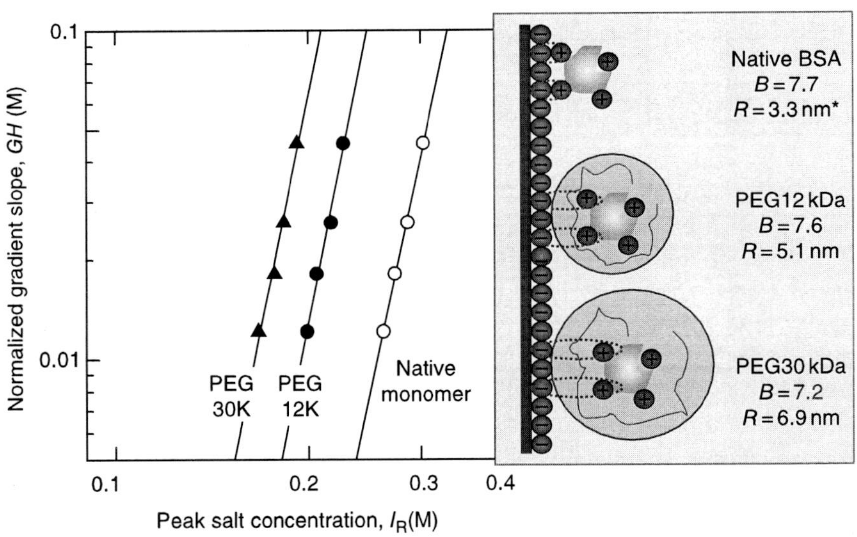

Figure 4.27 GH–I_R curves for PEGylated bovine serum albumin (BSA), the B values and the calculated hydrodynamic radius. The hydrodynamic radius R values were calculated by the Fee and Van Alstine equation [54]. *Source*: Abe et al. [55]. Reproduced with permission of Wiley.

(Figure 4.30). In other words, the B value is much larger than those for proteins even for small DNA molecules. The I_R values are also much higher compared with proteins (Figure 4.30). The B values for small DNAs are almost the same for different IEC columns as shown in Figure 4.31.

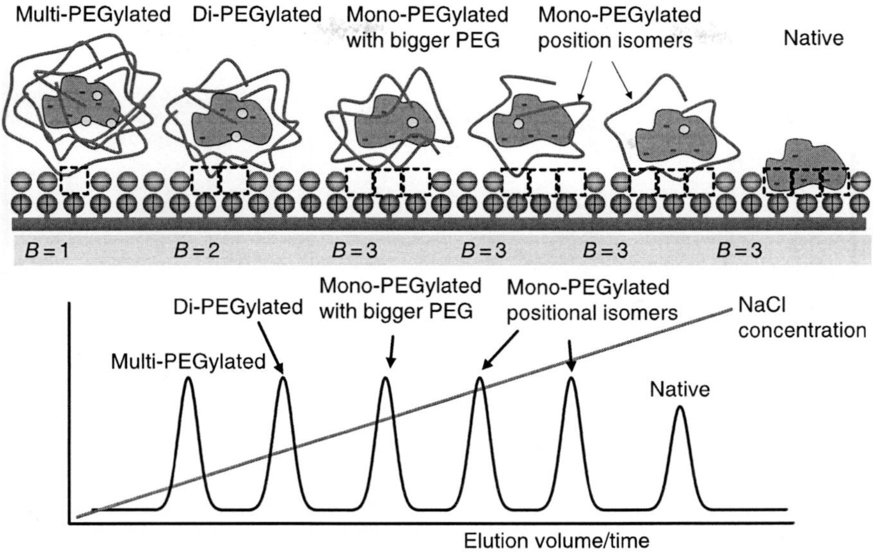

Figure 4.28 Possible retention mechanism of PEGylated proteins in linear gradient elution IEC. *Source*: Abe et al. [55]. Reproduced with permission of Wiley. (*See insert for color representation of the figure.*)

Because of such high B values, DNAs are eluted in a very narrow salt concentration window as shown in the trajectory in Figure 4.32. This is the reason why the peak became narrower with increasing DNA size, which corresponds to larger B values and higher I_R values. Because the monolithic chromatography media are based on the convection-aided separation, the slow pore diffusion due to large DNA molecule does not exist. This is another reason why the peak became sharper for larger DNA molecules.

4.5 Summary

In this chapter, our simplified models based on the mechanistic models for LGE were explained. The basic principle and the applications were given.

For the prediction of the peak retention time in LGE, GH–I_R curves are sufficient. K–I curves derived from GH–I_R curves can be used for the prediction of the peak retention time in SE.

When the column size or the operating conditions are changed, the resolution (separation performance) can be tuned with the aid of the dimensionless group Y_m (Yamamoto Number). It is also possible to optimize the separation in terms of the separation time and the buffer consumption.

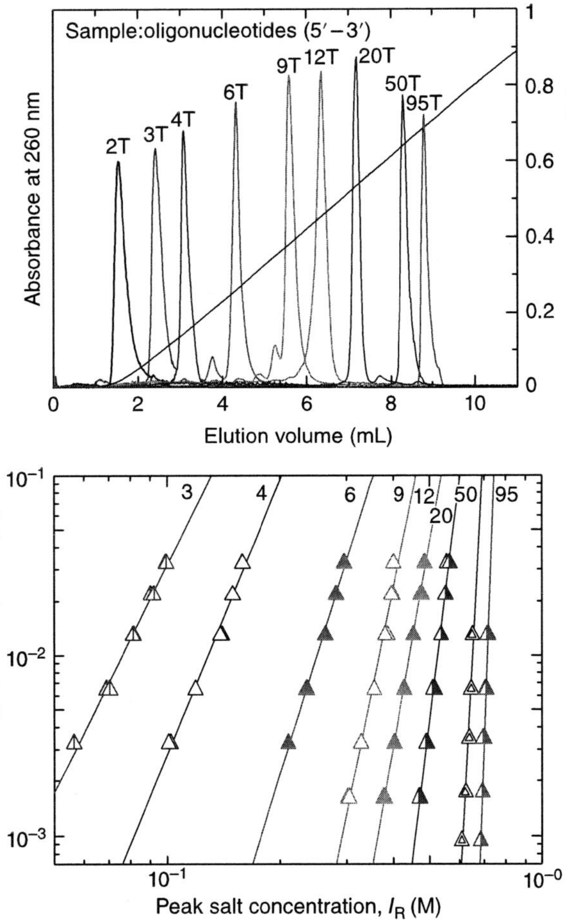

Figure 4.29 LGE elution curves and GH–I_R curve of DNAs (poly-T oligo DNA). *Source:* Yamamoto et al. [58]. Reproduced with permission of Elsevier.

More rigorous analysis is possible by using HETP for LGE curves. The binding site values B obtained from GH–I_R curves are also helpful for understanding the binding (adsorption) mechanism. Mostly we showed LGE–IEC with a salt gradient. As already mentioned in the text, it is possible to use the models (methods) for other modes of chromatography such as HIC and RPC. The model was extended to pH gradient elution IEC and mixed-mode chromatography [61, 62].

For process-scale chromatography, often an overloading condition is chosen in order to increase the productivity. As shown in our previous chapters [6, 20],

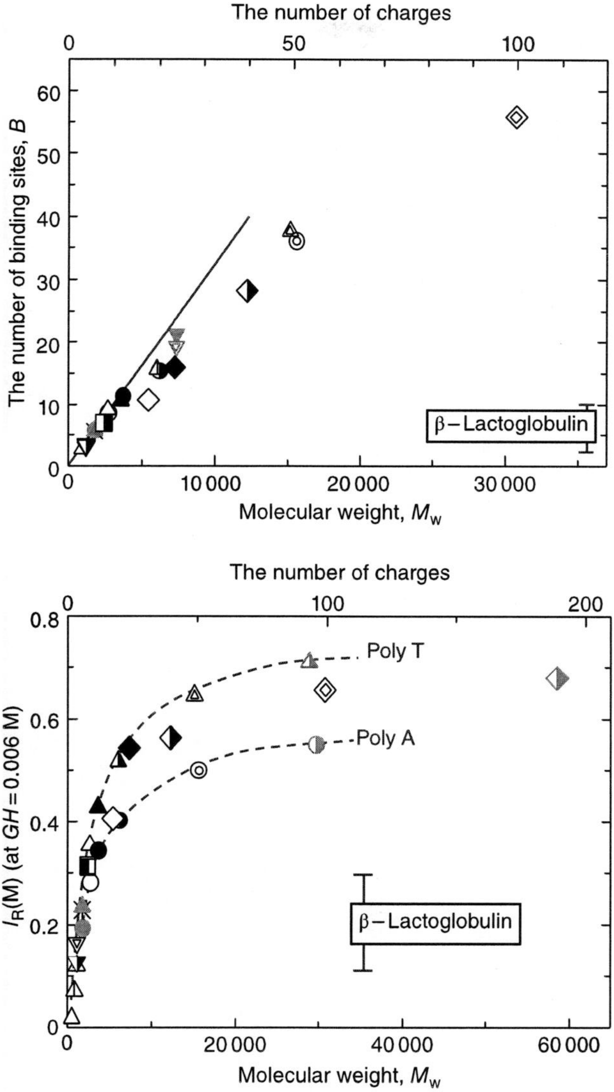

Figure 4.30 B and I_R as a function of DNA M_w or charges. *Source*: Yamamoto et al. [58]. Reproduced with permission of Elsevier.

GH–I_R curves can be prepared even for relatively overloaded conditions. It is recommended that the parameters needed are first obtained under low-loading conditions. Such parameters can be employed for numerical simulations of overloaded elution behavior.

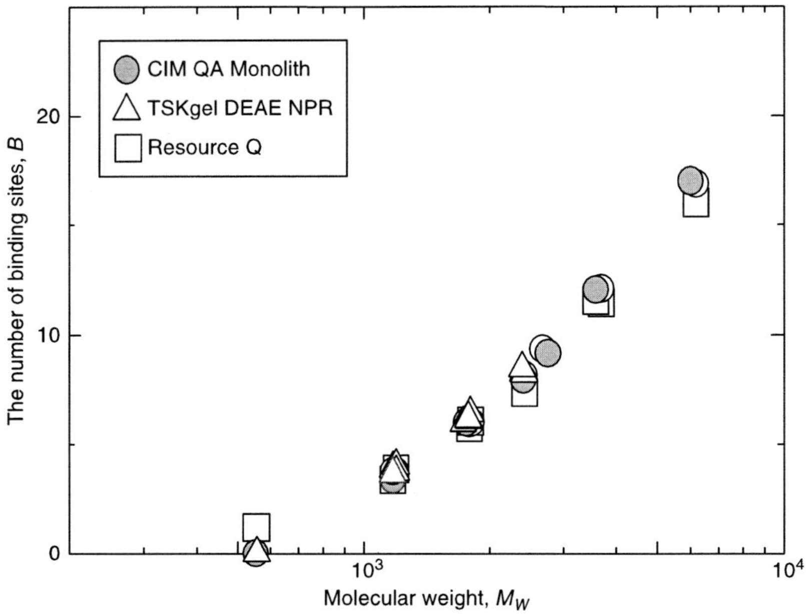

Figure 4.31 *B* as a function of DNA M_w for different IEC columns. *Source*: Yamamoto et al. [59]. Reproduced with permission of Elsevier.

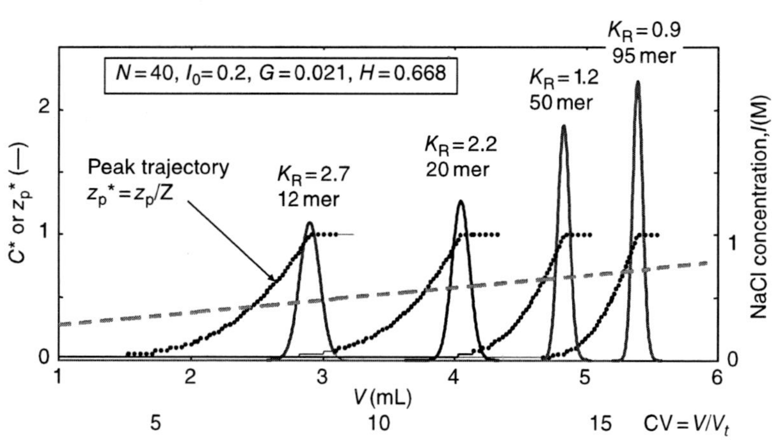

Figure 4.32 Simulated linear gradient elution curves of DNA. *Source*: Yamamoto et al. [60]. Reproduced with permission of Elsevier.

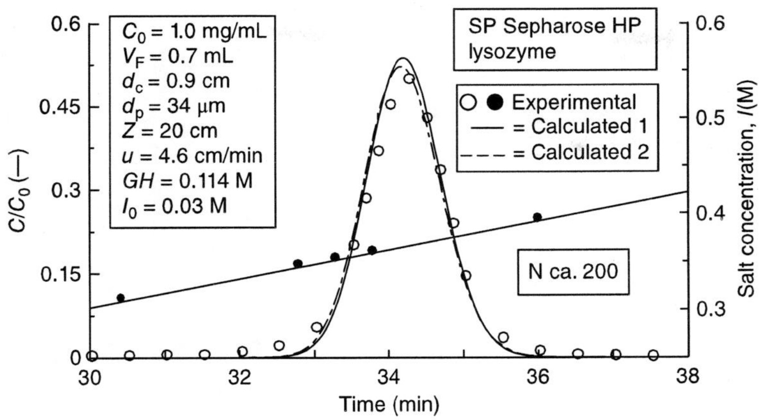

Figure 4.A.1 Numerically calculated linear gradient elution curve. *Source*: Yamamoto [26]. Reproduced with permission of Wiley.

Appendix 4.A Mechanistic Models for Chromatography

It is not very difficult to simulate chromatograms or elution curves numerically because high-speed, large-memory personal computers are readily available. However, it is still preferred to have a single zone-spreading parameter, which is sometimes called an overall dispersion coefficient or an effective overall volumetric mass transport coefficient. Different models containing such a single zone-spreading parameter give the identical elution curve in LGE provided that the plate number N is set to be equal. Therefore, the HETP-u relation is important and useful as it is directly used to calculate the plate number or the zone-spreading parameter.

In Figure 4.A.1, the experimental elution curve is compared with several calculated results. A single-lumped (or overall) zone-spreading parameter (N) was determined from the experimental $(\mathrm{HETP})_{\mathrm{LGE}}$-$u$ and GH–I_R curves. The numerically calculated elution curves agree well with the experimental results. It is also promising that the calculated curve by a simple model that does not need the numerical calculation is also in good agreement with the experimental one.

Appendix 4.B Distribution Coefficient and Binding Sites [20]

4.B.1 Ion Exchange Reaction

The ion exchange reaction or electrostatic interaction between a protein having Z_p charged sites and ionized (ion exchange) group in the presence of a counterion (salt) is described as follows (Figure 4.26):

$$\mathrm{P} + Z_p\overline{\mathrm{S}} \overset{K_e}{\Leftrightarrow} \overline{\mathrm{P}} + Z_p\mathrm{S} \tag{B1}$$

where P is the protein and S is the counterion (salt). The over bar indicates the concentration in the stationary phase. The equilibrium coefficient K_e is given by

$$K_e = \frac{[\bar{P}][S]^{Z_p}}{[P][\bar{S}]^{Z_p}} \tag{B2}$$

The application of this law of mass action to IEC of proteins was first presented by Boardman and Partridge [63]. Later a more elaborate treatment was developed by Regnier and his coworkers [42, 64] as the stoichiometric displacement model (SDM).

Assuming that the activity coefficient is constant and close to unity, the following equation is derived [6]:

$$K_e = \frac{C_q}{C}\left(\frac{I}{I_q}\right)^{Z_p} \tag{B3}$$

C_q is the protein concentration in the stationary phase (or the protein concentration bound to the ion exchange group), and C is the protein concentration in the liquid phase (the mobile phase in chromatography). I is the salt (ion) concentration in the liquid phase (the mobile phase in chromatography), and I_q is that in the stationary-phase (or the salt (ion) concentration bound to the ion exchange group).

The total concentration in the stationary phase is governed by the ion exchange capacity Λ, which is related to I_q and C_q as follows:

$$\Lambda = I_q + Z_p C_q \tag{B4}$$

As proteins are large molecules, it is likely that the protein in the stationary phase may hide or shield some ion exchange groups, which are not involved in the protein binding. Based on this consideration, a concept of shielded charge or the steric factor (S_F) has been introduced by Cramer's group [65] as the steric mass action (SMA) formula. Here S_F (in their papers, the symbol σ is commonly used) is the average number of sites, which are sterically shielded by the protein. Then, the total capacity becomes

$$\Lambda = I_q + (S_F + Z_p)C_q \tag{B5}$$

$$K_e = \frac{C_q}{C}\left(\frac{I}{\Lambda - (S_F + Z_p)C_q}\right)^{Z_p} \tag{B6}$$

When the protein concentration is low, $\Lambda = I_q$, the distribution coefficient $K_q = C_q/C$ is given by

$$K_q = \frac{C_q}{C} = \left(K_e \Lambda^{Z_p}\right) I^{-Z_p} \tag{B7}$$

The number of binding sites, Z_p, is also referred as the characteristic charge in the SMA model (the symbol ν is used in their papers). We use the symbol B for the number of binding sites.

4.B.2 Distribution Coefficient

The stationary phase consists of pores and solid surface, which can be described as

$$C_s = \varepsilon_P C_P + (1-\varepsilon_P)C_q = \varepsilon_P C_P + \rho_P q \tag{B8}$$

where ε_P is the particle porosity, C_P is the concentration in the pore, C_q is the concentration on the solid surface, ρ_P is the particle density, and q is the concentration in kg-protein–kg-adsorbent. C_P and C_q are given as follows:

$$C_P = K_P C \tag{B9}$$

$$C_q = K_{pq} C_P \tag{B10}$$

Then, C_s is described as

$$C_s = \{[\varepsilon_P + (1-\varepsilon_P)K_{pq}]K_P\}C = [(\varepsilon_P K_P) + (1-\varepsilon_P)K_{pq}K_P]C = KC \tag{B11}$$

$$K = [\varepsilon_P + (1-\varepsilon_P)K_{pq}]K_P = (\varepsilon_P K_P) + (1-\varepsilon_P)K_{pq}K_P = K_{SEC} + K_{ads} \tag{B12}$$

K_{SEC} is the distribution coefficient due to size exclusion (=K_{crt} the critical distribution coefficient).

When Equation B7 is inserted to K_{ads}, the following equation is derived:

$$K = (K_e \Lambda^{Z_P}) I^{-Z_P} + K_{SEC} = AI^{-B} + K_{SEC} \tag{B13}$$

In many studies, the capacity factor k' rather than K is determined from V_R:

$$V_R = V_o'(1+k') \tag{B14}$$

Here, V_o' is the peak elution volume of a non-retained (adsorbed) but fully permeable solute. Namely, a small solute that can use all the pore volume and does not interact with the ligand (ion exchange group, ligand spacers) or the polymer solid surface is employed for determining V_o'. k' is related to K as follows:

$$V_o' = V_o + V_p = V_t\varepsilon + V_s\varepsilon_P = V_t\varepsilon + V_t(1-\varepsilon)\varepsilon_P = V_t[\varepsilon + (1-\varepsilon)\varepsilon_P] \tag{B15}$$

$$V_t - V_o' = V_t - V_t\varepsilon - V_t(1-\varepsilon)\varepsilon_P = V_t(1-\varepsilon)(1-\varepsilon_P) \tag{B16}$$

$$V_R = V_t\varepsilon + V_t(1-\varepsilon)\varepsilon_P + V_t[\varepsilon + (1-\varepsilon)\varepsilon_P]k' \quad \text{(B17)}$$

Therefore,

$$V_t(1-\varepsilon)K = V_t(1-\varepsilon)\varepsilon_P + V_t[\varepsilon + (1-\varepsilon)\varepsilon_P]k' \quad \text{(B18)}$$

$$K = \varepsilon_P + \left[\frac{\varepsilon}{(1-\varepsilon)} + \varepsilon_P\right]k' \quad \text{(B19)}$$

In most experimental investigations, K_P is considered to be unity. However, this may cause another contradiction for size exclusion [66]. The particle porosity ε_P must be constant. K_P reflects the ratio of the pore volume accessible for a solute.

As pointed out by Hearn [67], the determination of ε_P must be carefully examined. If the sample probe is not small enough, the value is smaller than the true ε_P value due to size exclusion. If there is some interaction between the probe and the stationary phase, the value may be larger than the true k' value.

References

1. Carta, G., & Jungbauer, A. (2010). *Protein Chromatography: Process Development and Scale-up*, John Wiley & Sons, Weinheim.
2. Guiochon, G., Felinger, A., & Shirazi, D. G. (2006). *Fundamentals of Preparative and Nonlinear Chromatography*, Academic Press, Boston.
3. Ladisch, M. R. (2001). *Bioseparations Engineering*, Wiley-Interscience, New York.
4. LeVan, M. D., Carta, G., & Yon, C. (2008). Adsorption and ion exchange. In *Perry's Chemical Engineers' Handbook*, 8th Ed., McGraw-Hill, New York.
5. Sherwood, T. K., Pigford, R. L., & Wilke, C. R. (1975). *Mass Transfer*, McGraw-Hill, New York.
6. Yamamoto, S., Nakanishi, K., & Matsuno, R. (1988). *Ion-Exchange Chromatography of Proteins*, Marcel Dekker, New York.
7. Arnold, F. H., Blanch, H. W., & Wilke, C. R. (1985a). Analysis of affinity separations: I: Predicting the performance of affinity adsorbers. *Chemical Engineering Journal*, 30, B9–B23.
8. Arnold, F. H., Blanch, H. W., & Wilke, C. R. (1985b). Analysis of affinity separations: II: The characterization of affinity columns by pulse techniques. *Chemical Engineering Journal*, 30, B25–B36.
9. Arnold, F. H., Blanch, H. W., & Wilke, C. R. (1985c). Liquid chromatography plate height equations. *Journal of Chromatography A*, 330, 159–166.
10. Van Deemter, J. J., Zuiderweg, F. J., & Klinkenberg, A. V. (1956). Longitudinal diffusion and resistance to mass transfer as causes of nonideality in chromatography. *Chemical Engineering Science*, 5(6), 271–289.

11. Kubin, M. (1965). Beitrag zur theorie der chromatographie. *Collection of Czechoslovak Chemical Communications*, 30, 1104–1118.
12. Kucera, E. (1965). Contribution to the theory of chromatography. Linear non-equilibrium elution chromatography. *Journal of Chromatography*, 19, 237–248.
13. Bird, R. B., Stewart, W. E., & Lightfoot, E. N. (2002). *Transport Phenomena*, 2nd Ed., John Wiley & Sons, Inc, New York.
14. Harison, R. G., Todd, P., Rudge, S. R., & Petrides, D. P. (2003). *Bioseparation Science and Engineering*, Oxford University Press, New York.
15. Colby, C. B., O'Neill, B. K., & Middelberg, A. P. J. (1996). A modified version of the volume-averaged continuum theory to predict pressure drop across compressible packed beds of Sepharose Big-Beads SP. *Biotechnology Progress*, 12 (1), 92–99.
16. Soriano, G. A., Titchener-Hooker, N. J., Ayazi-Shamlou, P. (1997). The effects of processing scale on the pressure drop of compressible gel supports in liquid chromatographic columns. *Bioprocess Engineering*, 17, 115–119.
17. Stickel, J. J., & Fotopoulos, A. (2001). Pressure-flow relationships for packed beds of compressible chromatography media at laboratory and production scale. *Biotechnology Progress*, 17, 744–751.
18. Yamamoto, S., Nomura, M., & Sano, Y. (1987a). Resolution of proteins in linear gradient elution ion-exchange and hydrophobic interaction chromatography. *Journal of Chromatography A*, 409, 101–110.
19. Yamamoto, S., & Ishihara, T. (1999). Ion-exchange chromatography of proteins near the isoelectric points. *Journal of Chromatography A*, 852, 31–36.
20. Yamamoto, S. (2005). Electrostatic interaction chromatography process for protein separations: impact of engineering analysis of biorecognition mechanism on process optimization. *Chemical Engineering & Technology*, 28, 1387–1393.
21. Yamamoto, S. (1995). Plate height determination for gradient elution chromatography of proteins. *Biotechnology and Bioengineering*, 48, 444–451.
22. Xiao, Y., Rathore, A., O'Connell, J. P., & Fernandez, E. J. (2007). Generalizing a two-conformation model for describing salt and temperature effects on protein retention and stability in hydrophobic interaction chromatography. *Journal of Chromatography A*, 1157, 197–206.
23. Yamamoto, S., Hakoda, M., Oda, T., & Hosono, M. (2007b). Rational method for designing efficient separations by chromatography on polystyrene–divinylbenzene resins eluted with aqueous ethanol. *Journal of Chromatography A*, 1162, 50–55.
24. Yamamoto, S., Nomura, M., & Sano, Y. (1987b). Adsorption chromatography of proteins: determination of optimum conditions. *AIChE Journal*, 33, 1426–1434.
25. Snyder, L. R., & Stadalius, M. A. (1986). High-performance liquid chromatography separations of large molecules: a general model. *High-Performance Liquid Chromatography*, 4, 195–312.

26. Slaff, G. (1993). Chromatography column scale-up in an industrial environment, paper presented at Biopharm Conference '93, Cambridge, MA, June 1993.
27. Carta, G., Ubiera, A. R., & Pabst, T. M. (2005). Protein mass transfer kinetics in ion exchange media: measurements and interpretations. *Chemical Engineering & Technology*, 28, 1252–1264.
28. Ishihara, T., & Yamamoto, S. (2005). Optimization of monoclonal antibody purification by ion-exchange chromatography: application of simple methods with linear gradient elution experimental data. *Journal of Chromatography A*, 1069(1), 99–106.
29. Ishihara, T., Kadoya, T., & Yamamoto, S. (2007). Application of a chromatography model with linear gradient elution experimental data to the rapid scale-up in ion-exchange process chromatography of proteins. *Journal of Chromatography A*, 1162, 34–40.
30. Rüdt, M., Gillet, F., Heege, S., Hitzler, J., Kalbfuss, B., & Guélat, B. (2015). Combined Yamamoto approach for simultaneous estimation of adsorption isotherm and kinetic parameters in ion-exchange chromatography. *Journal of Chromatography A*, 1413, 68–76.
31. Yamamoto, S., & Kita, A. (2005). Theoretical background of short chromatographic layers: optimization of gradient elution in short columns. *Journal of Chromatography A*, 1065, 45–50.
32. Hosono, M., Maeda, R., Yoshimoto, N., & Yamamoto, S. (2012). Rational method for designing efficient chromatography processes based on the iso-resolution curve. *Chemical Engineering & Technology*, 35, 198–203.
33. Yamamoto, S., Nakanishi, K., Matsuno, R., & Kamikubo, T. (1983). Ion exchange chromatography of proteins—prediction of elution curves and operating conditions. I. Theoretical considerations. *Biotechnology and Bioengineering*, 25, 1465–1483.
34. Yamamoto, S., & Kita, A. (2006). Rational design calculation method for stepwise elution chromatography of proteins. *Food and Bioproducts Processing*, 84, 72–77.
35. Watler, P. D., Kaltenbrunner, O., Feng, D., & Yamamoto, S. (2002). Engineering aspects of ion-exchange chromatography. *Scale-up and Optimization in Preparative Chromatography: Principles and Biopharmaceutical Applications* (Eds. A. S. Rathore, & A. Velayudhan), Dekker, New York, pp. 123–171.
36. Yamamoto, S., Watler, P. K., Feng, D., & Kaltenbrunner, O. (1999). Characterization of unstable ion-exchange chromatographic separation of proteins. *Journal of Chromatography A*, 852, 37–41.
37. Kaltenbrunner, O., Jungbauer, A., & Yamamoto, S. (1997). Prediction of the preparative chromatography performance with a very small column. *Journal of Chromatography A*, 760, 41–53.
38. Gallant, S. R., Kundu, A., & Cramer, S. M. (1995). Optimization of step gradient separations: consideration of nonlinear adsorption. *Biotechnology and Bioengineering*, 47(3), 355–372.

39. Kelley, B. D., Tobler, S. A., Brown, P., Coffman, J. L., Godavarti, R., Iskra, T., Switzer, M., & Vunnum, S. (2008). Weak partitioning chromatography for anion exchange purification of monoclonal antibodies. *Biotechnology and Bioengineering*, 101, 553–566.
40. Suda, E. J., Thomas, K. E., Pabst, T. M., Mensah, P., Ramasubramanyan, N., Gustafson, M. E., & Hunter, A. K. (2009). Comparison of agarose and dextran-grafted agarose strong ion exchangers for the separation of protein aggregates. *Journal of Chromatography A*, 1216(27), 5256–5264.
41. Yamamoto, S., & Ishihara, T. (2000). Resolution and retention of proteins near isoelectric points in ion-exchange chromatography. Molecular recognition in electrostatic interaction chromatography. *Separation Science and Technology*, 35, 1707–1717.
42. Kopaciewicz, W., Rounds, M. A., Fausnaugh, J., & Regnier, F. E. (1983). Retention model for high-performance ion-exchange chromatography. *Journal of Chromatography*, 266, 3–21.
43. GE-Healthcare Bio-Sciences AB. (2004). *Ion Exchange Chromatography: Principles and Methods*, GE Healthcare Life Sciences, Uppsala.
44. Scopes, R. K. (1994). *Protein Purification, Principles and Practice*, 3rd Ed., Springer-Verlag, New York.
45. Yoshimoto, N., Minakuchi, K., Itoh, D., Isakari, Y., & Yamamoto, S. (2013a). High-throughput process development methods for chromatography and precipitation of proteins: advantages and precautions. *Engineering in Life Sciences*, 13, 446–455.
46. Fogle, J., Cano, T., Wong, M., Hoang, H., & von Hirschheydt, T. (2010). Retention versus resolution: optimizing aggregate removal in ion exchange chromatography, paper presented at the Recovery of Biological Products XIV, August 1–6, Lake Tahoe.
47. Yoshimoto, N., Nishijima, Y., Akbarzadehlaleh, P., Fujii, S., Abe, M., & Yamamoto, S. (2008). Micro-plate based monolithic ion-exchange chromatography for high throughput protein purification process design. *Journal of Chemical Engineering of Japan*, 41, 200–205.
48. Pabst, T. M., Suda, E. J., Thomas, K. E., Mensah, P., Ramasubramanyan, N., Gustafson, M. E., Hunter, A.(2009). Binding and elution behavior of proteins on strong cation exchangers. *Journal of Chromatography A*, 1216, 7950–7956.
49. Suda, E., Ubiera, A., & Göklen, K. (2011). Biopharm downstream platform development: resolution and pore structure characterization of high capacity ion exchangers for the removal of product-related variants, paper presented at PREP Symposium 2011, July 10–13, Boston, MA.
50. Ubiera, A., Levy, N., Loladze, V., Zhang, Y., Branco, C., & Göklen, K. (2014). Accelerating the early-phase portfolio: platform or first-intent experimental approaches for cation-exchange chromatography and the removal of aggregates, paper presented at ACS 2014 National Meeting, BIOT, March 16–20, Dallas, TX.

51. Urmann, M., Graalfs, H., Joehnck, M., Jacob, L. R., & Frech, C. (2010). Cation-exchange chromatography of monoclonal antibodies: characterization of a novel stationary phase designed for production-scale purification. *MAbs*, 2, 395–404.
52. Urmann, M., Hafner, M., Frech, C. (2011). Influence of protein and stationary phase properties on protein–matrix-interaction in cation exchange chromatography. *Journal of Chromatography A*, 1218, 5136–5145.
53. Yoshimoto, N., Isakari, Y., Itoh, D., Podgornik, A., & Yamamoto, S. (2015). *Biotechnology Journal*, 10(12), 1929–1934.
54. Fee, C. J., & Van Alstine, J. M. (2004). Prediction of the viscosity radius and the size exclusion chromatography behavior of PEGylated proteins. *Bioconjugate Chemistry*, 15(6), 1304–1313.
55. Abe, M., Akbarzaderaleh, P., Hamachi, M., Yoshimoto, N., & Yamamoto, S. (2010). Interaction mechanism of mono-PEGylated proteins in electrostatic interaction chromatography. *Biotechnology Journal*, 5(5), 477–483.
56. Yoshimoto, N., Isakari, Y., Itoh, D., & Yamamoto, S. (2013b). PEG chain length impacts yield of solid-phase protein PEGylation and efficiency of PEGylated protein separation by ion-exchange chromatography: Insights of mechanistic models. *Biotechnology Journal*, 8, 801–810.
57. Yamamoto, S., Fujii, S., Yoshimoto, N., & Akbarzadehlaleh, P. (2007a). Effects of protein conformational changes on separation performance in electrostatic interaction chromatography: Unfolded proteins and PEGylated proteins. *Journal of Biotechnology*, 132, 196–201.
58. Yamamoto, S., Yoshimoto, N., Tarmann, C., & Jungbauer, A. (2009a). Binding site and elution behavior of DNA and other large biomolecules in monolithic anion-exchange chromatography. *Journal of Chromatography A*, 1216, 2616–2620.
59. Yamamoto, S., Nakamura, M., Tarmann, C., & Jungbauer, A. (2007c). Retention studies of DNA on anion-exchange monolith chromatography: binding site and elution behavior. *Journal of Chromatography A*, 1144, 155–160.
60. Yamamoto, S., Yoshimoto, N., & Nishizumi, Y. (2009b). Theoretical background of monolithic short layer ion-exchange chromatography for separation of charged large biomolecules or bioparticles. *Journal of Chromatography A*, 1216, 2612–2615.
61. Kluters, S., Hafner, M., von Hirschheydt, T., & Frech, C. (2015). Solvent modulated linear pH gradient elution for the purification of conventional and bispecific antibodies: modeling and application. *Journal of Chromatography A*, 1418, 119–129.
62. Lee, Y. F., Schmidt, M., Graalfs, H., Hafner, M., & Frech, C. (2015). Modeling of dual gradient elution in ion exchange and mixed-mode chromatography. *Journal of Chromatography A*, 1417, 64–72.
63. Boardman, N. K., & Partridge, S. M. (1955). Separation of neutral proteins on ion-exchange resins. *Biochemical Journal*, 59, 543–552.

64. Regnier, F. E., & Mazsaroff, I. (1987). A theoretical examination of adsorption processes in preparative liquid chromatography of proteins. *Biotechnology Progress*, 3, 22–26.
65. Gallant, S. R., Vunnum, S., & Cramer, S. M. (1996). Optimization of preparative ion-exchange chromatography of proteins: linear gradient separations. *Journal of Chromatography A*, 725(2), 295–314.
66. Janson, J.-C., & Jonsson, J.-A. (1998). Introduction to chromatography. In *Protein Purification* (Eds. J.-C. Janson, & L. Ryden), 2nd Ed., Wiley-VCH, New York, 43.
67. Hearn, M. T. (1998). High-resolution reversed-phase chromatography of polypeptides and proteins. In *Protein Purification* (Eds. J.-C. Janson, & L. Ryden), 2nd Ed., Wiley-VCH, New York, 239.

5

Development of Continuous Capture Steps in Bioprocess Applications

Frank Riske and Tom Ransohoff

BioProcess Technology Consultants, Woburn, MA, USA

5.1 Introduction

What is a continuous process? Cooney and Konstantinov [1] defined continuous in the following way: "A unit operation is continuous if it is capable of processing a continuous flow input for prolonged periods of time with a minimal internal hold volume. The output can be continuous or discretized in small packets produced in a cyclic manner. A process is continuous if it is composed of integrated (physically connected) continuous unit operations with zero or minimal hold volume in between." To emphasize that all the unit operations are continuous and integrated, such processes are also referred to as fully continuous or end-to-end continuous. A hybrid process is composed of both batch and continuous unit operations.

Why are bioprocess scientists interested in pursuing continuous chromatography (CC) when batch chromatography has been successfully used for the last 60 years? Historically, the conversion of batch systems to continuous systems has resulted in more efficient processes that show lower costs [2]. The use of continuous systems in manufacturing operations has a long history that stretches back to the late 1800s. The chemical and pharmaceutical industries have used simulated moving bed (SMB) technology for a number of years to separate binary mixtures of small compounds and drug enantiomers [3]. Recently, more complex mixtures and proteins have been separated on SMB and SMB-like systems [4–6].

Proponents of CC in bioprocessing expect an improvement in capital efficiency and in operating cost of goods sold (COGS) due to a reduction in the resin, buffer volume, and time needed to process the same mass of material as in a batch system among other factors [7, 8]. CC systems are efficient because multiple operations are occurring at the same time on columns that have a high resin loading capacity (close to the static capacity). Target product that is not captured by the loading column is captured by a second in-line column. The continuous rapid processing of perfusion harvest reduces the protein residence time (R_T), which benefits relatively unstable molecules like factor VIII.

Skeptics point out that the higher productivity of CC versus batch systems may be counterbalanced by a higher rate of failure due to system complexity. The SMB-like systems contain 2–12 columns, and each additional column adds additional valves, lines, detectors, and tanks. Furthermore, Kaltenbrunner suggests that placing CC into a mature commercial process may not have a compelling COGS benefit, especially for MAb that are produced at a few hundred dollars per gram [9]. The best utility for CC may be in the clinical production of MAb in situations where the very expensive Protein A (PA) resin may only be used for several cycles. In this case, substantial savings are realized by using less resin more efficiently.

Although traditional SMB systems are well suited for separating binary and ternary mixtures of small compounds, they are not optimally designed to separate a protein of interest from the complex mixture of materials that are found in a clarified cell culture harvest. During the last decade, several multicolumn chromatography (MCC) systems have been developed to specifically meet the needs of purification of biomolecules using adsorptive chromatography, such as capture of product from harvested cell culture supernatant [10–13]. This paper will focus on the development and practical aspects of implementation of a continuous capture step in a biopharmaceutical process.

5.2 Economic Rationale for Continuous Processing

The rationale for considering continuous processing for the manufacture of a wide range of products is well described in other chapters in this book (e.g., see Chapter 15) and in many other sources (cf [14–16]), so our focus in this section will be on highlighting key benefits that will impact manufacturing practice and drive improved process economics specifically for biopharmaceuticals. These benefits include the following:

Process intensification/improved facility utilization. This is perhaps one of the most important benefits to continuous processing in biopharmaceutical manufacturing. The improved utilization of equipment—by minimizing nonproductive time—enables the use of significantly smaller equipment at all stages of the process. This reduction in equipment scale will lead to reduced manufacturing facility size, lowering capital costs and time of construction for new capacity.

Enables single-use equipment for larger processes/improved manufacturing flexibility. A highly related benefit to the process intensification described previously is that facilities using single-use equipment at the maximum currently practicable scale (e.g., ~2000 L bioreactors, ~60 cm single-use chromatography columns) will be able to produce far more material if operating continuous or semicontinuous processes instead of batch processes. As a result, manufacturing flexibility (i.e., the ability to move processes between facilities to "change over" quickly between production campaigns of different products) will be enhanced as well.

Reduced water and consumables usage. For certain unit operations, such as a capture chromatography step, operation in continuous mode will reduce water and chromatography resin usage. This is due to the ability to run the step at higher loading capacities at equivalent yields when operating a continuous, countercurrent multicolumn process as compared to a batch process [17]. In addition to economic benefits, this benefit may have added importance in applications where capacity is limited by buffer or product tank volume.

Improved process control and understanding. For those who are used to operating batch processes, it is somehow counterintuitive that continuous or semicontinuous operation of these unit operations would improve process control. In fact, continuous processes—by nature—are fully and continuously controlled, which should not only result in a reduction in labor demand but should also enhance product quality control. In addition to this, one should expect that enhanced process control will result in fewer deviations. This is also reflected in the occurrence of incidents and accidents in continuous processes of more mature process industries, the vast majority of which happen during transient operations, such as start-up and shutdown. Furthermore, the ability to more rapidly extend process understanding should result from the increased ability to obtain data from continuously running unit operations.

More rapid availability of product. As we improve our ability to operate processes continuously, the ability to rapidly start-up a continuous processing line for a new product (or production campaign) may allow us to deliver product more quickly than is possible with batch processes. The operation of a fully continuous processing line where unit operations are operated in parallel, with only brief lags between start-up of sequential operations, may allow for compression of the time to the first availability of purified drug substance.

More efficient utilization of labor. Particularly in conjunction with emerging facility concepts, such as flexible, modular facilities (e.g., GE's FlexFactory line) along with improvements in isolator technologies, it is possible to envision operation of multiple processes in the same hall being operated by the same labor team. While there are many obstacles (i.e., regulatory acceptability and ensuring adequate segregation) to implementing this mode of operation, it is reasonable to expect that this will be possible in the future. Of course, this is possible with both batch and continuous operations; however, there are several reasons that continuous processes are better aligned to this operating mode:

1) Smaller processes enable operation of multiple processing lines in a smaller footprint.
2) While batch processes are characterized by periods of heavy intensity and long periods of downtime/low activity, continuous processing operations are active at a more consistent, predictable level.

5.3 Developing a Continuous Capture Step

The capture of a protein of interest from a cell culture harvest is a critical process step. The purpose of the capture column is to reduce process volume and to separate the target biomolecule from a surrounding environment of proteases, glycosidases, and other materials, which may adversely affect protein stability. The capture step is often the largest process column as well as the longest operation, due to high load volumes and low binding capacity. Relatively unstable proteins will benefit from a rapid capture and exchange into a buffer that will stabilize the target. In cases where the molecule is relatively stable (e.g., MAb), and the capture resin is very expensive (PA), one of the primary purposes for using an MCC system is to reduce resin and production costs by increasing resin capacity.

The interaction between a resin and a biomolecule is governed by thermodynamics and mass transfer kinetics, which are related to the intrinsic characteristics of both the molecule and resin. Factors such as the size and distribution of the resin particles, the resin pores, the functional groups, the characteristics of the solid and liquid phase, and the protein structure, size, shape, charge, hydrophobicity, and surface character all affect this interaction. The interaction between resin and proteins has been described for single proteins or simple mixtures of a few well-characterized proteins, but not for complex biological feeds where multiple materials compete for the resin binding sites that reside primarily within the resin pores [18–21]. Although the modeling of the molecular interaction between biomolecules and ligands is rapidly improving, these programs are not yet sophisticated or powerful enough to *a priori* determine an appropriate capture resin for a target protein present in a complex mixture. This is because many of the low and high molecular weight materials produced by the cells, and secreted into the media, are not fully characterized. Furthermore, a number of ligand–protein interactions are needed to bind the protein to a traditional resin such as an ion exchanger or hydrophobic interaction column.

However, the process scientist does not need a deep mathematical fundamental understanding of resin–protein interaction to design a capture step for complex mixtures. The shape of the resin breakthrough curve and the breakthrough point reflects the competition of the harvest materials for the resin binding sites. A mathematical transformation of the breakthrough curve is used for both single-column (batch) chromatography (SCC) and MCC development. A typical breakthrough curve is shown in Figure 5.1.

The initial step for both SCC and MCC development is to experimentally screen resins for target capacity and selectivity (i.e., target separation from other cell culture components). The choice of resins to examine is often based on an understanding of the stability and the molecular characteristics of the biomolecule (e.g., pI, molecular weight, glycosylation pattern). Initial screening may be done in high-throughput (e.g., plate-based) systems to obtain equilibrium binding data (e.g., see Chapter 11). Experimentally derived binding curves and small-scale studies (that examine recovery and impurity removal) are then used to sort the chosen resins until there is a single resin left. The goal is to narrow the field to several resins, which have a high binding capacity, high target recovery, and a reasonable removal of impurities. Critical to being able to whittle down the resins is an assay that can quantify the target in both the load and column effluent.

The approach to the development of a capture resin for a platform process, such as a MAb, is significantly less complex than a non-MAb recombinant protein. PA is the capture resin of choice for many MAb, Fc-Fusion proteins, and antibody-like molecules. The resin screening process is often limited to an evaluation of several PA resins in order to pick the one that shows the highest dynamic capacity at the desired linear flow rate. In some cases only the dynamic capacity is determined because a company may use a single PA resin for all their products. The static and dynamic binding curves, at several linear flow rates, are sufficient to move forward in designing a capture step.

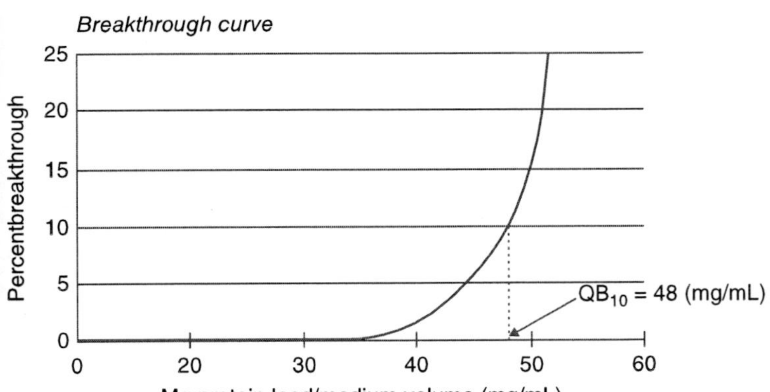

Figure 5.1 Determining dynamic binding capacity. *Source*: Image courtesy GE Healthcare Uppsala, Günter Jagschies.

For non-MAb proteins, a wide variety of capture resins are possible. These include ion exchange, hydrophobic, metal chelate, and mixed mode resins to name a few types. The current state of the art is to use high-throughput robotic systems to screen multiple resins and resin conditions to pick the appropriate candidates [22–25]. Typically, orthogonal resins, and several buffer ions and concentrations, are screened in batch mode (96 well plates), in dynamic mode (small columns), or both [26].

High-throughput systems can next be used to develop the wash, elution, and cleaning buffers. The resin is loaded to about 50–90% of their dynamic capacity, and a number of different buffers and conditions are examined. The candidates are then moved to the 5–10 mL scale on an automated chromatography system such as the AKTA avant (GE Healthcare, Piscataway, NJ) to evaluate dynamic binding performance. The chromatography conditions are refined using the AKTA system. It's at this point that the processes for MCC and SCC somewhat diverge. For SCC the step is essentially complete although it may be further refined to maximize the resin capacity, target recovery, and secondarily target purity. For MCC, the information derived from dynamic binding curves and the development of the step is used to move forward to continuous process design.

The dynamic binding curves are affected by the target concentration, the mass transfer properties of the target and resin, and the column R_T. An increasing linear flow rate decreases the MCC binding capacity. A 1% target breakthrough on the dynamic binding curve represents only a fraction of the total (i.e., static equilibrium) resin binding capacity (Figure 5.2, A). In MCC the column continues to be loaded (A + B) until it is close to the static binding capacity of the column (A + B + C). The MCC binding capacity in relation to

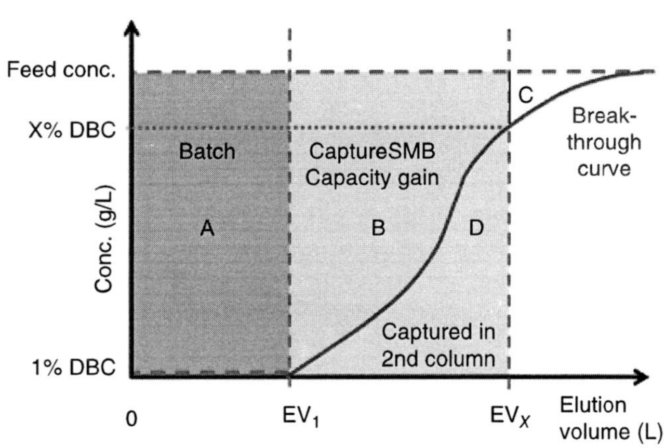

Figure 5.2 Loading a column in periodic counter-current chromatography (PCC) versus SCC. *Source*: Image courtesy of Thomas Műller-Späth.

the static binding capacity can be expressed as (A + B)/(A + B + C) and the % binding gain of MCC versus SCC as (A + B)/A × 100. The total target capacity in MCC can be 30–40% higher than SCC because the target breakthrough (Fig. 5.4(a)) is mostly captured on subsequent in-line column(s) in the MCC system.

5.4 The Operation of MCC Systems

Lab scale MCC systems have been used to purify a number of molecules including MAb, bi-specific Ab, peptides, enzymes, other recombinant proteins, and nanoparticles [27]. MCC systems are marketed by Pall (Westborough, MA), Novasep (Pompey, France), Semba (Madison, WI), and GE Healthcare (Westborough, MA). They support anywhere from 3 to 16 columns and are designed to reduce resin and buffer usage versus SCC and thereby improve productivity and the COGS. A different type of MCC system is the two-column multicolumn countercurrent solvent gradient purification (MCSGP) system (ChromaCon, Switzerland). It is designed as a polishing step for difficult separations and can be run with linear gradients. In this system the fractions containing overlapping product and impurities are recycled from one column onto the other while product is retained on both columns. The system continuously cycles between the two columns until the desired target purity is achieved. Recently, software has been developed for using this system to capture MAb from cell culture harvest [28].

While the designs of the specific systems differ, all MCC systems intensify the chromatography process by increasing equipment utilization. In part, this is achieved by directing multiple process solutions (i.e., load, equilibration, wash, elution, and/or regeneration buffers) to columns in the system simultaneously as opposed to delivery of one solution at a time in a batch system. This requires a different approach to system design as shown in Figure 5.3, depicting the cassette style valve used by Tarpon's BioSMB system to address this requirement.

Although the number of columns may differ among MCC systems, they all maximize resin utilization by having two or more columns in-line during the load phase [29, 30]. For illustrative purposes a three-column MCC operation (PCC, GE Healthcare) is described later [31–33] (Figure 5.4). At the start of PCC, the first column is in the loading phase, and the second and third columns are in the equilibration phase. The absorbance at 280 nm (A_{280}) at the first column outlet will rise from baseline when nonbinding UV absorbing material flows through to waste. The absorbance will climb to a plateau and remain at that plateau, as long as all the target protein is bound by the column. The second column is brought on-line when the absorbance starts rising from the plateau, indicating that the target is breaking through the first column. At this point, the first and second columns are in-line and the third column is idle.

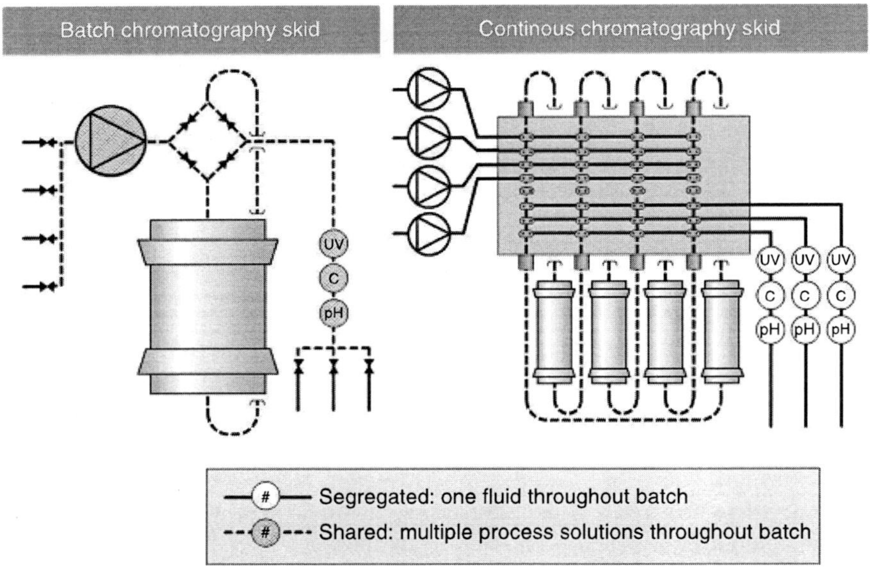

Figure 5.3 Management of process chromatography solutions in a batch chromatography system (left) and a continuous system (right). *Source*: Courtesy of Marc Bisschops.

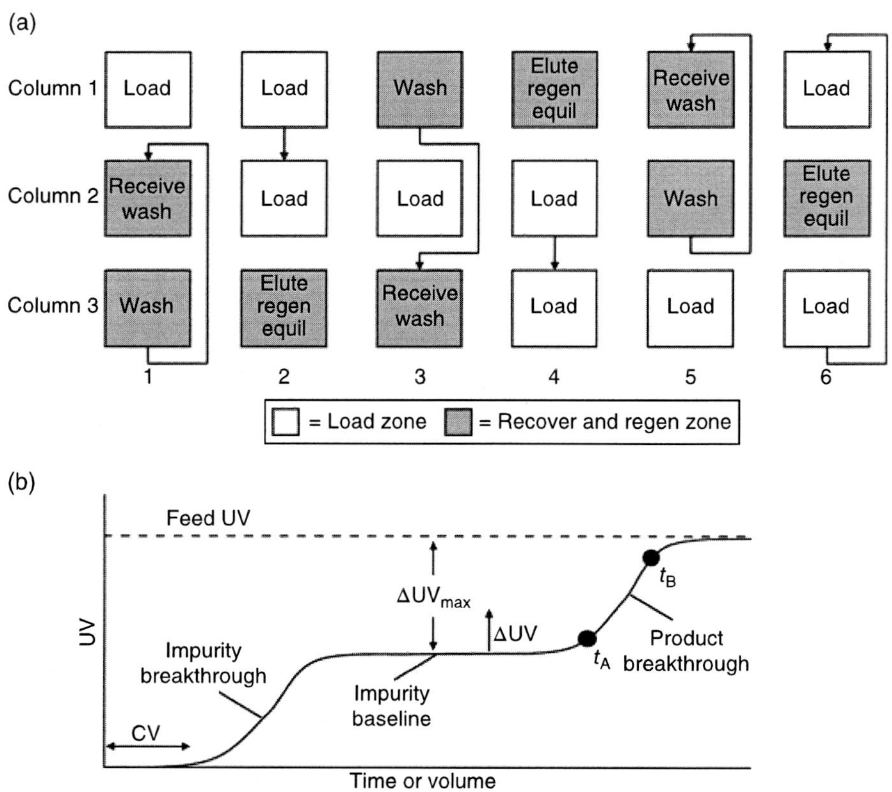

Figure 5.4 Operation of a three-column PCC system. (a) Switching sections of a PCC cycle operation. (b) UV monitoring of the PCC cycle.

The two columns remain in the loading phase until the first column approaches its (experimentally determined) static binding capacity. The column is not loaded further because the small gain in target binding is offset by the time required to further load the column (the binding curve is flattening out). At this point column 1 switches out of the load mode to undergo the remainder of the purification cycle (wash, elution, regeneration, and equilibration), and column 2 becomes the primary load column. Column 3 is put in-line with column 2 when target starts breaking through column 2, as indicated by the increase in UV. Column 2 is then loaded to saturation at which point it is disconnected from column 3 and undergoes the remainder of the column cycle. The process described for columns 1 and 2 and columns 2 and 3 is repeated for columns 3 and 1. The overall system column load flow is 1, 1 + 2, 2, 2 + 3, 3, 3 + 1, 1.

To maximize target recovery the column in the load mode is washed with one column volume of equilibration buffer prior to going through the remainder of the column cycle. The target present in the feed and eluate lines, and other interstitial spaces, is thereby captured by the second in-line column. The program for the MCC should be written such that there is at least one column available to be placed in-line prior to the loading column reaching saturation in order to receive this wash.

The switch from one column to another in the PCC system can be based on a set absorbance at 280 nm, a rapid product measurement tool such as UPLC, or the ΔUV system (Figure 5.4b). In this PAT system there is a UV monitor on the feed line and one on the column effluent. When the effluent absorbance plateaus, the ΔUV between the feed line and effluent represents the UV contribution of all materials in the media except the target. When the absorbance rises from that plateau, the target is breaking through the first column and a second column is brought on-line (tA). At tB, the first column is close to saturation. The column is washed with one volume of buffer, disconnected from column 2, and then washed, eluted, regenerated, and equilibrated. In other MCC systems (e.g., Tarpon's BioSMB), there are always multiple columns in the load zone, so developing this switching strategy is not necessary.

5.5 Modeling MCC Operation

An MCC capture step is designed based on a series of experimental binding curves and based on an understanding of the process requirements (e.g., mass to process in a given time frame). Basic modeling can be used to determine the number and size of columns for an MCC capture step. Langmuir isotherms are generated at the low and high target concentration range (in harvest) to understand the effect of harvest variability on column performance. As long as the target concentration is in the linear portion of the Langmuir curve, the MCC operation will be relatively insensitive to concentration changes, and similar

K_{MCC} (MCC binding capacity) values will be generated. Next, the target midpoint concentration (if in the linear portion of the curve) is used to generate experimental breakthrough curves at different R_T. An operational linear flow rate is chosen based on the shape of the dynamic binding curves and the K_{MCC}. The R_T defines the time an MCC column will be in the loading mode. The R_T in MCC is less than in SCC because two columns are used to capture the load; in essence the column height and R_T are doubled for two columns in the load zone.

A calculation for determining the number of MCC columns for a given process is shown in Equation 5.1. The necessary number of columns is based on the R_T and K_{MCC} values derived from the target breakthrough curves, the target concentration in the feed (C_F), and the time necessary to perform all the column steps after column loading (T_R).

$$N > 2 + \left(\frac{T_R \times C_F}{R_T \times K_{MCC}} \right) \tag{5.1}$$

Both T_R and R_T are variable inputs, whereas K_{MCC} is dependent on both R_T and C_F (which is fixed). The number of MCC columns increases with increasing C_F and T_R. Conversely, the number of columns decreases with a longer R_T and a higher K_{MCC}. Choosing a high binding capacity rigid resin, and developing a short T_R cycle, will lead to more efficient processes due to a reduction in both the size and number of columns.

In order for PCC to operate appropriately, T_L (the time necessary to perform the loading step) must be greater than T_R [34]. Since there are two columns in-line during the loading phase, the relationship is $T_R/N{-}2 < T_L$. With a three-column MCC such as the GE PCC system, $N{-}2 = 1$ and T_L must be $>T_R$. Since the column load is often shorter than the time to perform all the other column operations, a solution must be found to allow continuous operation. That solution is the addition of columns to the system to accommodate the T_R and T_L disparity. In essence, T_R decreases because it is divided by a greater N.

Besides N, the volume (V_C) of the MCC columns needs to be determined. First the necessary flow rate (D_r) is calculated based on the volume of media to process (V_{total} (L)) and the time required to complete that processing (T; Eq. 5.2).

$$D_r = \frac{V_{total}}{T} \tag{5.2}$$

Vc is then equal to D_r multiplied by the retention time (R_T; Eq. 5.3).

$$V_c = D_r \times R_T \tag{5.3}$$

The power of MCC can be illustrated by contrasting the time needed for processing a harvest that has been concentrated 10× by TFF. The 10× reduction in harvest volume results in about a 3× improvement in SCC processing time, given that the column is used and assuming that the loading capacity is comparable. The time savings are limited to T_L as T_R remains the same. The operation is sequential and T_R cannot start until T_L is completed. In MCC, a $10\times C_F$ reduces T_L as well. In contrast to SCC, T_R is also reduced because the number of columns is increased to accommodate the shorter T_L. T_R is reduced by N (Eq. 5.4). The total processing time improvement is greater than threefold because T_L and T_R are not sequential activities. In a properly designed MCC, the loading of the next column is not dependent on the completion of the first column cycle.

$$\text{Total column time} = \left(N \times T_L\right) + \left(\frac{T_R}{N}\right) \quad (5.4)$$

To some extent higher feed stream concentrations can be accommodated without increasing N by operating the system using less than theoretically optimal flow rates at different stages of the process. For example, increasing T_L (lower flow rates), decreasing T_R (higher flow rates), or a combination of both will decrease N if the changes are sufficient in magnitude. The column length can also be increased to increase T_L. However, the columns should not be longer than 10–15 cm because the system height and pressure double when two columns are in-line during the loading and initial wash phase.

As described elsewhere, more detailed models based on Langmuir binding curves may also be developed to optimize and refine designs of MCC systems [27], but the straightforward calculations described previously will allow for a reasonably sized MCC capture step that will enable users to realize the benefits of continuous operation.

5.6 Processing Bioreactor Feeds on a Capture MCC

The volume to be processed, the titer, or both often increase during the developmental lifecycle of a biotherapeutic. To accommodate the increasing target mass and volume, a process is scaled-up. In SCC scale-up typically involves increasing the column diameter while keeping all other factors the same; however, other approaches may also apply (see Chapter 7). In scaling up an MCC capture step, several strategies can be employed. These include adding additional columns, increasing cycle number and run time, decreasing T_R and increasing T_L, and/or increasing the column height and diameter (or volume). However, at some point the MCC piping may not be capable of handling the increased flow rates that are needed. The modeling described previously can aid in determination of the optimal strategy for scale-up of an MCC operation.

The appropriate column packing in SCC is important for system performance. Packing is also important when running multiple columns; an optimally running MCC depends on column uniformity. The columns should be packed to the same compression factor and show a comparable HETP and Af. In many systems, the switch from one column to another is automatic and based on a time or volume trigger. Column packing differences can lead to suboptimal MCC performance as the columns may be over- or underloaded. Furthermore, the column performance may decrease with increased cycling, and the rate of decrease may be different if columns are packed differently. Lastly, fluctuations in the feed concentration will also affect MCC performance if, for example, the upstream section of the process is also continuous. The influence of these factors will be magnified with an increasing number of columns in the system.

The decision to develop an MCC capture step should be made with the overall process design in mind and after conducting a risk/benefit and cost analysis. What is the nature of the process before and after the capture step? Discontinuous or continuous? And if continuous for how many steps? For example, if the next step is continuous, the discontinuous eluates from the capture MCC need to be synchronized with the loading of the columns on a second MCC or a properly sized surge tank established. This "connectivity" adds complexity to the system and scientific and engineering challenges. An assessment of MCC should examine the productivity gain and cost reduction weighed against a higher risk of system failure due to increased complexity. Another consideration in an analysis of MCC versus SCC is the higher costs of MCC production skids, the higher costs to maintain these systems (parts and labor), and increased consumable costs.

Lab to pilot scale MCC systems have been used to process material from perfusion and batch reactors. Typically, several days to two weeks of batch or fed-batch reactor harvest is collected for relatively stable molecules, such as MAb. Typically, the entire volume of the reactor is processed on the MCC after the cells and debris were removed. A majority of the MCC reports have been performed with MAb [6, 21, 31, 35].

The most complicated MCC situation is the direct connection of a perfusion reactor to the capture system [36, 37]. In this conception the perfusion flow rate must be closely matched to the MCC flow rate. Typically, a surge vessel is present between the bioreactor and MCC in order to accommodate slight differences in the perfusion and MCC flow rates. The joining of the bioreactor and MCC system introduces a number of design and operating constraints and limits the operating ranges for both systems. Perhaps the most challenging aspect of this concept is that the bioreactor and MCC must be connected and aseptically maintained for relatively long periods of time.

A perfusion bioreactor may run for weeks or months. If connected directly to the perfusion bioreactor, the MCC system, including its columns, must be sterilized prior to its connection to the reactor. After the reactor and MCC system are aseptically connected, the contamination of any part of the system contaminates the entire system. Furthermore, the reactor to column harvest feed lines cannot be treated with base until the campaign is completed because harvest is constantly being produced. Microbial contamination is difficult to avoid especially when complicated systems are running for long periods of time. In batch systems, which only require sanitary operation, microbial contaminations are a serious problem. Biologics production facilities average a batch failure every 40.6 weeks and lose 7.6% of all batches [38]. The greatest reported single source of failure in large volume facilities was from contamination (30%) followed by operator error, equipment failure, and specification failure. These four categories are accounted for 80% of all process failures. Once microbes gain access to a process, they can grow quickly. Bacterial doubling times are often orders of magnitude less (<0.5 h) than mammalian cells (18–48 h) under favorable environmental conditions such as a bioreactor harvest. The ingress of bacteria into a bioreactor typically results in the bacteria overgrowing the culture and killing the mammalian cells. Fungi tend to grow slower than bacteria, but over the course of several days, they can also overtake the cells in a reactor. Lastly, once in a system bacteria can be difficult to remove because they often form biofilms [36, 39] or, in the case of Gram negatives, endospores [35].

A potentially more robust and facile approach for operating a perfusion reactor and MCC is to decouple both systems by using manifolds and disposable bags or collection vessels. The bioreactor would be maintained aseptically, and the MCC would be operated in a sanitary fashion. When a bag or vessel is filled, it would be aseptically disconnected from the bioreactor, transported to the MCC, and attached to a manifold connected to the unit. There would always be one container feeding the MCC when a second filled container is attached. The physical break between the bioreactor and the MCC limits the sterile barrier to the bioreactor, simplifies MCC design and operation, and allows far greater range and flexibility of MCC flow rates during different stages of a process. The simpler MCC design reduces the cost of the unit and the system footprint. The amount of spent media to collect before starting to purify the biomolecule would depend on the flow rate of the bioreactor and MCC and the stability of the molecule. The more unstable a molecule, the more important it becomes to minimize the time that the target remains in harvest. A half to full day of perfused harvest might be enough to provide process robustness and for the ability to fully clean the entire skid once a day.

5.7 The Future of MCC

The appropriateness and use of MCC in capture steps and in other parts of the downstream process may depend on a number of factors including the molecular characteristics and stability of the target molecule, the feed titer and product amount required, and the facility design and intention (multipurpose or dedicated).

The decision will also depend on the durability, flexibility, and reliability of the commercial MCC systems. Although there is a cGMP system available, no one has yet disclosed the use of MCC in a clinical or commercial production.

Are MCC systems more flexible than SCC systems? If the bioreactor and capture systems are disconnected, there are a number of levers to accommodate process changes. However, flexibility decreases when the system is connected to a reactor as the MCC flow rate must match the perfusion flow rate.

The durability of the commercial MCC units is an open question as there is no history of use in commercial or clinical settings. Multiple daily column operations will occur for weeks or months in perfusion systems. The easiest way to test robustness is to run scale-down systems for extended periods of time. Unfortunately, the available small-scale systems may not model the production system. To collect the necessary data, it may be necessary to conduct multiple pilot runs using the scale equipment.

Several reports have been published that extend the continuous barrier post MCC [37, 40, 41]. The overall footprint for a continuous process may be smaller than a conventional process due to a reduction in the number of tanks and other unit operations. However, the footprint of MCC and SCC skids may be similar. Although a column will be smaller, each column will add valves, lines, and detectors that increase the size of the system. CIP and SIP capabilities (for an aseptic skid) will also add additional piping to the skid.

Perhaps the future of bioprocessing is not just a decision between MCC and SCC but rather a decision on downstream approach. When a substantially greater quantity of target is needed, will a process be scaled-up or will the process be scaled-out? In a scaled-out or modular approach, the purification train is copied as many times as necessary to handle the increased product demand. This simplifies the learning curve and the regulatory pathway, as the purification lines are identical. As industry gains more experience with MCC and other forms of continuous processing, this approach is likely to become more commonly used throughout industry.

References

1. Cooney, C. and Konstantinov, K. 2014. White paper on continuous processing, Continuous manufacturing symposium, May 20–21, Cambridge, MA.
2. Utterback, J. M. 1994. *Mastering the dynamics of innovation*. Boston, MA, Harvard Business School Press.
3. Fuchs, G. et al. 1992. Optical isomers purification with the simulated moving bed technology: experimental and theoretical approaches. In Proceeding of the 9th International Symposium on preparative and industrial chromatography and related techniques, Basel, CH, 392–402.

4. Mun, S. et al. 2003. Optimal design of a size-exclusion tandem simulated moving bed for insulin purification. *Ind. Eng. Chem. Res.* 42, 1977–1993.
5. Buhlert, K. et al. 2009. Construction and development of a new single-column simulated moving bed system on the laboratory scale. *J. Chromatogr. A* 1216, 8778–8786.
6. Thömmes, J. et al. 2003. Protein A affinity simulated moving bed chromatography. Recovery of Biological Products, XI, Banff, Alberta, Canada, September 14–19.
7. Juzza, M. et al. 2000. Preparative-scale separation by simulated moving bed of biocatalytically produced regioisomeric lactones. *Trends Biotechnol.,* 18, 108.
8. Stock, R. et al. 2015. The potential impact of continuous processing on the practice and economics of biopharmaceutical manufacturing. *Continuous Processing in Pharmaceutical Manufacturing,* Ed. G. Subramanian. Wiley-VCH, Weinheim, pp. 479–493.
9. Kaltenbrunner, O. 2014. Continuous chromatography: the good, the bad and the unexpected. Recovery of Biologics XVI, Rostock, Germany.
10. Műller-Späth, T. et al. 2008. Chromatographic separation of three monoclonal antibody variants using multicolumn countercurrent solvent gradient purification (MCSGP). *Biotechnol. Bioeng.* 100, 1166–1177.
11. Muller-Späth, T. et al. 2009. Role of cleaning-in-place in the purification of Mab supernatants using continuous cation exchange chromatography. *Sep. Sci. Technol.* 44, 1–26.
12. Műller-Späth, T. et al. 2010. Two step capture and purification of IgG2 using multicolumn countercurrent solvent gradient purification (MCSGP). *Biotechnol. Bioeng.* 107, 974–984.
13. Műller-Späth, T. et al. 2010. Increasing the activity of monoclonal antibody therapeutics by continuous chromatography (MCSGP). *Biotechnol. Bioeng.* 107, 652–662.
14. Van Walsem H.J. and Thompson M.C. 1997. Simulated moving bed in the production of lysine. *J. Biotechnol.* 59, 127–132.
15. Miller L. et al. 1999. Chromatographic resolution of the enantiomers of a pharmaceutical intermediate from the milligram to the kilogram scale. *J. Chromatogr. A* 849, 309–317.
16. Trout B. and Bisson W. Dec 2009. Continuous manufacturing of small molecule pharmaceuticals: the ultra lean way of manufacturing, presented at MIT Leaders for Global Operations Conference; Cambridge, MA.
17. Bisschops M. and Brower M. 2013. The impact of continuous multicolumn chromatography on biomanufacturing efficiency. *Pharm Bioprocess.* 1(4), 361–372.
18. DePhillips, P. and Lenhoff, A.M. 2000. Pore size distributions of cation-ex-change adsorbents determined by inverse size-exclusion chromatography. *J. Chromatogr. A* 883, 39–54.

19. Gritti, F. and Guiochon, G. 2010. A protocol for the measurement of all the parameters of the mass transfer kinetics in columns used in liquid chromatography. *J. Chromatogr. A*, 1217, 5137–5151.
20. Hahn, R. et al. 2005. Comparison of protein A affinity sorbents II. Mass transfer properties. *J. Chromatogr. A* 1093, 98–110.
21. Yan, Y. and Lenhoff, A.M. 2006. Pore size distributions of ion exchangers and relation to protein binding capacity. *J. Chromatogr. A* 1126, 107–119.
22. Hubbuch, J. 2012. Editorial: high-throughput process development. *Biotechnol. J.* 7, 1185.
23. Łacki, K.M. 2012. High-throughput process development of chromatography steps: advantages and limitations of different formats used. *Biotechnol. J.* 7, 1192–1202.
24. Sisodiya, V.N. et al. 2012. Studying host cell protein interactions with monoclonal antibodies using high throughput protein A chromatography. *Biotechnol. J.* 7, 1233–1241.
25. Li, X. et al. 2012. Self-packed filter plates: a good alternative for pre-packed filter plates for developing purification processes for therapeutic proteins. *Biotechnol. J.* 7, 1269--1276.
26. Sanaie, N. et al. 2012. Applying high-throughput methods to develop a purification process for a highly glycosylated protein. *Biotechnol. J.* 7, 1242–1255.
27. Jungbauer, A. 2013. Continuous downstream processing of biopharmaceuticals. *Trends Biotechnol.* 31 (8), 479–492.
28. Ströhlein, G. et al. 2007. Continuous processing: the multicolumn countercurrent solvent gradient purification process. *Biopharm Int.* February, S42–S48.
29. Carta, G. and Perez-Almodovar, E.X. 2010. Productivity considerations and design charts for biomolecule capture with periodic countercurrent adsorption systems. *Sep. Sci. Technol.* 45, 149–154.
30. Arve, B.H. and Liapis, A.I. 1988. Biospecific adsorption in fixed and periodic countercurrent beds. *Biotechnol. Bioeng.* 32, 616–627.
31. Mahajan, E. et al. 2012. Improving affinity chromatography resin efficiency using semi-continuous chromatography. *J. Chromatogr. A*, 1227, 154–162.
32. Warikoo, V. et al. 2012. Integrated continuous production of recombinant therapeutic proteins. *Biotechnol. Bioeng.* 109, 3018–3029.
33. Bryntesson, M. et al. 2011. Chromatography method. Assigned to GE Healthcare, US patent 7901581 B2.
34. Godawat, R. et al. 2012. Periodic counter-current chromatography—design and operational considerations for integrated and continuous purification of proteins *Biotechnol. J.* 7, 1496–1508.
35. Davis, B. D. et al. 1973. The enteric bacilli and similar Gram-negative bacteria. *Microbiology*, Ed. M McCarthy. Harper & Row, Hagerstown, MD, pp. 753–789.

36. Flemming, H. C. and Wingender, J. 2010. The biofilm matrix Nature reviews. *Microbiology*, 8, 623–633.
37. Godawat, R. 2013. Integrated and fully continuous processing of recombinant proteins—from cell culture media to purified drug substance. PREP, July 17–23, Boston, MA.
38. Langer, E. S. 2008. Biotech facilities average a batch failure every 40.6 weeks. *BioProcess Int.* 6, 28–30.
39. Lazar, V. 2011. Quorum sensing in biofilms—how to destroy the bacterial citadels or their cohesion/power? *Anaerobe*, 17, 280–285.
40. Xenopoulos, A. 2013. A new, integrated, continuous purification process template for monoclonal antibodies. Integrated continuous biomanufacturing conference. October 20–24, Barcelona, Spain.
41. Daszkowski, T. 2013. Continuous processing in biotech production: an alternative to a modern single use, batch facility? Integrated continuous biomanufacturing conference. October 20–24, Barcelona, Spain.

6

Computational Modeling in Bioprocess Development

Francis Insaidoo[1*], *Suvrajit Banerjee*[2*], *David Roush*[1], *and Steven Cramer*[2]

[1] Biologics and Vaccines, Process Development and Engineering, Merck Research Labs, Kenilworth, NJ, USA
[2] Department of Chemical and Biological Engineering, Rensselaer Polytechnic Institute, Troy, NY, USA

6.1 Linkage of Chromatographic Thermodynamics (Affinity, Kinetics, and Capacity)

Modeling of the adsorption behavior of proteins and the thermodynamics of process chromatography is covered in detail in the other chapters of this book (e.g., see Chater 2). Hence the focus of this chapter will be to detail the connections between biophysics and molecular modeling to the thermodynamics (e.g., ΔG) and site-specific interactions associated with protein–ligand interactions.

Retention in chromatography is a macroscopic measurement representing the average of multiple binding sites, conformations, and avidities [1] coupled with key transport properties (e.g., intra- and interparticle diffusion and dispersion) [2, 3]. For complex biological molecules, the linkage from the macroscopic retention to avidity is provided by the stoichiometric displacement model (SDM) [4]. The SDM allows one to calculate the net number of charges or effective avidity between a protein and an ion exchange (IEX) surface.

For example, an overview of the SDM (full derivation in Kopciewicz et al. [4]) is provided to illustrate the relationship between the capacity factor and the equilibrium constant for the binding reaction corrected for the phase ratio as follows:

$$K_\mathrm{b} = \frac{P_\mathrm{b}\left(aD_\mathrm{o}\right)^Z\left(bC_\mathrm{i}\right)^Z}{PC_\mathrm{i}D_\mathrm{b}^Z} \tag{6.1}$$

* Contributed equally to the authorship of the chapter.

where K_b is the binding constant, P_b is the bound protein concentration, P is the maximum binding capacity of the resin for the protein, a and b are parameters to account for valency and activity of the ions, D_o is the displacing ion concentration in the mobile phase, C_i is the counterion concentration, D_b is the concentration of the displacing ions on the surface, and Z is number of charges associated with the adsorption–desorption process.

The chromatographic distribution coefficient, K_i, or the ratio of the protein concentration in the stationary-phase C_s to that of the protein concentration in the mobile phase, C_m, can also be expressed as follows:

$$K_i = K_b\left(\frac{(D_{bi})^z}{(a)^Z(b)^Z(D_o)^Z(C_i)^Z}\right)(1-f) \quad (6.2)$$

where f is the fraction of the surface that is occupied by bound protein, C_i is counterion concentration, D_o is the displacing concentration of the mobile phase in moles/m^2, and a and b account for adjustment of valence resulting from the displacing power of an ion proportional to its ionic strength.

Equation 6.3 describes the relationship between k' and chromatographic distribution coefficient, K_i,

$$k' = K_i\left(\frac{A_s}{V_m}\right) \quad (6.3)$$

where A_s is the effective surface area of the media and V_m is the mobile-phase volume. Therefore, k' is a function of the overall interaction free energy, ΔG, for the protein binding event on the chromatographic surface. By measuring the change in k' as a function of elution buffer concentration, one can calculate the number of apparent interactions between the protein and ligand. This approach has been applied to protein binding to AEX resins [5–8]. The approach can also be applied to other protein–ligand interactions including immobilized metal affinity chromatography (IMAC)—Johnson and Arnold [9]. The approach was also extended to displacement chromatography with a specific application to a monolithic stationary phase operated in preparative mode [10]. Hence, in theory the SDM approach could be utilized to design specific binding sites for a given protein to achieve a specific interaction free energy (e.g., optimize the number of interactions). Although this approach combined with the use of site-directed mutants can be used to determine specific interaction sites between proteins and chromatographic media, the apparent number of interactions determined using this approach is often fractional making interpretation challenging.

Multiple parameters (e.g., mobile-phase composition, pH, temperature) can impact the free energy of binding including entropic (liberation of waters of hydration) and enthalpic (e.g., conformational changes) contributions. Thermodynamic components of the retention can be determined by measuring k' as function of temperature and performing an approximate van't Hoff analysis (plot of log k' vs. reciprocal of absolute temperature). A linear relationship allows one to determine the relative levels of entropic, ΔS, and enthalpic, ΔH, contributing to the protein binding to a chromatographic media [7, 11, 12]. A specific example of the use of a nonlinear van't Hoff analysis coupled with isothermal titration calorimetry (ITC) to estimate conformational changes associated with hydrophobic interaction chromatography (HIC) are presented in the literature [13, 14]. Some researchers have applied a combination of ITC and crystallography to discern the individual contributions of the various types of protein–ligand interactions (e.g., H bonds, electrostatics) [15].

Interaction energies between ligands and protein–protein complexes can be measured directly via ITC [16–20]. Subsequent researchers have developed techniques and methodologies for measuring competitive binding of ligands to proteins utilizing ITC [21]. ITC allows for measurement of an overall interaction free energy (ΔG) and an associated average dissociation constant or $k_{d,avg}$ for the multiple protein–ligand interactions. Computational models [22] can allow for direct calculation of interaction free energies (ΔG) for a specific motif and ligand from which dissociation constant (k_d) can be determined. Hence theoretically one should be able to rank order relative retention based on a combination of affinities and/or free energies associated with a wide range of interactions.

Multiple examples exist in the literature linking key specific parameters to retention in various systems for normal phase, reverse phase, and IEX systems based on hydrophobic, electrostatic interactions, etc. Commercially available packages (e.g., DryLab—Molnar Institute) allow for an estimation of relative retention based on a few key parameters. This approach has proven successful for relatively small molecules (e.g., steroids—Snyder et al. [23]) and for molecules with significantly different properties (e.g., pI, hydrophobicity).

Although the aforementioned techniques allow one to estimate relative retention in analytical (linear) chromatography conditions, process chromatography is often performed in a preparative mode (including nonlinear isotherms) where a significantly more complex set of interactions exist. For example, kinetics of adsorption may differ over the length of a column and within a chromatography bead. Some insights into the complexity of pore geometry and potential impact on adsorption phenomena have been determined via transmission electron microscopy (TEM) by Yao et al. [24]. Confocal microscopy [25–29] has provided some key insights into these phenomena and has been essential in understanding differences in uptake kinetics as a function of time, ligand density, and mobile-phase composition. However, this technique does not allow one to directly determine the individual contributions of affinity (linked to site-specific interaction energy) and avidity (overall impact of multiple ligand interactions from different regions of the molecule and binding heterogeneity to the chromatographic media at various ligand densities) to retention. Direct measurements of the impact of patches and charge distribution in the literature are relatively limited but clearly demonstrate the challenges of current experimental techniques [30].

In theory, one should be able to calculate the affinity of a ligand and a protein *in silico* and utilize that information to determine relative retention time in chromatography. Although significant progress has been made in that area, summarized in this chapter, gaps remain in a detailed understanding of the individual contributions of affinity, avidity, and conformational changes that may contribute to protein chromatographic retention.

6.2 Binding Maps and Coarse-Grained Modeling

Molecular dynamics (MD) simulations provide the most realistic estimate of free energies but are computationally expensive. Coarse-grained modeling can speed up this process by making certain appropriate approximations. Previous coarse-grained modeling efforts include simulation of protein-adsorbent electrostatic interactions. Initial attempts to study the relationship between chromatographic retention behavior and properties of the protein and the adsorbent utilized simplified representations of the protein, such as a charged surface [31] or a sphere having a centered charge equivalent to the net charge of the protein [32]. However, it has been shown experimentally that charge heterogeneity on the protein molecule plays a major role in chromatographic separation of similarly charged proteins [33], including the impact of changing a single residue [34] and in attractive interactions between protein and adsorbent with like charges [35, 36].

The key is to provide a linkage between the experimentally observed phenomena and the computational modeling efforts. For example, simulation of the interactions of an AEX surface with cytochrome b_5 revealed the presence of a preferred contact region in agreement with experimental data on site-directed mutants of cytochrome b_5 [37]. Modeling of complex polymer surfaces including the dynamics of interactions (e.g., chain collapse), associated counterions, and water structure is an area that requires significantly more investment to ensure accurate quantification of interaction energy.

The SDM [38] along with various extensions [39] can be applied for the analysis of the effects of salt concentration on protein retention. Due to lack of mechanistic details in such models, the model parameters, though defined with a physical basis, are difficult to interpret. Other models have been utilized to predict the binding of proteins to chromatographic surfaces. The charge regulated slab model was used to evaluate the salt dependence of proteins in cation exchange chromatography [40]. The Poisson–Boltzmann (PB) equation was used to efficiently predict the ionic strength dependence of retention [41]. A more rigorous form of the PB equation, that is, the nonlinear PB equation, was used for predicting the equilibrium constants for protein IEX in terms of protein size, ζ potentials, and electrolyte concentration [42]. Roth et al. [43] have examined the electrostatic contribution to the energy and entropy of protein adsorption, while Dismer and Hubbuch have used MD simulations to develop a protocol to find the preferred binding orientation for proteins on IEX surfaces [44].

Multiple force fields are available and are used extensively for evaluating the binding energy of a protein and a ligand or protein–surface interaction. To extend these to include bulk solvent, especially to calculate the free energies of binding, required a continuum solvent model such as that obtained through solving the PB equation [45]. Using this approach, it is possible to apply a coarse-grained representation of water and estimate free energies of binding following a continuum electrostatics approach. The hydrophobic component of the free energy can be estimated using surface area models [46] in concert with van der Waals interactions. Another force field is the Hummer's hydrophobic force field [47], which is based on the potential of mean force expansions and actually takes into consideration water structure around hydrophobic atoms.

For certain applications, e.g., flexible and non-globular proteins, it is desirable to also perform MD simulations and obtain free energies of binding using coarse-grained force fields. In these cases, net free energy is obtained as an ensemble average of the sum of electrostatic, hydrophobic, and dynamic components. The molecular mechanics/Poisson–Boltzmann solvent accessible surface area (MM/PBSA) force field [48] is best suited for this purpose and provides a method to approximate free energies from equilibrium MD simulations. The approach is computationally more efficient in certain applications than using traditional rigorous free energy calculation methods such as Thermodynamic Integration, Free Energy Pertubation, Umbrella Sampling or enhanced sampling methods like Replica Exchange, Metadynamics, etc. [49–55].

One specific example that has been studied extensively in the literature both theoretically and experimentally is IEX chromatography. The PB equation has to be solved in order to ascertain the electrostatic potential (EP) at each point in the system. The integration over all of space of the EP multiplied by the charge densities of atoms yields the electrostatic free energy of charging the system to its existing form starting from an uncharged system [56]. Hence, generation of the EP is a key step in determining electrostatic free energies. Further discussion of the PB equation is provided in Section 6.2.2. Here, the various software available for solving the PB equation is discussed. Delphi [57], GRASP [58], MEAD [59], UHBD [60], and PBEQ [61] module in CHARMM [62] are based on the finite difference method and are used to evaluate electrostatic free energies of charging biomolecules and free energy differences of biomolecules in bound and unbound states. The finite element method has also been applied to design the APBS software [63]. The APBS package has gained immense popularity in evaluating EPs and electrostatic free energies of biomolecules. The PARSE force field is very efficient in predicting EPs starting from accurately parameterized partial charges and is based on implicit solvent calculations [64], which result in some approximations. Gabdoulline and Wade devised a shorter method to calculate EP without having to do full PB calculations [65]. Their approach involves assignment of effective charges to protein atoms in a dielectric medium the same as that of the solvent and is implemented in their software SDA [66]. Such effective charges are capable of reproducing the EP around the protein. This approach effectively solves the dielectric–counterion boundary layer challenge at the protein–solvent interface with an approximation. For solvation free energy calculations, the free energy of solvation is generally quantified by the sum of two energy terms, for example, the polar and nonpolar solvation energies using an implicit solvation model [67, 68]—an approach that has been applied extensively in docking simulations [69, 70].

6.2.1 Protein–Surface Interaction Maps

6.2.1.1 Binding Maps and Preferred Binding Orientations

This example has been adapted from the work of Freed and Cramer [71] and shows how the combination of force fields can yield insights into the strength of binding of proteins on IEX chromatographic surfaces and the various preferred binding orientations of proteins on these surfaces. The force field developed by Gabdoulline and Wade [65] (mentioned in the previous section) was used in that work wherein the effective charges were assigned by fitting charges to an EP field generated by APBS. This force field was coupled with the Lennard–Jones interaction term for van der Waals interactions. Since electrostatics is the main driving force in IEX, only the aforementioned electrostatic and van der Waals interaction terms were considered to determine their ability to predict chromatographic separations.

Figure 6.1a shows how the various orientations of the protein are generated by rotating the protein (lysozyme) in two different axes of rotation φ and θ. Lysozyme is parameterized using the PARSE force field, and the IEX surface is modeled as composed of point discrete charges (−1 in this case) at fixed distance of separation from each other. The interaction energies are calculated using the coarse-grained force fields giving rise to two-dimensional binding maps as a function of the two angles of rotation and distance away from the IEX surface (Figure 6.1b). The preferred orientations of binding can be found by locating the favorable free energy hotspots on the binding map (Figure 6.1c).

6.2.1.2 Comparison with Chromatography Experiments

A similar procedure was followed with cold shock protein B (cspB) and its mutants as the one described in Section 6.3.1 with lysozyme. Chromatographic retention time differences between mutants of cspB and native cspB are shown in Figure 6.2. Also shown is the number of points on binding maps generated

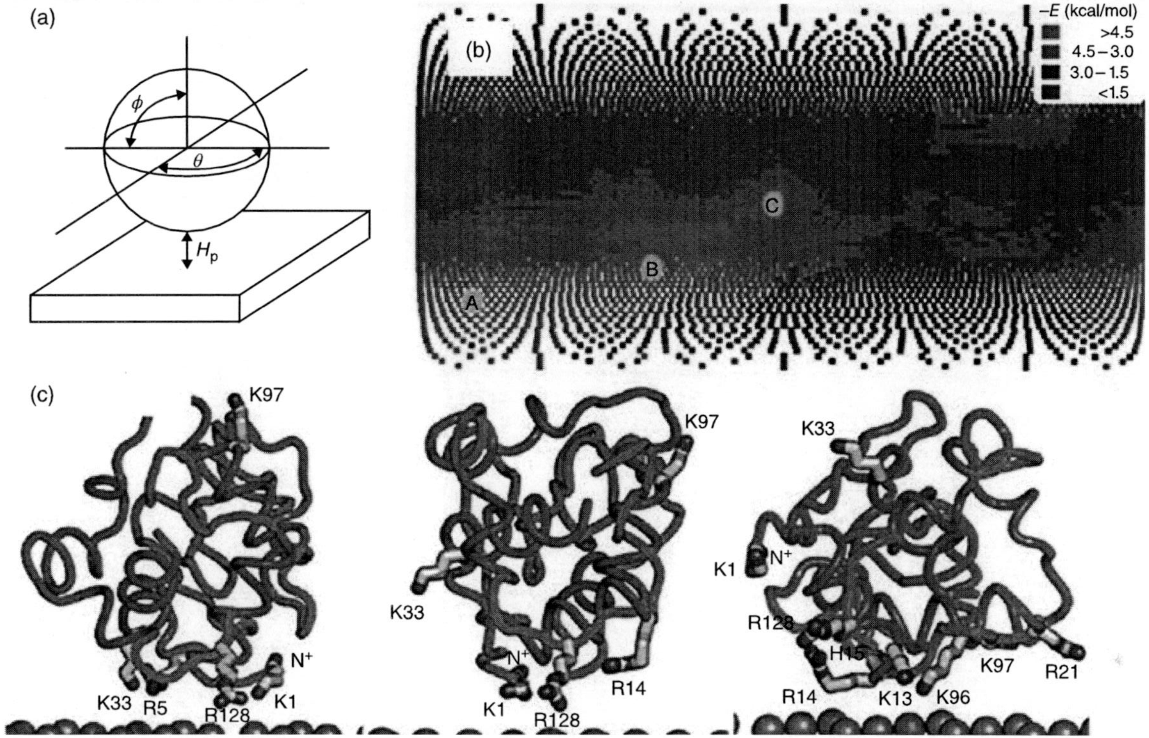

Figure 6.1 (a) Schematic diagram showing the three parameters, which define the orientation of a protein with respect to the chromatographic surface (φ, θ, and H_P). (b) Protein-surface interaction map of lysozyme on a model ion exchange surface. (c) Preferred binding orientations of lysozyme on the surface as determined from the hotspots on the interaction map in (b). *Source*: Freed and Cramer [71]. Reproduced with permission of American Chemical Society. (*See insert for color representation of the figure.*)

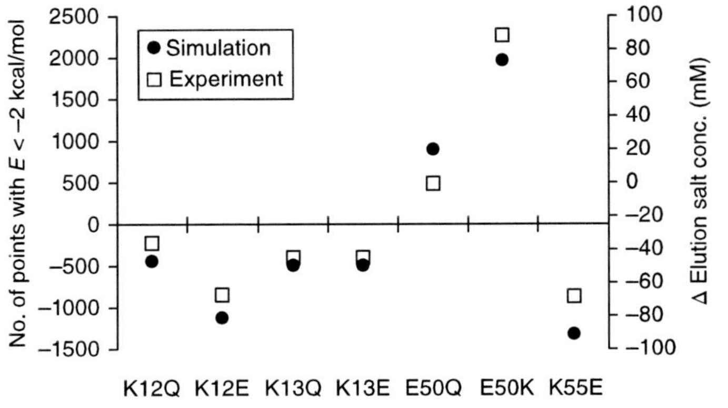

Figure 6.2 Comparison of retention data from chromatography and number of highly favorable orientations obtained from protein–surface interaction maps for various mutants of cold shock protein B on an ion exchange surface. *Source*: Freed and Cramer [71]. Reproduced with permission of American Chemical Society.

for these proteins with theoretical electrostatic free energy of binding less (more favorable) than −2 kcal/mol. This qualitative agreement between retention times and predicted binding free energies illustrates how coarse-grained modeling can help predict retention behavior in these systems and aid bioprocess development.

6.2.1.3 Effects of Salt and Inclusion of the Hydrophobic Effect

The effects of salt can be calculated by the PB equation, which utilizes the salt concentration of mobile ions in the solution as inputs. These calculations have the power to predict theoretical retention factors at various salt concentrations and can assist in designing separations. Hydrophobic effects can also be incorporated by an extra term in the free energy of binding calculations, the Hummer's hydrophobic force field [47]. The MM/PBSA method has also been reported in the literature to account for desolvation effects [72]. Inclusion of solvation terms can also facilitate prediction of protein binding in multimodal chromatography. A major impediment in such calculations is determining the synergy of electrostatic and hydrophobic interactions in the binding process. However, free energy calculations from all atom MD simulations (i.e., atomistic detail) with explicit counterions and water molecules can be used to calibrate such models.

6.2.2 Characterization of Chemical Heterogeneities on Protein Surfaces

In some cases (e.g., resolution of product-related variants), it is important to identify the regions on a protein surface, which can play a major role in binding. It has been shown previously that there is an appreciable effect of chemical heterogeneities on the surface of the protein in chromatographic separation processes [73]. It has been mentioned in Section 6.2.1 (Coarse-Grain modeling) that charge heterogeneity on the protein molecule plays a major role in chromatographic separation of similarly charged proteins [33] and attractive interactions between protein and adsorbent with like charges [35, 36]. Similarly, patches of hydrophobicity/philicity can influence the hydrophobic driving forces, which play an important role in colloidal and biological self-assembly phenomena [74–76]. Presence of hydrophobic patches on proteins can also influence assembly/protein aggregation to a significant extent [77]. The predominantly employed experimental method to directly measure hydrophobicity of a surface is through droplet contact angle measurements, which are not possible on nanoscopic objects like proteins [78]. Hence, special mapping tools or alternative indirect approaches are required to quantify electrostatic and hydrophobic patches on the surface of a protein molecule.

6.2.2.1 Electrostatic Patches

The principal method to quantify strength of electrostatic patches on a protein molecule is by calculating the EP around the protein in three-dimensional space and plotting the contour of this potential on the surface. The way to achieve this is by solving the PB equation. The PB equation [56] is as follows:

$$\vec{\nabla}\left[\varepsilon(\vec{r})\vec{\nabla}\psi(\vec{r})\right] = -4\pi\rho^{\mathrm{f}}(\vec{r}) - 4\pi\sum_i c_i^{\infty} z_i q \exp\frac{-z_i q\psi(\vec{r})}{k_{\mathrm{B}}T}\lambda(\vec{r}) \tag{6.4}$$

where $\varepsilon(\vec{r})$ is the dielectric constant, $\psi(\vec{r})$ is the EP, $\rho^{\mathrm{f}}(\vec{r})$ represents the molecular charges of the protein, c_i^{∞} is the concentration of ion i at an infinite distance from the protein, z_i is the valency of the ith ion, q is the proton charge, k_{B} is the Boltzmann constant, T is the absolute temperature, and $\lambda(\vec{r})$ describes accessibility to ions at point $\vec{r}$. The algorithms and software available to solve for the EP have been discussed earlier in this chapter. From Equation 6.4 it can be seen that it is possible to generate the EP field around the protein as a function of concentration and type of ions in solution, temperature, and different ionization states of side chains. The evaluated EP is then plotted on the solvent accessible surface of the protein to depict patches of positive and negative potentials.

6.2.2.2 Hydrophobic Patches

Hydrophobic patches can be identified by using the equilibrium ratio of concentration in oil to water (the partition coefficient) [79]. Values can be calculated based on the K_{p} data and can be used to generate hydropathy values to amino acids. However, hydropathy values alone cannot describe the regions highly prone to hydrophobic association. Hence, special techniques are available in literature for this purpose. A method based on local weighted averaging of hydropathy indices is available, known as spatial aggregation propensity (SAP) maps [80]. The SAP value of every atom is given as follows:

$$(\mathrm{SAP})_{\mathrm{atom}\ i} = \sum_{\substack{\text{Simulation}\\ \text{average}}} \left\{ \sum_{\substack{\text{Residues with at}\\ \text{least one side}\\ \text{chain atom}\\ \text{within } R\\ \text{from atom } i}} \left(\frac{\text{SAA of side chain atoms within radius } R}{\text{SAA of side chain atoms of fully exposed residue}} \times \text{residue hydrophobicity} \right) \right\} \tag{6.5}$$

SAA stands for solvent accessible area, and residue hydrophobicity is usually adapted from a hydropathy scale (e.g., see Ref. [81]) after suitable normalization. The cutoff distance R is a parameter to this equation. The more positive the SAP value, the more hydrophobic is the atom. There are other methods to map hydrophobicity on protein surface-based water density fluctuations [82], which have been shown to be the signature of hydrophobicity of a surface [75, 83–85]. Also, the chemical and topographical heterogeneities of a protein surface lead to ligand shape and size dependence of hydrophobic association. This property has also been used to map hydrophobicity [86].

6.2.2.3 Using Protein Surface Characterization Techniques to Explain Protein–Ligand Binding in NMR Spectroscopy

Srinivasan et al. [87] showed how two multimodal ligands with similar functional groups but different geometrical presentations of such groups interact very differently with the same protein. They used nanoparticle coated with self-assembled monolayers (SAMs) presenting these ligand headgroups, which were made to interact with ubiquitin.

Figure 6.3a and b show the binding of "Capto" and "Nuvia" multimodal cation exchangers to ubiquitin. Figure 6.3c and d show SAP and EP maps of ubiquitin, respectively. It was demonstrated that the Capto ligand that had greater surface exposure of a phenyl moiety along with a carboxylic acid moiety bound ubiquitin very specifically at the overlap of high hydrophobicity and positive EP. However, for the Nuvia ligand, the phenyl group is buried away from solvent, and only the carboxylic acid moiety is exposed to solvent. Hence, the interaction mainly consists of non-specific electrostatic interactions and extended over the entire region of the protein having positive EP. Thus, these surface mapping techniques can aid in fundamental understanding of protein–ligand binding processes.

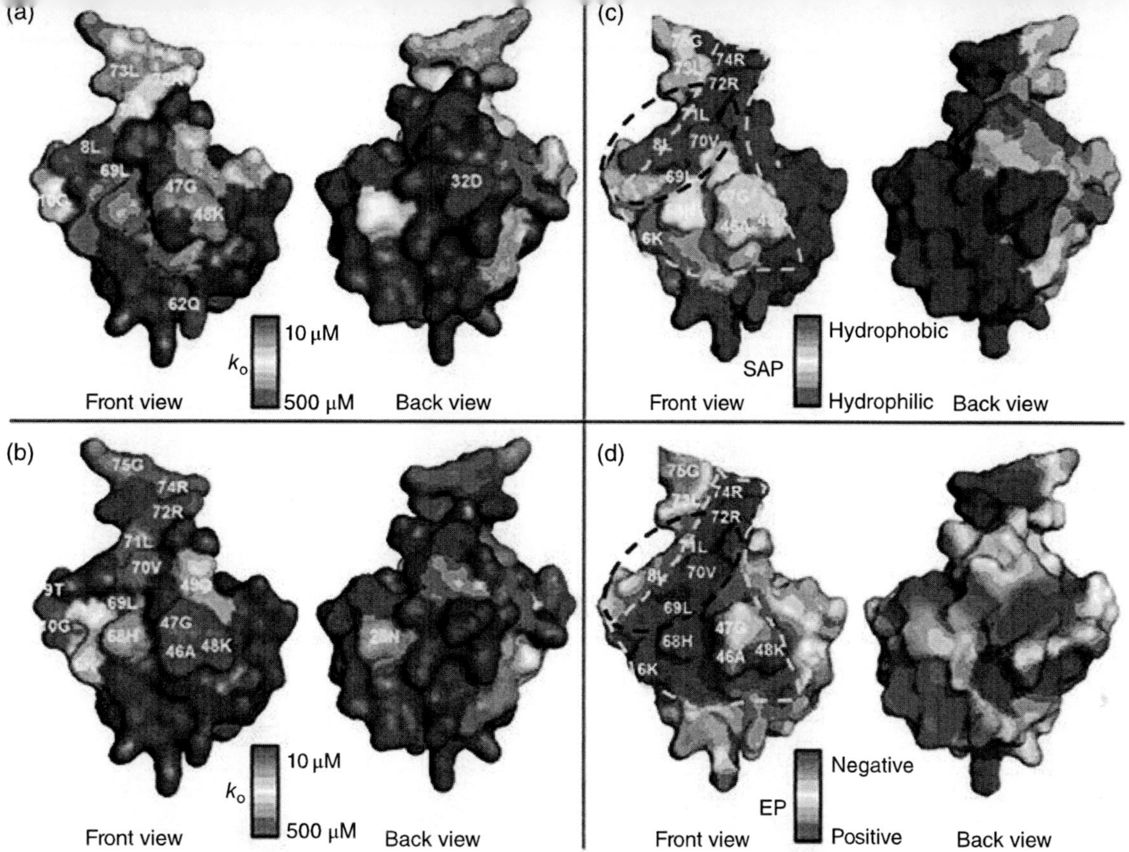

Figure 6.3 Multimodal ligand surface binding residues determined by NMR with color-coded dissociation constants for (a) "Capto ligand" surface and (b) "Nuvia ligand" surface. (c) SAP map and (d) EP map for ubiquitin. *Source:* Srinivasan et al. [87]. Reproduced with permission of American Physical Society Publications. (*See insert for color representation of the figure.*)

6.3 QSPR for Either Classification or Quantification Prediction

The *a priori* prediction of protein adsorption behavior has been a long-standing goal in several fields. A potentially powerful approach for predicting protein chromatographic behavior in a number of systems is the use of quantitative structure–property relationship (QSPR) modeling. In QSPR, the objective is to correlate the observed experimental data such as protein salt elution concentration, adsorption isotherm parameters, or free energy changes (the response-dependent variables) with specific physicochemical properties of the proteins (the independent variables). In order to carry this out, the protein properties are represented by a set of molecular property "descriptors" (or features), which are calculated from protein crystal structure data or appropriate homology models. These descriptors can include those available from commercial software packages such as molecular operating environment (MOE) [88] and Schrödinger [89] as well as more custom-designed protein surface descriptors developed by different investigators. Experimental chromatographic data is then obtained using an appropriate protein library. These protein libraries can include sets of commercially available proteins with a range of properties and/or more homologous sets of proteins. Machine-learning methods are then employed using the chromatographic data in concert with the molecular descriptors to generate a mathematical model to predict the response of each molecule in the dataset (e.g., elution salt concentration or retention volume) from the descriptors provided. There are many machine-learning techniques available (i.e., decision trees, random forest, neural networks): Multiple Linear Regression (MLR), Partial Least Square (PLS) regression, and Support Vector Models (SVM) regression methodologies that can be employed in these QSPR models [90, 91]. Models that are properly trained will have a high correlation with the actual response, will be verified to be robust to small changes in the dataset, and will not be overfitted in that they can differentiate false data from true response values. The SVM method of nonlinear regression has many important properties, including an effective avoidance of overfitting through capacity control and regularization [92]. These features improve its ability to build models using large numbers of molecular property descriptors with relatively few experimental results in the training set. An important aspect of QSPR model generation is the validation of model performance. Since QSPR models can be prone to overfitting, it is critical that they be verified to ensure that the model can predict the activity of new molecules and that it can reject false response values. The most rigorous validation technique is an external test set. In this method, a certain percentage of data is withheld during the training of the model and is then used to compare the accuracy of the model's predictions to the known response. While this is useful for large datasets where data can be removed without compromising the quality of the model, it is difficult to do when the dataset is limited or heterogeneous. Another key component in the generation of models can be the feature selection step, which will prune the number of molecular descriptors down to a smaller set that can adequately model the behavior of a given system. In addition to the predictive abilities of these models, the molecular descriptors selected during model generation can also help to identify potential mechanisms of chromatographic affinity and selectivity in these systems. Some of the advances that have occurred over the past fourteen years in the use of QSPR for predicting protein chromatographic behavior in a number of systems are summarized in Section 6.3.1.

6.3.1 QSPR Models for Ion Exchange Chromatography

Mazza et al. [93, 94] first presented QSPR models for the *a priori* prediction of protein retention in IEX systems. In that work, a genetic algorithm feature selection was combined with a PLS modeling approach, and the molecular descriptors were computed using protein crystal structures of commercially available proteins. The resulting QSPR models were well-correlated, and the predictive power of these models was demonstrated using proteins not included in the derivation of the models. Importantly, these models were able to predict selectivity reversals observed with two different stationary-phase materials. Song et al. [95] developed QSPR models for protein retention in AEX systems using topological, subdivided surface area, and transferable atom equivalent (TAE) electron-density-based descriptors computed using molecular connectivity patterns and crystal structure geometries. In addition, that work employed SVM regression to obtain predictive models by first using a sparse linear SVM for feature selection followed by an ensemble of nonlinear SVM regression models combined using bootstrap aggregation for the predictions. In addition, a visualization scheme (star plots) was employed to display the relative importance of each selected descriptor in the final set of models. Tugcu et al. [96] studied the effect of different salt types on protein retention and selectivity in AEX systems and developed predictive QSPR models. That work demonstrated that in addition to being predictive, proper interpretation of the QSPR models could provide insights into the particular elution trends for the proteins in the presence of these different salts.

Ladiwala et al. [97] expanded this line of investigation to the prediction of protein adsorption isotherm parameters and Gibbs free energy changes in IEX systems using the SVM modeling technique. Molecular descriptors selected during model generation were examined to gain insights into the important physicochemical factors influencing stoichiometry, equilibrium, steric effects, and binding affinity in protein IEX systems. The *a priori* prediction of protein isotherm parameters rather than the retention times (as was done in previous work) now enabled the coupling of QSPR models with column models to predict a range of IEX column operations. As proof of concept, this work showed that this multiscale modeling approach could successfully be used for directly predicting the chromatographic separation of a test set of proteins in the linear gradient mode of operation using the isotherm parameters obtained from the QSAR models. Malmquist et al. [98] developed a set of novel pH-dependent descriptors to predict protein retention in IEX chromatography at various pH values. Yang et al. [99] expanded this line of investigation by employing a refined set of pH-dependent predictions to estimate the steric mass action isotherm parameters of a binary test set of proteins at a pH not employed in the training set and by employing the predicted parameters in a column model to estimate the column behavior at this new pH.

6.3.2 QSPR Models for Hydrophobic Interaction Chromatography (HIC)

QSPR models have also been developed for predicting protein affinity and selectivity in HIC systems. Ladiwala et al. [100] examined the effect of stationary-phase resin chemistry and protein physicochemical properties on protein binding affinity in HIC using linear gradient chromatography and QSPR modeling. This work was carried out for a set of model proteins on four different HIC resins having different backbones and ligand chemistries. The retention data exhibited significant differences in protein binding affinity, not only across the phenyl and butyl ligand chemistries but also for the different backbone chemistries found in this series of resins. QSPR models were developed using traditional molecular descriptors based on protein crystal structure and primary sequence information as well as a set of new hydrophobicity descriptors based on the solvent accessible protein surface area. The results indicated that these models were successfully able to capture and predict the selectivity changes observed in these systems. Furthermore, the new descriptors resulted in physically interpretable models, which provided insights into the factors influencing protein affinity in these different HIC systems. Chen et al. [101] extended this work by developing a two-step methodology for the prediction of protein retention time in linear gradient HIC systems. QSPR models were generated to predict isocratic retention parameters and the resulting parameters then used to calculate protein gradient retention times. Chen et al. [102] further investigated the retention of proteins in the HIC systems using the preferential interaction theory in concert with QSPR modeling. In that work, models were successfully generated for predicting the water molecules released upon protein binding, and the selected descriptors provided insights into the protein physicochemical properties, which influenced protein affinity in a HIC system. Chen et al. [103] also developed classification models of HIC adsorption behavior in order to establish appropriate bioprocessing conditions without the need for high salt concentrations. Parallel batch screening experiments were carried out with 26 proteins using five resins with different ligand chemistries, ligand densities, and backbone chemistries to elucidate conditions for different salt HIC operations, and recursive partitioning (i.e., classification and regression tree (CART) analysis) was used to explore the interactions of specific descriptors on protein adsorption and to generate groups of proteins with similar adsorption behavior.

6.3.3 QSPR Models for Hydroxyapatite Chromatography

QSPR models have also been developed for hydroxyapatite chromatography (CHT) [104]. In that paper, chromatographic data was generated using a sodium chloride gradient in the presence of different phosphate concentrations, and the results were analyzed using various QSPR modeling techniques. Using a new set of molecular descriptors, QSPR classification models were generated to provide improved understanding and prediction of protein binding in CHT systems. An important advance from this work was the development of a new class of descriptors, which described synergistic binding with both metal chelation and cation exchange interactions on the ångström length scale and which was found to play a key role for protein binding in all of the models developed for CHT.

6.3.4 QSPR Models for Multimodal Chromatography

QSPR modeling has also been employed to correlate the physicochemical properties and structural components of multimodal IEX ligands with their ability to bind proteins under high salt conditions [105]. In that work, a set of molecular descriptors was calculated based on the structures of multimodal cation exchange ligands, and the resulting molecular descriptors were used to generate QSPR models for characterizing the ligand molecules and for predicting the protein chromatography behavior. Importantly, this work demonstrated that this approach was able to predict the performance of test ligands not included in the generation of the models and that the selected descriptors provided insight into the important ligand physicochemical and structural characteristics required for protein binding under high salt conditions. Interestingly, the results indicated that while aromatic rings played an important role in promoting protein binding under high salt conditions, moieties associated with intermediate hydrophobicity (e.g., aliphatic side chain) or the presence of hydrogen bond donors (e.g., NH and OH) tended to suppress the binding. This work demonstrated that ligand-based QSPR models may have utility for the *a priori* design of mixed-mode chromatographic systems for protein separations.

Hou et al. [106] developed QSPR models to predict protein retention in multimodal anion-exchange (AEX) systems. In that work linear gradient experiments were carried out with a multimodal anion exchanger, and the results were compared with those obtained with a traditional AEX material. Proteins were more strongly retained on the multimodal resin, and a different elution pattern was obtained as compared with AEX. Predictive QSPR models were developed, and the molecular descriptors selected during model generation suggested that multiple interaction modes contributed to the stronger retention and different elution patterns observed in the multimodal system. Finally, mobile-phase modifiers such as ethylene glycol, urea, and arginine were shown to be able to impart unique selectivity trends in multimodal chromatography.

Buyel et al. [107] demonstrated that QSPR can be used to facilitate process development for the chromatographic removal of tobacco host cell proteins. In that work, more than 100 tobacco proteins were identified by mass spectrometry, and their structures were reconstructed from X-ray crystallography, nuclear magnetic resonance (NMR) spectroscopy, and/or homology modeling data. The resulting three-dimensional models were used to calculate protein descriptors, and significant descriptors were selected to develop QSPR models for protein retention on anion, cation, and mixed-mode resins. The predicted protein retention profiles were compared with experimental results using crude tobacco protein extracts, and good agreement was attained for all resin systems.

6.4 All Atoms MD Simulations for Free Solution Studies and Surfaces

All atoms MD simulations provide a direct method to assess protein–ligand interactions [108]. The binding of ligands to proteins essentially occurs in a water-mediated environment. Hence, explicit representation of water molecules in a simulation helps in elucidating the exact mechanisms of binding (electrostatic, hydrophobic, hydrogen bonding, and van der Waals interactions) of various ligands to proteins. Although these simulations are computationally very expensive, the exponential growth in computing power over the past few decades have made it easier for such simulations to be performed for longer length and time scales. MD simulations also provide a key computational approach for comparing theory with experiments. These experimental techniques include NMR spectroscopy [109–111] and atomic force microscopy [112]. It is expected that ligand molecules would bind to proteins in a way that resembles drug binding to proteins, which is inherently characterized by strong interactions [113]. Also, the binding of fluid-phase modifiers to proteins can be nonspecific and hence is an example of weak interactions resembling those found in folding and unfolding of proteins [114]. Hence protein–ligand interactions in chromatography are characterized by a spectrum of possibilities ranging from strong to weak interactions.

Crystallographic and NMR structures have increased the number of structural snapshots available for most molecules. For novel biologics, homology modeling provides an additional source of structural information when empirical methods prove difficult. These static conformations and conformational changes report on structural changes occurring at a particular point in time. However, a comprehensive understanding of chromatographic processes relies on the dynamic processes that influence protein chromatographic ligand interaction. There is overwhelming evidence to support the fact that protein structure, dynamics, and function are strongly interconnected [115, 116]. To understand protein retention in chromatography is to understand the interplay between structure and dynamics and how these two parameters ultimately influence retention and separation. Structure–function relationships (e.g., static or constrained structures, use of average structures) can often be misleading because they suggest that one structural state is responsible for the resulting purification profile. Available results indicate that at the global minimum of a folded protein, there are still micro- or macroscopic conformational changes that account for function and interaction with ligands. These substates are sampled at different rates and are not always resolved in static structures [117].

Each binding interaction has an associated Gibbs free energy (ΔG), which consists of an enthalpic component (ΔH) and an entropic component (ΔS) as described in Equation 6.6. The basic understanding for protein interaction with a ligand can therefore be described as such: protein dynamics capture part of ΔS and static structure captures ΔH. In essence, these contributions are limited to not only the protein but the system being examined (solvent, ions, ionization energy, etc.). The formation of a complex (protein—(chromatographic affinity ligand)) can therefore result in either an increase or decrease in protein–ligand dynamics (entropy). These entropic effects (conformational changes/dynamics) are in most cases allowable states with energy barriers that can be overcome with time or thermal energy during an MD simulation [115]. In this section, a brief introduction to the fundamentals of MD simulations is presented first in order to make the reader familiar with its concepts followed by specific examples:

$$\Delta G = \Delta H - T\Delta S \tag{6.6}$$

6.4.1 Fundamentals about Molecular Dynamics Simulation

The potential energy of a system can be represented by a function, which includes both bonded and nonbonded interactions. Bonded interactions include stretching of bonds, altering of angles between bonds, and rotating a single bond (changing of dihedrals). Nonbonded interactions include intermolecular interactions, that is, electrostatic (Coulomb potential) and van der Waals interactions (usually Lennard–Jones potential). The type of function used for describing the potential energy along with the parameters in that function constitutes the force field. The potential energy changes with respect to equilibrium values of bonded interactions in the force field and with respect to distance of separation between two atoms for non-bonded interactions. It is important to note that to calculate long-range electrostatic interactions, the particle mesh Ewald method [118] (based on Ewald summation) is generally used. This approach is employed since the summation of the Coulomb potential energy function becomes a conditionally convergent series whose sum depends on the order of terms considered. It is also to be noted that parameters in force fields are mostly empirical because they are derived from different types of experimental data. Some of the most popular force fields in use are AMBER [119], CHARMM [62], and GROMOS [120]. This linkage of experimental data to parameterization is essential to ensure that the outcome of the MD is physically relevant.

The force on each atom is obtained by differentiating the potential energy function. Once the forces are obtained, the equations of motion are integrated to obtain the new position of the atoms. Thus, it is possible to track the evolution of the system with time. The first few nanoseconds are treated as equilibration of the system, and the remaining time is considered as the production run, which is used for analysis. Algorithms used to integrate the equations of motion include the Verlet, velocity Verlet, and LeapFrog algorithms. It is also very important to simulate under proper conditions of temperature and pressure. For this purpose, model barostats and thermostats are available. The most commonly used models for thermostats are Berendsen [121], Nosé–Hoover [122, 123], Langevin, and velocity rescaling thermostats. Berendsen and Parrinello–Rahman barostats [124] are two model barostats that are popularly used. A well-established rule for simulating biomolecular systems is to use Berendsen thermostat/barostat for relaxation and Nosé–Hoover thermostat/Parrinello–Rahman barostat for production simulations. Packages such as GROMACS [125–127], NAMD [128], and LAMMPS [129] have all these algorithms incorporated in them and are used extensively for performing MD simulations.

6.4.2 Protein Dynamics and Time Scale of Molecular Motion

The crystallographic coordinates of a protein describe the atomic x, y, and z coordinates at a particular time (t) in the energy landscape. Following the change in coordinates with time reveals the energy landscape and provides information on conformational changes required for activity. Since each atom in a protein complex can move in any direction (x, y, and z), for each microscopic change in time (δt), the energy landscape hence becomes multidimensional. The resulting structures can conceivably present different surface properties or binding sites to the ligand of interest. For a given protein molecule to go from one energy state or well (local minima) to another, it has to overcome an energy barrier, which varies depending on the required conformational change. Protein conformational changes are therefore characterized by the timescale of the fluctuations (kinetic) and the amplitude/direction of the fluctuation (structural). Each energy landscape is a function of a particular set of individual conditions of temperature, pressure, and solvent conditions. By manipulating these conditions in process development, the relative energy barriers and populations can be changed, which would influence the separation profile [115, 130].

Protein motion can be grouped in two main classes: collective motions (slow timescale motions) and local fluctuations (fast timescale motions). Slow timescale motions are therefore motions that occur between two kinetically distinct states. Most biological processes fall within this timescale of dynamics, including enzyme catalysis, signal transduction, and binding. Within each of the energy wells sampled on the slow timescale, proteins are not static, and the motions within these wells constitute fast timescale motion [115, 116, 130, 131]. Dynamics within the energy wells are referred to as fast timescale. These consist of a large ensemble of structurally similar states. There are two main

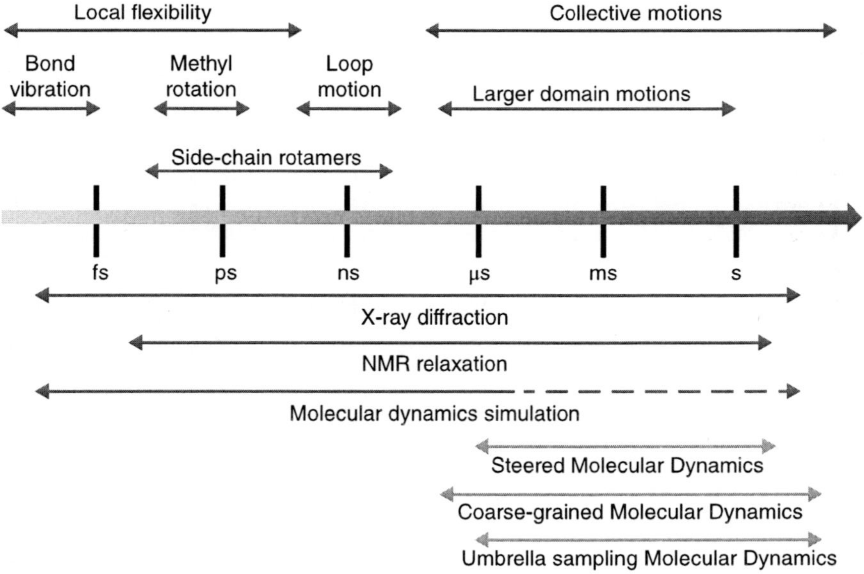

Figure 6.4 Currently available molecular dynamics methodologies and time scale of protein motions that can be detected using such an approach. Most of these methods impose assumptions and simplification of protein dynamics. Hence appropriate data processing is needed to distinguish the different modes of motion sampled and how the results replicate realistic protein dynamic properties. (*See insert for color representation of the figure.*)

classes of motion within this regime: side-chain motion (normally in the picoseconds time scale) and loop motions (normally in the nanosecond time scale). These types of motions are dictated by the structure of the protein molecule. For instance, a side chain in the core or part of a secondary structure is more likely to be restrained than the same residue in a loop region [115, 116, 130, 131]. A general classification of protein motions and experimental procedures aimed at sampling these motions is outlined in Figure 6.4. These methods can be grouped into two main classes: high-resolution (atomic-resolution) methods and low-resolution methods (e.g., NMR).

6.4.3 Representations of Proteins and Ligands

Proteins structures are readily available from the protein data bank (PDB: over 10 000 structures, March 2015). For proteins whose structure has not been determined, homology modeling can be performed leveraging similar (in sequence space) template structures from the PDB. Homology models require using parameters from a particular force field. The AMBER force field in conjunction with the TIP3P water model [132] is a typical basis for simulation of biomacromolecules, since typically the initial homology model requires some extensive relaxation or dynamics approach to generate a relevant model. The TIP3P model of water has been optimized in accordance with the AMBER force field and is highly effective in simulating protein structural changes accurately, whereas, for simulating protein ligand interactions, it is more important to use a water model, which can effectively reproduce the surface tension of water. The SPC/E [133] water model can achieve this to a higher accuracy though water models are the subject of continued research.

Ligand parameters are not readily available in literature. However, parameters for its constituent groups are available in various force fields. To be internally consistent during the course of a simulation, it is advisable to construct ligands using force field parameters that are the same as those used for constructing the topology of the protein.

6.4.4 Effect of Protein Amino Acid Mutation and Dynamics on Affinity Ligand Binding

Previous studies have demonstrated that a single amino acid change can significantly influence the dynamics and function of proteins [134, 135]. For instance, complementarity-determining region (CDR) loops of antibodies and T-cell receptors (TCR) are highly variable, and a single amino acid change in a CDR loop sequence can completely abolish the binding affinities to a cognate ligand or lead to cross-reactivity [136–138]. These functional and biophysical property changes result from dynamic and conformational changes upon amino acid mutations and/or ligand interaction. The dynamic properties of proteins can be either enhanced or attenuated by the buffer conditions, interacting ligands, or interactions with other proteins. These dynamical property differences can result in proteins adopting different structural conformations or functional states, leading to increased solubility, crystallization, precipitation, and function [139–143].

In process development, the binding and separation of proteins on affinity, IEX, or mixed-mode resins are governed by the same principles that influence the biochemical and biophysical properties described previously. Dynamics of chromatographic ligands and biologics to be separated thus bind to each other based on conformation and can be influenced by mutations as demonstrated by Insaidoo et al. [144]. Structure and dynamic properties also influence non-chromatographic purification and crystallization as an alternative to standard chromatographic purification. The differences in the binding affinity between the native protein and mutant variants can influence chromatographic retention and separation.

6.4.5 Protein–Ligand Docking and Molecular Dynamics Simulation

There are many experimental methods for determining the binding affinities of protein and ligand. Some of the more quantitative methods include surface plasmon resonance (SPR) [145], ITC [146], and NMR spectroscopy [147]. Yet, for rapid screening of potential protein–ligand interactions, molecular docking provides an excellent opportunity to screen ligands in a high-throughput platform. In drug discovery, computer-aided drug design (CADD) is used to dock small molecules into macromolecules to identify potential drug candidates. With improved algorithms and lower equipment cost, the use of CADD in the pharmaceutical industry is increasing [148–150]. As of 2014, there were about 130 different software programs and Web servers for docking and virtual screening [151]. In structure-based drug design, CADD is an essential component. Since the same principle of protein–ligand interaction is at play in process development, the use of CADD can enhance and facilitate the selection and optimization of affinity ligands for chromatographic separation. However, reported binding affinities from these computational methods should be interpreted and utilized with caution, unless proper care is taken to validate the results [152, 153].

An effective way to evaluate docking scores, in addition to experimental data, is to use MD simulation to refine the ranked candidates from the docking studies.

MD simulation provides information of molecular motion at different timescales and allows for visualization of conformational changes responsible for protein–protein and protein–ligand interaction [154]. Another advantage of MD simulation is that correlated motions within a protein can be observed. For fast timescale motion, which can influence protein–ligand interaction,

MD simulation is ideal. This is mainly due to the relatively short simulation time required and the lower energy barriers, allowing conformations to be sampled easily and bonds to be formed and broken. For process development and drug design, the binding affinities of a bound ligand can be observed and calculated as a function of time and conformational change. In that instance, the strength of the interaction (ΔG) between the ligand and protein can be evaluated and applied to process development.

With advances in computational power, improved force fields [49, 53] and new algorithms [49, 155], longer timescale (microsecond to millisecond and beyond) can now be sampled. Most of these longer timescale conformational changes can result in changes in the surface properties of molecules that are recognized by chromatographic ligands. Classical MD simulation cannot sample these conformational changes, and hence new methods with simplified force fields have been developed. Some of these methods include normal mode analysis [52], Gaussian network models [50], and replica exchange MD [55]. Another way to sample these slow timescale events is to accelerate or steer the simulation in a desired direction [51, 156, 157] or via use of umbrella sampling [54] (Figure 6.4). The combined effect of an increase in PDB structures and improvement in MD simulation methodologies has advanced the use of these methods, resulting in more accurate modeling results. However, a comprehensive interpretation is only achieved by validating MD simulation results with experimental data [49, 155, 158].

6.4.6 Free Ligand Simulations

A computationally efficient approach to probe protein–ligand interactions is to employ the so-called free ligand simulations. In these simulations the ligands are clipped off from their base matrices and dispersed freely in water. The protein is solvated in this aqueous solution of ligands, and the system is allowed to evolve with time to ascertain how the ligands interact with the protein. A potential challenge is that the free ligand can adopt different conformations in solution than on a constrained surface. Two very common methods that are available to quantify the strength of binding of ligands to proteins are (i) preferential interaction coefficients [159, 160] and (ii) radial distribution functions [159]. However, both methods provide an overall average free energy of binding and do not help to elucidate the binding hotspots on a protein or determine the strength of binding of the ligand to such hotspots relative to the nonbinding regions on the protein surface, which in turn can be used to define selectivity. In this section, two methods are presented, which have been used to overcome the problems presented by the preferential interaction coefficient and the radial distribution function based approaches: (i) a method based on enumeration of ligands near heavy atoms of proteins, that is, determining the probability of binding of the ligand to each heavy atom of the protein and (ii) a method based on spherical harmonics expansions, which provides an estimate of the density of ligands directly at or near the surface of the protein. An example of this approach is provided in Figure 6.5, which describes the interaction of a mixed-mode ligand (MMC) with a model protein ubiquitin.

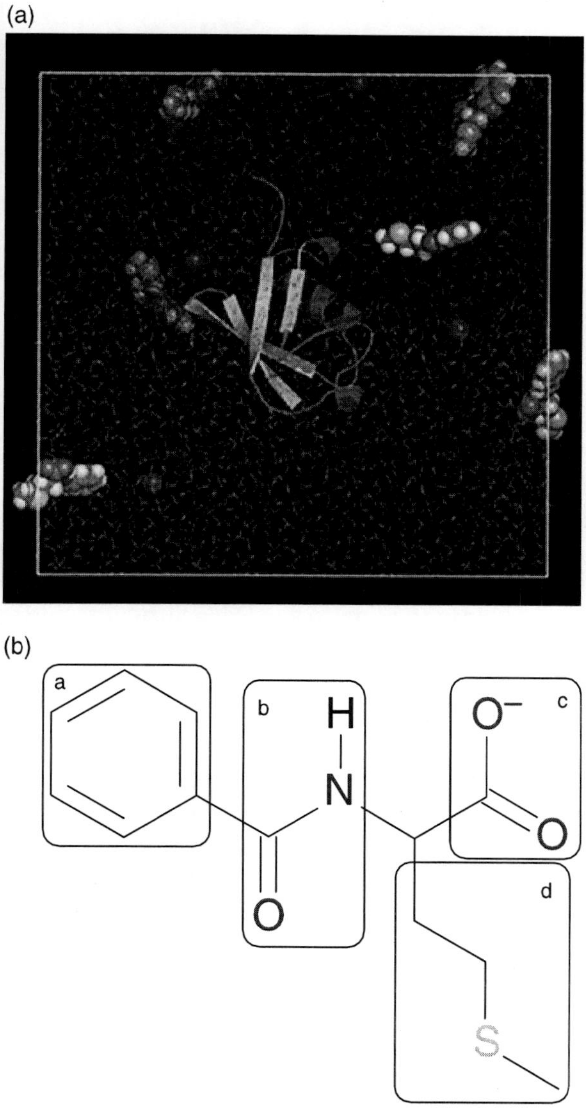

Figure 6.5 (a) MD snapshot of ubiquitin solvated in an aqueous solution of free Capto MMC ligands. (b) Various moieties present on Capto MMC. *Source*: Freed et al. [22]. Reproduced with permission of American Physical Society Publications. (*See insert for color representation of the figure.*)

Figure 6.5a shows a snapshot of an MD simulation of ubiquitin in a well-dispersed solution of Capto MMC multimodal ligands in water. Figure 6.5b shows the various moieties on the ligand (a: phenyl group, b: amide group, c: carboxylic acid group, d: thiol group), which represents a subset of potential types of interactions (electrostatic, hydrophobic, hydrogen bonding, π–π interactions) that could occur with a protein. The timescale of these simulations is determined by the diffusivity of the ligands in solution. Usually about 50 ns are sufficient to equilibrate the system, and timescales of the order of 10^2 ns are used for production runs. Cubic or rectangular boxes with periodic boundary conditions are generally used [161]. To ensure that pH is correctly considered, the protonation states of all ionizable side chains have to be taken care of carefully.

6.4.7 Analysis Techniques for Free Ligand Simulations

6.4.7.1 Cutoff-Based Probability of Binding

In this technique, distances between heavy atoms of ligands and proteins are calculated [22]. The entire ligand is considered to be "bound" if the distance between subunit center of mass (COM) and the nearest protein heavy atom (non-hydrogen) is less than a cutoff distance, which is typically around 5 Å. For distances less than 5 Å, there is usually no intervening water molecule between the two interacting groups. The binding of the entire ligand or its constituent moieties can be tracked using this approach. This process is performed over the entire simulation, and analysis of the average number of atoms of ligands/moieties on ligands is performed. Normalizing these counts with the total count allows one to determine the probability of binding for each heavy atom of the protein. When plotted using a color code, this technique yields a map of various hotspots where the ligand had bound to the protein.

Figure 6.6 shows the different regions of the protein ubiquitin that interacted with the entire ligand as well as with different moieties of the ligand. It can be seen clearly that although the ligand seems to be bound around a particular area on one face of the protein, the individual moieties of the ligand have particular regions on the protein surface, which they interrogate. Such information is very useful to gain fundamental insights into how the same moieties on a ligand interrogate different proteins depending on the chemical and topological heterogeneities of their surfaces.

6.4.7.2 Spherical Harmonics Expansion Approach to Quantify Distribution of Ligands

Spherical harmonics expansion is widely used in physics to rationally reduce the complexity of a problem [162, 163], for example, in spectroscopy [22], computer graphics [164], and medical imaging [165]. Hence, it can be leveraged to support

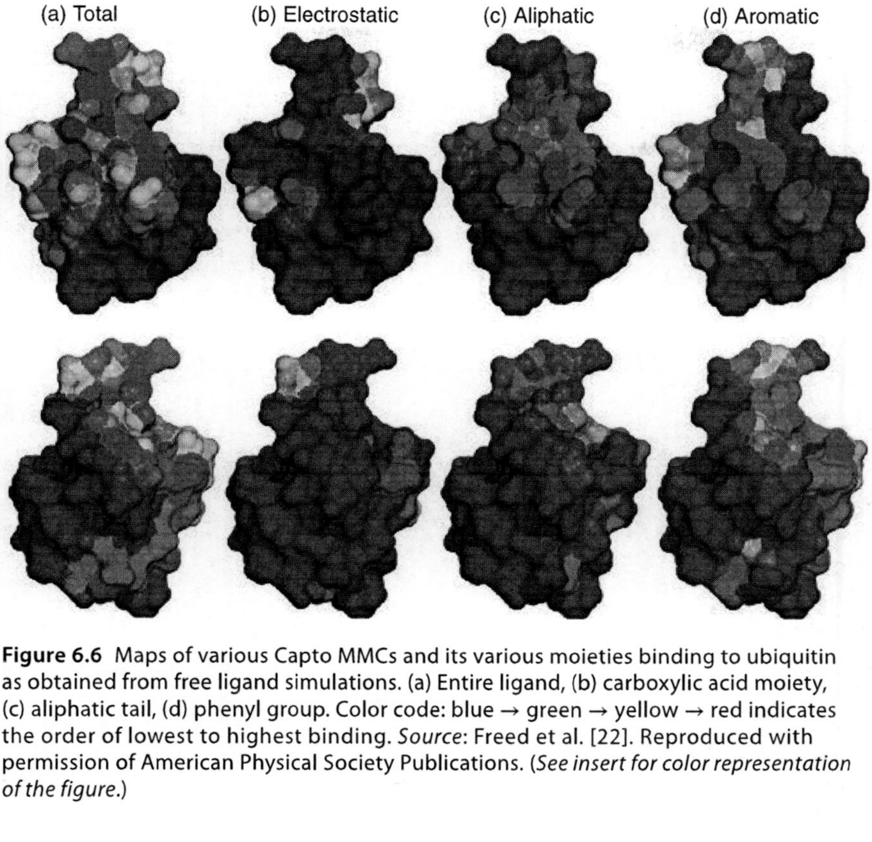

Figure 6.6 Maps of various Capto MMCs and its various moieties binding to ubiquitin as obtained from free ligand simulations. (a) Entire ligand, (b) carboxylic acid moiety, (c) aliphatic tail, (d) phenyl group. Color code: blue → green → yellow → red indicates the order of lowest to highest binding. *Source*: Freed et al. [22]. Reproduced with permission of American Physical Society Publications. (*See insert for color representation of the figure.*)

the base mapping tool for visualization and quantitative comparison of binding hotspots of ligands on protein surfaces. The basic foundation of this approach is explained as follows (for more details see Ref. [166]).

The three-dimensional density profile, $\rho(r)$, of the molecules of interest is expressed as an expansion in spherical harmonics truncated at the Nth order and described in Equations 6.7–6.9:

$$\frac{\rho(r)}{\rho_0} = \sum_{n=0}^{N} \sum_{m=0}^{n} \left[C_{nm}(r) Y_{nm}^{e}(\theta,\phi) + S_{nm}(r) Y_{nm}^{o}(\theta,\phi) \right] \tag{6.7}$$

Using

$$\rho(r) = \sum_{i=1}^{N_{\text{atoms}}} \delta(r - r_i) \tag{6.8}$$

And the orthonormality relation

$$\int_0^{2\pi}\int_0^{\pi}\left[Y_{\mathrm{nm}}^{e\text{ or }o}(\theta,\phi)\right]^2 \sin\theta d\theta d\phi = \frac{4\pi}{2(2n+1)}\frac{(n+m)!}{(n-m)!}(1\pm\delta_{m0}) \tag{6.9}$$

Coefficients $C_{\mathrm{nm}}(r)$ and $S_{\mathrm{nm}}(r)$ for atoms of interest are obtained from simulation trajectories and substituted into Equation 6.7 yielding a map of the ligand density distribution in three-dimensional space around the protein.

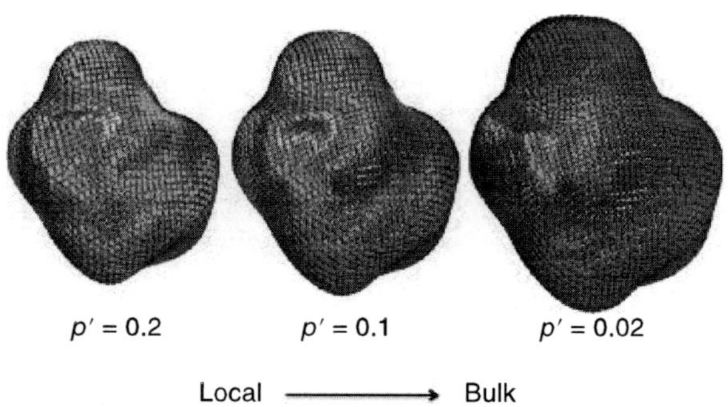

Figure 6.7 Spatial binding maps of Capto MMC at various distances away from the surface of ubiquitin. Maps are based on density of ligand (ρ') as evaluated by spherical harmonics analysis and normalized by bulk density of ligands. *Source*: Parimal et al. [162]. Reproduced with permission of American Chemical Society. (*See insert for color representation of the figure.*)

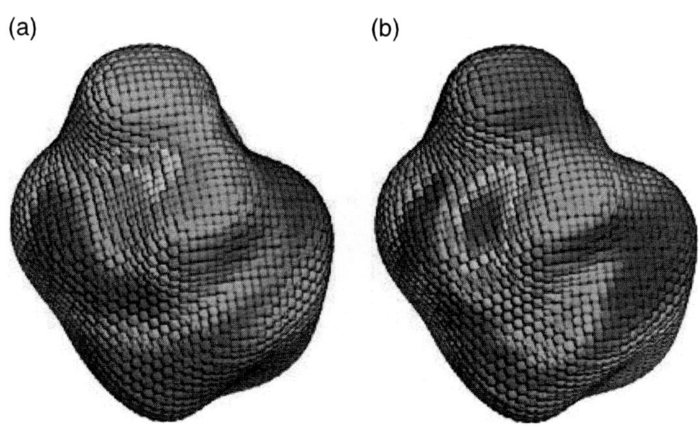

Figure 6.8 Differences in binding of benzene subgroup on Capto MMC (a) and benzene molecule (b) dispersed freely in aqueous solution with ubiquitin. *Source*: Parimal et al. [162]. Reproduced with permission of American Chemical Society. (*See insert for color representation of the figure.*)

Figure 6.7 shows the difference in ligand distribution around the protein (ubiquitin–Capto MMC system) with distance away from the protein surface. This is a powerful method to assess which moieties of the ligand contact the protein upon binding and shed light on differences in binding behavior and can also be performed with individual moieties of the ligand.

Figure 6.8 shows how the benzene group of Capto MMC prefers a different region for binding than just a benzene molecule (both ascertained from spherical harmonics analysis). This shows interesting behavior and demonstrates how the presence of additional groups in a ligand can alter the way a group interacts with a protein surface.

6.4.8 Comparisons of Free Ligand Simulations with Experiments

As discussed in detail previously in this chapter, these simulations can be directly compared with results from various biophysical experiments. Figure 6.9 shows the probability of binding of Capto MMC ligands to ubiquitin using method described in Section 6.4.7.1 previously [22]. The binding hotspots as identified by MD simulations are shown in rainbow scheme with violet to red indicating low to high binding. The overall agreement between the two methods (spherical harmonics and cutoff based, as described previously) is highly encouraging and demonstrates how experimental and computational

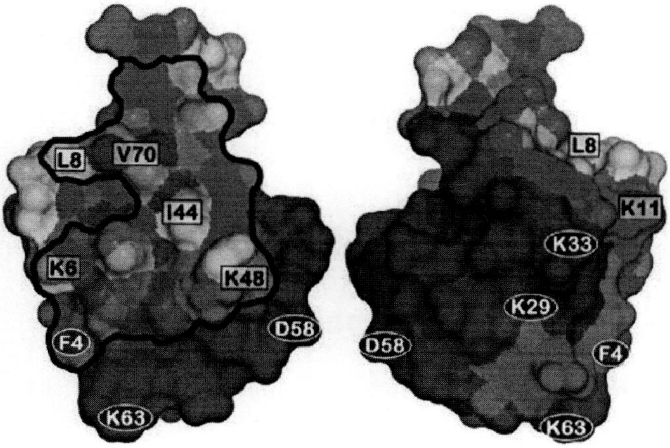

Figure 6.9 Comparison of binding hotspots as determined from MD simulations (color coding same as in Figure 6.6) and NMR experiments (region within the dark black line showed maximum binding in NMR). Residues marked in boxes showed changes in retention time in chromatographic experiments upon mutation, whereas those in ovals did not. *Source*: Freed et al. [22]. Reproduced with permission of American Physical Society Publications. (*See insert for color representation of the figure.*)

methods can complement one another. These results suggest that there is a preferred binding face of free Capto MMC ligands to ubiquitin. Such noteworthy agreement shows that MD simulations can greatly aid in fundamental understanding of protein ligand binding without having to perform a large number of biophysical and chromatography experiments, and only targeted experiments are necessary to validate key findings from MD simulations. The boxed residues are the ones that affected retention time upon mutation, whereas the ones in ovals did not affect [167], and the region within the black line indicates the binding region identified by NMR experiments [167].

6.4.9 Surface Simulations

The biggest challenge of computational modeling of chromatographic systems is to represent the stationary-phase surface having ligands immobilized on them. However, pore structure, curvature, and other effects make it difficult to actually create a representative surface. Controlled surfaces may overcome this difficulty. Specifically, the use of SAMs aids in studying interactions of proteins with surfaces presenting chromatographic ligand head groups [87]. The construction of SAMs has been done *in silico* for various fundamental studies using MD simulations [168–170]. Similar methods can be used to create SAMs presenting chromatographic ligands. Special simulation techniques such as umbrella sampling or metadynamics can be used to calculate the free energy of binding of specific faces of proteins to the ligand surface. This approach enables one to calculate face-specific binding of proteins to chromatographic ligand surfaces and also comparison of overall strengths of binding of various proteins to a surface by performing ensemble averaging overall binding orientations of the protein. These simulations are an excellent computational method to complement single-molecule force spectroscopy experiments, which also provide free energies of binding of proteins to surfaces [112].

6.5 Ensemble Average and Comparison of Binding of Different Proteins in Chromatographic Systems

Free energies obtained from the various simulation methods mentioned in Sections 6.3 and 6.4 are functions of various reaction coordinates (also referred to as order parameters or degrees of freedom). For instance, in coarse-grained protein surface free energy calculation methods, the calculated free energies are a function of the two angles of orientation and the distance away from the surface, whereas, in free ligand simulations, the free energy of binding is resolved in three-dimensional space around the protein (note: spherical harmonics analysis yields normalized densities of ligands around proteins.

The natural logarithm of these normalized densities is proportional to the free energy of binding at each location around the protein). From a statistical standpoint, the probability distribution of protein conformations on the surface correlates with the corresponding free energies of interaction [171]. Ensemble averaging these free energies' overall reaction coordinate yields such an average free energy of binding. In general, the weighting factor for a given orientation can be calculated according to the Boltzmann distribution, which describes the probability of a certain orientation depending on its energy,

$$p(r,\theta,\phi)=\frac{1}{Z}e^{-\beta\Delta G(r,\theta,\phi)} \tag{6.10}$$

where $p(r, \theta, \phi)$ is the probability for a given position and orientation. Z is the partition function, ΔG is the free energy associated with this particular position and orientation (evaluated theoretically from coarse-grained or MD simulations) and $\beta = 1/k_B T$ where k_B is the Boltzmann constant, and T is the absolute temperature. The partition function is expressed as

$$Z=\sum e^{-\beta\Delta G(r,\theta,\phi)} \tag{6.11}$$

Then the ensemble averaged free energy of binding $\langle\Delta G\rangle$ would be the sum of the weighted electrostatic interaction energies' overall possible positions and orientations. That is,

$$\langle\Delta G\rangle=\sum p(r,\theta,\phi)\Delta G(r,\theta,\phi) \tag{6.12}$$

This $\langle\Delta G\rangle$ can be calculated for various proteins interacting with chromatographic ligands using techniques mentioned in Sections 6.3 and 6.4 under different pH, temperatures, and fluid-phase conditions. Evaluation of such overall binding free energies would greatly facilitate design of chromatographic processes from *in silico* approaches as well as help explain retention behavior in these systems [172, 173].

6.6 Antibody Homology Modeling and Bioprocess Development

Antibodies have made a significant addition to the field of biologics discovery and development. The significant advantage for using antibodies as lead candidates in this class of biological therapeutics is the ability for these molecules to bind to a vast array of clinical targets and elicit significant therapeutic efficacy. In humans, there are five main isotypes of antibodies (IgA, IgD, IgE, IgG, and IgM) [174–176], of which only one (IgG) has been developed extensively for clinical use. IgG has four subtypes (IgG1, IgG2, IgG3, and IgG4) with different structural features and biophysical properties [177]. The inherent diversity between the different subtypes of IgG presents unique challenge for bioprocess development and formulation [178]. To advance and improve the efficiency of process development, the structure of these individual molecules and its impact on purification becomes critical.

IgG is composed of two polypeptide chains called light chain and heavy chains. The two chains fold into a general "Y"-shaped structure shown in Figure 6.10 with two fragment antigen-binding (Fab) and fragment crystallizable (Fc) regions. The Fab regions are connected to the Fc region by a linker composed of amino acids from the heavy chain and disulfide bonds. The Fab is a heterodimer consisting of light chain and heavy chain amino acids that fold into two domains (variable and constant domain) and linked by a disulfide bond. The variable domain has six flexible loops referred to as the CDR loops that are used for ligand recognition (antigen-binding site) and are responsible for the effector functions of most antibodies. Further, the CDR loops are highly variable, thus allowing for the diversity in ligand recognition and the biophysical properties of the molecule. The Fc region folds into two constant domains and consists of heavy chain amino acids only. Functionally, the Fc domain modu-

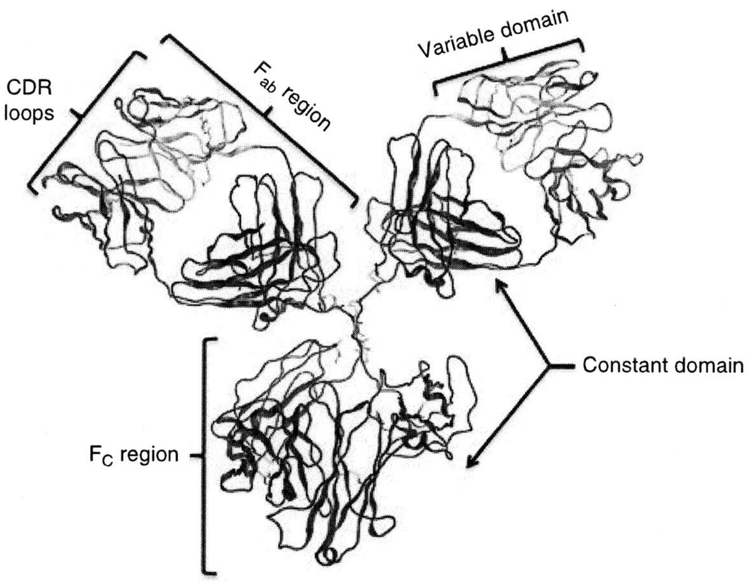

Figure 6.10 Structure of IgG_1 antibody illustrating the structural features. The Fab region is composed of both light and heavy chain sequence. The CDR loops are in purple, red, and brown. The Fc region is in blue and linked to the Fab with the purple linker. Disulfide bonds are shown in yellow. The ribbon diagram for IgG_1 was generated from PDB file 1IGY [179]. (*See insert for color representation of the figure.*)

lates immune cell activity by binding to a specific class of Fc receptors and other immune molecules such as complement proteins. In bioprocess development, the Fc region plays a critical role in the capturing step of antibody purification by binding to protein-A resin.

6.6.1 Molecular Modeling of Antibody Structures

Currently, there are about five full chain antibody structures in the PDB. In contrast, there are about 1000 Fc structures and 1700 Fab structures published in the PDB. The aforementioned statistics indicate that structure determination of full chain antibody is a challenging problem. This presents a significant bottleneck when multiple full chain antibody structures are needed to inform process development using a combination of empirical and computational modeling. The three-dimensional structure of an antibody provides relevant information about surface properties, liability regions that can undergo posttranslational modification, and other biophysical properties that can influence process development (e.g., p*I*, extinction coefficient, and solubility profile). Thus, for a high-throughput evaluation of multiple antibody candidates, homology modeling presents an excellent opportunity to probe some of these relevant biophysical properties [180].

One potential problem with the use of homology modeling in antibody structure determination is the lack of a vast array of full chain antibody structures. Yet, with the abundance of Fab and Fc structures, the limited full chain structures can be used as scaffold for the orientation of the final "Y"-shaped structure and still have a representative structure for computational modeling. Thus, a general strategy for antibody modeling would require that the molecule sequence be divided into the various fragments (two Fabs and one Fc). Then the next step is to utilize the BLAST algorithm (National Center for Biotechnology Information) to assess sequence against the nonredundant PDB for Fab and Fc structures that have a high level of sequence similarity with the target molecule. It is important at this stage to only focus on the framework region of the molecules and if possible BLAST just the CDR loops to find Fab structures that can be used to better model that region of the protein. Once PDB files are identified that best represent the molecule of interest, structural manipulation can begin, which will lead to a model structure of the target antibody. It is important to minimize the final structure to ensure phi and psi angles are within what is allowed for proteins (reference Ramachandran plot) [181]. The final step in this process, if a full chain antibody structure is desired, will be to find a full chain antibody close to the antibody class of interest and superimpose the structures to allow modeling of the linker sequence. Final steps will include making sure the appropriate disulfide bonds are formed and the structure is minimized (Figure 6.11).

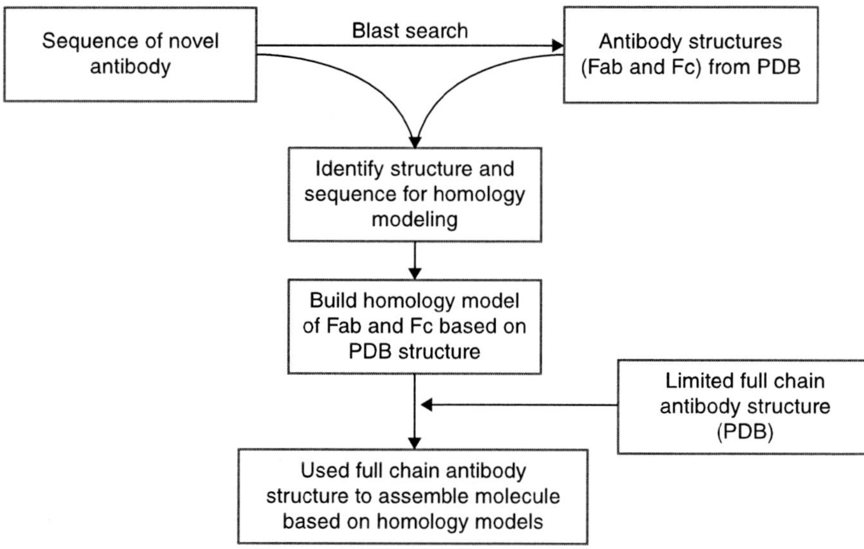

Figure 6.11 Flow diagram for the development of antibody homology structure. The major steps in the process of developing a homology model of antibodies based on initial protein sequence.

The aforementioned procedure is very time consuming and does not allow for an efficient approach to study antibodies in a high-throughput fashion. To this end, several companies including Chemical Computing Group (CCG) have developed algorithms to automate most of the steps required to build an antibody. In this chapter, CCG's MOE will be used as a case study in the automation of antibody homology modeling. Using the MOE graphic user interface, it is possible to paste in the sequence of any antibody, BLAST search against all solved antibody structures, and select the appropriate scaffold model. In addition, MOE allows the user to select the most appropriate CDR loops and set conditions for refinement and build models. Depending on the conditions defined, it is possible to generate several model structures and select the most appropriate based on any selection criteria. The homology- and antibody-building module in MOE contains an algorithm to access if significant structural errors exist in the homology model (e.g., clashes of atoms during and after energy minimization). The same module in MOE provides a rank ordering of homology models based on overall free energy. A similar approach can also be used to generate the full chain antibody structure. This algorithm saves time and is computationally more efficient, allowing for rapid screening of lead candidates and improvements in bioprocess development [182, 183].

6.6.2 Antibody Modeling and Bioprocess Development

Chromatographic ligands bind to proteins and separate variants based on binding affinities (protein retention), which can be modulated by mobile-phase composition (salt concentration, pH, and buffer components). The interaction between proteins and chromatographic ligands predominantly occurs via the surface residues of the protein. These surface residues define the surface properties of the protein (hydrophobic, negative, and positive patches). Thus, using computational modeling, chromatography ligands can be stack ranked based on their ability to bind proteins (retention) and differences in binding affinities that can result in better separation (column resolution or selectivity). Recently published data explored the correlation between computationally determined protein binding affinities (chromatographic affinity ligand) and experimental purification profile that revealed a similar trend [144].

The surface properties of a protein molecule are subject to pH and buffer conditions. With a well-resolved structure or homology model of any given protein, computational modeling can be used to evaluate the effect of pH and solvent on surface patches to aid process development. In a situation where multiple variants of a protein are possible, subtle changes in surface properties, resulting from amino acid misincorporation or degradation, can be exploited to improve resolution and separation. To illustrate this point, two publically available antibody structures (1IGY [179] (complete IgG1 structure) and 1IGT [184] (complete IgG2a structure)) were leveraged to demonstrate changes in biophysical properties that can influence purification. The two molecules have coordinate based p*I* of 7.85 and 7.72 respectively, with a sequence based p*I* of 6.64 and 6.62, respectively. Figure 6.12a illustrates the impact of increasing pH from 4.5 to 9.0; hydrophobic properties remain constant for both molecules, while negative and positive charges vary as a function of pH. The negative charge properties increased (Δ –ive for 1IGY of ~1.58% and ~1.94% for 1IGT), while positive charge properties decreased (Δ +ive for 1IGY of ~1.08% and ~1.87% for 1IGT) as a function of pH. When protein net charge and solubility properties are calculated as a function of pH, the profile looks very similar for both molecules, yet a closer look reveals that there are differences between the molecules from pH 4 to 8, which can be exploited for process development by adjusting pH condition and buffer ionic strength (Figure 6.12b).

6.7 Summary of Gaps and Future State

The current state of computational biophysics allows for determination of individual contributions of key interactions at an atomistic level. However, there remains a significant gap in the linkage of experimental techniques (typically macroscopic) to biophysical modeling. It is essential that these gaps be

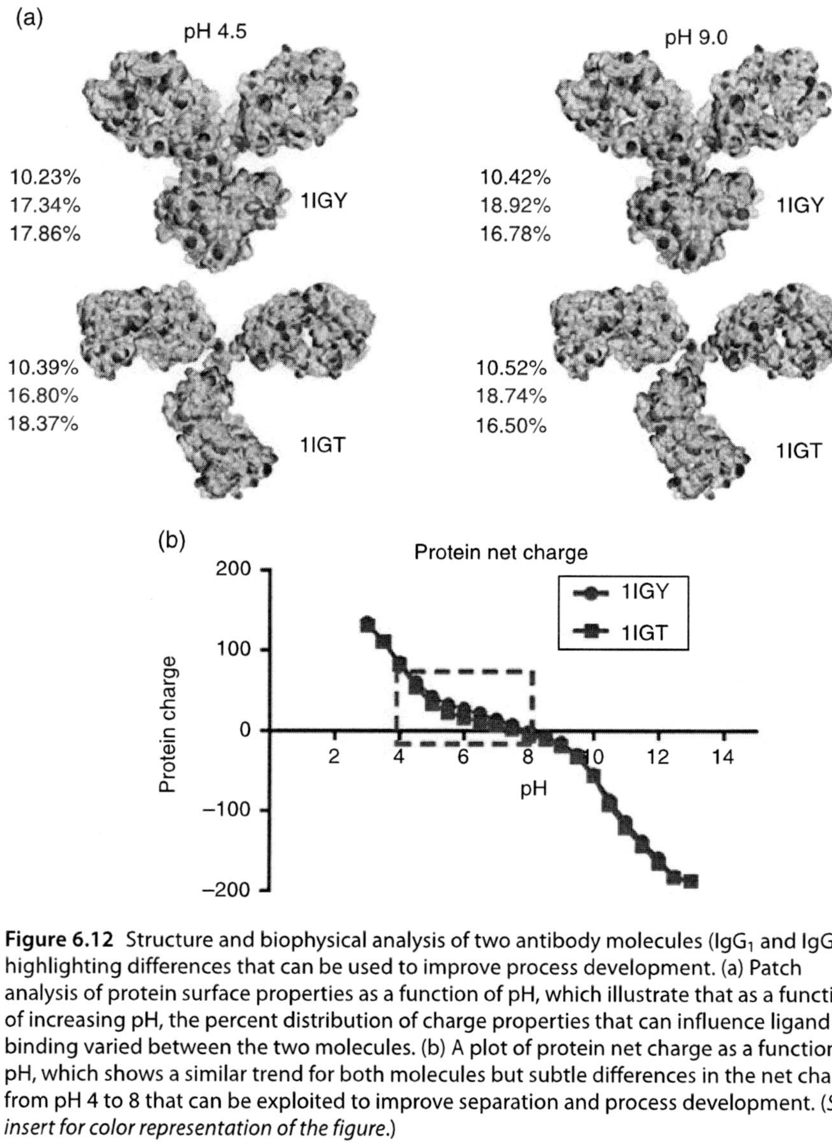

Figure 6.12 Structure and biophysical analysis of two antibody molecules (IgG_1 and $IgG2_a$), highlighting differences that can be used to improve process development. (a) Patch analysis of protein surface properties as a function of pH, which illustrate that as a function of increasing pH, the percent distribution of charge properties that can influence ligand binding varied between the two molecules. (b) A plot of protein net charge as a function of pH, which shows a similar trend for both molecules but subtle differences in the net charge from pH 4 to 8 that can be exploited to improve separation and process development. (*See insert for color representation of the figure.*)

closed in order to realize the potential for rational process design. An overview of the current modeling approaches, benefits, and liabilities is presented in Table 6.1 to guide the reader in selection of the most appropriate technique based on the specific scientific question.

One example of a gap is the ability to measure average quantities such as the number of charge interactions involved in a protein binding event via the SDM.

Table 6.1 Table outlining the cost and benefits of using the computational methods discussed previously in process development.

Method	Pros	Cons
Docking	Rapid read on binding affinities	Results can be misleading (false positive interactions) and require careful analysis
Docking plus MD simulation	Effective approach to identifying potential ligands	Computational cost of simulating thousands of protein–ligand complexes
Coarse-grained models	Efficient approach to estimate binding interaction energies	Loss of mechanistic information and accuracy possible (depending on method of application)
Free ligand simulation	Ability to evaluate multiple protein–ligand interactions	Free ligands may adopt nonrepresentative states in solution Avidity of interaction and ligand density may be difficult to model

SDM approach is indirect based on macroscopic measurements of protein–resin interactions, resulting in determination of apparent or net number of charges involved in an interaction. Hence it is not possible to readily determine if some of the charges are sterically excluded, nor the relative contributions, nor affinities of the individual interactions. The balance of avidity and affinity is an essential element in designing a separation process since these two parameters directly link to capacity and resolution. Additional experimental techniques with the fidelity to measure protein–ligand interactions at the individual level are essential.

Modeling of small protein–ligand complexes is readily feasible and can be performed using workstations or high-performance clusters. However, modeling of complex proteins (Mab or assemblies of Mab) requires extensive time on supercomputers. This gap in computational speed could be potentially be resolved by GPU in the future.

Computational modeling allows one to approach the problem from a different perspective specifically calculating the interaction energy between a ligand and protein potentially with an atomistic level of detail. However, it is extremely challenging to experimentally confirm these values directly and computationally inefficient to apply the atomistic level of detail to multiple interactions between a large protein (eg. Mab) and a ligand surface.

Computational techniques that allow one to mimic heterogeneity in the ligand distribution and the avidity of interaction could potentially be compared with experimental results of binding performed with media of known ligand heterogeneity and density.

Acknowledgment

We are grateful to the Merck Research Laboratories (MRL) postdoctoral research fellows program for supporting this work.

References

1. Narayanan, S. R., Preparative affinity chromatography of proteins. *Journal of Chromatography A* 1994, 658, 237–58.
2. Gritti, F., Guiochon, G., The van Deemter equation: Assumptions, limits, and adjustment to modern high performance liquid chromatography. *Journal of Chromatography A* 2013, 1302, 1–13.
3. Khirevich, S., Holtzel, A., Daneyko, A., Seidel-Morgenstern, A., Tallarek, U., Structure–transport correlation for the diffusive tortuosity of bulk, monodisperse, random sphere packings. *Journal of Chromatography A* 2011, 1218(37), 6489–97.
4. Kopaciewicz, W., Rounds, M. A., Fausnaugh, J., Reginer, F. E., Retention model for high-performance ion-exchange chromatography. *Journal of Chromatography* 1983, 266, 3–2.
5. Mazsaroff, I., Bischoff, R., Tice, P. A., Regnier, F. E., Influence of mobile phase pH on high-performance liquid chromatographic column loading capacity. Implications for the design of preparative protein separations. *Journal of Chromatography* 1988, 437(2), 429–35.
6. Mazsaroff, I., Cook, S., Regnier, F. E., Molecular orientation of immunoglobulin G at high concentration on an ion-exchange sorbent. *Journal of Chromatography* 1988, 443, 119–31.
7. Roush, D. J., Gill, D. S., Willson, R. C., Anion-exchange chromatographic behavior of recombinant rat cytochrome b5. Thermodynamic driving forces and temperature dependence of the stoichiometric displacement parameter Z. *Journal of Chromatography A* 1993, 653(2), 207–18.
8. Whitley, R. D., Wachter, R., Liu, F., Wang, N. H., Ion-exchange equilibria of lysozyme, myoglobin and bovine serum albumin. Effective valence and exchanger capacity. *Journal of Chromatography* 1989, 465(2), 137–56.
9. Johnson, R. D., Todd, R. J., Arnold, F. H., Multipoint binding in metal-affinity chromatography II. Effect of pH and imidazole on chromatographic retention of engineered histidine-containing cytochromes c. *Journal of Chromatography A* 1996, 725, (2), 225–35.
10. Ghose, S., Cramer, S. M., Characterization and modeling of monolithic stationary phases: Application to preparative chromatography. *Journal of Chromatography A* 2001, 928(1), 13–23.

11. Chester, T. L., Coym, J. W., Effect of phase ratio on van't Hoff analysis in reversed-phase liquid chromatography, and phase-ratio-independent estimation of transfer enthalpy. *Journal of Chromatography A* 2003, 1003(1–2), 101–11.
12. Takeuch, T., Kumaki, M., Ishi, D., Role of column temperature in open-tubular microcapillary liquid chromatography. *Journal of Chromatography* 1982, 235, 309–22.
13. Ueberbacher, R., Rodler, A., Hahn, R., Jungbauer, A., Hydrophobic interaction chromatography of proteins: Thermodynamic analysis of conformational changes. *Journal of Chromatography A* 2010, 1217(2), 184–90.
14. Chen, W., Huang, H., Lin, C., Lin, F., Chan, Y., Effect of temperature on hydrophobic interaction between proteins and hydrophobic adsorbents: Studies by isothermal titration calorimetry and the van't Hoff equation. *Langmuir* 2003, 19, 9395–403.
15. Baum, B., Muley, L., Smolinski, M., Heine, A., Hangauer, D., Klebe, G., Non-additivity of functional group contributions in protein–ligand binding: A comprehensive study by crystallography and isothermal titration calorimetry. *Journal of Molecular Biology* 2010, 397(4), 1042–54.
16. Falconer, R. J., Collins, B. M., Survey of the year 2009: Applications of isothermal titration calorimetry. *Journal of Molecular Recognition* 2010, 24(1), 1–16.
17. Falconer, R. J., Penkova, A., Jelesarov, I., Collins, B. M., Survey of the year 2008: Applications of isothermal titration calorimetry. *Journal of Molecular Recognition* 2010, 23(5), 395–413.
18. Gill, D. S., Roush, D. J., Shick, K. A., Willson, R. C., Microcalorimetric characterization of the anion-exchange adsorption of recombinant cytochrome b5 and its surface-charge mutants. *Journal of Chromatography A* 1995, 715(1), 81–93.
19. Leavitt, S., Freire, E., Direct measurement of protein binding energetics by isothermal titration calorimetry. *Current Opinion in Structural Biology* 2001, 11(5), 560–6.
20. Pierce, M. M., Raman, C. S., Nall, B. T., Isothermal titration calorimetry of protein–protein interactions. *Methods* 1999, 19, 213–21.
21. Sigurskjold, B. W., Exact analysis of competition ligand binding by displacement isothermal titration calorimetry. *Analytical Biochemistry* 2000, 277(2), 260–6.
22. Freed, A. S., Garde, S., Cramer, S. M., Molecular simulations of multimodal ligand–protein binding: Elucidation of binding sites and correlation with experiments. *The Journal of Physical Chemistry B* 2011, 115(45), 13320–7.
23. Snyder, I. R., Dolan, J. W., Lommen, D. C., Drylab® computer simulation for high-performance liquid chromatographic method development I. Isocratic elution. *Journal of Chromatography A* 1989, 485, 65–89.

24. Yao, Y., Czymmek, K. J., Pazhianur, R., Lenhoff, A. M., Three-dimensional pore structure of chromatographic adsorbents from electron tomography. *Langmuir*, 2006, 22(26), 11148–57.
25. Hubbuch, J., Linden, T., Knieps, E., Thommes, J., Kula, M. R., Mechanism and kinetics of protein transport in chromatographic media studied by confocal laser scanning microscopy. Part II. Impact on chromatographic separations. *Journal of Chromatography A* 2003, 1021(1–2), 105–15.
26. Ljunglof, A., Bergvall, P., Bhikhabhai, R., Hjorth, R., Direct visualisation of plasmid DNA in individual chromatography adsorbent particles by confocal scanning laser microscopy. *Journal of Chromatography A* 1999, 844(1–2), 129–35.
27. Ljunglof, A., Larsson, M., Knuuttila, K. G., Lindgren, J., Measurement of ligand distribution in individual adsorbent particles using confocal scanning laser microscopy and confocal micro-Raman spectroscopy. *Journal of Chromatography A* 2000, 893(2), 235–44.
28. Bowes, B. D., Lenhoff, A. M., Protein adsorption and transport in dextran-modified ion-exchange media. III. Effects of resin charge density and dextran content on adsorption and intraparticle uptake. *Journal of Chromatography A* 2011, 1218(40), 7180–8.
29. Wu, Y., Abraham, D., Carta, G., Particle size effects on protein and virus-like particle adsorption on perfusion chromatography media. *Journal of Chromatography A* 2015, 1375, 92–100.
30. Chung, W. K., Evans, S. T., Freed, A. S., Keba, J. J., Baer, Z. C., Rege, K., Cramer, S. M., Utilization of lysozyme charge ladders to examine the effects of protein surface charge distribution on binding affinity in ion exchange systems. *Langmuir* 2009, 26(2), 759–68.
31. Staahlberg, J., Joensson, B., Horvath, C., Theory for electrostatic interaction chromatography of proteins. *Analytical Chemistry* 1991, 63(17), 1867–74.
32. Roth, C. M., Lenhoff, A. M., Electrostatic and van der Waals contributions to protein adsorption: Computation of equilibrium constants. *Langmuir* 1993, 9(4), 962–72.
33. Chicz, R. M., Regnier, F. E., Single amino acid contributions to protein retention in cation-exchange chromatography: Resolution of genetically engineered subtilisin variants. *Analytical Chemistry* 1989, 61(18), 2059–66.
34. Ramsden, J., Roush, D., Gill, D., Kurrat, R., Willson, R., Protein adsorption kinetics drastically altered by repositioning a single charge. *Journal of the American Chemical Society* 1995, 117(33), 8511–6.
35. Rounds, M. A., Regnier, F. E., Evaluation of a retention model for high-performance ion-exchange chromatography using two different displacing salts. *Journal of Chromatography A* 1984, 283, 37–45.
36. Lesins, V., Ruckenstein, E., Patch controlled attractive electrostatic interactions between similarly charged proteins and adsorbents. *Colloid and Polymer Science* 1988, 266(12), 1187–90.

37. Roush, D. J., Gill, D. S., Willson, R. C., Electrostatic potentials and electrostatic interaction energies of rat cytochrome b5 and a simulated anion-exchange adsorbent surface. *Biophysical Journal* 1994, 66(5), 1290.
38. Boardman, N., Partridge, S., Separation of neutral proteins on ion-exchange resins. *Biochemical Journal* 1955, 59(4), 543.
39. Melander, W. R., El Rassi, Z., Horváth, C., Interplay of hydrophobic and electrostatic interactions in biopolymer chromatography: Effect of salts on the retention of proteins. *Journal of Chromatography A* 1989, 469, 3–27.
40. Hallgren, E., Kálmán, F., Farnan, D., Horváth, C., Ståhlberg, J., Protein retention in ion-exchange chromatography: Effect of net charge and charge distribution. *Journal of Chromatography A* 2000, 877(1), 13–24.
41. Jönsson, B., Ståhlberg, J., The electrostatic interaction between a charged sphere and an oppositely charged planar surface and its application to protein adsorption. *Colloids and Surfaces B: Biointerfaces* 1999, 14(1), 67–75.
42. Bowen, W. R., Sharif, A. O., Long-range electrostatic attraction between like-charge spheres in a charged pore. *Nature* 1998, 393(6686), 663–5.
43. Roth, C. M., Sader, J. E., Lenhoff, A. M., Electrostatic contribution to the energy and entropy of protein adsorption. *Journal of Colloid and Interface Science* 1998, 203(1), 218–21.
44. Dismer, F., Hubbuch, J., *In silico* prediction of protein binding using molecular dynamic simulations: A future tool for accelerated process development. *Journal of Bioscience and Bioengineering* 2009, 108, S60.
45. Honig, B., Nicholls, A., Classical electrostatics in biology and chemistry. *Science* 1995, 268(5214), 1144–9.
46. Levy, R. M., Zhang, L. Y., Gallicchio, E., Felts, A. K., On the nonpolar hydration free energy of proteins: Surface area and continuum solvent models for the solute-solvent interaction energy. *Journal of the American Chemical Society* 2003, 125(31), 9523–30.
47. Hummer, G., Hydrophobic force field as a molecular alternative to surface-area models. *Journal of the American Chemical Society* 1999, 121(26), 6299–305.
48. Kuhn, B., Gerber, P., Schulz-Gasch, T., Stahl, M., Validation and use of the MM-PBSA approach for drug discovery. *Journal of Medicinal Chemistry* 2005, 48(12), 4040–8.
49. Adcock, S. A., McCammon, J. A., Molecular dynamics: Survey of methods for simulating the activity of proteins. *Chemical Reviews* 2006, 106(5), 1589–615.
50. Haliloglu, T., Bahar, I., Erman, B., Gaussian dynamics of folded proteins. *Physical Review Letters* 1997, 79(16), 3090–3.
51. Isralewitz, B., Gao, M., Schulten, K., Steered molecular dynamics and mechanical functions of proteins. *Current Opinion in Structural Biology* 2001, 11(2), 224–30.

52. Ma, J., Karplus, M., Ligand-induced conformational changes in ras p21: A normal mode and energy minimization analysis. *Journal of Molecular Biology* 1997, 274(1), 114–31.
53. Mackerell, A. D., Jr., Empirical force fields for biological macromolecules: Overview and issues. *Journal of Computational Chemistry* 2004, 25(13), 1584–604.
54. Raux, B., The calculation of the potential of mean force using computer-simulations. *Computer Physics Communications* 1995, 91(1–3), 275–82.
55. Sugita, Y., Okamoto, Y., Replica-exchange molecular dynamics method for protein folding. *Chemical Physics Letters* 1999, 314(1–2), 141–51.
56. Fogolari, F., Brigo, A., Molinari, H., The Poisson–Boltzmann equation for biomolecular electrostatics: A tool for structural biology. *Journal of Molecular Recognition* 2002, 15(6), 377–92.
57. Klapper, I., Hagstrom, R., Fine, R., Sharp, K., Honig, B., Focusing of electric fields in the active site of Cu–Zn superoxide dismutase: Effects of ionic strength and amino-acid modification. *Proteins: Structure, Function, and Bioinformatics* 1986, 1(1), 47–59.
58. Nicholls, A., Graphical representation and analysis of surface properties. *Surfaces* 1993, 5, 2.4.
59. Bashford, D., An object-oriented programming suite for electrostatic effects in biological molecules. An experience report on the MEAD project, In: *Scientific Computing in Object-Oriented Parallel Environments*, Yutaka Ishikawa, Rodney R. Oldehoeft, John V. W. Reynders, Marydell Tholburn, eds. Berlin/New York: Springer 1997; pp 233–40.
60. Madura, J. D., Briggs, J. M., Wade, R. C., Davis, M. E., Luty, B. A., Ilin, A., Antosiewicz, J., Gilson, M. K., Bagheri, B., Scott, L. R., Electrostatics and diffusion of molecules in solution: Simulations with the University of Houston Brownian Dynamics program. *Computer Physics Communications* 1995, 91(1), 57–95.
61. Im, W., Beglov, D., Roux, B., Continuum solvation model: Computation of electrostatic forces from numerical solutions to the Poisson–Boltzmann equation. *Computer Physics Communications* 1998, 111(1), 59–75.
62. Brooks, B. R., Bruccoleri, R. E., Olafson, B. D., States, D. J., Swaminathan, S., Karplus, M., CHARMM: A program for macromolecular energy, minimization, and dynamics calculations. *Journal of Computational Chemistry* 1983, 4(2), 187–217.
63. Baker, N. A., Sept, D., Joseph, S., Holst, M. J., McCammon, J. A., Electrostatics of nanosystems: Application to microtubules and the ribosome. *Proceedings of the National Academy of Sciences* 2001, 98(18), 10037–41.
64. Tan, C., Yang, L., Luo, R., How well does Poisson–Boltzmann implicit solvent agree with explicit solvent? A quantitative analysis. *The Journal of Physical Chemistry B* 2006, 110(37), 18680–7.
65. Gabdoulline, R. R., Wade, R. C., Effective charges for macromolecules in solvent. *The Journal of Physical Chemistry* 1996, 100(9), 3868–78.

66. Gabdoulline, R. R., Wade, R. C., Simulation of the diffusional association of barnase and barstar. *Biophysical Journal* 1997, 72(5), 1917.
67. Sitkoff, D., Sharp, K. A., Honig, B., Accurate calculation of hydration free energies using macroscopic solvent models. *The Journal of Physical Chemistry* 1994, 98(7), 1978–88.
68. Still, W. C., Tempczyk, A., Hawley, R. C., Hendrickson, T., Semianalytical treatment of solvation for molecular mechanics and dynamics. *Journal of the American Chemical Society* 1990, 112(16), 6127–9.
69. Degliesposti, G., Portioli, C., Parenti, M. D., Rastelli, G., BEAR, a novel virtual screening methodology for drug discovery. *Journal of Biomolecular Screening* 2011, 16(1), 129–33.
70. Lindström, A., Edvinsson, L., Johansson, A., Andersson, C. D., Andersson, I. E., Raubacher, F., Linusson, A., Postprocessing of docked protein–ligand complexes using implicit solvation models. *Journal of Chemical Information and Modeling* 2011, 51(2), 267–82.
71. Freed, A. S., Cramer, S. M., Protein–surface interaction maps for ion-exchange chromatography. *Langmuir* 2011, 27(7), 3561–8.
72. Tong, H. F., Lin, D. Q., Zhang, Q. L., Wang, R. Z., Yao, S. J., Molecular recognition of Fc-specific ligands binding onto the consensus binding site of IgG: Insights from molecular simulation. *Journal of Molecular Recognition* 2014, 27(8), 501–9.
73. Regnier, F. E., The role of protein structure in chromatographic behavior. *Science* 1987, 238(4825), 319–23.
74. Dill, K. A., Dominant forces in protein folding. *Biochemistry* 1990, 29(31), 7133–55.
75. Chandler, D., Interfaces and the driving force of hydrophobic assembly. *Nature* 2005, 437(7059), 640–7.
76. Tanford, C., Contribution of hydrophobic interactions to the stability of the globular conformation of proteins. *Journal of the American Chemical Society* 1962, 84(22), 4240–7.
77. Giasson, B. I., Murray, I. V., Trojanowski, J. Q., Lee, V. M.-Y., A hydrophobic stretch of 12 amino acid residues in the middle of α-synuclein is essential for filament assembly. *Journal of Biological Chemistry* 2001, 276(4), 2380–6.
78. Granick, S., Bae, S. C., Chemistry. A curious antipathy for water. *Science (New York, NY)* 2008, 322(5907), 1477–8.
79. Ghose, A. K., Viswanadhan, V. N., Wendoloski, J. J., Prediction of hydrophobic (lipophilic) properties of small organic molecules using fragmental methods: An analysis of ALOGP and CLOGP methods. *The Journal of Physical Chemistry A* 1998, 102(21), 3762–72.
80. Chennamsetty, N., Voynov, V., Kayser, V., Helk, B., Trout, B. L., Prediction of aggregation prone regions of therapeutic proteins. *The Journal of Physical Chemistry B* 2010, 114(19), 6614–24.
81. Black, S. D., Mould, D. R., Development of hydrophobicity parameters to analyze proteins which bear post- or cotranslational modifications. *Analytical Biochemistry* 1991, 193(1), 72–82.

82. Acharya, H., Vembanur, S., Jamadagni, S. N., Garde, S., Mapping hydrophobicity at the nanoscale: Applications to heterogeneous surfaces and proteins. *Faraday Discussions* 2010, 146, 353–65.
83. Patel, A. J., Varilly, P., Chandler, D., Fluctuations of water near extended hydrophobic and hydrophilic surfaces. *The Journal of Physical Chemistry B* 2010, 114(4), 1632–7.
84. Huang, D. M., Chandler, D., The hydrophobic effect and the influence of solute-solvent attractions. *The Journal of Physical Chemistry B* 2002, 106(8), 2047–53.
85. Sarupria, S., Garde, S., Quantifying water density fluctuations and compressibility of hydration shells of hydrophobic solutes and proteins. *Physical Review Letters* 2009, 103(3), 037803.
86. Patel, A. J., Garde, S., Efficient method to characterize the context-dependent hydrophobicity of proteins. *The Journal of Physical Chemistry B* 2014, 118(6), 1564–73.
87. Srinivasan, K., Parimal, S., Lopez, M. M., McCallum, S. A., Cramer, S. M., Investigation into the molecular and thermodynamic basis of protein interactions in multimodal chromatography using functionalized nanoparticles. *Langmuir* 2014, 30(44), 13205–16.
88. Vilar, S., Cozza, G., Moro, S., Medicinal chemistry and the molecular operating environment (MOE): Application of QSAR and molecular docking to drug discovery. *Current Topics in Medicinal Chemistry* 2008, 8(18), 1555–72.
89. Gao, L., Zu, M., Wu, S., Liu, A.-L., Du, G.-H., 3D QSAR and docking study of flavone derivatives as potent inhibitors of influenza H1N1 virus neuraminidase. *Bioorganic and Medicinal Chemistry Letters* 2011, 21(19), 5964–70.
90. Afantitis, A., Melagraki, G., Sarimveis, H., Koutentis, P. A., Igglessi-Markopoulou, O., Kollias, G., A combined LS-SVM & MLR QSAR workflow for predicting the inhibition of CXCR3 receptor by quinazolinone analogs. *Molecular Diversity* 2010, 14(2), 225–35.
91. Shen, Q., Jiang, J.-H., Jiao, C.-X., Shen, G.-L., Yu, R.-Q., Modified particle swarm optimization algorithm for variable selection in MLR and PLS modeling: QSAR studies of antagonism of angiotensin II antagonists. *European Journal of Pharmaceutical Sciences* 2004, 22(2), 145–52.
92. Evgeniou, T., Pontil, M., Poggio, T., Regularization networks and support vector machines. *Advances in Computational Mathematics* 2000, 13(1), 1–50.
93. Mazza, C. B., Sukumar, N., Breneman, C. M., Cramer, S., Prediction of protein retention in ion-exchange systems using molecular descriptors obtained from crystal structure. *Analytical Chemistry* 2001, 73(22), 5457–61.
94. Mazza, C. B., Rege, K., Breneman, C. M., Sukumar, N., Dordick, J. S., Cramer, S. M., High-throughput screening and quantitative structure–efficacy relationship models of potential displacer molecules for ion-exchange systems. *Biotechnology and Bioengineering* 2002, 80(1), 60–72.

95. Song, M., Breneman, C. M., Bi, J., Sukumar, N., Bennett, K. P., Cramer, S., Tugcu, N., Prediction of protein retention times in anion-exchange chromatography systems using support vector regression. *Journal of Chemical Information and Computer Sciences* 2002, 42(6), 1347–57.
96. Tugcu, N., Song, M., Breneman, C. M., Sukumar, N., Bennett, K. P., Cramer, S. M., Prediction of the effect of mobile-phase salt type on protein retention and selectivity in anion exchange systems. *Analytical Chemistry* 2003, 75(14), 3563–72.
97. Ladiwala, A., Rege, K., Breneman, C. M., Cramer, S. M., A priori prediction of adsorption isotherm parameters and chromatographic behavior in ion-exchange systems. *Proceedings of the National Academy of Sciences of the United States of America* 2005, 102(33), 11710–5.
98. Malmquist, G., Nilsson, U. H., Norrman, M., Skarp, U., Strömgren, M., Carredano, E., Electrostatic calculations and quantitative protein retention models for ion exchange chromatography. *Journal of Chromatography A* 2006, 1115(1), 164–86.
99. Yang, T., Sundling, M. C., Freed, A. S., Breneman, C. M., Cramer, S. M., Prediction of pH-dependent chromatographic behavior in ion-exchange systems. *Analytical Chemistry* 2007, 79(23), 8927–39.
100. Ladiwala, A., Xia, F., Luo, Q., Breneman, C. M., Cramer, S. M., Investigation of protein retention and selectivity in HIC systems using quantitative structure retention relationship models. *Biotechnology and Bioengineering* 2006, 93(5), 836–50.
101. Chen, J., Yang, T., Cramer, S. M., Prediction of protein retention times in gradient hydrophobic interaction chromatographic systems. *Journal of Chromatography A* 2008, 1177(2), 207–14.
102. Chen, J., Yang, T., Luo, Q., Breneman, C. M., Cramer, S. M., Investigation of protein retention in hydrophobic interaction chromatographic (HIC) systems using the preferential interaction theory and quantitative structure property relationship models. *Reactive and Functional Polymers* 2007, 67(12), 1561–9.
103. Chen, J., Luo, Q., Breneman, C. M., Cramer, S. M., Classification of protein adsorption and recovery at low salt conditions in hydrophobic interaction chromatographic systems. *Journal of Chromatography A* 2007, 1139(2), 236–46.
104. Hou, Y., Morrison, C. J., Cramer, S. M., Classification of protein binding in hydroxyapatite chromatography: Synergistic interactions on the molecular scale. *Analytical Chemistry* 2011, 83(10), 3709–16.
105. Yang, T., Malmquist, G., Johansson, B.-L., Maloisel, J.-L., Cramer, S., Evaluation of multi-modal high salt binding ion exchange materials. *Journal of Chromatography A* 2007, 1157(1), 171–7.
106. Hou, Y., Cramer, S. M., Evaluation of selectivity in multimodal anion exchange systems: *A priori* prediction of protein retention and examination

of mobile phase modifier effects. *Journal of Chromatography A* 2011, 1218(43), 7813–20.

107. Buyel, J., Woo, J., Cramer, S., Fischer, R., The use of quantitative structure–activity relationship models to develop optimized processes for the removal of tobacco host cell proteins during biopharmaceutical production. *Journal of Chromatography A* 2013, 1322, 18–28.
108. Frenkel, D., Smit, B., *Understanding Molecular Simulation: From Algorithms to Applications*. San Diego: Academic Press 2001; Vol. 1.
109. Harner, M. J., Frank, A. O., Fesik, S. W., Fragment-based drug discovery using NMR spectroscopy. *Journal of Biomolecular NMR* 2013, 56(2), 65–75.
110. Vuignier, K., Schappler, J., Veuthey, J.-L., Carrupt, P.-A., Martel, S., Drug–protein binding: A critical review of analytical tools. *Analytical and Bioanalytical Chemistry* 2010, 398(1), 53–66.
111. Renaud, J.-P., Delsuc, M.-A., Biophysical techniques for ligand screening and drug design. *Current Opinion in Pharmacology* 2009, 9(5), 622–8.
112. Friddle, R. W., Battle, K., Trubetskoy, V., Tao, J., Salter, E. A., Moradian-Oldak, J., De Yoreo, J. J., Wierzbicki, A., Single-molecule determination of the face-specific adsorption of Amelogenin's C-terminus on hydroxyapatite. *Angewandte Chemie* 2011, 123(33), 7683–7.
113. Jorgensen, W. L., The many roles of computation in drug discovery. *Science* 2004, 303(5665), 1813–8.
114. Canchi, D. R., Jayasimha, P., Rau, D. C., Makhatadze, G. I., Garcia, A. E., Molecular mechanism for the preferential exclusion of TMAO from protein surfaces. *The Journal of Physical Chemistry B* 2012, 116(40), 12095–104.
115. Henzler-Wildman, K., Kern, D., Dynamic personalities of proteins. *Nature* 2007, 450(7172), 964–72.
116. Henzler-Wildman, K. A., Lei, M., Thai, V., Kerns, S. J., Karplus, M., Kern, D., A hierarchy of timescales in protein dynamics is linked to enzyme catalysis. *Nature* 2007, 450(7171), 913–6.
117. Eisenmesser, E. Z., Bosco, D. A., Akke, M., Kern, D., Enzyme dynamics during catalysis. *Science* 2002, 295(5559), 1520–3.
118. Essmann, U., Perera, L., Berkowitz, M. L., Darden, T., Lee, H., Pedersen, L. G., A smooth particle mesh Ewald method. *The Journal of Chemical Physics* 1995, 103(19), 8577–93.
119. Cornell, W. D., Cieplak, P., Bayly, C. I., Gould, I. R., Merz, K. M., Ferguson, D. M., Spellmeyer, D. C., Fox, T., Caldwell, J. W., Kollman, P. A., A second generation force field for the simulation of proteins, nucleic acids, and organic molecules. *Journal of the American Chemical Society* 1995, 117(19), 5179–97.
120. Scott, W. R., Hünenberger, P. H., Tironi, I. G., Mark, A. E., Billeter, S. R., Fennen, J., Torda, A. E., Huber, T., Krüger, P., van Gunsteren, W. F., The GROMOS biomolecular simulation program package. *The Journal of Physical Chemistry A* 1999, 103(19), 3596–607.

121. Berendsen, H. J., Postma, J. P. M., van Gunsteren, W. F., DiNola, A., Haak, J., Molecular dynamics with coupling to an external bath. *The Journal of Chemical Physics* 1984, 81(8), 3684–90.
122. Evans, D. J., Holian, B. L., The Nose–Hoover thermostat. *The Journal of Chemical Physics* 1985, 83(8), 4069–74.
123. Nosé, S., A unified formulation of the constant temperature molecular dynamics methods. *The Journal of Chemical Physics* 1984, 81(1), 511–9.
124. Parrinello, M., Rahman, A., Crystal structure and pair potentials: A molecular-dynamics study. *Physical Review Letters* 1980, 45(14), 1196.
125. Berendsen, H. J., van der Spoel, D., van Drunen, R., GROMACS: A message-passing parallel molecular dynamics implementation. *Computer Physics Communications* 1995, 91(1), 43–56.
126. Hess, B., Kutzner, C., Van Der Spoel, D., Lindahl, E., GROMACS 4: Algorithms for highly efficient, load-balanced, and scalable molecular simulation. *Journal of Chemical Theory and Computation* 2008, 4(3), 435–47.
127. Pronk, S., Páll, S., Schulz, R., Larsson, P., Bjelkmar, P., Apostolov, R., Shirts, M. R., Smith, J. C., Kasson, P. M., van der Spoel, D., GROMACS 4.5: A high-throughput and highly parallel open source molecular simulation toolkit. *Bioinformatics* 2013, 29(7), 845–54.
128. Nelson, M. T., Humphrey, W., Gursoy, A., Dalke, A., Kalé, L. V., Skeel, R. D., Schulten, K., NAMD: A parallel, object-oriented molecular dynamics program. *International Journal of High Performance Computing Applications* 1996, 10(4), 251–68.
129. Plimpton, S., Fast parallel algorithms for short-range molecular dynamics. *Journal of Computational Physics* 1995, 117(1), 1–19.
130. Palmer, A. G., 3rd, NMR characterization of the dynamics of biomacromolecules. *Chemical Reviews* 2004, 104(8), 3623–40.
131. Boehr, D. D., Dyson, H. J., Wright, P. E., An NMR perspective on enzyme dynamics. *Chemical Reviews* 2006, 106(8), 3055–79.
132. Jorgensen, W. L., Chandrasekhar, J., Madura, J. D., Impey, R. W., Klein, M. L., Comparison of simple potential functions for simulating liquid water. *The Journal of Chemical Physics* 1983, 79(2), 926–35.
133. Vega, C., De Miguel, E., Surface tension of the most popular models of water by using the test-area simulation method. *The Journal of Chemical Physics* 2007, 126(15), 154707.
134. Peng, T., Zintsmaster, J. S., Namanja, A. T., Peng, J. W., Sequence-specific dynamics modulate recognition specificity in WW domains. *Nature Structural and Molecular Biology* 2007, 14(4), 325–31.
135. Stefl, S., Nishi, H., Petukh, M., Panchenko, A. R., Alexov, E., Molecular mechanisms of disease-causing missense mutations. *Journal of Molecular Biology* 2013, 425(21), 3919–36.

136. Garcia, K. C., Teyton, L., Wilson, I. A., Structural basis of T-cell recognition. *Annual Review of Immunology* 1999, 17, 369–97.
137. Armstrong, K. M., Piepenbrink, K. H., Baker, B. M., Conformational changes and flexibility in T-cell receptor recognition of peptide-MHC complexes. *Biochemistry Journal* 2008, 415(2), 183–96.
138. Pan, Y., Yuhasz, S. C., Amzel, L. M., Anti-idiotypic antibodies: Biological function and structural studies. *FASEB Journal* 1995, 9(1), 43–9.
139. Borbulevych, O. Y., Insaidoo, F. K., Baxter, T. K., Powell, D. J., Jr., Johnson, L. A., Restifo, N. P., Baker, B. M., Structures of MART-126/27-35 Peptide/HLA-A2 complexes reveal a remarkable disconnect between antigen structural homology and T cell recognition. *Journal of Molecular Biology* 2007, 372(5), 1123–36.
140. Insaidoo, F. K., Borbulevych, O. Y., Hossain, M., Santhanagopolan, S. M., Baxter, T. K., Baker, B. M., Loss of T cell antigen recognition arising from changes in peptide and major histocompatibility complex protein flexibility: Implications for vaccine design. *Journal of Biological Chemistry* 2011, 286(46), 40163–73.
141. Smejkal, B., Agrawal, N. J., Helk, B., Schulz, H., Giffard, M., Mechelke, M., Ortner, F., Heckmeier, P., Trout, B. L., Hekmat, D., Fast and scalable purification of a therapeutic full-length antibody based on process crystallization. *Biotechnology and Bioengineering* 2013, 110(9), 2452–61.
142. Becker, H. D., Kern, D., *Thermus thermophilus*: A link in evolution of the tRNA-dependent amino acid amidation pathways. *Proceedings of the National Academy of Sciences of the United States of America* 1998, 95(22), 12832–7.
143. Becker, H. D., Roy, H., Moulinier, L., Mazauric, M. H., Keith, G., Kern, D., *Thermus thermophilus* contains an eubacterial and an archaebacterial aspartyl-tRNA synthetase. *Biochemistry* 2000, 39(12), 3216–30.
144. Insaidoo, F. K., Rauscher, M. A., Smithline, S. J., Kaarsholm, N. C., Feuston, B. P., Ortigosa, A. D., Linden, T. O., Roush, D. J., Targeted purification development enabled by computational biophysical modeling. *Biotechnology Progress* 2015, 31(1), 154–64.
145. Olaru, A., Bala, C., Jaffrezic-Renault, N., Aboul-Enein, H. Y., Surface plasmon resonance (SPR) biosensors in pharmaceutical analysis. *Critical Reviews in Analytical Chemistry* 2015, 45(2), 97–105.
146. Kabiri, M., Unsworth, L. D., Application of isothermal titration calorimetry for characterizing thermodynamic parameters of biomolecular interactions: Peptide self-assembly and protein adsorption case studies. *Biomacromolecules* 2014, 15(10), 3463–73.
147. Ghitti, M., Musco, G., Spitaleri, A., NMR and computational methods in the structural and dynamic characterization of ligand-receptor interactions. *Advances in Experimental Medicine and Biology* 2014, 805, 271–304.
148. Warren, G. L., Do, T. D., Kelley, B. P., Nicholls, A., Warren, S. D., Essential considerations for using protein–ligand structures in drug discovery. *Drug Discovery Today* 17(23–24), 1270–81.

149. Zheng, M., Liu, X., Xu, Y., Li, H., Luo, C., Jiang, H., Computational methods for drug design and discovery: Focus on China. *Trends in Pharmacological Sciences* 2013, 34(10), 549–59.
150. Pei, J., Yin, N., Ma, X., Lai, L., Systems biology brings new dimensions for structure-based drug design. *Journal of the American Chemical Society* 2014, 136(33), 11556–65.
151. Chen, Y. C., Beware of docking! *Trends in Pharmacological Sciences* 2015, 36(2), 78–95.
152. Friesner, R. A., Banks, J. L., Murphy, R. B., Halgren, T. A., Klicic, J. J., Mainz, D. T., Repasky, M. P., Knoll, E. H., Shelley, M., Perry, J. K., Shaw, D. E., Francis, P., Shenkin, P. S., Glide: A new approach for rapid, accurate docking and scoring. 1. Method and assessment of docking accuracy. *Journal of Medicinal Chemistry* 2004, 47(7), 1739–49.
153. Hou, X., Du, J., Zhang, J., Du, L., Fang, H., Li, M., How to improve docking accuracy of AutoDock4.2: A case study using different electrostatic potentials. *Journal of Chemical Information and Modeling* 2013, 53(1), 188–200.
154. Berendsen, H. J., Hayward, S., Collective protein dynamics in relation to function. *Current Opinion in Structural Biology* 2000, 10(2), 165–9.
155. Scheraga, H. A., Khalili, M., Liwo, A., Protein-folding dynamics: Overview of molecular simulation techniques. *Annual Review of Physical Chemistry* 2007, 58, 57–83.
156. Hamelberg, D., Mongan, J., McCammon, J. A., Accelerated molecular dynamics: A promising and efficient simulation method for biomolecules. *Journal of Chemical Physics* 2004, 120(24), 11919–29.
157. Isralewitz, B., Baudry, J., Gullingsrud, J., Kosztin, D., Schulten, K., Steered molecular dynamics investigations of protein function. *Journal of Molecular Graphics and Modelling* 2001, 19(1), 13–25.
158. McCammon, J. A., Gelin, B. R., Karplus, M., Dynamics of folded proteins. *Nature* 1977, 267(5612), 585–90.
159. Kang, M., Smith, P. E., Preferential interaction parameters in biological systems by Kirkwood–Buff theory and computer simulation. *Fluid Phase Equilibria* 2007, 256(1), 14–9.
160. Shukla, D., Shinde, C., Trout, B. L., Molecular computations of preferential interaction coefficients of proteins. *The Journal of Physical Chemistry B* 2009, 113(37), 12546–54.
161. Yeh, I.-C., Hummer, G., System-size dependence of diffusion coefficients and viscosities from molecular dynamics simulations with periodic boundary conditions. *The Journal of Physical Chemistry B* 2004, 108(40), 15873–9.
162. Parimal, S., Cramer, S. M., Garde, S., Application of a spherical harmonics expansion approach for calculating ligand density distributions around proteins. *The Journal of Physical Chemistry B* 2014, 118(46), 13066–76.
163. Garde, S., Hummer, G., García, A. E., Pratt, L. R., Paulaitis, M. E., Hydrophobic hydration: Inhomogeneous water structure near nonpolar

molecular solutes. *Physical Review E* 1996, 53(5), R4310.
164. Westin, S. H., Arvo, J. R., Torrance, K. E., Predicting reflectance functions from complex surfaces. Computer Graphics (SIGGRAPH '92 Proceedings), 26, 255–264, July 1992, ACM.
165. Alexander, D., Barker, G., Arridge, S., Detection and modeling of non-Gaussian apparent diffusion coefficient profiles in human brain data. *Magnetic Resonance in Medicine* 2002, 48(2), 331–40.
166. Arfken, G. B., *Mathematical Methods for Physicists.* Waltham, MA: Academic Press 2013.
167. Chung, W. K., Freed, A. S., Holstein, M. A., McCallum, S. A., Cramer, S. M., Evaluation of protein adsorption and preferred binding regions in multimodal chromatography using NMR. *Proceedings of the National Academy of Sciences of the United States of America* 2010, 107(39), 16811–6.
168. Godawat, R., Jamadagni, S. N., Garde, S., Characterizing hydrophobicity of interfaces by using cavity formation, solute binding, and water correlations. *Proceedings of the National Academy of Sciences of the United States of America* 2009, 106(36), 15119–24.
169. Jamadagni, S. N., Godawat, R., Garde, S., How surface wettability affects the binding, folding, and dynamics of hydrophobic polymers at interfaces. *Langmuir* 2009, 25(22), 13092–9.
170. Shenogina, N., Godawat, R., Keblinski, P., Garde, S., How wetting and adhesion affect thermal conductance of a range of hydrophobic to hydrophilic aqueous interfaces. *Physical Review Letters* 2009, 102(15), 156101.
171. McQuarrie, D., *Statistical Mechanics.* Sausalito, CA: University Science Books 2000; Vol. 12, p 641.
172. Yao, Y., Lenhoff, A. M., Electrostatic contributions to protein retention in ion-exchange chromatography. 1. Cytochrome C variants. *Analytical Chemistry* 2004, 76(22), 6743–52.
173. Dismer, F., Hubbuch, J., 3D structure-based protein retention prediction for ion-exchange chromatography. *Journal of Chromatography A* 2010, 1217(8), 1343–53.
174. Underdown, B. J., Schiff, J. M., Immunoglobulin A: Strategic defense initiative at the mucosal surface. *Annual Review of Immunology* 1986, 4, 389–417.
175. Geisberger, R., Lamers, M., Achatz, G., The riddle of the dual expression of IgM and IgD. *Immunology* 2006, 118(4), 429–37.
176. Chen, K., Xu, W., Wilson, M., He, B., Miller, N. W., Bengten, E., Edholm, E. S., Santini, P. A., Rath, P., Chiu, A., Cattalini, M., Litzman, J., Bussel, B., Huang, B., Meini, A., Riesbeck, K., Cunningham-Rundles, C., Plebani, A., Cerutti, A., Immunoglobulin D enhances immune surveillance by activating antimicrobial, proinflammatory and B cell-stimulating programs in

basophils. *Nature Immunology* 2009, 10(8), 889–98.

177. Oxelius, V. A., Pandey, J. P., Human immunoglobulin constant heavy G chain (IGHG) (Fcgamma) (GM) genes, defining innate variants of IgG molecules and B cells, have impact on disease and therapy. *Clinical Immunology* 2013, 149(3), 475–86.
178. Ravetch, J. V., Bolland, S., IgG Fc receptors. *Annual Review of Immunology* 2001, 19, 275–90.
179. Harris, L. J., Skaletsky, E., McPherson, A., Crystallographic structure of an intact IgG1 monoclonal antibody. *Journal of Molecular Biology* 1998, 275(5), 861–72.
180. Hanke, A. T., Ottens, M., Purifying biopharmaceuticals: Knowledge-based chromatographic process development. *Trends in Biotechnology* 2014, 32(4), 210–20.
181. Ramachandran, G. N., Ramakrishnan, C., Sasisekharan, V., Stereochemistry of polypeptide chain configurations. *Journal of Molecular Biology* 1963, 7, 95–9.
182. Almagro, J. C., Beavers, M. P., Hernandez-Guzman, F., Maier, J., Shaulsky, J., Butenhof, K., Labute, P., Thorsteinson, N., Kelly, K., Teplyakov, A., Luo, J., Sweet, R., Gilliland, G. L., Antibody modeling assessment. *Proteins* 2011, 79(11), 3050–66.
183. Maier, J. K., Labute, P., Assessment of fully automated antibody homology modeling protocols in molecular operating environment. *Proteins* 2014, 82(8), 1599–610.
184. Harris, L. J., Larson, S. B., Hasel, K. W., McPherson, A., Refined structure of an intact IgG2a monoclonal antibody. *Biochemistry* 1997, 36(7), 1581–97.

7

Chromatographic Scale-Up on a Volume Basis

Ernst B. Hansen

Mathematical Modelling, Novo Nordisk A/S, Bagsværd, Denmark

7.1 Introduction

The topic of scale-up for chromatographic process steps can be divided into two main areas:

1) Scale independent: The principle of scale-up dealing with the process parameters and how to change them (or not) during scale-up
2) Scale dependent: Practical issues such as column packing and hardware design

This chapter deals with the principle of scale-up from the laboratory to a pilot facility and into production. Notes on the practical issues can be found at the end of the chapter.

7.1.1 The Rigidity of Linear Scale-Up

The scale-up of a chromatographic separation is in principle a simple procedure since the process parameters are scalable in a linear fashion. The process is scaled by increasing the column diameter while keeping the bed height, the velocity, and the volumes of the different phases (measured in column volumes) constant. The problem with this simple and well-established concept, usually referred to as *linear scale-up*, is the lack of flexibility. For practitioners, especially those dealing with large-scale operations, it is no secret that chromatography columns come in discrete sizes regarding the diameter. Combined with a constant bed height, this means that the column volume is also a discrete variable.

In production scale the diameter and the number of columns are the only design variables we have available. As a consequence we most often cannot choose the column volume that matches the desired capacity. We are faced with the problem that for a chosen diameter we need, for instance, 2.2 columns in order to meet the capacity demands. The solution is either to invest in a third column or go back to the lab and try to optimize the process in order to squeeze it into two columns. Both solutions can be rather costly either in investment or in development time. Further, we have no guarantee that the optimization will actually succeed (not to speak of the documentation burden if it does). Even when we manage to meet the right capacity, we still risk encountering the same problem at a later date if we want to increase the production capacity.

In pilot scale the problem is even more pronounced since we must use existing equipment and have maybe two or three options for the column diameter and thus the volume. The lack of flexibility of the column volume can increase the number of cycles on each column and thereby the duration of the pilot campaign (where the chromatography steps are often the bottleneck). For products with limited stability, a processing time that varies with the scale is not attractive.

At the lower end of the scale, in the development phase, we are met with similar problems. Knowing that the process will end up in production some day and the bed height must be kept constant, we aim for a suitable bed height (like 20 cm) when developing the chromatography in lab scale. In order to avoid packing problems and flow irregularities (the so-called *wall effects*), a certain column diameter (~1 cm) is required, and we therefore end up with a lower volume limit around perhaps 15–20 mL in order to have a representative process step. Especially in the early phase of development where the product quantities are limited, this can be a substantial volume that reduces the number of trials, and we therefore risk a delay in development. A volume in this range is also certainly not desirable for special tasks like virus validation.

7.1.2 Increasing the Flexibility

Fifteen years ago we revisited the scaling principles of chromatography in order to find a more flexible way of scaling up (or down). From a conceptual point of view, the solution was quite simple and does not involve any new theory. By scaling the process based on a constant contact time (residence time), we can, within limits, introduce the column bed height as a process design variable rather than a fixed parameter. Since the bed height is a continuous variable, as opposed to the diameter, we can use it to obtain exactly the volume we need in any scale. After proof of concept in lab scale and later in pilot scale, the new concept has become an established approach in our company, which has proven extremely valuable over the entire range from development in lab scale to commercial production.

Our experience from interactions with companies, including producers of chromatography equipment, and with the regulatory authorities has shown that this is apparently still a new territory. In the following we will present more than 10 years of experience in chromatographic scale-up on a volume basis. This chapter will include the theoretical background and proof of concept as well as industrial cases with large-scaling factors. Examples include IEX, HIC, and RPC processes for the separation of both peptides and large proteins. Additional information can be found in Refs. [1] and [2].

7.2 Theoretical Background

In order to understand why this concept works, we take a look at a theory. For simplicity we consider the number of theoretical plates as a measure for the column performance. Using the general theory of chromatography in the form of mass and flux balances for the column leads to the same result. A brief outline is given at the end of this section.

7.2.1 Separation Performance: The Lower Limit

From the theory of chromatography, we know that the height of a theoretical plate depends on the linear velocity. This is known as the Van Deemter equation, which in a simplified form can be expressed as

$$H = A + Cv$$

where H is the plate height (cm), A and C are constants, and v is the linear velocity (cm/h). The number of theoretical plates is found simply as the ratio between the bed height and the plate height:

$$N = \frac{L}{H}$$

The number of theoretical plates is indicative for the separation (performance), and it is in principle this number we need to keep constant when scaling up (or down). In the traditional linear scale-up concept, a constant plate number is obtained by fixing the bed height and the velocity (i.e., the plate height) when scaling up. This leaves only the column diameter as a design variable when increasing the column volume during scale-up. As columns come in discrete sizes, this approach does not provide much flexibility.

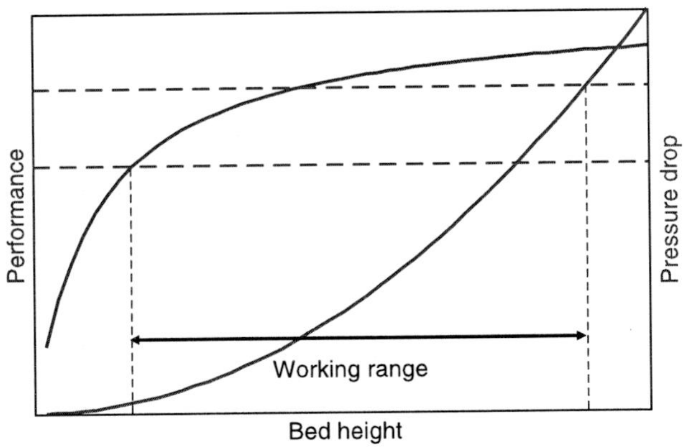

Figure 7.1 Scale-up on a volume basis. The line to the left shows how the performance (plate number) increases with the bed height and reaches an asymptotic level. The line to the right shows the quadratic increase of the pressure drop with the bed height. The horizontal dashed lines are the lower limit for the performance and the upper limit for the pressure drop. The intercept of the solid and dashed lines gives the limits of the working range for the bed height.

The basis of scaling on a volume basis is to rearrange the two equations into

$$N = \frac{1}{(A/L) + CQ}$$

where $Q = v/L$ is now the flow rate in CV/h (column volumes per hour). From this expression there are two important observations: (i) the plate number, at a fixed flow rate in CV/h, increases with the bed height and (ii) beyond a certain bed height, where $A/L \ll CQ$, the plate number becomes independent of the bed height. So, if we fix the flow rate (in CV/h), we know that in increasing the bed height, we will get the same or a better performance and that at some point the increase is insignificant. Instead of keeping the bed height and thus the plate number constant, we define a minimum bed height, corresponding to a minimum plate number, beyond which the performance is satisfactory as long as the flow rate is kept constant in CV/h corresponding to a constant residence time (see Figure 7.1).

7.2.2 Pressure Restriction: The Upper Limit

The bed height is limited upward by the pressure drop over the column. At a certain value of the bed height, the pressure restrictions are met, either from the resin or from the column limitations (hardware). The pressure drop over a packed bed can be expressed as

$$P = \alpha vL = \alpha QL^2$$

where α is an empirical constant. The first expression is the usual linear relationship between pressure, velocity, and height. In the second expression we have again substituted the velocity by the flow rate (CV/h), and from this expression we see that at a fixed flow rate the pressure drop increases by the square of the bed height. This relation together with a maximum allowed pressure drop provides an upper limit for the bed height at a fixed flow rate:

$$L_{\max} = \sqrt{\frac{P_{\max}}{\alpha Q}}$$

where $P_{\max}$ is the pressure restriction for the relevant scale.

7.2.3 Design Window

We now have a design window for the bed height with a lower limit from the separation performance and an upper limit from the pressure drop. Figure 7.1 illustrates the design window.

Within the design window we obtain a satisfactory product quality. The cycle time, the buffer volume (CV), and the productivity (g/L h) are constant, while the capacity (g/h) is proportional to the bed height.

7.2.4 General Theory

This part is optional and mainly intended for those who find the previous explanation too simple, are more used to differential equations, or think that this concept is not valid for the highly nonlinear case of preparative chromatography.

The differential mass balance for chromatography is in dimensionless form:

$$\frac{\partial c}{\partial x} - \frac{1}{\mathrm{Pe}}\frac{\partial^2 c}{\partial x^2} + \varepsilon\frac{\partial c}{\partial V} + (1-\varepsilon)\frac{\partial s}{\partial V} = 0$$

where x is the dimensionless position (z/L) and V is the dimensionless volume (V/V_{col}), which is the same as dimensionless time (tv/L). From left to right the balance is the sum of convection, axial dispersion, accumulation in the mobile phase, and accumulation in the stationary phase (pore volume in the particles comprising the pore liquid and the adsorbed state). The accumulation in the stationary phase is described by mass transfer using the linear driving force approximation:

$$\frac{\partial s}{\partial V} = \mathrm{St}_\mathrm{p}\left(c - c^*\right)$$

In this case we have chosen pore diffusion as the main driving force. This means we use a pore diffusivity (mass transfer coefficient) and that the driving force is expressed in the liquid phase concentration.

The mass balance looks complicated but is in fact a straightforward compilation of the phenomena occurring in the column. The mass transfer expression looks simple but is in fact just one of many possible expressions. The flux from the mobile phase into the pore phase and the adsorbed state can be described using several mechanisms such as external film mass transfer (mobile to particle surface), pore diffusion, solid-state diffusion (driving force in adsorbed concentration), and even adsorption kinetics (for affinity resins). Luckily for our purpose it doesn't really matter which mechanism we choose. The important thing to recognize is that in order to get the same solution of the equations, we must keep the Peclet number (Pe) and the Stanton number (St) constant:

$$\mathrm{Pe} = \frac{vL}{D_f + D_a} = \frac{vL}{\lambda v d_\mathrm{p}} = \frac{L}{\lambda d_\mathrm{p}}$$

where λ is parameter related to packing and d_p is the particle diameter. For a given process step, the Peclet number is proportional to the bed height.

$$\mathrm{St}_\mathrm{p} = \frac{K_\mathrm{p} L}{v} = \frac{K_\mathrm{p}}{Q}$$

The p in St_p indicates that this is the Stanton number based on pore diffusion. As discussed previously other definitions can apply in which case the mass transfer coefficient K_p is replaced by a similar coefficient. For a given process step, the Stanton number is inversely proportional to the flow rate or proportional to the residence time ($1/Q$).

The Peclet number relates to the A/L term in the Van Deemter equation, while the Stanton number relates to the CQ term. Instead of keeping Pe and St constant, we can define a minimum St number and a minimum Pe number, above which the separation will be satisfactory. Again, we do not need to know either Pe or St. We just need to identify a flow rate and a minimum bed height. Keeping the flow rate constant and staying above the minimum bed height, we then know that the Stanton number will be constant and that the

Peclet number will be the same or higher. Using a more physical or mechanistic approach, we have reached the same result as for the more simple plate number considerations.

7.3 Proof of Concept Examples

The following examples are lab studies performed as proof of concept for scaling on a volume basis. Different types of chromatography are used, and the height has been varied significantly in each case (Figures 7.2, 7.3, and 7.4).

The final example (Figure 7.5) demonstrates that there is indeed an actual lower limit for the bed height at a given flow rate and we must not use autopilot and assume that everything is fine as long as the residence time is the same.

The examples are in good agreement with theory and show that the concept is a general approach, not specific to the type of adsorption. In some cases there is no difference between bed heights, and in other cases we see a slight improvement when increasing the bed height.

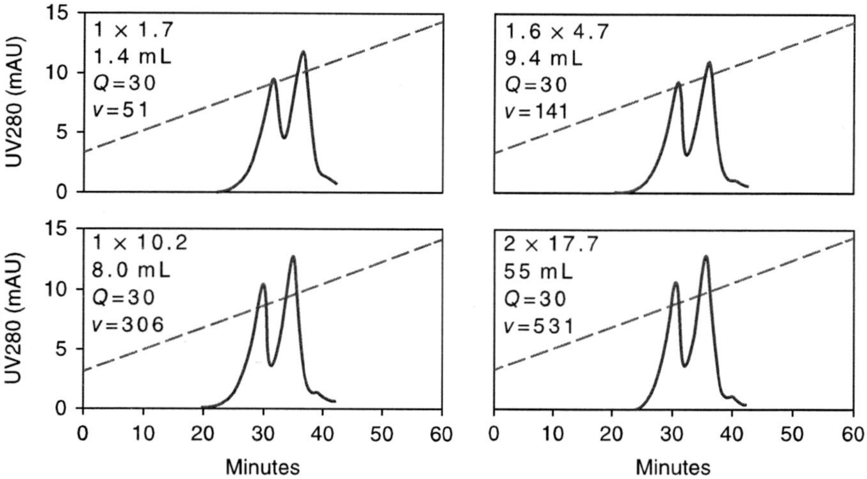

Figure 7.2 Separation of β-lactoglobulin A and B by AIEX on Source 30Q [3]. Dimensions are column diameter × bed height (cm). The dashed line is the gradient. The flow rate is constant at $Q = 30$ CV/h, while the velocity (v in cm/h) and the bed height (cm) changes by a factor of 10. The separation for $L = 1.7$ cm is slightly less than for the remaining bed heights in agreement with theory. *Source*: Mollerup and Pedersen [3]. Reproduced with permission of International Symposium on the separation of Proteins, Peptides & Polynucleotides.

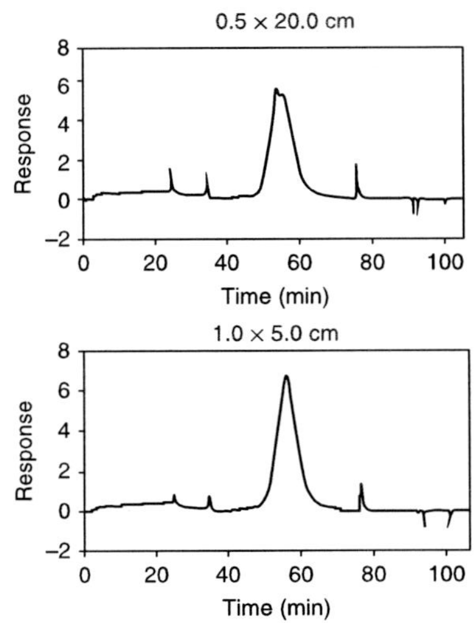

$Q = 40$ CV/h

Col. dim.	Velocity	Impurities (ppm)	
$D \times L$ (cm)	cm/h	No.1	No.2
0.5 × 20.0	800	4.7	4.0
1.0 × 5.0	200	4.8	4.9
1.6 × 2.0	80	5.0	4.1

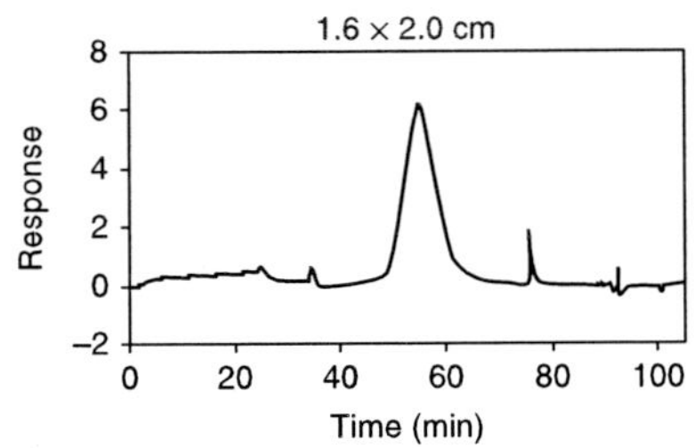

Figure 7.3 Purification of rFVII by AIEX [4]. The flow rate is constant at $Q = 40$ CV/h, while the velocity and the bed height change by a factor of 10. The data for the two main impurities show identical results. The calculated pressure drop is changed by a factor of 100. *Source*: Christensen [4]. Reproduced with permission of American Chemical Society.

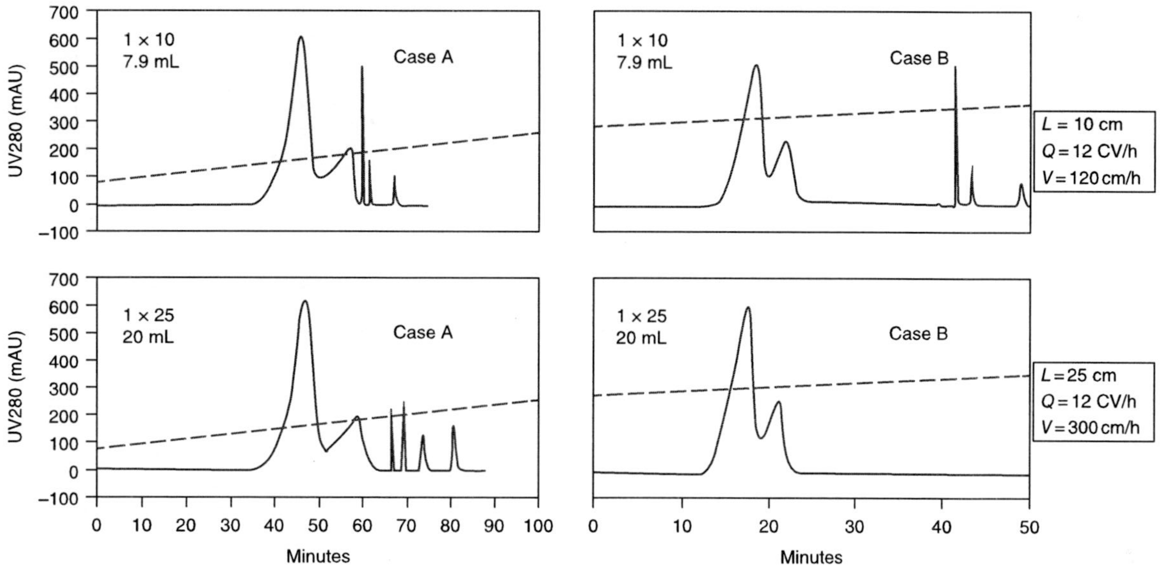

Figure 7.4 Separation of human insulin (HI) and deamidated HI on RPC [5]. Two cases (different gradients) are each at two bed heights (10 and 25 cm). The flow rate is constant at $Q = 12$ CV/h, while the velocity and the bed height changed by a factor of 2.5. The separation is the same with a slightly increased performance for from $L = 10$ to $L = 25$ in case A. *Source*: Hansen [5]. Reproduced with permission of RPC/HIC.

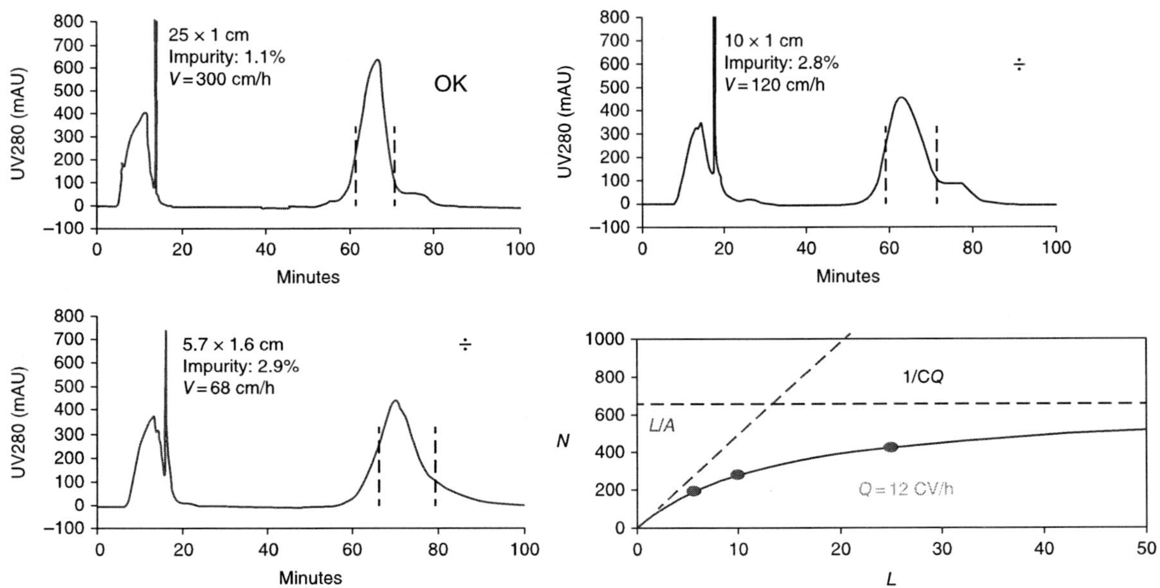

Figure 7.5 Purification of a recombinant peptide using RPC and bed heights of 25, 10, and 5.7 cm [6]. The flow rate was constant at $Q = 12$ CV/h. The impurity level was too high when the bed height was 10 cm and below. A plot of the corresponding plate numbers versus the bed height (bottom right) shows how the assumption of $A/L \ll CQ$ does not hold in this case. A minimum bed height of 15 cm was later established.
Source: Kornbeck and Hansen [6]. Reproduced with permission of RPC/HIC. (*See insert for color representation of the figure.*)

7.4 Design Applications: How to Scale up from Development Data

7.4.1 Industrial Cases

In this section we present applications of the scale-up concept in actual production cases including basic scale-up, optimization, and actual process design of multiple steps (Figures 7.6, 7.7, 7.8, and 7.9).

7.4.2 Process Design: Multiple Steps

This example is from the process design of a new facility for the production of a recombinant peptide. Three steps (A, B, C) are considered, each with a fixed set of process parameters (load, flow, gradients) and a given yield. In the original design there were two columns for each step each with a bed height of 25 cm, a standard value at the time. The problem was that using this design, we were not able to meet the capacity demand because of a bottleneck in step B. A third column was considered for step B, which is costly both in terms of equipment and footprint. At this time the concept of scaling on a volume basis was emerging out of development, and this case therefore became the first major production design using the new concept.

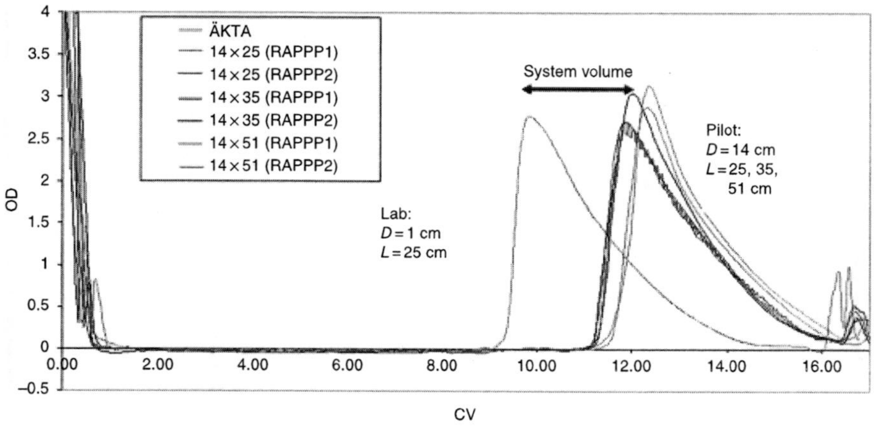

Figure 7.6 Scale-up of an RPC step for a recombinant peptide including a scaling factor of ~400 and the bed height × 2 (S. Kidal and L. Sejergaard, Private Communication). All the analytical data were well within specification. A slight improvement in purity was observed with increasing bed height. The difference in retention volume is due to the different dead volumes in the systems. *Source*: S. Kidal and L. Sejergaard, Private Communication. Reproduced with permission of Wiley.

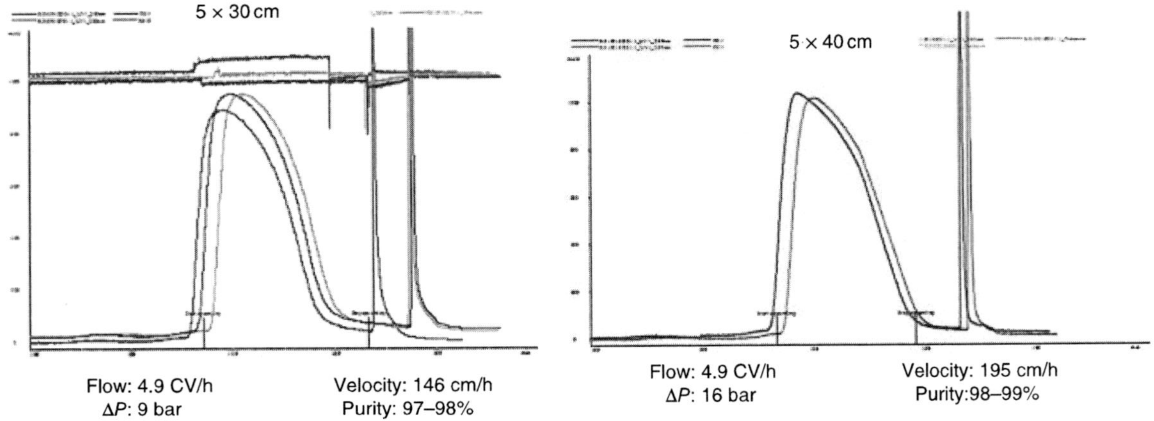

Figure 7.7 Increasing the production capacity for human insulin on RPC by 33% by increasing the bed height from 30 to 40 cm at constant flow rate (validation at intermediate scale, $D = 5$ cm (A. Laursen and E. Hansen, Private Communication)). A slight increase in purity is observed with increased bed height. As a bonus this increases the robustness of the step. *Source*: A. Laursen and E. Hansen, Private Communication. Reproduced with permission of *European Journal of Interdisciplinary Studies*.

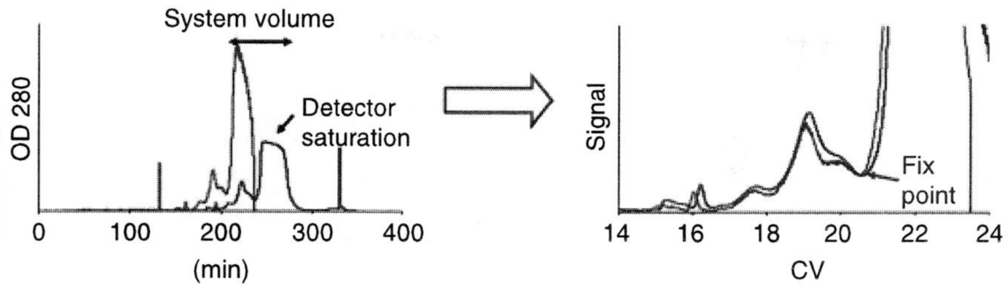

Figure 7.8 Scale-up of an IEX step for a recombinant peptide including a scaling factor of ~550 and the bed height from 25 cm (lab, black) to 35 cm (pilot, gray) (S. Kidal and O.E. Jensen, Private Communication). The chromatograms show big differences (left figure) due to a difference in the system volume and saturation of the UV detector in the lab run (black). For better comparison the two profiles are aligned in the valley of the separation (right), and this shows that there is only a small difference in separation, with the pilot profile (gray) being slightly sharper. No significant difference was found in the analytical results. *Source*: S. Kidal and O.E. Jensen, Private Communication. Reproduced with permission of Elsevier.

The three steps had the following design window:

Step A (IEC): $L = 10–31$ cm
Step B (RPC): $L = 20–42$ cm
Step C (RPC): $L = 20–46$ cm

The required capacity is set to 100%. Four scenarios were considered each using two columns per step (Figure 7.10):

I: The initial design with $L = 25$ cm for all columns. This gave a capacity of 95% with step B as bottleneck.
II: Adjustment of the bed heights to meet capacity. This gave bed heights of 23, 26, and 20 cm. Note that the bed height is increased for step B to resolve the bottleneck but lowered for step A and C. The capacity is 100% and there are no bottlenecks.
III: What is the maximum capacity? We increase the capacity by increasing the bed heights of all three steps until we reach the maximum bed height for one of them. In this scenario we reach a capacity of 132% with bed heights of 31, 35, and 24 cm and step A as bottleneck.
IV: Redesign of step C that had a low yield. From lab studies it was known that the yield could be increased from 75 to 90% if the load was lowered to 50% of the set point. In order to meet capacity, the column would then have to be much longer. This scenario could produce 159% of the required capacity. The bed heights were 29, 33, and 46 cm with step C as bottleneck.

In the end scenario II was chosen with scenario III as an alternative if capacity requirements increased. Although technically attractive, scenario IV was disregarded due to the amount of documentation required.

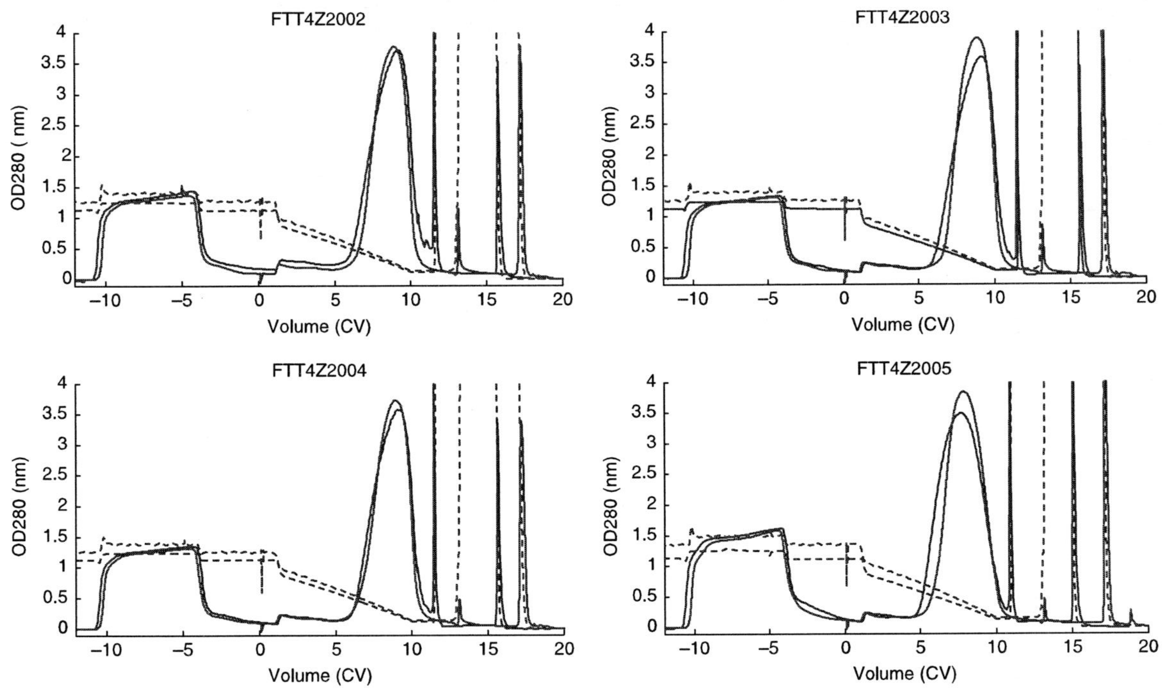

Figure 7.9 Scalability study of a recombinant protein purification using HIC (T.B. Hansen, Private Communication). OD280 versus volume (CV) for four batches run in production (48 L, 16.5 cm bed height) and simultaneously in the laboratory
(58 mL, 8.2 cm bed height). Elution starts at $V = 0$. The scaling factor is 1208 with 2× for the bed height. Still, both the chromatograms and the analytical results are identical. *Source*: T.B. Hansen, Private Communication. Reproduced with permission of Novo Nordisk. (*See insert for color representation of the figure.*)

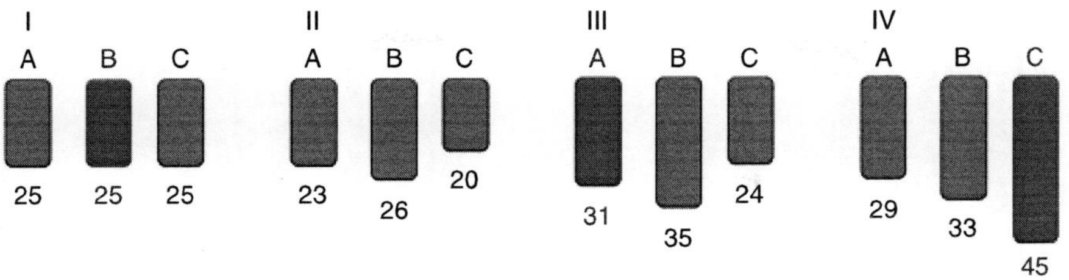

Figure 7.10 Process design for three columns in four scenarios. Numbers are the bed height and the columns IB, IIIA, and IVC are the bottleneck (E. Hansen and I. Christiansen, Private Communication) in the four scenarios. *Source*: E. Hansen and I. Christiansen, Private Communication. Reproduced with permission of Copenhagen.

This example illustrates that one should be careful using the expression *bottleneck*. Unless you design for maximum capacity (scenario III or IV), there shouldn't be a bottleneck. It is also noteworthy how we can build in extra capacity from the beginning, without reserving room for an extra column, simply by choosing the column dimensions so that the bed height can be increased if required. In this case we have 32% extra capacity build in by starting with scenario II.

7.5 Discussion

Using the concept of scaling on a volume basis, we have transformed the bed height from being an important parameter defining the performance to being a design parameter that we adjust to meet capacity requirements in terms of the column volume. Together with the column diameter, this gives us a continuous range for the column volume and therefore a greatly increased flexibility at all scales. The bed height is not important for the product quality, as long as we know the column volume and can adjust the process volumes and flows accordingly. This makes packing less stressful as we do not need to hit a narrow range for the bed height.

So is this a revolution in scaling? No. Is it useful and easy to apply? Yes. If you ask people working with continuous reactors, scaling with the residence time will not lift an eyebrow. If you ask people working with water purification (Amberlite-type resins), they will tell you to keep the residence time constant. This is also true for biotech people in recovery designing capture steps. On the other hand, if you ask biochemists or pharmacists with a background coming from analytical chromatography, they will tell you that the bed height is important. This is also the case if you read the guidelines from for example, *GE Healthcare*. So, how can this be? Who got it wrong? The answer is not surprisingly "it depends." More precisely the answer is in the rearranged Van Deemter equation:

$$N = \frac{1}{(A/L) + CQ}$$

When the contribution from axial dispersion, A/L, is much larger than the contribution from mass transfer, CQ, the plate number is proportional to the bed height, $N = L/A$, and independent of the flow rate. This is the case for small molecules with high diffusivities and for small chromatographic particles, both making the C term small, possibly combined with a low flow rate or low bed height. The case of small particles is definitely true for HPLC analysis. Having used chromatography mainly for analysis, you will "know" that the bed height is important. At the opposite end, when $CQ \gg A/L$ only the flow rate is important, as you will "know" having worked with capture steps (or water purification) using large particles to reduce the pressure drop. Figure 7.11 is an example comparing the analytical to the preparative case. It is normal that the A term is proportional to the particle diameter while the C term is proportional to the diameter squared. The parameters here are calculated as $A = 0.0048 \times \text{dp}$ and $C = 4 \times 10^{-7} \times \text{dp}^2$, with the particle diameter dp in μm. This corresponds to working with the same protein using two different particle sizes. In this case we use 5 μm particles and $Q = 15$ CV/h for the analysis and 50 μm particles and $Q = 30$ CV/h for the preparative case. This gives a factor of 10 for A and a

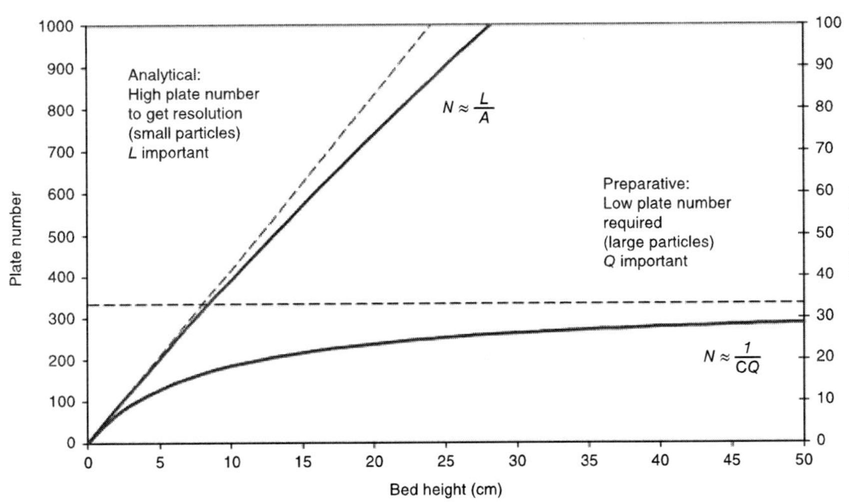

Figure 7.11 Plate number for typical analysis (left) and preparative mode (right). Dashed lines are L/A (left) and $1/CQ$ (horizontal, right), respectively. For the preparative case the particle diameter is 10× and the flow rate 2× compared to the analytical case.

factor of 100 for C (200 for CQ) between the two cases. As a consequence we have around 30 plates for the preparative case and around 1000 for the analytical case.

The example illustrates why we can have different experience working with chromatography depending on which application we have worked with. It also highlights the importance of knowing *why* and not only *how*. Scaling on a volume basis is not difficult to understand or to apply. Still, it took years for the concept to get accepted within our company. The example in Figure 7.11 helps to understand why.

7.6 Recommendations

This section includes a guide to using scale-up on a volume basis during process development. In addition we have included a list of the most common challenges encountered during scale-up. These challenges are not related to the present scale-up concept but apply in general when changing the volume significantly. We present them for convenience to prevent mistakes when you test the scale-up concept.

The flexibility provided by the volumetric scale-up is mainly used for seamlessly adjusting the column volume at scale. This can be in a factory design, deciding which columns to purchase and in a pilot context where the column volume can be adjusted from batch to batch based on the batch size. In the succeeding text we describe how to apply the scale-up principle based on development data and list the main design applications, followed by actual case studies related to production.

7.6.1 How to Scale up from Development Data

The concept of scaling on a volume basis is easy to apply and consists basically of two steps:

1) *Determine the minimum bed height at the relevant flow rate.* This is done in the laboratory. Often we are not very interested in finding the actual minimum value possible (~edge of failure), and therefore we just test the minimum relevant bed height. We would normally not want to pack columns shorter than 10 cm in production scale, so therefore there is no need to test columns shorter than $L = 10$ cm. Exceptions are capture steps and when dealing with virus validation where using a small column in the laboratory is essential and a low bed height therefore relevant. In this case an effort is put into showing scalability down to a lower bed height.
2) *Know the pressure limits in production scale.* Once you know the maximum pressure you can calculate the maximum bed height from the pressure

equation presented in *Theoretical Background*. Preferably the calculation is based on data from the relevant column at scale in order to get the most accurate pressure-flow correlation. If this is not possible, you will have to use lab data and assume the same pressure-flow correlation. The maximum bed height is calculated from a reference run as

$$L_{\max} = L_{R}\sqrt{\frac{P_{\max}Q_{R}}{P_{R}Q}}$$

where R indicates the reference run. Notice that the reference run does not have to be at the same flow rate. If you got a pressure drop of 2 bars when running a 15 cm column at 10 CV/h and the maximum pressure you can have is 3 bars with 8 CV/h as set point for the flow rate, the calculation is

$$L_{\max} = 15\times\sqrt{\frac{3\times10}{2\times8}} = 20 \text{ cm}$$

The pressure limitations are sometimes obtained from the resin limitations and sometimes from the column limitations, whichever is the lowest. The maximum bed height can thus depend on the hardware used and is therefore an equipment (scale)-dependent parameter.

Note that we do not need to determine the constants of the plate number or the pressure drop correlations. We only need to determine a minimum bed height at the relevant flow rate in the laboratory, know the pressure limits in production scale, and have a reference run.

Our process description will read something like:

Flow rate: 8 CV/h
Load: 7–10 g/L
Bed height: >10 cm*

*Bed height is limited upward by the pressure drop.

We do not operate with a set point for the bed height. The bed height is a design variable adjusted to meet the capacity requirements. We have tested the lower limit for the bed height as part of development. The upper limit is not fixed in centimeter but is determined from the pressure. In this way we have "validated" the upper limit based on theory.

7.6.2 The Real Challenges of Scale-Up

There are several challenges to deal with when scaling a chromatographic process. As you hopefully have come to recognize the principle of scale-up is not one of them. The real issues of scale-up are related to column packing and to equipment dimensions or design. Here are few of the most important:

Column packing. The packing pressure depends on the bed height to diameter ratio L/D. Packing a slim column requires much more pressure than a wide (pancake) column, because a lot of the pressure drop comes from the wall support. Therefore, if you pack using the same pressure in large and small scale, you get different densities, and if you are unlucky you will crush the resin in the large (wide) column. We have on several occasions realized that the high pressure resin operated at 30–50 bars in the lab could only withstand 10 bars in large scale. In other words the maximum pressure, which we use to define the maximum bed height, is definitely a scale-dependent parameter. This is something we would like resin vendors to understand some day and provide us with relevant data. For a better understanding of the pressure restrictions for resins at different aspect ratios, the paper by Stickel and Fotopoulos [7] is highly recommended.

Dead volume The dead volume or system volume is the volume in the surrounding piping and solvent mixer (for gradients). Not knowing your dead volumes at the different scales can lead to serious errors. You can end up trying to adjust the retention volume by adjusting the gradient, because you think your buffers have the wrong composition, when in fact it is the dead volumes that are different (measured in CV). You must know your dead volumes and correct them at both end of the scale.

Mixer volume. Having a large mixer volume will apart from the volume delay itself also smear the gradient leading to an error in retention and possibly decrease the separation. You should keep your mixer volume as low as possible, preferably less than 0.1 CV.

References

1. Kidal, S. and Elvang, O., Using volumetric flow to scale up chromatographic processes. *BioPharm International*, 2006, 19: 34–43.
2. Al-Jibbouri, S., Scale-up of chromatographic ion-exchange processes in biotechnology. *Journal of Chromatography A*, 2006, 1116: 135–142.
3. Mollerup, J. and Pedersen, L., Scale-up of chromatographic separations in ion-exchange chromatography, poster presented at ISPPP, Ljubljana, Slovenia, November 5–8, 2000.
4. Christensen, J., Chromatographic scale-up at constant residence time, presentation at American Chemical Society 216th National Meeting, Boston, MA, August 23–27, 1998.

5. Hansen, E., Scale-up on a volume basis, presentation at third RPC/HIC, Athens, Greece, February 5–9, 2001.
6. Kornbeck, C. and Hansen, E., On the design of a preparative chromatography step, poster presented at third RPC/HIC, Athens, Greece, February 5–9, 2001.
7. Stickel, J. J. and Fotopoulos, A., Pressure-flow relationships for packed beds of compressible chromatography media at laboratory and production scale. *Biotechnology Progress*, 2001, 17: 744–751.

8

Scaling Up Industrial Protein Chromatography

Where Modeling Can Help

Chris Antoniou, Justin McCue, Venkatesh Natarajan, Jörg Thömmes, and Qing Sarah Yuan

Biogen, Cambridge, MA, USA

8.1 Introduction

Industrial-scale purification of biopharmaceuticals relies heavily on chromatographic separations, mostly driven by the superior resolution that chromatography offers over alternative methods. Despite this widespread utilization, development of industrial protein chromatography continues to be based on experimental approaches; most attempts to improve process development have focused on generating more experimental data faster through microscale high-throughput experimentation and a statistical evaluation of the generated data. The substantial academic work in the modeling of protein separations has not yet found its way into industrial process development. This is on one hand surprising since by now the fundamental equations underlying the separations are well understood, and there are many examples of models for different separation mechanisms that describe protein separations quite well [1]. On the other hand, these modeling approaches often only explain experimental behavior "after the fact," while *a priori* prediction of chromatographic separations of proteins still seems elusive, despite the excellent advances made through quantitative structure–property relationship (QSPR) approaches [2], see also Chapter 6. As the biomanufacturing industry matures and manufacturing becomes more industrialized, models of chromatographic separations will certainly find their way into both the development space and control strategies.

Another aspect of chromatographic theory that is not utilized in industrial practice sufficiently is the crucial link between small-scale data and routine manufacturing at process scale. Scale-up of chromatographic protein separation still appears to be regarded as somewhat of an art, which is based on a few time-tested rules and requires substantial experience by manufacturing and its support functions. The largely empirical approach to scaling up protein chromatography is most evident in the way column packing is approached, in terms of how columns are packed as well as the quality of packing is measured and controlled. This has in turn influenced how hardware manufacturers have designed large-scale chromatography columns.

A final aspect of long-term commercial biomanufacturing is the influence of variations in raw material quality on process stability. Chromatographic separations are not exempt from such issues. The most critical raw material in these steps is the stationary phase, whose properties will influence the separation. Typically, these questions are addressed by requiring very tight specifications of the incoming raw materials; often neglected is the power of theoretical understanding of the impact of variations and opportunities to use models to build operating spaces within which conditions exist to mitigate variations in raw materials.

Our goal in this contribution is to show that scaling up protein chromatography to relevant process scale does not have to be an art but can be supported by theoretical considerations. In the course of the chapter, we will first explore why column packing is such an important criterion that has to be consistent across scales. We will then discuss how models can be utilized to predict column packing across scales and to perform packing consistently in an industrial environment. From there we will progress to using computational fluid design as a tool to understand and predict the impact of column hardware, specifically fluid distribution systems, on operations at industrial scale. We conclude the chapter with a discussion of how theoretical models can be used to mitigate the impact of resin lot-to-lot variability in a quality by design (QbD) approach. We hope that this brief overview can contribute to the continued maturation of biopharmaceutical manufacturing from art to true industrial operation.

8.2 Packing Quality: Why and How to Ensure Column Packing Quality Across Scales

8.2.1 Impact of Packing Quality on Separations

It is well established that the packing quality of a chromatography column can affect the separation performance. Martin and Synge [3] modeled the chromatography column as a series of mixing tanks and developed the concept of number of theoretical plates (N) and height equivalent to a theoretical plate (HETP). These have become the customary measurements for column efficiency and packing quality. While N is directly related to column resolution for analytical-scale chromatography (which utilize chromatography adsorbents with particle sizes ≤ 10 μm), for preparative-scale chromatography, the impact of N on column resolution is less prominent because intraparticle diffusion is usually the rate-limiting step for mass transfer. Nonetheless, when chromatography is used to separate multiple species that possess similar chemical and physical properties, the separation performance could be sensitive to packing quality. Therefore, HETP is still a useful indicator of packing quality and is thus monitored routinely to ensure consistent and reproducible separation performance.

The impact of packing quality on protein separations can be modeled using well-established mass transfer equations coupled with appropriate adsorption and desorption isotherms. Equations 8.1 and 8.2 represent intraparticle diffusion and macroscopic mass transfer in a chromatography column [4]:

$$\frac{\partial q_i}{\partial t} = D_{\mathrm{eff},i}\left(\frac{\partial^2 q_i}{\partial r^2} + \frac{2}{r}\frac{\partial q_i}{\partial r}\right) \tag{8.1}$$

$$\tau\frac{\partial c_i}{\partial t} + L\frac{\partial c_i}{\partial z} + 3\frac{1-\varepsilon}{\varepsilon}\frac{k_{\mathrm{f}}\tau}{R}\left(c_i - c_i^s\right) = \frac{L^2}{\mathrm{Pe}}\frac{\partial^2 c_i}{\partial z^2} \tag{8.2}$$

In the previous set of equations, c_i and q_i refer to the liquid- and solid-phase concentration of each solute species, respectively; $D_{\mathrm{eff},i}$ is the effective intraparticle diffusion coefficient; τ is the residence time; L is the packed bed height; R is the adsorbent particle radius; k_{f} is the external film mass transfer coefficient; ε is the porosity of the packed bed; and Pe is the Peclet number. The column residence time (τ) is a function of the column length and fluid-phase interstitial velocity (u), while the Peclet number is related to the column length, interstitial velocity, and dispersion coefficient (D_{a}) [5, 6].

$$\tau = \frac{L}{u} \tag{8.3}$$

$$\mathrm{Pe} = \frac{uL}{D_{\mathrm{a}}} \tag{8.4}$$

The Peclet number is a measure of macroscopic flow dispersion in the column and is directly impacted by flow inhomogeneity. As such, a number of correlations exist between N and Pe [6, 7]. Under normal operating conditions for preparative chromatography, Pe is significantly greater than 1, and the following equation suffices:

$$\mathrm{HETP} \cong \frac{2L}{\mathrm{Pe}} \tag{8.5}$$

In Equations 8.1 and 8.2, values for intraparticle diffusion coefficient $D_{\text{eff},i}$ and external film mass transfer coefficient k_f can be estimated using published correlations [8]. Equilibrium isotherms for each species can be determined by fitting batch adsorption and desorption data at different concentrations to appropriate isotherm equations for different solute species [9]. Thus, Equations 8.1 and 8.2 can be solved simultaneously for a given set of operating conditions (L, u) for both the load and elution phases, and the effect of plate height on column performance can be assessed by varying the Peclet number. "Close–close" Danckwerts boundary conditions [5, 10], in which dispersion is limited to inside the column and plug flow is assumed at the column inlet and outlet, are typically applied. The initial condition for column loading is $c_i = q_i = 0$, while the initial condition for column elution is the c_i and q_i profiles in the column at the end of the loading and wash steps.

McCue et al. [4] modeled the column packing impact on separation efficiency of a hydrophobic interaction chromatography (HIC) column, which is commonly used for monomer and aggregate separations in the purification of protein therapeutics [11–13]. Their results demonstrated that column packing quality can impact the separation performance for preparative-scale chromatographic processes. Specifically, an increase in N from 25 to 300 improved the column performance in terms of both product yield and aggregate clearance (Figure 8.1). In the particular system studied, the performance levels at about $N = 150$, which corresponds to an HETP of ~0.1 cm. Furthermore, at longer residence times, interparticle dispersion becomes the dominant mass transfer limitation. Therefore, column performance is more sensitive to changes in N with increasing residence time.

The simulation approach described in this section can be widely applied to a variety of modes of chromatographic separations other than HIC. Such simulations can be used as a predictive tool to understand the relationship between column packing quality, process throughput, and product purity prior to process scale-up for preparative-scale manufacturing processes.

It is important to note that the simulation presented here excludes any extra-column flow dispersion. Extra-column flow dispersion, which includes contributions from the column flow adaptor and skid piping in industrial-scale chromatography equipment, can sometimes contribute significantly to the measured HETP values. Extra-column flow dispersion effects are discussed in Section 8.2.2.

8.2.2 Predicting Packing Quality Across Scales

As column efficiency can impact its performance, it is important to understand packing quality across scales. There are a number of different column packing methods, with advantages and disadvantages existing for each method.

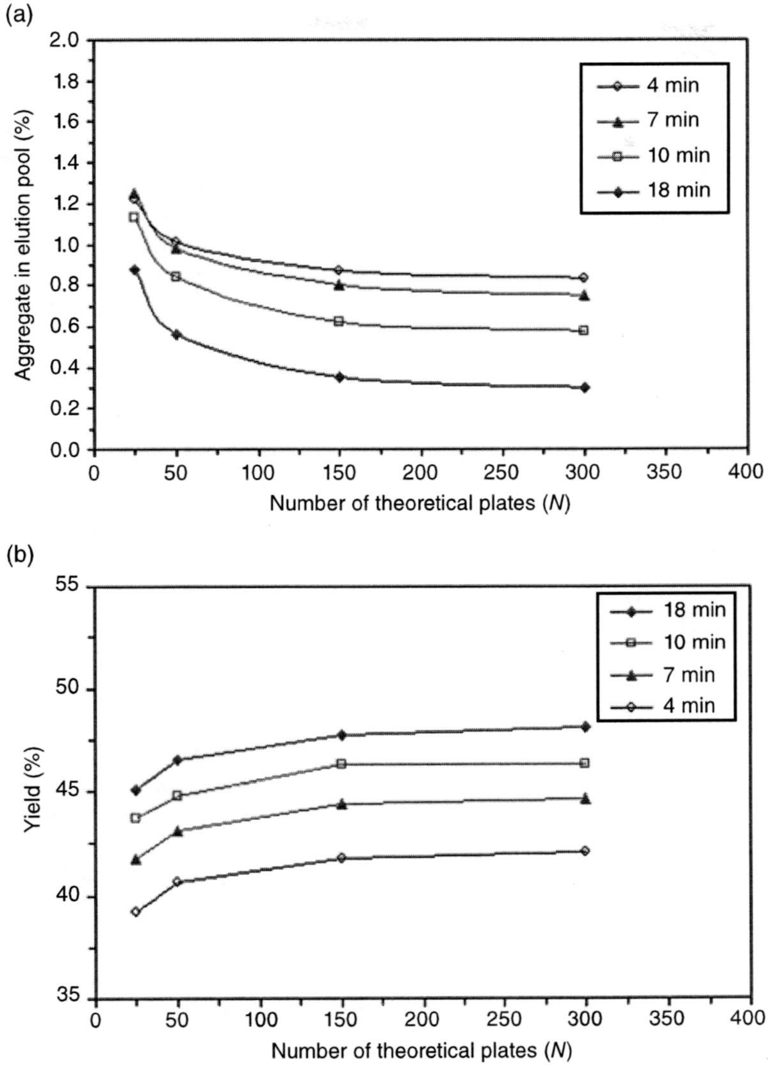

Figure 8.1 Model simulations of the aggregate levels (a) and yield (b) in the elution pool of HIC columns packed with increasing number of theoretical plates (N). *Source*: McCue et al. [4]. Reproduced with permission of Elsevier.

For industrial-scale columns, dynamic axial compression and ram packing are among the most prevalent packing methods [14].

During axial compression packing, the column is first filled with slurry containing the chromatography media homogeneously suspended in the packing fluid. The slurry is allowed to settle by gravity before the top flow adaptor is installed, and the bed is further compressed by flow of liquid through the column. Once the bed is considered packed, the flow adapter is moved to the top of the packed bed and sealed. During ram packing, the top adaptor is lowered against the top of the resin bed, and the column is packed using only mechanical compression.

Sarker et al. [15] and Stanley et al. [16] carried out experiments to study the consolidation behavior of several conventional chromatography media during axial compression, and found that the packing density varies with the nature of the solid phase, the applied pressure, and the number of packing cycles the resins have gone through. The nature of the fluid phase is also important to packing density and homogeneity. Furthermore, the kinetics of bed consolidation under pressure can be significantly different for irregular and spherical particles. Irregular particles tend to consolidate slowly by chaotic steps of random height at random time intervals, whereas smooth spherical particles tend to consolidate rapidly with smooth, monotonic height decrease [17].

It is important to note that when discussing packing inhomogeneity, chromatographers are primarily concerned about radial inhomogeneity, which increases macroscopic dispersion and, consequently, the HETP value. One special case of radial inhomogeneity is the wall effect. The local porosity at the wall region can be significantly different from that in the center of the column. This is why it is important to consider both the axial and radial dimensions when modeling the column packing process.

Column packing is essentially a process of pressure filtration of concentrated slurry. Kynch's model for the behavior of gravity settling [18] sets the basis for most current studies. In contrast to gravitational settling, few mathematical analyses exist for pressure filtration. Landman et al. [19, 20] extended some simple linear models for constant liquid expression rate and constant applied pressure to incorporate the yielding solid rheological properties of flocculated suspensions. Later, Keener et al. [21–23] formulated a model able to describe the pressure–flow properties of several adsorbent types subject to both mechanical and flow compression. However, all of these models are one-dimensional and assume radial uniformity. Hence they cannot be used to predict scale effects on flow distribution within the column.

Yuan [14] developed a two-dimensional theoretical model to calculate the local axial and radial displacements during the column packing process for a range of column aspect ratios. Their results show that wall friction is the dominant force providing support for the bed in small diameter columns. As the column diameter increases, the wall effects gradually diminish toward the center of the column. In addition, the compressive and shear stress profiles in the column indicated that plastic deformation, that is, deformation caused by particles sliding against each other, will begin around the wall, resulting in these regions to be formed first.

Östergren et al. [24–26] developed a Cauchy elastic two-dimensional model to describe the pressure–flow behavior using both liquid flow and mechanical compression packing. Material nonlinearity resulting from particle compression was accounted for by expressing modulus of elasticity as an empirical function of particle porosity. Parameters for the empirical modulus were measured experimentally. The two-dimensional model predictions for pressure–flow behavior agree with lab-scale experimental data, but comparison over a range of column dimensions, scales, and different adsorbent types was not evaluated.

McCue et al. [27] used a two-dimensional model to describe the pressure–flow behavior of compressible adsorbent phases for protein chromatography at different temperatures and column scales. Using a single fitted value for the empirical modulus parameters, the model was applied to describe the pressure–flow behavior of three commonly used adsorbents packed using both fluid flow and mechanical compression. Simulation results were in agreement with experimental data and accurately predicted the pressure–flow and compression behavior of three adsorbents over a range of column scales and operating temperatures. The practical value of this study is that it can potentially improve the accuracy of the column scale-up process, allowing the use of limited laboratory-scale data to predict column behavior in large-scale applications.

Column packing can be modeled as the interaction of two coexisting pseudo-continuous phases: the assembly of packing particles and the percolating fluids. For the particle phase, the momentum balance is written as

$$-\nabla \cdot [\sigma] = \nabla P \tag{8.6}$$

$$\nabla P = \nabla p - \rho g \tag{8.7}$$

where σ is the stress tensor of the solid phase, P is the fluid pressure, ρ is the fluid density, and g is the gravitational constant.

The pressure gradient can be determined from Darcy's law:

$$-\frac{1}{\eta}\kappa \nabla P = V \tag{8.8}$$

where η is the viscosity of the fluid, κ is the column permeability, and V is the superficial linear velocity. The column permeability can be further expressed as a function of interstitial void fraction (μ), particle diameter (d_p), and an empirical constant (C), using the Blake–Kozeny model:

$$\kappa = \frac{d_p^2 \mu^3}{C(1-\mu)^2} \tag{8.9}$$

For steady-state linear elastic deformation, the stress (σ)–strain (ε) relationship can be approximated by Hooke's law:

$$\sigma = D\varepsilon \tag{8.10}$$

where D is a elasticity matrix defined by material properties, which include the modulus of elasticity (E) and the Poisson ratio (ν). Furthermore, ε is related to the displacement (u) through the following equation:

$$\varepsilon_{i,j} = \frac{1}{2}\left(\frac{\partial u_i}{\partial x_j} + \frac{\partial u_j}{\partial x_i}\right) \tag{8.11}$$

For the fluid phase, continuity equation for steady state, one-phase incompressible flow is

$$\nabla \cdot \left(V\right) = 0 \tag{8.12}$$

Combing Equation (8.8) with Equation (8.12) leads to

$$\nabla \cdot \left[-\frac{1}{\eta} \kappa \nabla P \right] = 0 \tag{8.13}$$

Equations 8.6 and 8.13 form a system of differential equations, with pressure gradient linking the liquid-phase pressure drop with solid-phase deformation. Equations 8.10 and 8.11 further connect the solid phase stress to displacement and the material properties. Physical properties of the absorbent particles (particle diameter, modulus of elasticity, Poisson ratio) and settling beds (wall friction coefficient and bed porosity) can be measured through independent experiments.

The pressure–flow and bed compression behavior over a broad range of column scales (5–200 cm ID) and settled bed heights (15, 30 cm) was examined in [27] during dynamic axial compression packing and after the bed is packed. In the case of packing process, the simulation results predicted a shift in the pressure–flow curve and critical velocity over the entire range of column diameters, especially notable at larger bed heights (Figure 8.2). Critical velocity continued to decrease, even for column diameters > 50 cm. The dependence of critical velocity on scale can be explained by considering wall friction force. For small columns (<5 cm ID), the wall friction extends over nearly the entire column radius. As the column diameter is increased, the wall support impacts a smaller portion of the column radius, resulting in a particle displacement gradient across the column radius. As a result, greater compression is required to achieve the critical velocity for intermediate column diameters. As the column diameter is further increased (>50 cm), the wall effect only impacts the outer rim of the column, resulting in a homogeneous core. Therefore, less compression was required to obtain the critical velocity for both small and large diameter

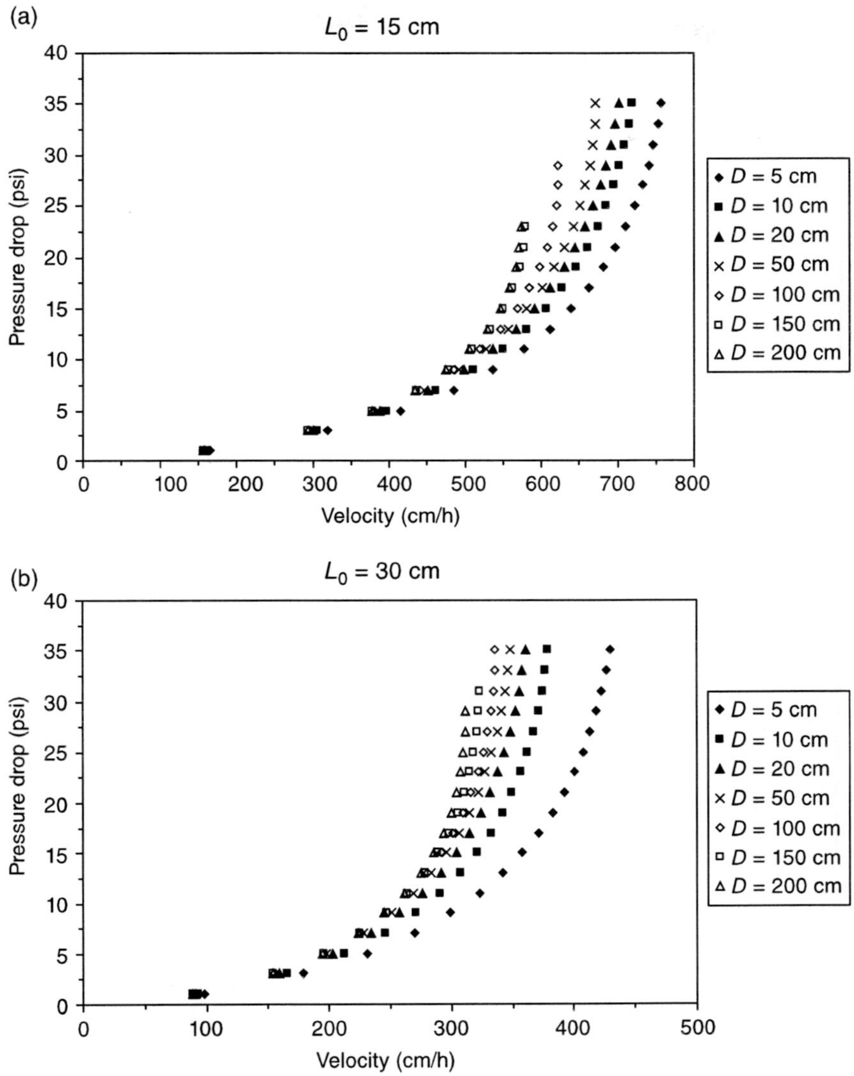

Figure 8.2 Model predictions for the pressure–flow behavior of Phenyl Sepharose 6FF at different scales for gravity-settled bed heights of (a) 15 cm and (b) 30 cm. *Source*: McCue et al. [4]. Reproduced with permission of Elsevier.

columns, where the displacement profiles were the most uniform over a column cross section (Figures 8.3 and 8.4). In addition to column diameter, the simulations also showed that the pressure–flow profile, critical velocity, and packed bed void fraction were dependent on the bed height. Thus, both column bed height and diameter may need to be considered when scaling up the column

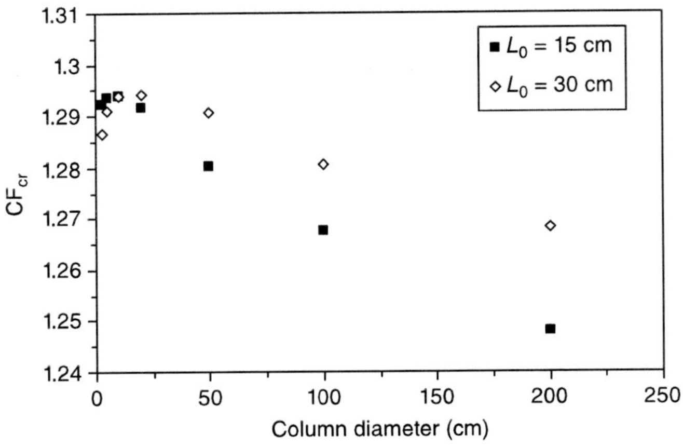

Figure 8.3 Column packing model predictions for Phenyl Sepharose 6FF showing the effect of column diameter (D) and gravity-settled bed height (L_0) on the compression factor at critical velocity (CF_{cr}) at the column center. *Source*: McCue et al. [4]. Reproduced with permission of Elsevier.

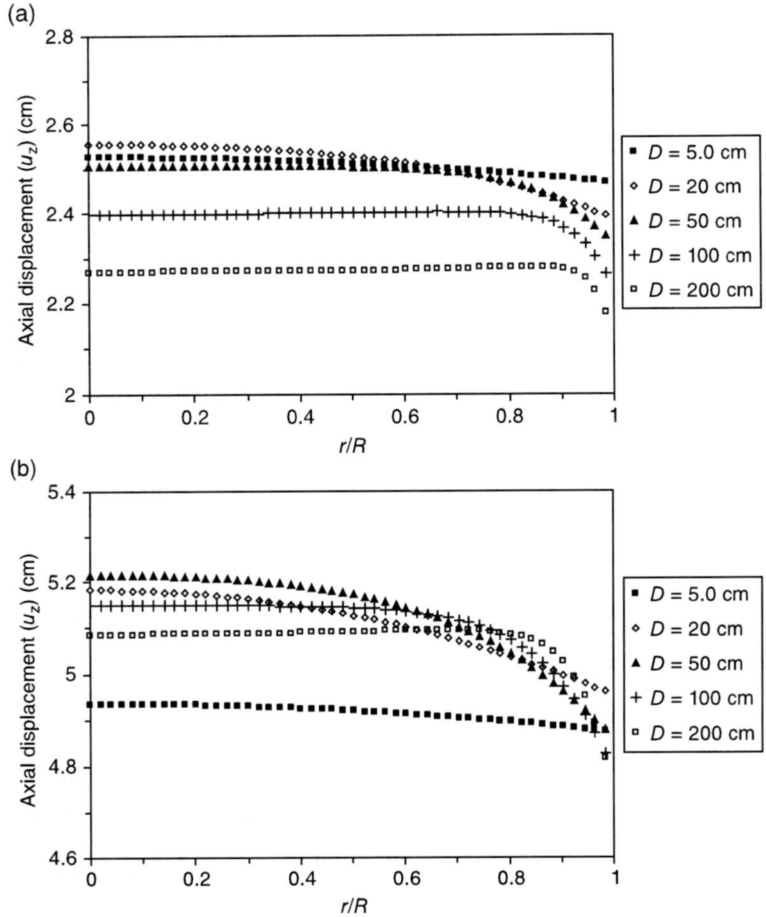

Figure 8.4 Column packing predictions for Phenyl Sepharose 6FF showing the effect of column radius (r) on the particle axial displacement profile for two different gravity-settled bed heights (L_0). (a) L_0 = 15 cm. (b) L_0 = 30 cm. *Source*: McCue et al. [4]. Reproduced with permission of Elsevier.

packing. The study also showed that model predictions were less sensitive to changes in the wall friction coefficient and Poisson ratio of the particle but more sensitive to settled bed void fraction, permeability constant, and particle diameter.

In the case of a packed bed, the simulations predicted the pressure–flow behavior approached a constant profile for large columns. This behavior is different from the packing process due to the inlet boundary condition that prevented particle displacement.

The simulation results were compared against experiments and found to be in good agreement with experimental measurements. This study demonstrates that simulation is a useful tool for guiding large-scale column packing, in which full-scale experimentation may be costly to perform.

8.3 Process Equipment: Using CFD to Describe Effects of Equipment Design on Column Performance

Chromatography processes are scaled up using scale-invariant process parameters, such as number of column volumes (CV), superficial velocity (u_s), and/or residence time (L/u_s, where L is the bed height) (see also Chapter 7). Typically, the bed heights are maintained constant (within tolerance) during scale-up, and the column diameter (D) is increased to accommodate the larger volumes. Thus, the aspect ratios (L/D) encountered at process scale can be significantly lower than those at lab scale. The differing aspect ratios can affect the packing process, the hydraulic behavior of and the flow distribution in the packed bed. It is very important for the chromatographer to understand these effects to ensure a robust and successful scale-up of a chromatography process.

The effect of scale on the hydraulic behavior of compressible resins has received a great deal of attention from practitioners. Both empirical [28] and theoretical [29] approaches have been employed to develop models to reliably predict pressure–flow behavior of compressible resins on scale-up. The hydraulic behavior at scale is typically a function solely of the resin compressibility and is relatively insensitive to the hardware configuration as long as the inlet and outlet piping sizes have been designed appropriately for the operational flow rates. The flow distribution in process-scale columns, on the other hand, may be affected by design features such as the number of inlets and outlets, size of anti-jets, header design [30, 31], location and number of retaining bolts, and the sealing design [32]. Thus, it is critical to understand how the particular design a vendor has selected impacts the resulting flow distribution in the packed bed.

Information on the axial and radial flow distribution within a packed bed is usually obtained by performing dye tests. Such tests involve injecting a finite volume of a colored dye solution into the column, stopping the flow when the dye is expected to be approximately halfway down the column, and excavating the packed bed to visually observe the dye pattern in the bed. The shape of the dye pattern would be indicative of the flow distribution within the packed bed. Such tests, though highly informative, are resource intensive and may not be feasible with all resin types as the dye needs to be unretained in the resin.

Computational fluid dynamics (CFD) is a numerical modeling tool to analyze problems involving fluid flow and associated transport phenomena. In the bioprocess industry, CFD has been employed to provide information on flow distribution in process-scale columns [32–34]. The tool provides similar and complementary data to dye tests while eliminating experimental complications. An overview of the CFD process can be found elsewhere [35].

In this case study, CFD was employed to evaluate the effect of two commercially available header designs on column performance and packed bed cleanability. The simulations were performed using a customized version of Fluent 6.3. Details of the numerical modeling can be found elsewhere [34]. The resin employed in this case study was a commercially available, compressible agarose resin. The two different designs that were considered were the flat header design from eastern rivers and the "ribbed" design from Pall Corporation. The flat design has four inlets and utilizes anti-jets at these inlets to prevent the incoming fluid momentum from transferring directly into the packed bed and to facilitate the radial distribution of the flow. In this design, the cross-sectional area available for the fluid within the header is radially constant. The "ribbed" header design, by contrast, has a single inlet with radially oriented channels that continually decrease in cross-sectional area as the fluid moves from the center toward the outer wall in order to maintain a constant velocity in the channel.

8.3.1 Model Verification

When using any modeling tool to develop and/or better understand a process, it is always recommended that the model be verified or validated against experimental data. This ensures that a model is capable of capturing the salient features of a process and, consequently, is reliable for the task at hand. In this case study, the simulation results were verified against two experimentally derived metrics—dye profiles obtained in a dye injection test and an unretained tracer peak profile.

A dye injection study was performed on a 1.4 m column with the "ribbed" header design to visualize the flow distribution in the packed bed. The column was packed using an optimized axial compression packing method to a bed height of 17.5 cm. The procedure consisted of equilibrating the column with 100 mM sodium hydroxide (NaOH), injecting a volume of dye (1 g/L phenol red in 100 mM NaOH) =2% of the CV, followed by a chase with 72 L of 100 mM NaOH.

Subsequently, another injection of the dye was made to introduce two distinct bands for later inspection. Following this second injection, the column was flushed further with ~140 L of 100 mM NaOH, and the flow was stopped when the first dye trace was expected to be close to the bottom flow cell. The fluid velocity during all steps was 60 cm/h. Finally, the packed bed was excavated to visualize the flow distribution. These were compared with the profiles predicted by the CFD simulations, the results of which are illustrated in Figure 8.5.

Figure 8.5 clearly indicates that the CFD model captures the salient features of the flow distribution in the actual packed bed. The presence of the central valve and the sealing design at the wall result in lower flow in these regions. The resultant "smile" is clearly evident in both the experimental and simulation results. Furthermore, the presence of retaining bolts to secure the mesh to the header results in streaking of the dye profile (Figure 8.5, top right). This experimental observation is well captured by the simulation (Figure 8.5, top left).

Finally, a simulated tracer profile was compared against an actual tracer profile from a 1.4 m column with the flat header design packed with the same resin to a bed depth of 13.7 cm (Figure 8.6). A ~1% CV tracer injection of 1.5 M sodium chloride (NaCl) against a background of phosphate buffer (pH 7.3) with 100 mM NaCl was made at a fluid velocity of 100 cm/h. The close agreement

Figure 8.5 The "smile" at the column center and the wall is clearly evident in the pictures at the bottom. (Top right) this figure illustrates the trailing dye profiles due to the retaining bolts. Both these features are captured in the simulation results (top left). *Source*: Johnson et al. [34]. Reproduced with permission of Wiley. (*See insert for color representation of the figure.*)

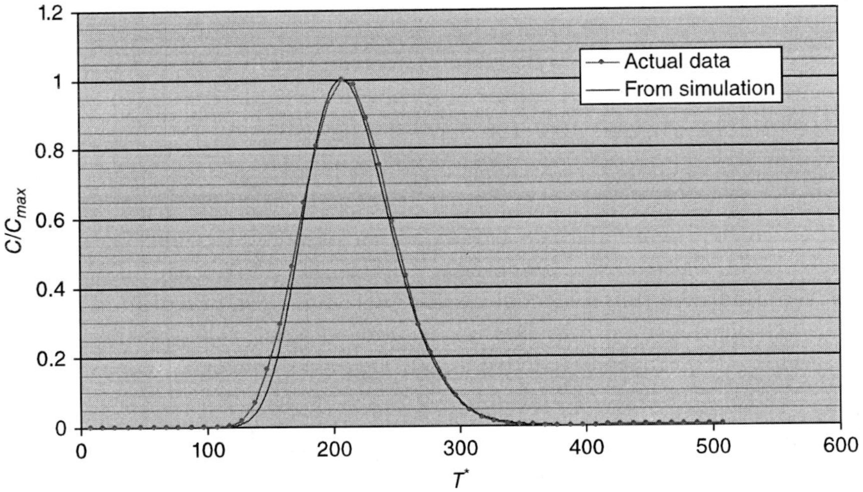

Figure 8.6 Comparison of actual and simulated tracer profiles in a 1.4 m column with a flat header design. The qualification fluid velocity was 100 cm/h. *Source*: Johnson et al. [34]. Reproduced with permission of Wiley.

between the simulated and experimental tracer peaks further validates the use of the model.

Once the model was verified, multiple scenarios were simulated to assess the difference between the two header designs.

Figure 8.7 illustrates the velocity contours in the packed bed for the two designs. The packed bed permeability is assumed to be spatially homogenous and the same in both header designs. As is evident, the flow distribution is relatively uniform over a large portion of the bed with both designs. However, there are qualitative differences between the two designs. In the case of the flat header design, reduced velocities are observed largely in the areas immediately below and above the slurry valves (Figure 8.7, top). On the other hand, in the case of the "ribbed" design, the areas of reduced velocity are located near the walls (Figure 8.7, bottom).

Figure 8.8 illustrates the pressure contours within the two columns at varying superficial velocities. Slightly higher pressure drops are encountered in the ribbed design (~3 to 4%). However, from an operational point of view, the increased pressures are not expected to negatively impact the operating velocities identified in small-scale experiments.

As mentioned previously, the two header designs yielded qualitatively different flow distributions within the packed bed. In order to gain insight on the effect of these differences on column quality (as measured by HETP and As) and solution transitions, the CFD simulations were rerun with the dispersion coefficient

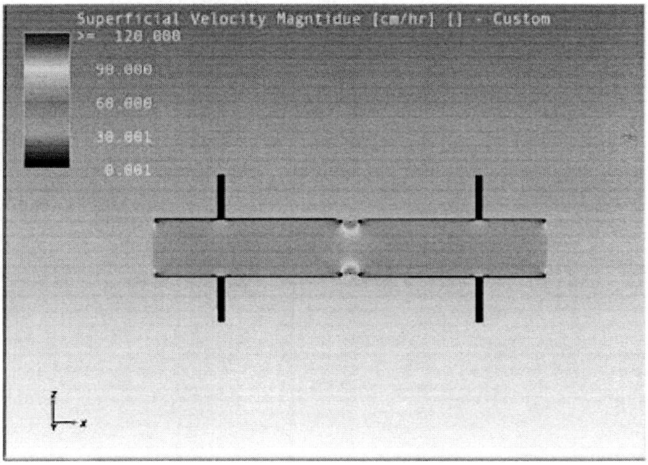

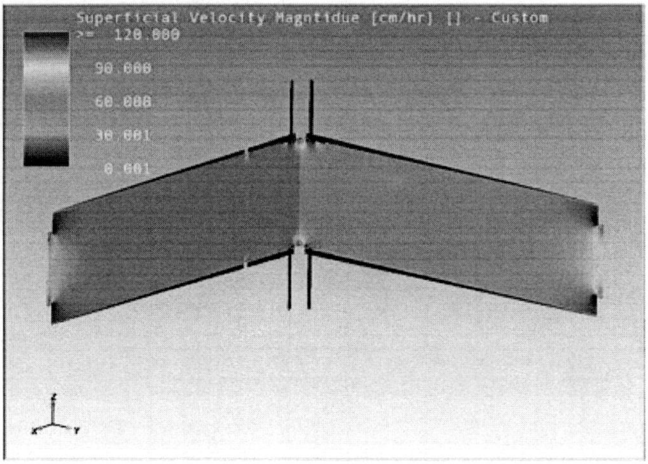

Figure 8.7 Velocity contours in the packed bed for the flat (top) and ribbed (bottom) designs. *Source*: Johnson et al. [34]. Reproduced with permission of Wiley.

set to zero. Figure 8.9 compares the unretained tracer peaks obtained from both geometries. The figure clearly indicates that the flow distribution resulting from the "ribbed" design yields a tight, symmetrical peak. On the other hand, significant tailing was observed in the case of the flat design.

Figure 8.10 compares the transition curves obtained with the two designs. The ribbed design yields a sharper transition as expected from the pulse input results. The tracer uptake within the packed bed is illustrated in Figure 8.11 for the two header designs. Even after 1.25 CVs, a significant fraction of the central portion of the packed bed has low levels of tracers (<10%) in the flat design.

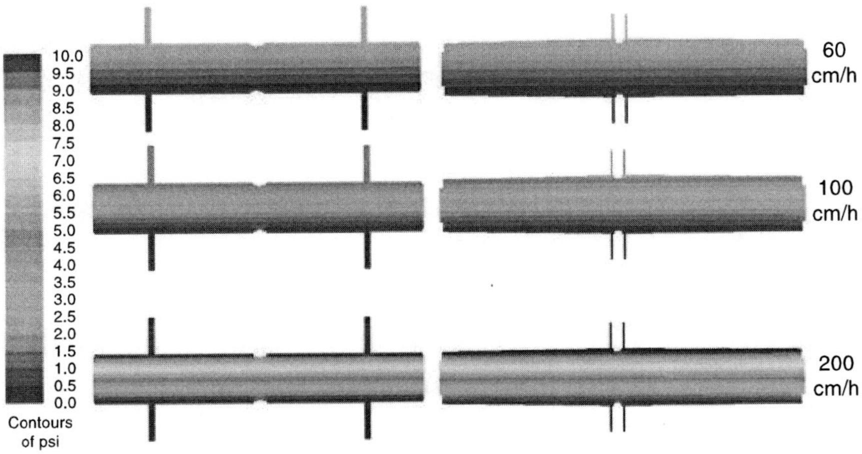

Figure 8.8 Pressure contours in the packed bed for the flat (left) and ribbed (right) designs. (*See insert for color representation of the figure.*)

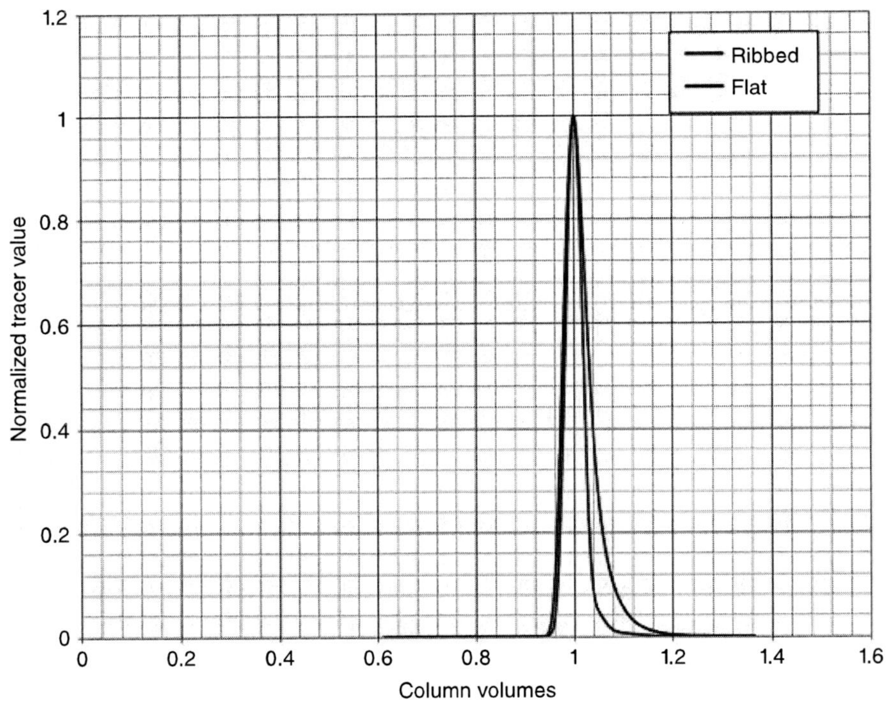

Figure 8.9 Tracer profiles obtained with the "ribbed" (narrow peak) and flat (broader peak) designs. The dispersion coefficient was set to zero in both cases. *Source*: Johnson et al. [34]. Reproduced with permission of Wiley.

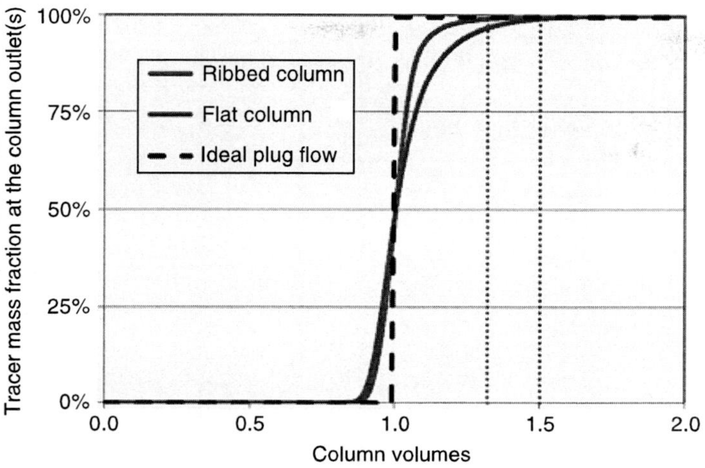

Figure 8.10 Tracer transition profiles at the column outlet(s) obtained with the "ribbed" (steeper curve) and flat designs. The ribbed design achieves 99% tracer saturation at the outlet after 1.32 CV compared with 1.50 CV for the flat design. *Source*: Johnson et al. [34]. Reproduced with permission of Wiley.

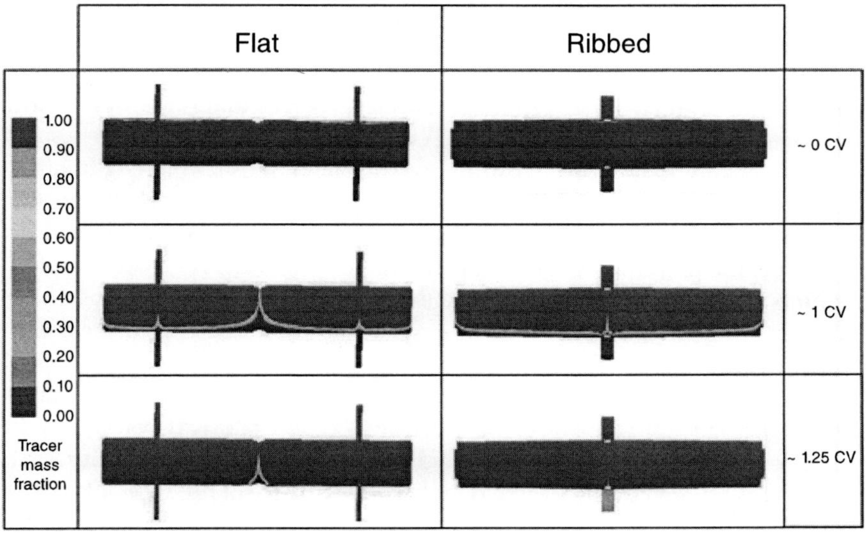

Figure 8.11 Tracer mass fraction contours on the $x = 0$ plane of the packed bed at times $t = 0$ CV (top), $t = 1$ CV (middle), and $t = 1.25$ CV (bottom) obtained with the "ribbed" and flat designs. Ribbed design contours utilize mirror imaging across the symmetry surfaces of the model to enable more equal comparisons of the two designs. *Source*: Johnson et al. [34]. Reproduced with permission of Wiley.

Consider a cleaning step in the chromatography process. Figures 8.10 and 8.11 indicate that only after 1.50 CV would the entire packed bed be at >99% of the inlet concentration of the cleaning agent. On the other hand, with the "ribbed" design, such a condition would be attained after 1.32 CV. Now, the typical duration of a cleaning step in a chromatography process is ~5 CV. Thus, the "ribbed" header design allows for incrementally higher effective cleaning agent contact time throughout the packed bed.

In summary, the CFD simulations were able to demonstrate tangible differences between the two column designs. The flow distribution in the packed bed (even when considered to be homogenous) was found to be qualitatively different in the two designs. The ribbed design yields a flow distribution that yields faster saturation times (potentially reducing cleaning solution volumes) and whose contribution to the overall HETP is lower than that of the flat design. The case study provides insight into the various process column design features that could have important implications for scaling up a chromatography process and should be carefully considered when selecting a column vendor.

8.4 Long-Term Column Operation at Scale: Impact of Resin Lot-to-Lot Variability

In the large-scale manufacturing of the biologic therapeutics, resin lot variability is another challenge to maintain product purity consistency. The binding properties of product-related impurity species, such as undesirable isoforms, and low or high molecular weight species are often very similar to the desired product, making separation performance extremely sensitive to slight variation in process conditions or adsorbent binding properties [36]. The impact of resin lot-to-lot variability on separation performance is also highly dependent on the specific protein separation. Therefore, it is difficult and costly to select resin lots that are only suitable for a specific separation. Manufactures of the biologic therapeutics must also develop strategies to mitigate resin variability during process development.

The impact of resin lot-to-lot variability on separation performance can be simulated using Equations 8.1 and 8.2 from Section 8.2, with variability in ligand density reflected in the shape of the equilibrium adsorption isotherms. McCue et al. studied the aggregate removal performance of different lots of the same HIC resin with slightly different ligand densities [37]. Batch adsorption experiments showed that under binding conditions, the different resin lots had the same values for the binding strength and static capacity for both the desired product and impurity species within experimental error. However, under elution conditions, both binding strength constants and the static capacity for the weakly bound impurities can vary from lot to lot.

The predictive models can significantly reduce the development effort for the control strategy to mitigate resin lot variability. Typically, the desired product yield and impurity content in the eluate pool is defined during the process development stage. If the impact of resin variability is understood and the new resin lot ligand density is known, the protein manufacturer can potentially change the elution conditions within a predefined range to ensure that the eluate pool is consistent despite the resin lot change. Although most protein manufacturing processes today are not designed with such considerations, the industry is beginning to shift its focus on controlling input parameters to controlling performance output via QbD initiatives [38, 39].

8.5 Closing Remarks

Through more than 30 years of biomanufacturing, chromatographic separations have been a core process step in almost every industrial manufacturing process. Although the dramatic increase in process productivity has triggered many discussions about the ability of chromatography to keep step with productivity demands, there is very little evidence of chromatography losing its role as a core element of manufacturing. We predict that protein chromatography models will become an important part of process development and process control in the future. Today's processes are still dominated by empirical approaches to development and control, but we see already today an important role of models in scale-up and long-term robustness of industrial protein chromatography. Our contribution outlined four cases of using modeling approaches to strengthen scale-up and operation of protein chromatography in an attempt to demonstrate that the maturation of biomanufacturing can and should include modeling as one tool to increase predictability and robustness.

References

1. E. J. Close, J. R. Salm, D. G. Bracewell and E. Sorensen, "Modeling of industrial biopharmaceutical multicomponent chromatography," *Chemical Engineering Research and Design*, vol. 92, pp. 1304–1314, 2014.
2. Y. Hou and S. M. Cramer, "Evaluation of selectivity in multimodal anion exchange systems: a priori prediction of protein retention and examination of mobile phase modifier effects," *Journal of Chromatography A*, vol. 1218, pp. 7813–7820, 2011.
3. A. Martin and R. Synge, "A new form of chromatogram employing two liquid phases 1. a theory of chromatography 2. Application to the micro-determination of the higher monoamino-acids in proteins," *Biochemistry Journal*, vol. 35, pp. 1358–1368, 1941.

4. J. McCue, P. Engel and J. Thommes, "Modeling the effects of column packing quality and residence time changes on protein monomer/aggregate separation," *Journal of Chromatography A*, vol. 1216, pp. 4895–4901, 2009.
5. H. Fogler, *Elements of Chemical Reaction Engineering*, 2nd ed., Engelwood Cliffs, NJ, Prentice Hall, 1992.
6. J. Deemter, F. Zuiderweg and A. Klinkenberg, "Longitudinal diffusion and resistance to mass transfer as causes of nonideality in chromatography," *Chemical Engineering Science*, vol. 5, pp. 271–289, 1956.
7. A. Athalye, "Effects of high sample loading in nonadsorptive protein chromatography". PhD thesis. University of Wisconsin–Madison, Madison, 1993.
8. S. Foo and R. Rice, "On the prediction of ultimate separation in parametric pumps," *AIChE Journal*, vol. 21, pp. 1149–1158, 1975.
9. J. McCue, P. Engel, A. Ng, R. Macniven and J. Thommes, "Modeling of protein monomer/aggregate purification and separation using hydrophobic interaction chromatography," *Bioprocess and Biosystems Engineering*, vol. 31, pp. 261–275, 2008.
10. O. Levenspiel and K. Bischoff, "Patterns of flow in chemical process vessels," *Advances in Chemical Engineering*, vol. 4, pp. 95–198, 1964.
11. C. Shepard and A. Tiselius, "The chromatography of proteins. The effect of salt concentration and pH on the adsorption of proteins to silica gel," *Discussions of the Faraday Society*, vol. 7, pp. 275–285, 1949.
12. M. Hearn, "Physicochemical factors in polypeptide and protein purification and analysis by high-performance liquid chromatographic techniques: current status and challenges for the future," In *Handbook of Bioseparation*, S. Ahuja, Ed., New York, Academic Press, pp. 71–229, 2000.
13. J. Queiroz, C. Tomaz and J. Cabral, "Hydrophobic interaction chromatography of proteins," *Journal of Biotechnology*, vol. 87, pp. 143–159, 2001.
14. Q. Yuan, *Improving Flow Distribution in Chromatographic Columns*, Madison, University of Wisconsin–Madison, 1999.
15. M. Sarker, A. Katti and G. Guiochon, "Consolidation of the packing material in chromatographic columns under dynamic axial compression. II. Consolidation and breakage of several packing materials," *Journal of Chromatography A*, vol. 719, pp. 275–289, 1996.
16. B. Stanley, M. Sarker and G. Guiochon, "Consolidation of the packing material in chromatographic columns under dynamic axial compression. IV. Mechanical properties of some packing materials," *Journal of Chromatography A*, vol. 741, pp. 175–184, 1996.
17. E. J. Fernandez, C. A. Grotegut, G. W. Braun, K. J. Kirschner, J. R. Staudaher, M. L. Dickson and V. L. Fernandez, "The effects of permeability heterogeneity on miscible viscous fingering: a three-dimensional magnetic resonance imaging analysis," *Physics of Fluids*, vol. 7, pp. 468–477, 1995.
18. G. Kynch, "A theory of sedimentation," *Transactions of the Faraday Society*, vol. 48, pp. 166–176, 1952.

19. K. Landman, C. Sirakoff and L. White, "Dewatering of flocculated suspensions by pressure filtration," *Physics of Fluids A*, vol. 3, pp. 1495–1509, 1991.
20. K. Landman and W. Russel, "Filtration at large pressures for strongly flocculated suspensions," *Physics of Fluids A*, vol. 5, pp. 550–560, 1993.
21. R. Keener, J. Maneval, K. Ostergren and E. Fernandez, "Mechanical deformation of compressible chromatographic columns," *Biotechnology Progress*, vol. 18, pp. 587–596, 2002.
22. R. Keener, J. Maneval and E. Fernandez, "Toward a robust model of packing and scale-up for chromatographic beds. 1. Mechanical compression," *Biotechnology Progress*, vol. 20, pp. 1146–1158, 2004.
23. R. Keener, J. Maneval and E. Fernandez, "Toward a robust model of packing and scale-up for chromatographic beds. 2. Flow packing," *Biotechnology Progress*, vol. 20, pp. 1159–1168, 2004.
24. K. Ostergren and C. Tragardh, "Numerical study of 2-dimensional compaction, flow, and dispersion in a chromatographic column," *Numerical Heat Transfer Part A*, vol. 32, pp. 247–265, 1997.
25. K. Ostergren, C. Tragardh, G. Enstad and J. Mosby, "Deformation of a chromatographic bed during steady state liquid flow," *AIChE Journal*, vol. 44, pp. 2–12, 1998.
26. K. Ostergren and C. Tragardh, "Modeling and analysis of axial flow through and compression of a non-rigid chromatographic bed," *Chemical Engineering Journal*, vol. 72, pp. 153–161, 1999.
27. J. McCue, D. Cecchini, C. Chu, W. Liu and A. Spann, "Application of a two-dimensional model for predicting the pressure-flow and compression properties during column packing scale-up," *Journal of Chromatography A*, vol. 1145, pp. 89–101, 2007.
28. J. J. Stickel and A. Fotopoulos, "Pressure-flow relationships for packed beds on compressible media at laboratory and production scale," *Biotechnology Progress*, vol. 17, pp. 744–751, 2001.
29. E. J. Fernandez, R. N. Keener, J. E. Maneval and R. A. Hart, "Advancement in the modeling of pressure-flow for the guidance of development and scale-up of commercial-scale biopharmaceutical chromatography," *Journal of Chromatography A*, vol. 1190, pp. 127–140, 2008.
30. J. Moscariello, G. Purdom, J. Coffman, T. W. Root and E. N. Lightfoot, "Characterizing the performance of industrial-scale columns," *Journal of Chromatography A*, vol. 908, pp. 131–141, 2001.
31. Q. S. Yuan, A. Rosenfeld, T. W. Root, D. J. Klingenberg and E. N. Lightfoot, "Flow distribution in chromatography columns," *Journal of Chromatography A*, vol. 831, pp. 149–165, 1999.
32. N. Pathak, C. Norman, S. Kundu, S. Nulu, M. Durst and Z. Fang, "Modeling flow distribution in large-scale chromatographic columns with computational fluid dynamics," *Bioprocess International*, vol. 6, pp. 74–81, 2008.

33. Z. Fang, "Large-scale chromatography columns, modeling flow distribution," In *Encyclopedia of Industrial Biotechnology, Bioprocess, Bioseparation, and Cell Technology*, M.C. Flickinger, Ed., Hoboken, John Wiley & Sons, Inc, pp. 3123–3140, 2010.
34. C. Johnson, V. Natarajan and C. Antoniou, "Evaluating two process scale chromatography column header designs using CFD," *Biotechnology Progress*, vol. 30, no. 4, pp. 837–844, 2014.
35. Z. Fang, "Applying computational fluid dynamics technology in bioprocesses," *Biopharm International*, vol. 23, pp. 38–44, 2010.
36. H. Aono, I. Iliescu, D. Cecchini, S. Wood and J. T. McCue, "Mitigation of chromatography adsorbent lot performance variability through control of buffer solution design space," *Journal of Chromatography A*, vol. 1318, pp. 198–206, 2013.
37. J. McCue, P. Engel and J. Thommes, "Effect of phenyl sepharose ligand density on protein monomer/aggregate purification and separation using hydrophobic interaction chromatography," *Journal of Chromatography A*, vol. 1216, pp. 902–909, 2009.
38. US Food and Drug Administration. "Guidance for Industry, Q8 Pharmaceutical Development," 2006.
39. US Food and Drug Administration. "Guidance for Industry, Q8 Pharmaceutical Development (Revision 1)," 2009.

9

High-Throughput Process Development

Silvia M. Pirrung and Marcel Ottens

Department of Biotechnology, Delft University of Technology, Delft, the Netherlands

9.1 Introduction to High-Throughput Process Development in Chromatography

The increasingly large quantities, in which some of the industrial proteins are needed, together with advances in, for instance, molecular engineering lead to higher titers produced during upstream processing. As a result, the bottleneck in biopharmaceutical protein production is being shifted toward downstream processing. In order to purify this increased amount of product, larger facilities are needed which implies higher investment and operating costs due to an increased consumption of buffers and other consumables [1]. Another challenge for the development of purification processes resulting from the optimization toward high titers in upstream processing is a modification of the broth composition regarding type and concentration of impurities [2].

Furthermore, the regulatory initiative quality by design (QbD) asks for a systematic and efficient approach during process development to explore the design space, which is defined by QbD as the permissible range of operating conditions assuring product quality, on the basis of critical process parameters (CPP), key process parameters (KPP), and critical quality attributes (CQA) as can be seen in Figure 9.1 and in Chapter 1. An extensive description of QbD principles and how they can be applied to biopharmaceuticals can be found elsewhere [3]. In order to do so, a comprehensive knowledge about the underlying mechanisms would be beneficial.

As a result, simple trial and error approaches cannot be applied anymore which forces the biopharmaceutical industry to look into new strategies for improving purification processes. One of these strategies that is being used more and more during the last 10 years is high-throughput process development (HTPD) [4].

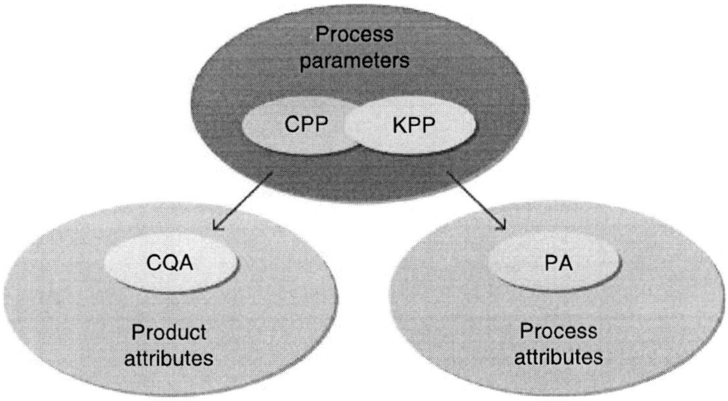

Figure 9.1 Here, CQA that define the needed product quality are identified; an example could, for instance, be viral clearance or product aggregation. Subsequently, process parameters, CPP, whose variation could have an effect on these CQA leading to a high risk of falling outside the design space, are determined. Additionally, other process parameters, KPP, that are not influencing the CQA but do influence the process attributes, PA, which is, for instance, yield, have to be determined as well for a full description of QbD design space.

These techniques can be seen as combinations of high-throughput experimentation (HTE), which is characterized by miniaturization, automation and parallelization, and mechanistic modeling of the chromatographic separations.

In a recent survey regarding process development in biomanufacturing, most respondents thought that the most important prerequisite for an easier process development is having a *sufficient understanding of performance driving parameters* and *having applicable scale-down models available* [5]. This clearly shows that the industry is aware of the need for high-throughput techniques in process development and is adopting this in their practices as showcased, for instance, in the academic–industrial conference series on HTPD [6].

The efforts here are mainly focused on chromatographic purification steps, since these are still the most important steps during biopurification [7], which can be underlined by the fact that downstream processes in large-scale biomanufacturing use an average of three chromatographic steps [5]. Optimization via HTPD can be done on different aspects of the chromatography unit. The optimal resin out of the huge amount available in the market today differing from each other in particle size, ligand type, and pressure flow characteristics, among others, has to be found while ensuring suitability for scale-up and reusability. Another aspect to be optimized is the binding of the protein to the resin where often design of experiment (DoE) approaches are used to find buffer conditions that allow optimal binding. Finding the optimal elution conditions is another important aspect considered during optimization of a chromatographic step. Finally, the robustness of the process has to be good enough for manufacturing [8].

9.2 Process Development Approaches

The process design space in chromatographic operations is rather big; choices between a vast number of media, operating conditions, chromatography modes, and their order have to be made. The optimum in this design space where the cost is at its minimum has to be found by still satisfying all constraints such as purity, quality, yield, and so on. This challenging task has been tackled with many different approaches, which are described in the following [9]. Which part of the design space some of these approaches can cover is visualized in Figure 9.2. However, one has to keep in mind that the more process design space is to be covered, the more information and resources are needed for developing a biopurification process.

9.2.1 Trial and Error Approach

In traditional purification development, due to heavy feed material demands only a few column experiments can be used to investigate different operating parameters in the process design space, which results in a process with technically and economically suboptimal conditions [10].

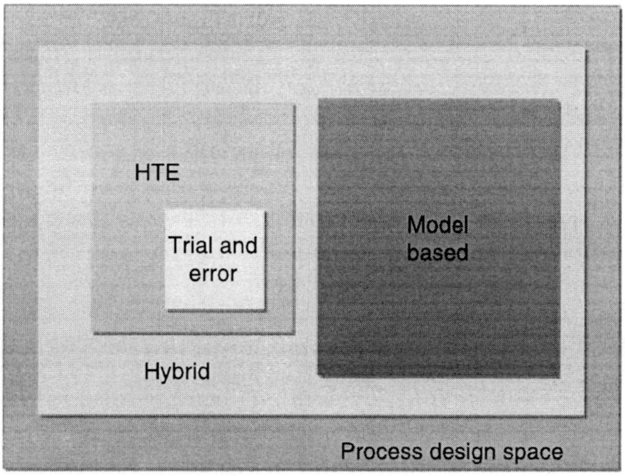

Figure 9.2 Using the trial and error approach, just a small part of the process design space can be explored due to the inefficient nature and the high consumption of resources during this approach. HTE allows the investigation of a bigger part of the design space while utilizing the same amount of resources due to its miniaturized experiments. Having a trustworthy mechanistic model available enables investigation of the design space by mere in silico simulation and, thus, eliminates the limitation due to resources. Combining mechanistic models with HTE techniques, the so-called hybrid approach, makes it possible to tailor your model to your experimental system and, hence, to cover an even bigger part of the design space.

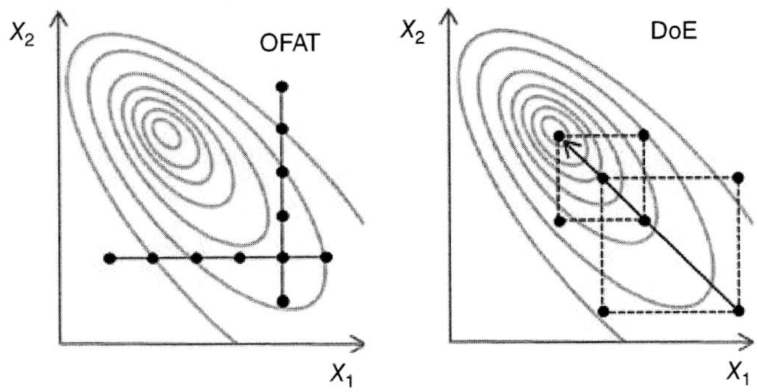

Figure 9.3 In the OFAT approach, one variable (x_1 or x_2) is changed at a time which neglects all interactions between these variables and can thus lead to a suboptimal point. In the DoE approach, on the other hand, x_1 and x_2 are varied simultaneously, which allows to capture dependencies between the two and hence enables finding the direction of the true optimum (based on Mandenius and Brundin [11]).

9.2.1.1 One Factor at a Time (OFAT)

One method often used in trial and error approaches is the OFAT. Like the name implies, the OFAT method just changes one factor at a time while other factors are kept constant and, thereafter, draws conclusion from the resulting effects. This method can lead to a suboptimal solution since dependencies between variables are being neglected as can be seen in Figure 9.3. Additionally, it is a material-consuming and time-intensive method since many experiments have to be performed to investigate several factors.

9.2.1.2 DoE

A more efficient method is DoE, since different factors are considered simultaneously as shown in Figure 9.3. Therefore, less resources are needed to explore a bigger part of the process design space compared to OFAT. Additionally, results obtained are more accurate since more observations are being made to describe one effect [12]. How multivariate techniques can be applied to chromatographic separations is extensively described in a review [13].

9.2.2 Expert Knowledge-Based Process Development

Asenjo et al. use an approach to select a purification step in a rational manner. Their approach relies on the application of expert knowledge based on large-scale production in a system that contains hundreds of different logical rules and methods. These rules are supposed to imitate an expert's reasoning. This system then uses data of the main contaminant proteins out of databases. In this way, the whole sequence of downstream process operations can be selected [14–16]. The disadvantage in this approach is that data about each protein contaminant has to be available in one of the databases used. Subsequently, they used different models to simulate the chromatographic unit operations and, thus, further optimize their entire process [17, 18].

9.2.3 High-Throughput Experimentation

In the beginning of developing a protein purification process, there is usually little information available about the proteins and the contaminants inside the crude protein mixture; hence, finding the optimal conditions for a process requires many experiments, for instance, in order to characterize the mixture or to screen for suitable chromatographic media. To accelerate this time-consuming process, high-throughput techniques are being commonly applied for their high level of miniaturization, automation, and parallelization.

Therefore, the main advantage of using this kind of system is that it allows a fast investigation of many chromatographic conditions while just using a small amount of material. By additionally taking into account accurate mass transfer and kinetic models, the column performance can be quantitatively foreseen [19]. As a net result, the time for having a functioning process is shortened, and other issues concerning the process can be addressed. Another advantage is that if a product fails during late clinical trials, at that moment, comparably less money has been spent as would have been spent in a traditional process development approach [10]. A review on different formats used in HTE such as miniature columns or 96-well plates and their applications can be found elsewhere [20].

In a purely experimental approach, optimization is usually achieved by DoE and subsequent empirical modeling by response surface analysis. This has been shown to be significantly inferior to an approach based on mechanistic modeling [21]. Additionally, using just HTE during process development does not guarantee that the optimal process is being selected in the end [9]. However, HTE can perfectly be used in industrial process development as a screening tool prior to further optimizing to assess the operating space as shown, for instance, in research at Biogen Idec and Merck [22, 23]. Furthermore, isotherm and kinetic parameters can be determined in a fully automated manner [24]. A reported downside of HTE is that in some applications just solutions of pure proteins can be applied to generate an accurate result. Additionally, the small working volume may limit further analytics [25]. Often the analytics used themselves are low-throughput techniques. Thus, analytics may become the bottleneck in HTE. A strategy that addresses this issue has been proposed [26]. An example for a detailed protocol on how to build an HTPD platform can be found in Ref. [8].

9.2.4 Model-Based Approaches

Models can be used to aid process development especially by reducing the time and material consumption involved. Instead of real experiments, different conditions can be tested by mere *in silico* simulation, and thus, a much wider range of operating conditions can be investigated. Another advantage is that mechanistic models are built on fundamental understanding of the process which has been made necessary to obtain by the QbD initiative.

Current modeling approaches on different aspects—the column, process parameters, and the process—are outlined in the following.

9.2.4.1 Modeling of a Chromatography Column

In the modeling of a chromatography column, there are several levels of complexity [27]. Often one-dimensional models such as the equilibrium transport-dispersive model are used for their speed in computing and often sufficient ability of predicting chromatograms [28, 29]. Another advantage here is that fewer parameters are needed. This model can describe the chromatographic behavior of components in the liquid phase by the following balance-based partial differential equation [30]:

$$\frac{\partial c_{\mathrm{p,i}}}{\partial t}+F\frac{\partial q_{\mathrm{p,i}}}{\partial t}=-v\frac{\partial c_{\mathrm{p,i}}}{\partial x}+D_{\mathrm{L,i}}\frac{\partial^2 c_{\mathrm{p,i}}}{\partial x^2} \tag{9.1}$$

where $c_{\mathrm{p,i}}$ is the protein concentration in the bulk phase, $q_{\mathrm{p,i}}$ is the protein concentration in the stationary phase, v is the interstitial velocity of the mobile phase calculated as $v=u/\varepsilon_{\mathrm{b}}$ where u is the superficial velocity and ε_{b} the bed porosity, and $D_{\mathrm{L,i}}$ is the axial dispersion coefficient. F is the phase ratio, defined as $F=(1-\varepsilon_{\mathrm{t}})/\varepsilon_{\mathrm{t}}$ in adsorption chromatography and $F=(\varepsilon_{\mathrm{t}}-\varepsilon_{\mathrm{b}})/\varepsilon_{\mathrm{b}}$ in size exclusion chromatography with the total porosity ε_{t}. The total porosity is calculated as $\varepsilon_{\mathrm{t}}=\varepsilon_{\mathrm{b}}+(1-\varepsilon_{\mathrm{b}})\varepsilon$. Figure 9.4 clarifies the definitions of different types of porosities and velocities inside a chromatographic column.

In the equilibrium transport-dispersive model, the mass transfer can be quantified by the liquid-film linear driving force approximation where a linear concentration profile in the boundary layer, a stagnant film surrounding the particle, can be assumed. Therefore, the external mass transfer can be expressed as

$$\frac{\partial q_{\mathrm{p,i}}}{\partial t}=k_{\mathrm{ov,i}}\left(c_{\mathrm{p,i}}-c_{\mathrm{p,i}}^{*}\right) \tag{9.2}$$

Here, $c_{\mathrm{p,i}}^{*}$ is the equilibrium concentration in the liquid phase which can be calculated using an appropriate isotherm description such as the mixed mode isotherm by Nfor et al. [30]. The overall mass transfer coefficient $k_{\mathrm{ov,i}}$ is defined as the series connection of the mass transfer resistance inside the stagnant film and the mass transfer resistance inside the pore. The effects occurring in a chromatography column as described by the equilibrium transport-dispersive model are depicted in Figure 9.4.

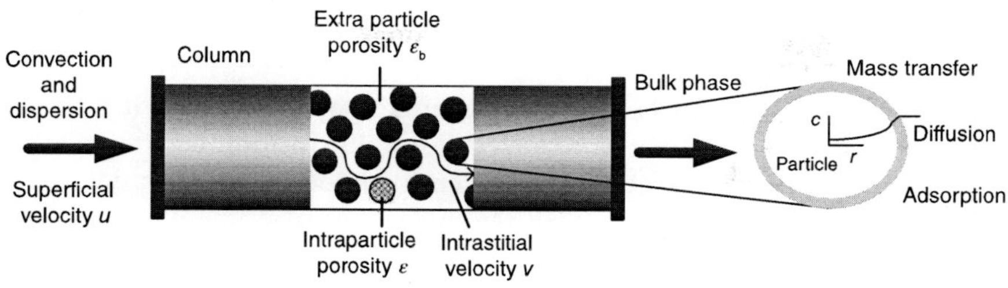

Figure 9.4 Schematic drawing of the different porosities, velocities, and effects inside a column as captured by the equilibrium transport-dispersive model including the mass transfer approximated via the liquid-film linear driving force.

Also two-dimensional models are being used. The most common one is the general rate model, which shows a high predictability [31–35]. This model is a detailed chromatography model taking radial mass transport inside the particle pores into account. Additionally, some attempts have been made to model a chromatographic column in three dimensions, thereby accounting for cross-sectional flow conditions and concentration distributions. An issue here is the large amount of beads inside a column (even in case of a micro column) that could not yet be reproduced in the modeling software [36]. A more detailed overview on models used for chromatography modeling can be found in relevant literature [37, 38] and elsewhere in this book.

These chromatographic models can, for example, be used for model-based resin selection [39] or to investigate control strategies and to evaluate the robustness of a process [28, 31, 40]. Models were also successfully applied for risk analysis of coupled process steps [41]. Also aggregate removal has been successfully predicted by mechanistic models [42]. Additionally, model-based single chromatographic unit operation optimization [43], as well as cascaded chromatographic process operation optimization have been realized [44].

9.2.4.2 Parameter Estimation

While using a mechanistic model to predict column performance, specific process parameters need to be determined beforehand in order to use them as input parameters for modeling. In bioprocesses, this can be an issue due to the ill-defined and complex nature of the cell broth. If the composition of the feed regarding at least the product and the main contaminants are known, some of these process parameters can be approximated by using engineering correlations such as the correlation by Young to determine the free diffusivity of globular proteins [45] or the extended Ogston model for the distribution coefficient in size exclusion chromatography [46]. Of course, parameters predicted by simple engineering correlations are not as accurate as the ones experimentally determined or predicted by a more advanced model, but in some cases they can help to predict the column performance sufficiently. In adsorptive chromatography techniques, it is difficult to predict all the parameters involved describing the binding in between protein and resin. There is quite some scientific effort focused on developing models that can predict these without using any experiments. One fully atomistic approach is to predict the resin–protein colloidal interactions based on the full three-dimensional protein structure using molecular dynamics simulations, which is a computationally rather expensive approach for large proteins [47]. Another noteworthy approach is based on quantitative structure–property relationships (QSPR). In this approach, properties of proteins such as pH-dependent atomic partial charges, here called descriptors, are calculated using their crystal structures and protein sequence. This allows to model the effect of protein surface properties on protein adsorption and, hence, the prediction of protein retention behavior in chromatographic columns [48–50]. Another promising approach utilizes a protein charge model that enables the simulation of resin–protein interaction while having the amino acid sequence of the proteins available and, therefore, the prediction of the retention factor [51]. However, detailed information about the investigated proteins, its amino acid sequence or even crystal structure, has to be known. Therefore, strictly model-based and at the same time accurate process development in adsorption chromatography is just possible while investigating the process for a well-known and defined feed. Computational prediction of protein–ligand interaction is not only of interest in academic research but also in industrial research as shown by the Research Laboratories located at Merck. Their work applied, for instance, an atomistic modeling approach to the separation of variants of glargine from the desired insulin for guiding the selection of ligands used in process development [52].

9.2.4.3 Modeling of a Chromatographic Process

Another area of chromatography modeling research focuses on not optimizing just the chromatographic column but instead on optimizing the whole sequence of chromatographic unit operations. Simple mathematical correlations describing two chromatography modes, ion exchange and hydrophobic interaction, which are based on charge densities and hydrophobicities of the proteins, respectively, are used to describe retention times in a chromatographic unit operation [53, 54]. Based on these, optimization models for different objectives such as purity and cost have been formulated [55, 56]. However, since very simple models that are based on triangles instead of actual peak shapes are used to describe the chromatographic behavior, it can be questionable how accurate the simulations describe the real process and, hence, how optimal the resulting process really is.

9.2.5 Hybrid Methods

A quickly emerging field in process development approaches is hybrid methods; they seem to be more and more recognized in academia and industry. A hybrid method is a combination of the previously outlined methods of HTE and detailed mechanistic modeling, and therefore, it is combining the advantages of both approaches. One advantage here is, for example, that scaling effects can be implemented in the mechanistic model and, thus, ensure scalability of the data obtained in HTE [57].

One concept on how high-throughput experimental data can be combined nicely with a model-based approach in order to optimize a single chromatographic step has been presented by Osberghaus et al. [58]. Firstly, parameters on column scale such as the porosity and dispersion are being determined experimentally on a high-throughput robotic platform. Secondly, parameters on the particle level concerning the adsorption to the resin are estimated [59]. At the same time, the model is being calibrated with experimental results and, thereafter as last step, the model is used to predict conditions for optimal elution experiments. Additionally, upscale predictive ability could be verified. However, the case study this approach was applied to consisted just of three model proteins and not of a crude unknown protein mixture.

More detailed examples, where the workflow in hybrid approaches is described, can be found in the following.

9.2.5.1 Parameter Estimation

A lot of research focuses on parameter estimation as already outlined in the modeling part since feed materials in bioprocessing are complex and ill defined. By using hybrid methods, the parameters can either be acquired via HTE, via modeling, or most likely via a combination of both depending on the type of parameter. A popular way of estimating model parameters is the inverse method. Here, the difference between experimentally determined elution profiles and simulated elution profiles is optimized with typically a least square minimization technique. The results are parameters that give the best fit of experimental to simulated data [60]. However, it has to be kept in mind that these parameters bear an uncertainty and that in case the experimental conditions are not accurately determined they can still be erroneous even if they result in a low residual [61]. Additionally, new HTE methods to determine model parameters by adapting mechanistic models to the specific experimental setup utilized in a robot arise as, for instance, shown by Traylor et al. in case of pore diffusivity [62]. Another nice example for parameter estimation describes how model input parameters for a complex lysate can be obtained via HTE and used for prediction of protein retention behavior as can be found in Kroner et al.; however, low-throughput analytics still seem to be an issue in this otherwise high-throughput approach [63, 64]. An overview of possible combinations between process design tools and the required sources for input data can be found in a review by Hanke and Ottens [57].

9.2.5.2 Process Optimization

Similarly to the mere modeling approach, the hybrid approach is also being applied to look at the complete process. However, approaches here are built on very detailed unit operation models, which lead to other limitations during modeling such as the high computational cost. Recently, the hybrid approach has been used to simulate and optimize a multiple-step chromatographic separation. Such an optimization can either be done sequentially or simultaneously. Sequentially has the advantage that a large computational problem can be solved as a sequence of several smaller subproblems and, thus, the problem becomes more manageable. In a recent approach, sequential modeling and optimization of chromatographic unit operations was successfully achieved, using parameters obtained from crude feedstocks, while taking into account the host cell proteins (HCP) [44]. This approach is described in detail in the following during the second example.

A simultaneous hybrid approach has been applied only to a two-step ion-exchange chromatographic process using model proteins [65]. Here, it was shown that the concerted approach can be superior to the sequential one which is reasonable since in a simultaneous approach possible trade-offs can all be taken into account at the same time which increases the chance of finding the optimal solution. However, in this sort of approach, computational costs are highly increased.

9.3 Case Descriptions

9.3.1 Optimization of a Single Chromatographic Purification Step

An application of a hybrid approach was shown to work in an industrial case for the separation of whey proteins in order to optimize a chromatographic unit operation [43]. A schematic overview on this approach can be seen in Figure 9.5.

The first step is to decide on which chromatographic unit operations to investigate for the separation. In this case, the choice was made to focus on ion-exchange chromatography (IEX) and hydrophobic interaction chromatography (HIC), but these could be expanded at will. As a next step, the operating window has to be established. Here, the crude mixture, which is in this case a

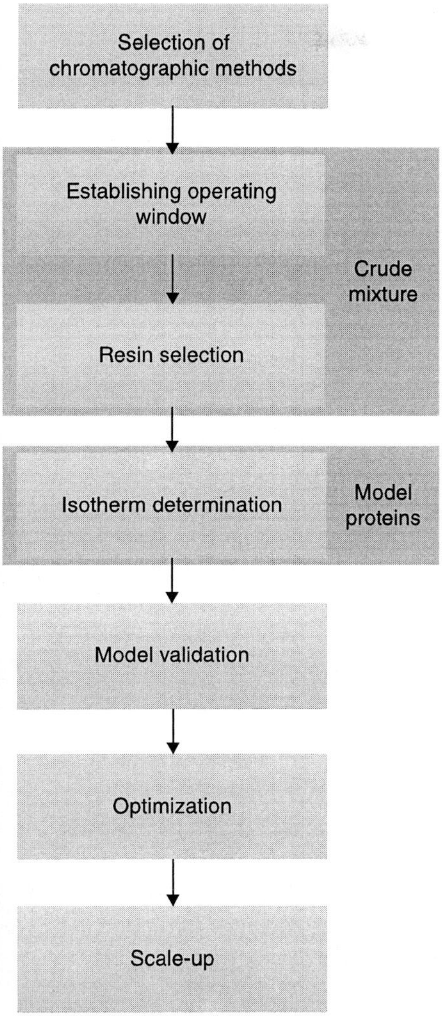

Figure 9.5 Workflow for applying the hybrid approach to a chromatographic separation.

milk serum, has been analyzed by different methods. In order to determine the protein composition of the crude mixture and the elution pH of the target protein, which is known to be a reliable indicator for protein retention, a pH gradient fractionation and subsequent analysis via gel electrophoresis and mass spectrometry was performed [66]. A scheme on this step excluding mass spectrometry is presented in Figure 9.6.

Additionally, protein precipitation behavior was studied to ensure stability of the product during operation. In this case, the precipitation results showed that cation-exchange chromatography (CEX) was not a reasonable option for

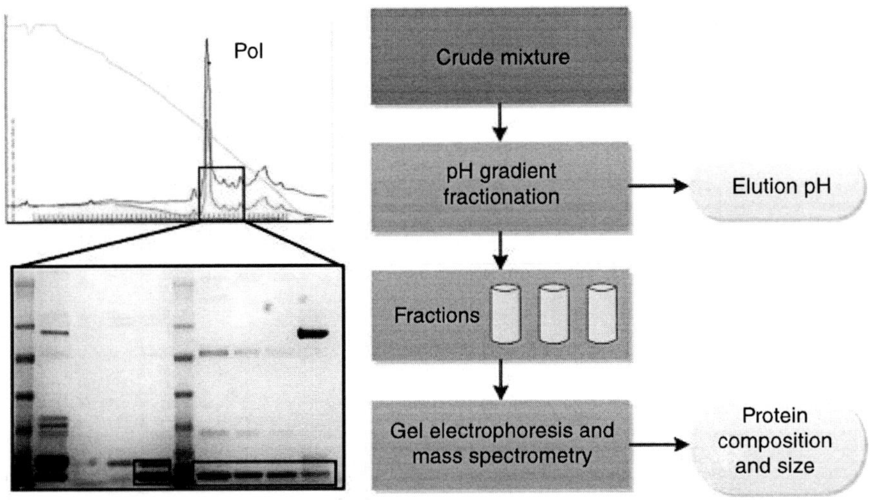

Figure 9.6 Scheme on the establishment of an operating window allowing pH selection. Upper part: pH gradient fractionation to determine elution pH of protein of interest (PoI)—fractions for gel electrophoresis are marked with a black box. Lower part: gel electrophoresis of the fractions retrieved from the pH gradient fractionation—PoI marked with a black box. *Source*: Nfor et al. [43]. Reproduced with permission of Docphin.

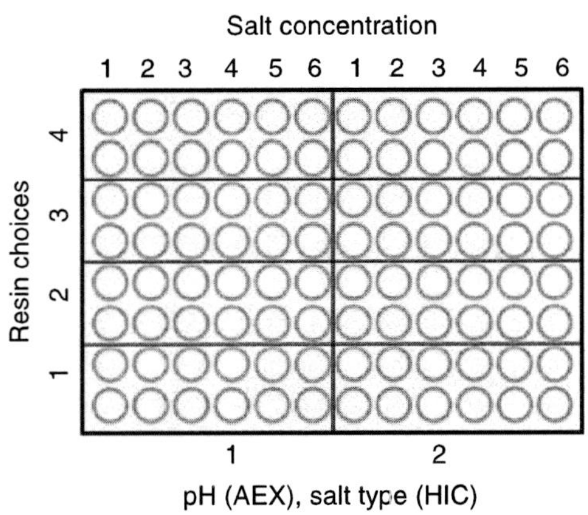

Figure 9.7 Possible plate configurations for HTE resin screening in batch uptake experiments; tested resins, salt concentrations, pH values in case of AEX, and salt type in case of HIC are varied.

this separation. For further experiments, multiple industrial-grade resins for anion exchange (AEX) and HIC were selected and evaluated using high-throughput batch uptake experiments based on a strategy considering various performance metrics such as selectivity and resin capacity as described by

Nfor et al. [39]. Possible 96-well plate configurations to be carried out during this type of resin screening are shown in Figure 9.7.

These remaining and selected resins and operating conditions were then used to determine adsorption isotherms over a wide range of salt, pH, and protein concentrations by HTE using pure protein solutions of the main proteins present in the milk serum. Subsequently, these isotherm parameters are used as input parameters for the column model to predict chromatograms. These predictions for AEX and HIC are compared to experimental runs in order to validate the model as can be seen in Figure 9.8a and b. After validation, the predictions generated by the model can be assumed to be accurate enough and the model can subsequently be used to optimize the chromatographic step regarding variables such as column loading and cut points of the product pool.

Finally, it could be shown that the optimization produced for lab scale could easily be scaled up to preparatory scale using the chromatography model by keeping the optimized operating variables constant and only changing the column dimensions (Figure 9.8c and d). In both cases, the simulation results showed very similar elution profiles with components eluting at around the same column volume (CV) during gradient elution. Moreover, peak resolution is significantly improved which is expected by using a wider column diameter. Therefore, a higher purity and yield of the chromatographic step can be achieved, which shows that column optimization should be done considering larger-scale columns to obtain accurate results regarding the highest available column loading and other variables without violating yield and purity constraints. Finally, optimal conditions were found via *in silico* process design of this chromatographic purification using step elution at preparative scale (data not shown).

9.3.2 Multiple-Column Process Design

A comprehensive example where a hybrid HTPD approach is nicely implemented in order to design and further optimize a downstream process for purification of an antibody without a protein A capture step, considering several chromatographic unit operations and starting from a *crude* protein mixture produced by a hybridoma cell culture, has been published by Ottens and coworkers [44]. The workflow in this approach is presented in Figure 9.9.

First, the purification task is being defined, which means that the CQA, such as HCP level and product stability, and PA, such as recovery yield, for the case have to be identified. Next, the best resin for each chromatography mode (in this case, ion exchange, hydrophobic interaction, and size exclusion chromatography (SEC) have been investigated) according to data and recommendations by the supplier is being selected. After that, the crude protein mixture is

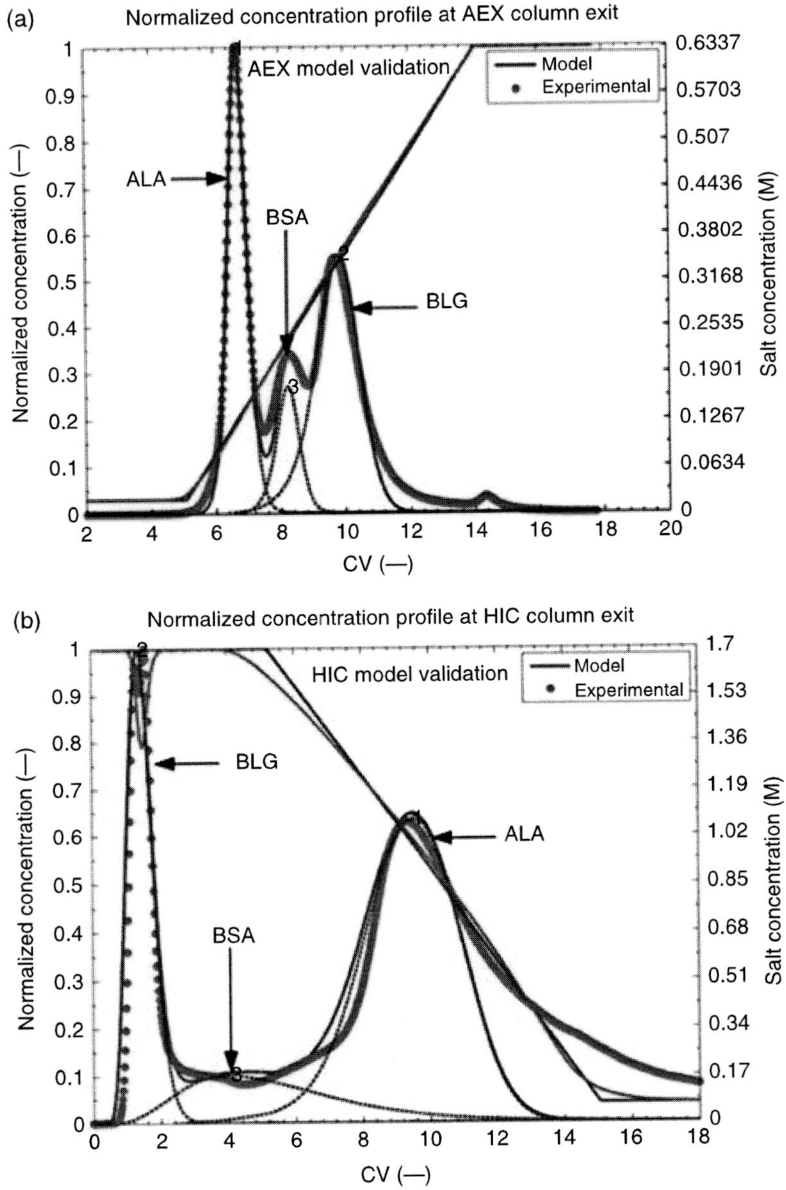

Figure 9.8 (a) and (b) Column model validation showing overlaid experimental and simulated elution profiles of serum proteins in AEX and HIC. Feed composition—1.5, 3.5, and 0.6 mg/mL of ALA, BLG, and BSA, respectively. ALA: molecular weight (Mw = 14.2 kDa, p*I* = 4.2–5.1; BLG: Mw = 18.3 kDa; p*I* = 5.2–5.4; BSA: Mw = 66–69 kDa, p*I* = 4.9–5.1). (c) and (d) Optimized column profiles. (c) Lab scale AEX: column volume = 1.0 mL; pH = 7.5; resin = Capto Q. Optimized conditions: gradient length = 15 CV; linear velocity = 400.7 cm/h, sample load = 5.0 CV (8% binding capacity); lower cut point = 11.7 CV, upper cut point = 15 7 CV. (d) Prep scale AEX (column volume = 35.3 L; column length = 50 cm; internal diameter = 30 cm) in linear gradient elution. The pH sample load and all other optimized conditions are the same as in (c). *Source*: Nfor et al. [43]. Reproduced with permission of Docphin.

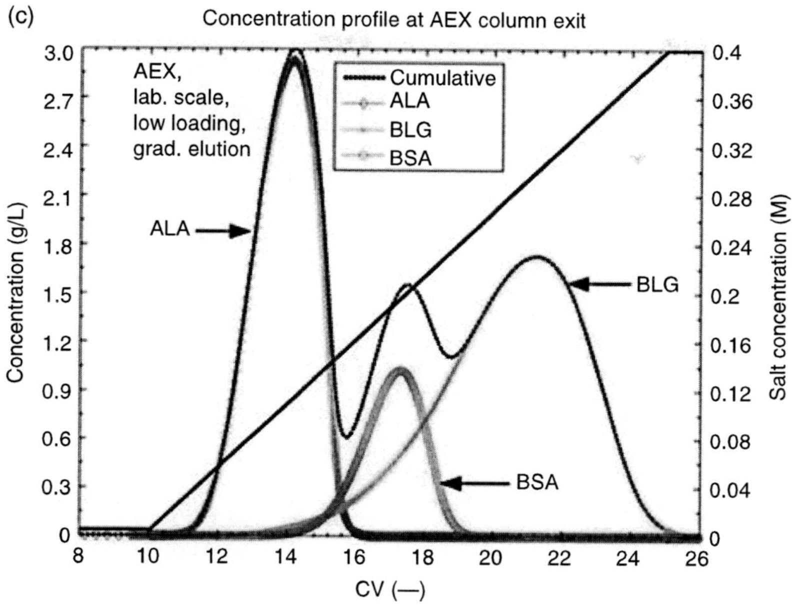

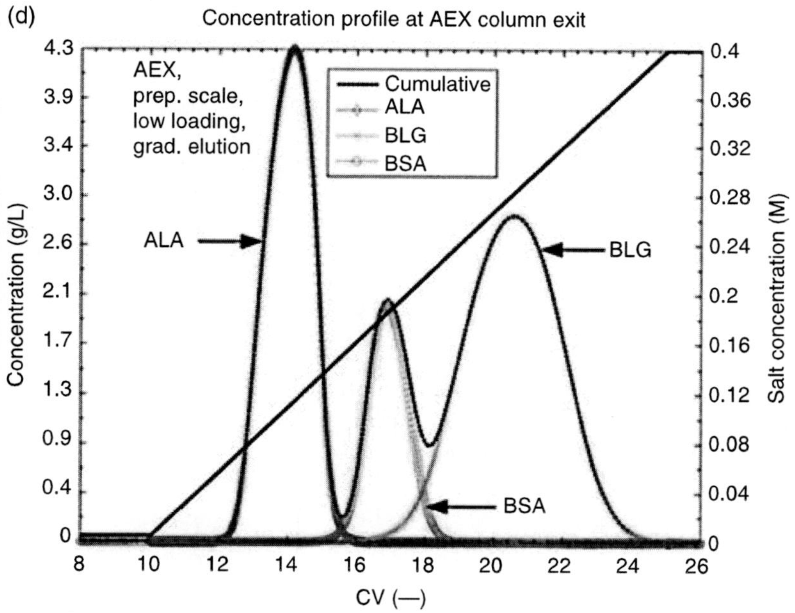

Figure 9.8 (Continued)

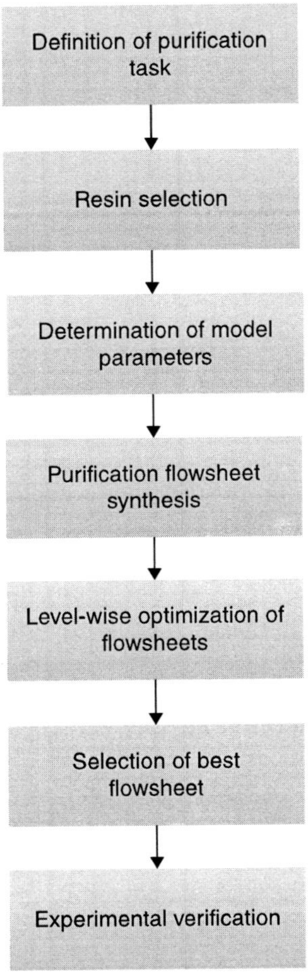

Figure 9.9 Workflow for applying the hybrid approach to cascaded process optimization.

being characterized by a multidimensional fractionation and characterization scheme by means of different chromatographic fractionation steps in combination with analytical methods such as mass spectrometry as depicted in Figure 9.10. In this way, crucial model parameters as, for instance, the physicochemical and thermodynamic properties of the protein product and the main contaminants required for the modeling of several chromatography modes are being determined in a fast and efficient manner [60].

These parameters can then be stored in a database together with the parameters describing the auxiliary materials such as the chromatographic media to enable utilization in future process development. As a next step, validated unit operation models as described in the section on modeling a chromatography column can use these parameters to simulate different process alternatives.

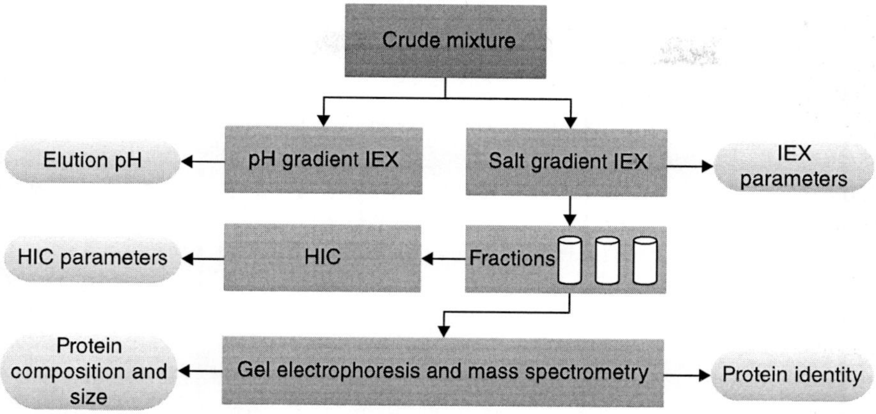

Figure 9.10 Complex fractionation and characterization scheme to determine important model parameters for crude protein mixtures. *Source*: Nfor et al. [60]. Reproduced with permission of Wiley.

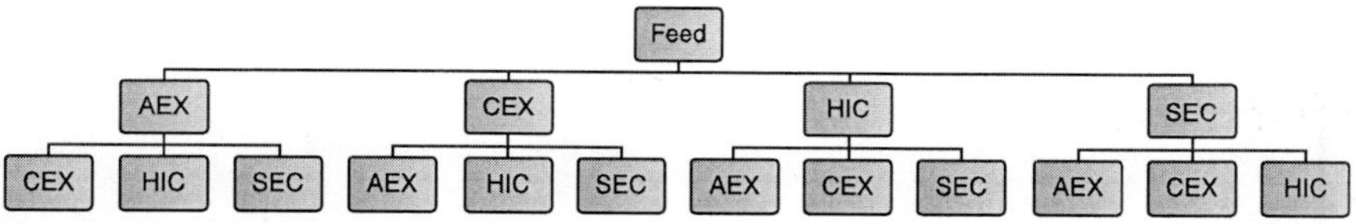

Figure 9.11 Tree diagram showing all possible combinations of different chromatographic operations (AEX, anion exchange; CEX, cation exchange; HIC, hydrophobic interaction; SEC, size exclusion) up to a level of two sequential units.

These process alternatives consist of cascaded chromatographic unit operations and can, for instance, be shown in a tree diagram (Figure 9.11). For simplification, just a two-level graphic has been chosen. The least promising of these alternatives are being discarded based on expert knowledge to simplify the optimization process. A mathematical algorithm is then applied to find the optimum process out of all these alternatives based on the selected objectives such as purity or operating cost. The optimization is executed by considering one level at a time; hence, it constitutes a sequential optimization.

9.4 Future Directions

Current downstream processing practices applied in the biopharmaceutical industry need to advance to be able to face the challenges resulting from higher cell culture titers and recent regulatory initiatives [67]. Applying high-throughput techniques and mechanistic modeling approaches during process development is a promising way to address some of these challenges.

Nonetheless, one part that still needs improvement in high-throughput techniques is the common usage of low-throughput analytics, which are then becoming the bottleneck regarding speed of development. Another issue is the huge amount of data that is being created which requires good data handling structures. Therefore, finding ways of achieving faster analysis and the handling of the produced results are still important areas for future research [4, 57].

One obvious trend is a higher degree of integration between different approaches as can be seen in the increasingly more applied hybrid approach. Another aspect where integration is becoming more relevant is the cascading of the chromatographic unit with other unit operations in the downstream process, which is likely to be taken even further by integrating the upstream process as well, which would allow for an overall optimization of the entire bio-production process. Optimization software tools based on accurate and complex mechanistic models using a flowsheeting approach for bioprocess development are fruitful ground for future research and development.

References

1. Low D, O'Leary R, Pujar NS. 2007. Future of antibody purification. *J Chromatogr B Analyt Technol Biomed Life Sci* 848(1):48–63.
2. Gronemeyer, P., Ditz, R., Strube, J. 2014. Trends in upstream and downstream process development for antibody manufacturing. *Bioengineering*, 1(4), 188–212.
3. Rathore, A.S., Winkle, H. (2009) Quality by design for biopharmaceuticals. *Nature Biotechnol* 27, 26–34.
4. Łącki KM. 2014. High throughput process development in biomanufacturing. *Curr Opin Chem Eng* 6:25–32.
5. Tran R, Lacki K, Davidson A, Sharma B, Titchener-Hooker N. 2014. Changing manufacturing paradigms in downstream processing and the role of alternative bioseparation technologies. *J Chem Technol Biotechnol* 89(10):1534–1544.
6. Hansen SK, Oelmeier SA. 2012. Meeting report: high-throughput process development—HTPD 2012: June 4–7, 2012, Avignon, France. *Biotechnol J* 7(10):1189–1191.
7. Bhambure R, Rathore AS. 2013. Chromatography process development in the quality by design paradigm I: establishing a high-throughput process development platform as a tool for estimating "characterization space" for an ion exchange chromatography step. *Biotechnol Prog* 29(2): 403–414.
8. Rathore AS, Bhambure R. 2014. High-throughput process development: I. Process chromatography. *Methods Mol Biol* 1129:29–37.

9. Nfor BK, Ahamed T, van Dedem GWK, van der Wielen LAM, van de Sandt EJAX, Eppink MHM, Ottens M. 2008. Design strategies for integrated protein purification processes: challenges, progress and outlook. *J Chem Technol Biotechnol* 83(2):124–132.
10. Titchener-Hooker NJ, Dunnill P, Hoare M. 2008. Micro biochemical engineering to accelerate the design of industrial-scale downstream processes for biopharmaceutical proteins. *Biotechnol Bioeng* 100(3):473–487.
11. Mandenius CF, Brundin A. 2008. Bioprocess optimization using design-of-experiments methodology. *Biotechnol Prog* 24(6):1191–1203.
12. Czitrom V. 1999. One-factor-at-a-time versus designed experiments. *Am Stat* 53(2):126–131.
13. Ferreira SL, Bruns RE, da Silva EG, Dos Santos WN, Quintella CM, David JM, de Andrade JB, Breitkreitz MC, Jardim IC, Neto BB. 2007. Statistical designs and response surface techniques for the optimization of chromatographic systems. *J Chromatogr A* 1158(1–2):2–14.
14. Asenjo J, Andrews B. 2004. Is there a rational method to purify proteins? From expert systems to proteomics. *J Mol Recognit* 17(3):236–247.
15. Leser EW, Lienqueo ME, Asenjo JA. 1996. Implementation in an expert system of a selection rationale for purification processes for recombinant proteins. *Ann N Y Acad Sci* 782:441–455.
16. Lienqueo ME, Asenjo JA. 2000. Use of expert systems for the synthesis of downstream protein processes. *Comput Chem Eng* 24(9–10):2339–2350.
17. Asenjo JA, Andrews BA. 2009. Protein purification using chromatography: selection of type, modelling and optimization of operating conditions. *J Mol Recognit* 22(2):65–76.
18. Sandoval G, Shene C, Andrews BA, Asenjo JA. 2010. Extension of the selection of protein chromatography and the rate model to affinity chromatography. *J Mol Recognit* 23(6):609–617.
19. Coffman JL, Kramarczyk JF, Kelley BD. 2008. High-throughput screening of chromatographic separations: I. Method development and column modeling. *Biotechnol Bioeng* 100(4):605–618.
20. Lacki KM. 2012. High-throughput process development of chromatography steps: advantages and limitations of different formats used. *Biotechnol J* 7(10):1192–1202.
21. Osberghaus A, Hepbildikler S, Nath S, Haindl M, von Lieres E, Hubbuch J. 2012c. Optimizing a chromatographic three component separation: a comparison of mechanistic and empiric modeling approaches. *J Chromatogr A* 1237:86–95.
22. Sanaie N, Cecchini D, Pieracci J. 2012. Applying high-throughput methods to develop a purification process for a highly glycosylated protein. *Biotechnol J* 7(10):1242–1255.

23. Welsh JP, Petroff MG, Rowicki P, Bao H, Linden T, Roush DJ, Pollard JM. 2014. A practical strategy for using miniature chromatography columns in a standardized high-throughput workflow for purification development of monoclonal antibodies. *Biotechnol Prog* 30(3):626–635.
24. Schuldt S, Schembecker G. 2013. A fully automated ad- and desorption method for resin and solvent screening. *Chem Eng Technol* 36(7):1157–1164.
25. Yoshimoto N, Minakuchi K, Itoh D, Isakari Y, Yamamoto S. 2013. High-throughput process development methods for chromatography and precipitation of proteins: advantages and precautions. *Eng Life Sci* 13(5):446–455.
26. Konstantinidis S, Kong S, Chhatre S, Velayudhan A, Heldin E, Titchener-Hooker N. 2012. Strategic Assay Selection for analytics in high-throughput process development: case studies for downstream processing of monoclonal antibodies. *Biotechnol J* 7(10):1256–1268.
27. Nilsson B. 2005. Aspects of modeling a preparative ion-exchange step for antibody purification. *Chem Eng Technol* 28(11):1367–1374.
28. Close EJ, Salm JR, Bracewell DG, Sorensen E. 2014. A model based approach for identifying robust operating conditions for industrial chromatography with process variability. *Chem Eng Sci* 116:284–295.
29. Javeed S, Qamar S, Seidel-Morgenstern A, Warnecke G. 2011. Efficient and accurate numerical simulation of nonlinear chromatographic processes. *Comput Chem Eng* 35(11):2294–2305.
30. Nfor BK, Noverraz M, Chilamkurthi S, Verhaert PD, van der Wielen LA, Ottens M. 2010. High-throughput isotherm determination and thermodynamic modeling of protein adsorption on mixed mode adsorbents. *J Chromatogr A* 1217(44):6829–6850.
31. Borg N, Brodsky Y, Moscariello J, Vunnum S, Vedantham G, Westerberg K, Nilsson B. 2014. Modeling and robust pooling design of a preparative cation-exchange chromatography step for purification of monoclonal antibody monomer from aggregates. *J Chromatogr A* 1359:170–181.
32. Gu TY, Iyer G, Cheng KSC. 2013. Parameter estimation and rate model simulation of partial breakthrough of bovine serum albumin on a column packed with large Q Sepharose anion-exchange particles. *Sep Purif Technol* 116:319–326.
33. Li ZG, Gu YS, Gu TY. 1998. Mathematical modeling and scale-up of size-exclusion chromatography. *Biochem Eng J* 2(2):145–155.
34. Nagrath D, Messac A, Bequette BW, Cramer SM. 2004. A hybrid model framework for the optimization of preparative chromatographic processes. *Biotechnol Prog* 20(1):162–178.
35. von Lieres E, Andersson J. 2010. A fast and accurate solver for the general rate model of column liquid chromatography. *Comput Chem Eng* 34(8): 1180–1191.

36. Schnittert S, Winz R, von Lieres E. Development of a 3D Model for Packed Bed Liquid Chromatography in Micro-Columns , UKSIM European Symposium on Computer Modeling and Simulation, Athens, Greece, November 25–27 2009. pp. 193–197.
37. Guiochon G, Felinger A, Shirazi DG, Katti AM. 2006. *Fundamentals of Preparative and Nonlinear Chromatography*: Elsevier Inc., Amsterdam
38. Schmidt-Traub H, Schulte M, Seidel-Morgenstern A. 2012. *Preparative chromatography*: John Wiley & Sons, Chichester.
39. Nfor BK, Zuluaga DS, Verheijen PJ, Verhaert PD, van der Wielen LA, Ottens M. 2011. Model-based rational strategy for chromatographic resin selection. *Biotechnol Prog* 27(6):1629–1643.
40. Westerberg K, Borg N, Andersson N, Nilsson B. 2012. Supporting design and control of a reversed-phase chromatography step by mechanistic modeling. *Chem Eng Technol* 35(1):169–175.
41. Westerberg K, Broberg-Hansen E, Sejergaard L, Nilsson B. 2013. Model-based risk analysis of coupled process steps. *Biotechnol Bioeng* 110(9):2462–2470.
42. Sejergaard L, Karkov HS, Krarup JK, Hagel AB, Cramer SM. 2014. Model-based process development for the purification of a modified human growth hormone using multimodal chromatography. *Biotechnol Prog* 30(5):1057–1064.
43. Nfor BK, Ripic J, van der Padt A, Jacobs M, Ottens M. 2012b. Model-based high-throughput process development for chromatographic whey proteins separation. *Biotechnol J* 7(10):1221–1232.
44. Nfor BK, Ahamed T, van Dedem GWK, Verhaert PDEM, van der Wielen LAM, Eppink MHM, van de Sandt EJAX, Ottens M. 2013. Model-based rational methodology for protein purification process synthesis. *Chem Eng Sci* 89:185–195.
45. Young ME, Carroad PA, Bell RL. 1980. Estimation of diffusion coefficients of proteins. *Biotechnol Bioeng* 22(5):947–955.
46. Bosma JC, Wesselingh JA. 2000. Partitioning and diffusion of large molecules in fibrous structures. *J Chromatogr B Biomed Sci Appl* 743(1–2):169–180.
47. Dismer F, Hubbuch J. 2010. 3D structure-based protein retention prediction for ion-exchange chromatography. *J Chromatogr A* 1217(8):1343–1353.
48. Chung WK, Hou Y, Freed A, Holstein M, Makhatadze GI, Cramer SM. 2009. Investigation of protein binding affinity and preferred orientations in ion exchange systems using a homologous protein library. *Biotechnol Bioeng* 102(3):869–881.
49. Hou Y, Cramer SM. 2011. Evaluation of selectivity in multimodal anion exchange systems: a priori prediction of protein retention and examination of mobile phase modifier effects. *J Chromatogr A* 1218(43):7813–7820.
50. Ladiwala A, Rege K, Breneman CM, Cramer SM. 2005. A priori prediction of adsorption isotherm parameters and chromatographic behavior in ion-exchange systems. *Proc Natl Acad Sci U S A* 102(33):11710–11715.

51. Guelat B, Delegrange L, Valax P, Morbidelli M. 2013. Model-based prediction of monoclonal antibody retention in ion-exchange chromatography. *J Chromatogr A* 1298:17–25.
52. Insaidoo FK, Rauscher MA, Smithline SJ, Kaarsholm NC, Feuston BP, Ortigosa AD, Linden TO, Roush DJ. 2015. Targeted purification development enabled by computational biophysical modeling. *Biotechnol Prog* 31(1):154–164.
53. Alvarez EV, Lienqueo ME, Pinto JM. 2001. Optimal synthesis of multistep protein purification processes. *Lat Am Appl Res* 31(4):373–381.
54. Lienqueo ME, Mahn A, Asenjo JA. 2002. Mathematical correlations for predicting protein retention times in hydrophobic interaction chromatography. *J Chromatogr A* 978(1–2):71–79.
55. Liu SS, Simaria AS, Farid SS, Papageorgiou LG. 2014. Optimising chromatography strategies of antibody purification processes by mixed integer fractional programming techniques. *Comput. Chem. Eng.* 68:151–164.
56. Polykarpou EM, Dalby PA, Papageorgiou LG. 2012. A novel efficient optimisation system for purification process synthesis. *Biochem Eng J* 67:186–193.
57. Hanke AT, Ottens M. 2014. Purifying biopharmaceuticals: knowledge-based chromatographic process development. *Trends Biotechnol* 32(4):210–220.
58. Osberghaus A, Drechsel K, Hansen S, Hepbildikler SK, Nath S, Haindl M, von Lieres E, Hubbuch J. 2012a. Model-integrated process development demonstrated on the optimization of a robotic cation exchange step. *Chem Eng Sci* 76:129–139.
59. Osberghaus A, Hepbildikler S, Nath S, Haindl M, von Lieres E, Hubbuch J. 2012b. Determination of parameters for the steric mass action model—a comparison between two approaches. *J Chromatogr A* 1233:54–65.
60. Nfor BK, Ahamed T, Pinkse MW, van der Wielen LA, Verhaert PD, van Dedem GW, Eppink MH, van de Sandt EJ, Ottens M. 2012a. Multi-dimensional fractionation and characterization of crude protein mixtures: toward establishment of a database of protein purification process development parameters. *Biotechnol Bioeng* 109(12):3070–3083.
61. Borg N, Westerberg K, Andersson N, von Lieres E, Nilsson B. 2013. Effects of uncertainties in experimental conditions on the estimation of adsorption model parameters in preparative chromatography. *Comput Chem Eng* 55:148–157.
62. Traylor SJ, Xu X, Li Y, Jin M, Li ZJ. 2014. Adaptation of the pore diffusion model to describe multi-addition batch uptake high-throughput screening experiments. *J Chromatogr A* 1368:100–106.
63. Kroner F, Elsasser D, Hubbuch J. 2013a. A high-throughput 2D-analytical technique to obtain single protein parameters from complex cell lysates for in silico process development of ion exchange chromatography. *J Chromatogr A* 1318:84–91.

64. Kroner F, Hanke AT, Nfor BK, Pinkse MW, Verhaert PD, Ottens M, Hubbuch J. 2013b. Analytical characterization of complex, biotechnological feedstocks by pH gradient ion exchange chromatography for purification process development. *J Chromatogr A* 1311:55–64.
65. Huuk TC, Hahn T, Osberghaus A, Hubbuch J. 2014. Model-based integrated optimization and evaluation of a multi-step ion exchange chromatography. *Sep Purif Technol* 136:207–222.
66. Ahamed T, Chilamkurthi S, Nfor BK, Verhaert PD, van Dedem GW, van der Wielen LA, Eppink MH, van de Sandt EJ, Ottens M. 2008. Selection of pH-related parameters in ion-exchange chromatography using pH-gradient operations. *J Chromatogr A* 1194(1):22–29.
67. Liu HF, Ma J, Winter C, Bayer R. 2010. Recovery and purification process development for monoclonal antibody production. *MAbs* 2(5):480–499.

10

High-Throughput Column Chromatography Performed on Liquid Handling Stations

Process Characterization and Error Analysis

Patrick Diederich[1] and Jürgen Hubbuch[2]

[1] *Peptide & Protein Purification, Novo Nordisk A/S, Bagsværd, Denmark*
[2] *Institute of Engineering in Life Sciences, Section IV: Biomolecular Separation Engineering, Karlsruhe Institute for Technology (KIT), Karlsruhe, Germany*

10.1 Introduction

Liquid chromatography (LC) is indispensible for the purification of therapeutic proteins. However, the development of chromatographic processes can be a challenging task. There is a variety of chromatography-based separation principles that can be applied and combined, and for each of them dozens of process parameters need to be optimized. Consequently, the required experimental effort for process development and process optimization can be huge. This is in particular true if only little is known on the molecular properties of target protein and contaminants, and heuristic rules do not apply. Even if process development scientists can employ a purification platform process or support the development by using mechanistic modeling, the number of experiments, the time needed for model parameter estimation and process characterization, and the material consumption can still be considerably high.

In response to these challenges, miniaturized chromatography formats have been developed within the last decade for the parallel and automated experimentation on liquid handling stations (LHSs). This has enabled high-throughput experimentation (HTE) within LC. Batch chromatography for resin screening and isotherm determination can be performed using 96-well filter plates and 384-well micro-plates [1–6]. Packed bed chromatography can be conducted in pipette tips filled with resin and miniaturized chromatography columns providing a fast way to obtain information on dynamic binding capacities and optimal binding and elution conditions [7–12]. Furthermore, HTE chromatography has been used for the initial determination of parameters required for mechanistic modeling and simulation of chromatographic processes [1, 13–16]. Particularly in early industrial process development, in which process development work is mainly driven by short timelines, HTE has been implemented. Here, the advantages of low sample volume and high throughput will outweigh the larger experimental errors, which are often inherent in HTE. However, if the intention is to use microliter-scale HTE column chromatography (in the following referred to as high-throughput column chromatography (HTCC)) for more than screening purposes, that is, as scale-down model for detailed process optimization, process characterization, and calibration of chromatography models, two categories of questions related to the quality of data arise. The questions in the first category deal with scale, that is, "What chromatography parameters can in fact be investigated in microliter scale compared with the liter scale?" "How is scale 'comparability' defined by development departments and regulatory authorities?" The questions in the second category deal with data quality, that is, "What are the main error sources in HTCC?" "Which effects are caused specifically by using a LHS and do not occur when using common chromatographic systems?" "To what extent can the accuracy be influenced by optimizing LHS settings and method parameters?" In this work, we have addressed questions of the second category, focusing on HTCC at a scale of 200 μL column volume (CV).

The study aims to generate novel insights into the performance of HTCC and develop a better understanding of the limitations while incorporating results of recently published studies on the uncertainty in HTE batch chromatography [6, 17]. We have studied the influence of LHS settings related to pipetting accuracy, working in micro-titer plates (MTPs), fractionation, fluid flow regime, and step gradient elution by employing a combination of experimental data and simulations. In combination with different levels of analytical errors, we aimed to identify the main sources of experimental noise and to quantitatively determine the correlation to the resulting data quality regarding peak characteristics such as retention times or peak width. The results also illustrate which HTCC settings require special attention, whereas others could be left out from extensive and time-consuming optimization without compromising the resulting data quality. A detailed knowledge on expected error ranges will help approaching issues from the aforementioned first category on scale comparability and on quality of model parameters estimated based on data from HTCC.

10.1.1 High-Throughput Column Chromatography: Method Review

In order to facilitate a better understanding of the presented objectives and results, this section will briefly review the experimental setup of HTCC while focusing on chromatography operated using salt gradient elution. The typical workflow in HTCC is displayed in Figure 10.1.

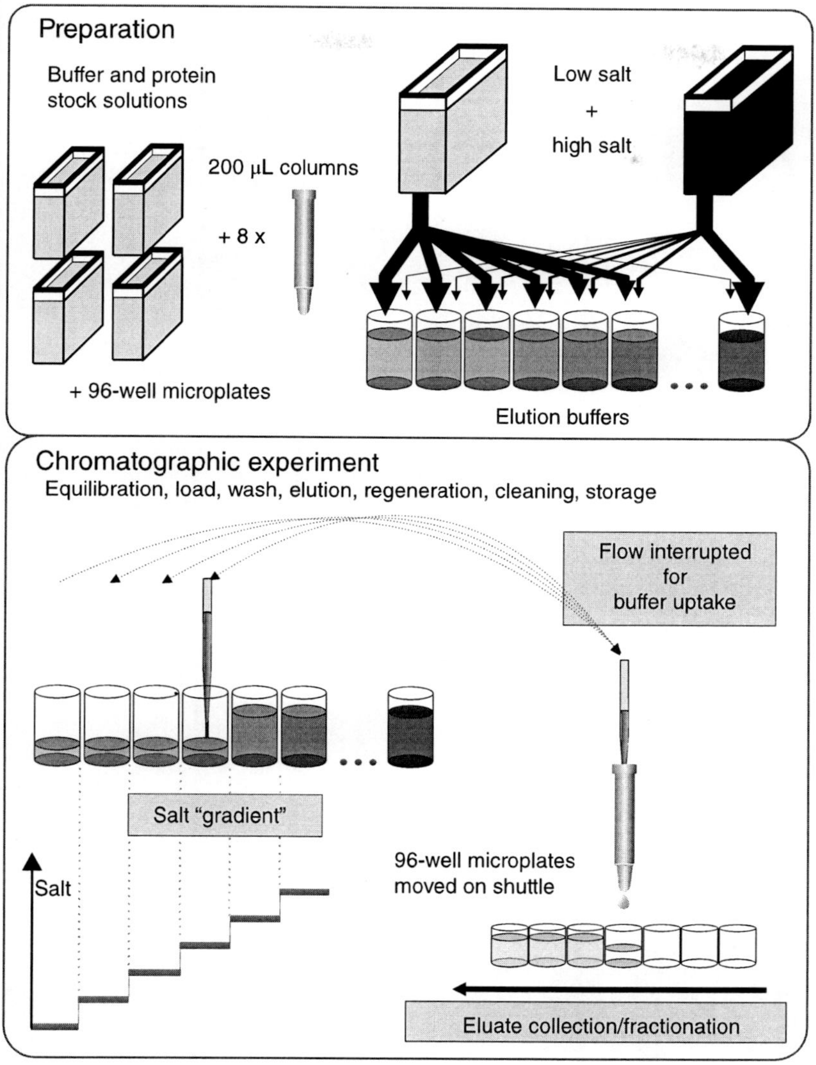

Figure 10.1 Sequence of HTCC process steps in case of gradient elution experiments. (*The figure continues in the following page.*)

The chromatography columns are fixed on the HTCC carrier, which is located above a plate shuttle. Further information including a description of the miniaturized columns used for HTCC are given in Refs. [9, 18]. The plate shuttle is a part of the plate storage and delivery system. On this shuttle, MTPs are moved with defined intervals underneath the columns that enable collection of column eluate droplets and fractionation hereof. The liquid flow on typically eight columns operated in parallel is realized using an eight-channel

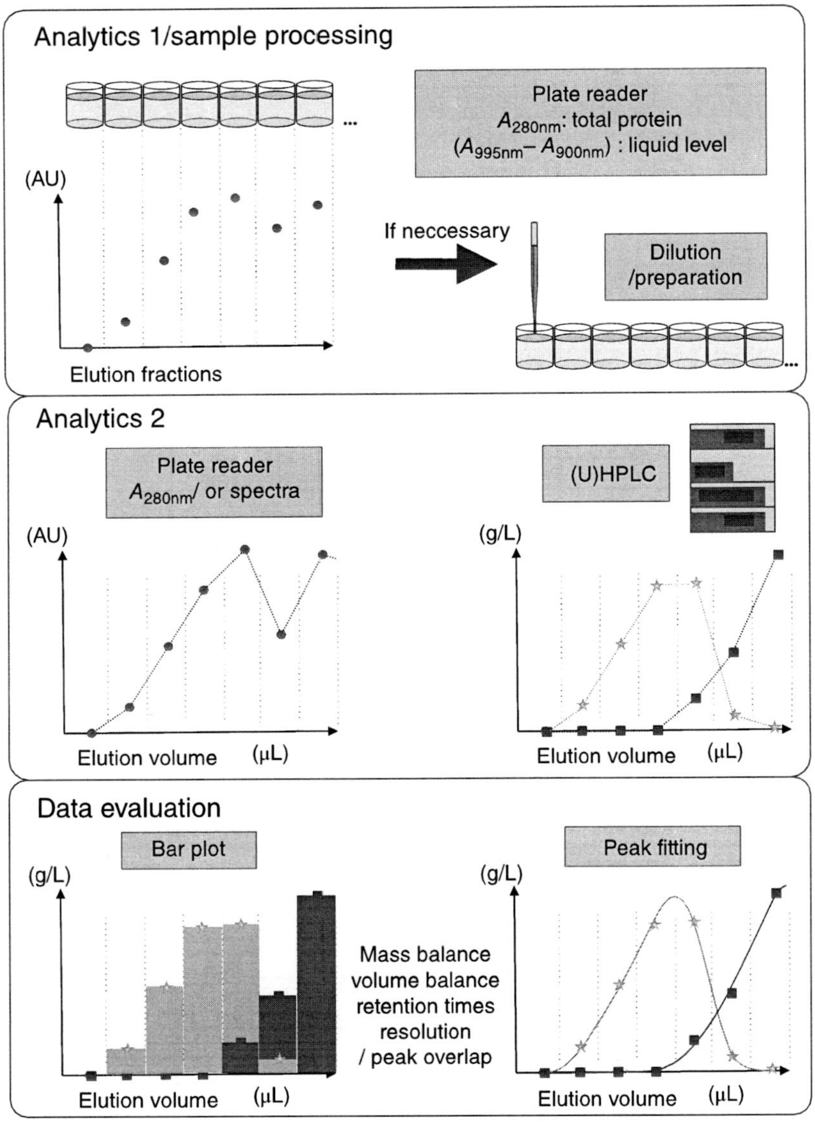

Figure 10.1 (Continued)

pipetting arm, which aspirates and dispenses the required process liquids such as buffers, protein solutions, regeneration, and cleaning solutions. Most commonly, the collected fractions are analyzed photometrically with a UV–VIS plate reader that is operated at-line fully integrated in the LHS. The photometric data are automatedly exported for subsequent data evaluation. The resulting "chromatogram" in fact is a scatter or bar plot with the width of the fraction volume. Peak fitting functions can be used to generate a "continuous" absorption signal versus the elution volume.

When considering the usage of HTCC, three major system characteristics must be kept in mind. First, the plunger size determines the maximal volume that can be dispensed without flow interruption. The most common size of the plunger in current format is 1 mL. Therefore, a sequence of dispense steps interrupted by aspiration steps must be performed in order to process the total volume required for dynamic chromatography experiments. Second, the minimal possible volumetric flow rate is given by the fixed plunger volume and the minimal step size for the plunger movement, for example, 0.83 µL/s for 1 mL plunger, which equals a residence time of 4 min when using 200 µL columns. Third, only one liquid at a time can be dispensed into a column and therefore inline gradient mixing is not possible. Hence, a gradient must be mimicked by a multiple step elution (step gradient). Preferably, this is done with buffers previously prepared by the LHS [12].

10.1.2 High-Throughput Column Chromatography: Error Sources

It has been demonstrated that chromatograms resulting from HTCC in 200 µL scale are comparable to chromatograms obtained with 2.5 or 1 mL used on a laboratory LC system [9, 12, 16]. However, HTCC features apparently higher noise levels when compared with the precision obtained on a laboratory-scale LC system [9, 16]. In particular, the variations in mass balances are relatively high. For example, a range of 86–104% (HPLC analysis) and 83–104% (photometric analysis) was stated in Hansen et al. [19] (four replicates). Treier et al. [12] stated a mass balance of 111% with a standard deviation (STD) of 10% (photometric analysis, eight replicates). No publications were found regarding a quantitative investigation of HTCC in order to identify the main contributors to the observed variances.

A cause-and-effect diagram is shown in Figure 10.2, in order to provide an overview of error sources potentially affecting the experimental outcomes. From this overview, LHS settings were identified, which are specific for HTCC when compared with column chromatography using standard laboratory LC systems. These settings were chosen for further investigation. Error sources that are similarly important to the experimental results when performing laboratory LC experiments such as false weighing of buffer components and proteins were not considered for further analysis.

It must be pointed out that parameters such as fraction volume and salt step height are parameters that need to be set in advance. These are self-evidently well controlled, whereas precision of pipetting can only be maximized to a certain extent by optimization of pipetting parameters. In the investigation,

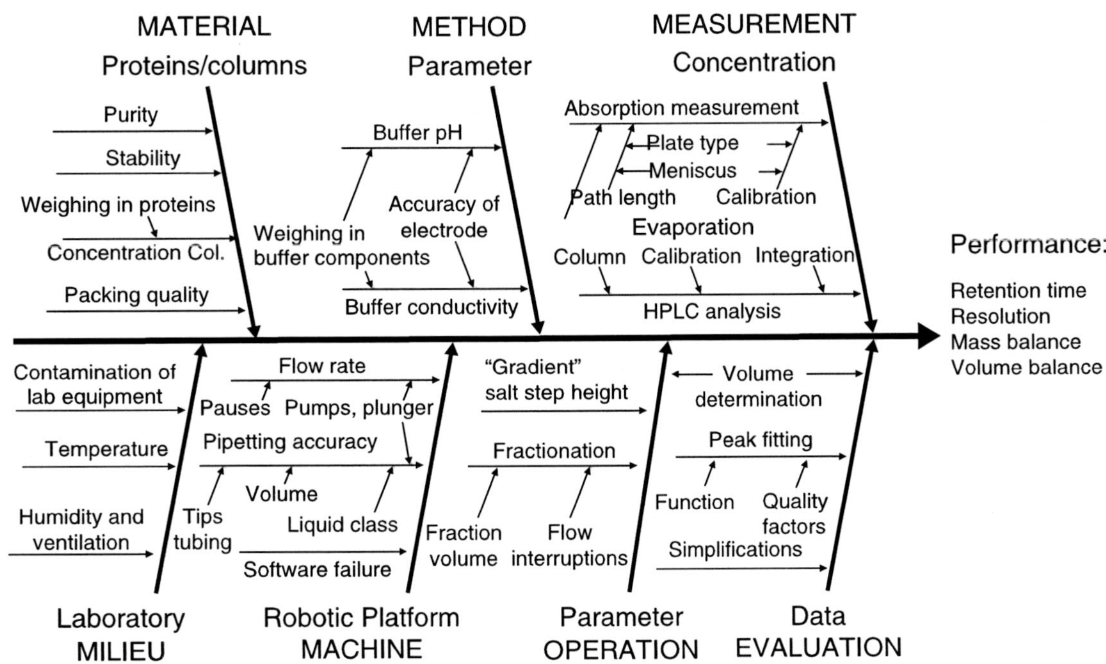

Figure 10.2 Fishbone diagram on potential parameters affecting the quality of HTCC experiments on a LHS.

the following process parameters were included and sorted by increasing degree of complexity and cumulation of possible error sources:

- Pipetting
- UV–VIS absorption measurements in 96-well MTP
- Fractionation volume (number of fractions)
- Discontinuous flow and needle movement in and out of the column
- Salt step height and flow interruptions in salt gradient elution experiments

10.2 Chromatographic Methods

10.2.1 HTCC Experiments

HTCC experiments were performed according to the general description given in Section 10.1.1 and Figure 10.1 with the following specifics: RoboColumns filled with SP Sepharose FF resin (Atoll GmbH, Weingarten, DE) and a CV of 200 μL (1 cm bed height) were used in both isocratic and gradient methods. Experiments were carried out on a Freedom EVO®200 station (Tecan, Crailsheim, DE), equipped with an eight-channel liquid handling arm (non-coated fixed tips), a centric gripper, a rotational shaker (Te-Shake), a plate delivery and storage devices (Te-Stack), and a chromatography unit (Te-Chrom). In all cases, the eluate was collected in fractions of 75 μL in half-area 96-well plates. Fractions were photometrically analyzed at wavelengths 280, 900, and 990 nm using the integrated plate reader. For selective protein quantification, reversed phase (RP) HPLC was performed on a UHPLC system as described in Ref. [20].

10.2.1.1 Isocratic Elution

Columns were loaded with 3.2 mg/mL_{CV} lysozyme (14.3 kDa) or avidin (68.0 kDa) dissolved in 100 mM phosphate buffer, pH 4.5. The limitation of the aspiration volume by the plunger size (commonly 1 mL) was removed by directly controlling the instrument via firmware commands. Thus, a volume of 3.2 mL elution buffer was aspirated into the system tubing, and a continuous dispension of 2.7 mL (=13.5 CV) buffer was performed interrupted in intervals of 0.9 mL for ~2 s.

10.2.1.2 Gradient Elution

As a model system, lysozyme and avidin were separated for loads of 1.06 and 3.20 mg/mL_{CV}, respectively. The application of salt gradients was mimicked by sequential pipetting of buffers with varying salt concentrations c_{salt} for each buffer calculated by Equation 10.1 with n being the total number of salt steps:

$$c_{salt} = c_{gradient,start} + \frac{c_{gradient,end} - c_{gradient,start}}{n} \cdot \frac{2i-1}{2}, \quad i = 1,\ 2,\ 3 \ldots n. \qquad (10.1)$$

The applied gradient lengths ranged from 8 to 28 CV for an increase in NaCl concentration from 150 to 675 mM in 100 mM phosphate buffer at pH 4.5. A buffer step length of 150 μL was kept constant for all gradient experiments. Thus, in the case of a salt gradient length of 18 CV, the salt step height in the robotic experiments corresponded to 21.9 mM.

10.2.2 Lab-Scale Experiments

For comparison to HTCC, gradient chromatography experiments were performed on an Äkta Purifier10 laboratory LC system from GE Healthcare (Uppsala, SE). Pre-packed mini-columns with a bed volume of 200 μL purchased from Atoll GmbH (Weingarten, DE) were used providing identical dimensions compared with the RoboColumns and allowing a close connection of the tubing. Salt gradient length and fraction volume were identical to the HTCC gradient experiments as described earlier.

10.3 Results and Discussion

The challenge of investigating the influence of each single factor lies in finding proper methods to distinguish between effects exclusively caused by the robotic method and the general effects of working in such small scale. The following result sections start with investigations of very basic effects and gain complexity with sequentially incorporating more aspects of gradient elution in HTCC.

10.3.1 Pipetting Accuracy

Pipetting accuracy is probably the first error source to be considered when discussing the performance of automated methods. In common control software, the researcher can adjust many pipetting parameters such aspiration and dispense speed as well as experimentally derived correlations ("liquid classes") between plunger movement and pipetted liquid volume for each type of liquid. However, optimization of a large number of liquid classes can become time consuming, especially if with each class a wide volume range needs to be covered accurately. Liquid class parameters and data on pipetting accuracy can be found in literature [12, 17, 21, 22] demonstrating that an optimization of the liquid class is necessary in order to reduce the deviation between nominal and true pipetted volume below 1%.

In Figure 10.3, the relative standard deviation (RSTD) is plotted as a function of the nominal pipetting volume when using optimized liquid classes for the pipetting of a phosphate buffer and a protein solution (1 g/L avidin). Two liquid handling setups were compared: first, using stainless steel (non-coated) tips

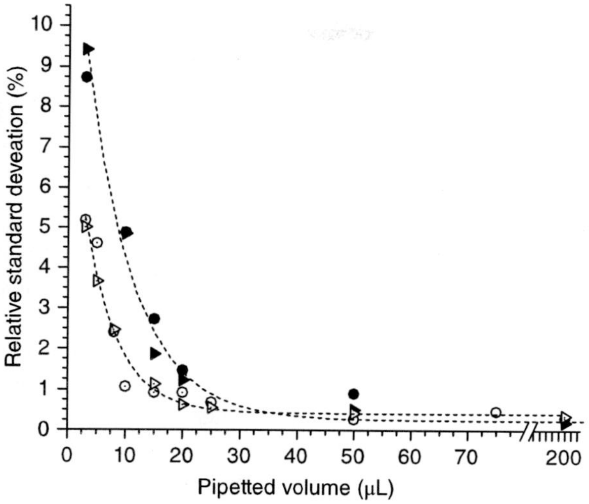

Figure 10.3 Relative standard deviation as a function of the pipetted volume for a LHS. -•-, 50 mM phosphate buffer, pH 7, with fixed tip, stainless steel non-coated, 1 mL syringe; -○-, 50 mM phosphate buffer, pH 7, with low volume fixed tip, stainless steel FEP-coated, 0.25 mL syringe; -▶-, 1 g/L avidin with fixed tip, stainless steel noncoated, 1 mL syringe; -▷-, 1 g/L avidin with low volume fixed tip, stainless steel FEP-coated, 0.25 mL syringe.

and a 1 mL plunger volume and second, using low volume stainless steel tips (FEP-coated) and a plunger volume of 0.25 mL.

The pipetting precision decreased exponentially with a decreasing pipetted volume, which is in concert with trends reported by Treier [12] and Osberghaus [17]. When using 1 mL plungers the precision decreased strongly from RSTD values <1% for volumes larger than ~25 μL to a RSTD of ~10% for pipetting <5 μL. With a reduced plunger volume, a RSTD of <1% was obtained for volumes larger than ~15 μL, and the precision was reduced to 5% when pipetting 3 μL. The data clearly demonstrates that low volumes should be avoided for sample dilution, that is, volumes lower than 15 μL. The average pipetting accuracy of all channels was 100% ± 1% when using optimized liquid classes calibrated in three volume ranges (2–15 μL, 15–55 μL, 55–200 μL) and for each needle tested in duplicate. In terms of accuracy, there was no significant difference obtained when comparing the two types of setups. The accuracy hence depends on the quality of the calibration, and therefore, only the random variance in pipetting was taken into account in this study.

10.3.2 Absorption Measurement in Micro-Titer Plates

Photometric analysis is probably the most suitable analytical technique for high-throughput screenings. It is very fast, cheap, label-free, and noninvasive and can be automated on LHSs in 96-well and 384-well plate formats. Most commonly, the absorption at a wavelength of 280 nm is used for the quantification of the protein content. Furthermore, the liquid level within a well and thus the liquid volume can be linearly correlated to the absorption difference of wavelengths ≥970 nm and 900 nm [12, 23].

The accurate determination of volume balances and of mass balances (if protein concentration is analyzed by absorption measurements) hence depends on the accuracy of the photometric measurement in MTPs and the quality of the calibration curves. The main differences of measurements in micro-plate wells compared with measurements using cuvettes are the variable path length. The dropwise collection of the elution volume during a HTCC experiment can lead to deviations in collected volume between fractions by one drop (~25 µL), and hence, the path length is not constant for all fractions. Furthermore, meniscus formation results in a decrease of the path length and, hence, in a decrease of the absorption value according to Lambert–Beer's law. The curvature of the meniscus and hence the path length depend on liquid/protein characteristics and protein concentration, all having an influence on the liquid surface tension [24].

10.3.2.1 Determination of Volume Based on Absorption Difference

A typical calibration curve obtained when measuring the absorption difference $A_h = A_{990nm} - A_{900nm}$ as a function of the pipetted volume in half-area and full-area MTPs is given in Figure 10.4a for a constant protein concentration of 0.6 g/L lysozyme and for a 100 mM phosphate buffer solution, pH 7.0. For both micro-plate types, A_h values were higher when using buffer compared with using protein solution with an offset being rather constant for all volumes. In Figure 10.4b, A_h obtained in half-area and full-area MTPs at volumes of 75 and 150 µL, respectively, is shown as a function of varying lysozyme concentrations. The absorption difference decreases exponentially with increasing protein concentration until a rather constant value is obtained for lysozyme concentrations higher than ~0.15 g/L. The formation of a meniscus, being the cause of a decreasing path length with increasing protein concentration, thus clearly affects the use of a single calibration curve with constant protein concentration as shown in Figure 10.4a and, thus, the calculation of liquid volume in MTPs based on path length determination via $A_{990nm} - A_{900nm}$. As the meniscus can change with the composition of the protein sample, the accurate determination of the volume using an univariate calibration becomes practically not feasible.

The different path lengths for a solution of 1 g/L lysozyme compared to the buffer blank resulted in deviations of predicted volumes of ~6 and ~16 µL in half-area MTP wells and full-area MTP wells, respectively. This equals an error of 8 and 11% for sample volumes of 75 and 150 µL. These values are in concert with data Treier et al. [12] reported as offsets between protein solutions and water (18 µL, full-area MTP).

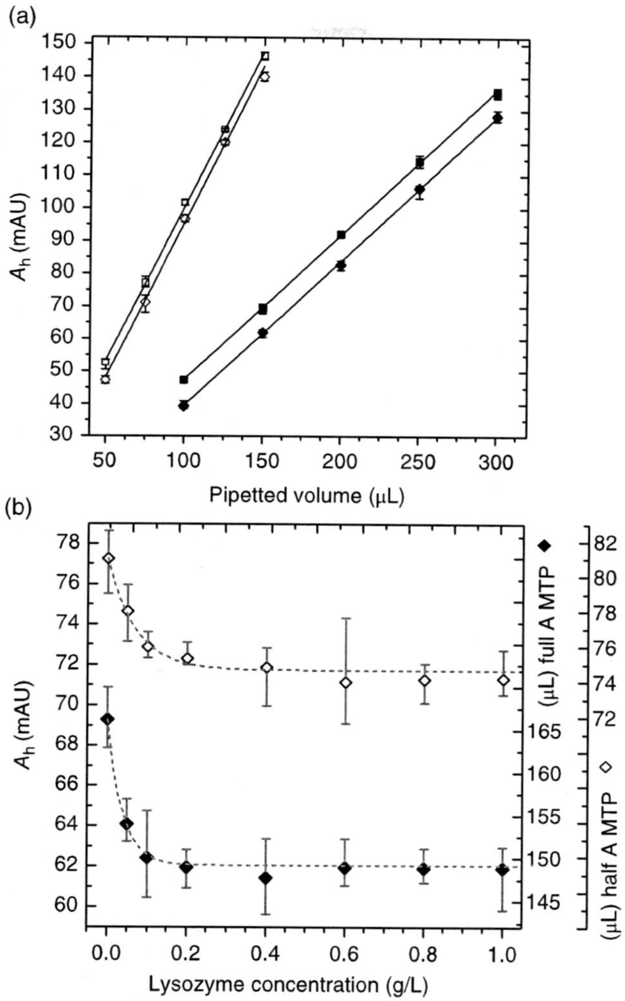

Figure 10.4 (a) Absorption difference $A_h = A_{990nm} - A_{900nm}$ as a function of pipetted volumes of 100 mM phosphate buffer (marker squares) and 0.6 g/L lysozyme (marker diamonds) in half-area MTPs (open marker) and full-area MTPs (filled marker). Fitted by linear regression, $R^2 > 0.998$ (eight replicates for each volume level). (b) Absorption difference $A_h = A_{990nm} - A_{900nm}$ obtained in half-area (-◇-) and full-area MTPs (-◆-) at nominal volumes of 75 and 150 μL, respectively, as a function of the lysozyme concentration in 100 mM phosphate buffer, pH 7.0. The *y*-axis to the right is obtained relying on a calibration curve established by using a lysozyme solution of 0.6 g/L as displayed in Figure 10.4a.

10.3.2.2 Determination of Protein Concentration Based on UV 280 nm

When using common cuvette measurements, the path length is constant and thus a fixed value when correlating A_{280nm} with protein concentration. In MTP measurements, however, the path length might increase with decreasing protein concentration, which is an effect not represented by a linear calibration curve for A_{280nm}. It may thus be advisable to correlate A_{280nm}/A_h to the protein concentration when establishing the calibration curve. Furthermore, it was observed that the constant value of a linear correlation between A_{280nm} and the protein concentration usually is higher than the absorption value of the buffer (data not shown). Therefore, one can conclude that the blank value should not be included in the calibration data set.

In conclusion, when using micro-plate absorption measurements for the determination of protein concentrations and fraction volumes, potential meniscus effects must be considered when setting up the calibration curves. Deviations from 100% in the total volume balance of an HTCC experiment can possibly be explained by meniscus effects rather than false pipetting.

10.3.3 Effect of Fractionation and Number of Fractions

The number of collected fractions is an important parameter to be set when performing HTCC. The fractionation using low volumes, on the one hand, results in more information on the peak shape and provides a higher resolution, which, in turn, provides the possibility of performing fractionation and peak cutting comparable to a large-scale process. On the other hand, the analytical effort obviously increases with a larger number of fractions, and the number of fractions might be limited by the minimum fraction volume required for subsequent analysis.

10.3.3.1 *In Silico* Fractionation Method

The influence of the chosen fraction size regarding the determination of the peak area and the retention time was investigated by using Monte Carlo (MC) simulations for two different, experimentally derived peak shapes, a rather symmetric peak and an asymmetric, strongly tailing peak (presented in Figures 10.6a and 10.7a, respectively). These peaks, serving as reference peaks, were *in silico* fractionated corresponding to fraction volumes ranging from 50 to 350 μL by using an exponential Gaussian hybrid function (EGH), which features excellent fitting of also strongly asymmetric peaks [25, 26]. The theoretical concentrations in each of these artificial fractions were then altered by adding experimental errors of different levels for the determination of volume and protein concentration as well as pipetting errors from potential dilution of fractions. These varied data points were then fitted again in order to calculate the retention time and the peak area. The simulation procedure is schematically illustrated in Figure 10.5.

All introduced errors were added to the initial concentration and volume value for each fraction ($y_{ref,frac,V}$ and $y_{ref,frac,c}$), resulting in a simulated data point defined by the modified volume $\tilde{y}_V$ and concentration $\tilde{y}_c$:

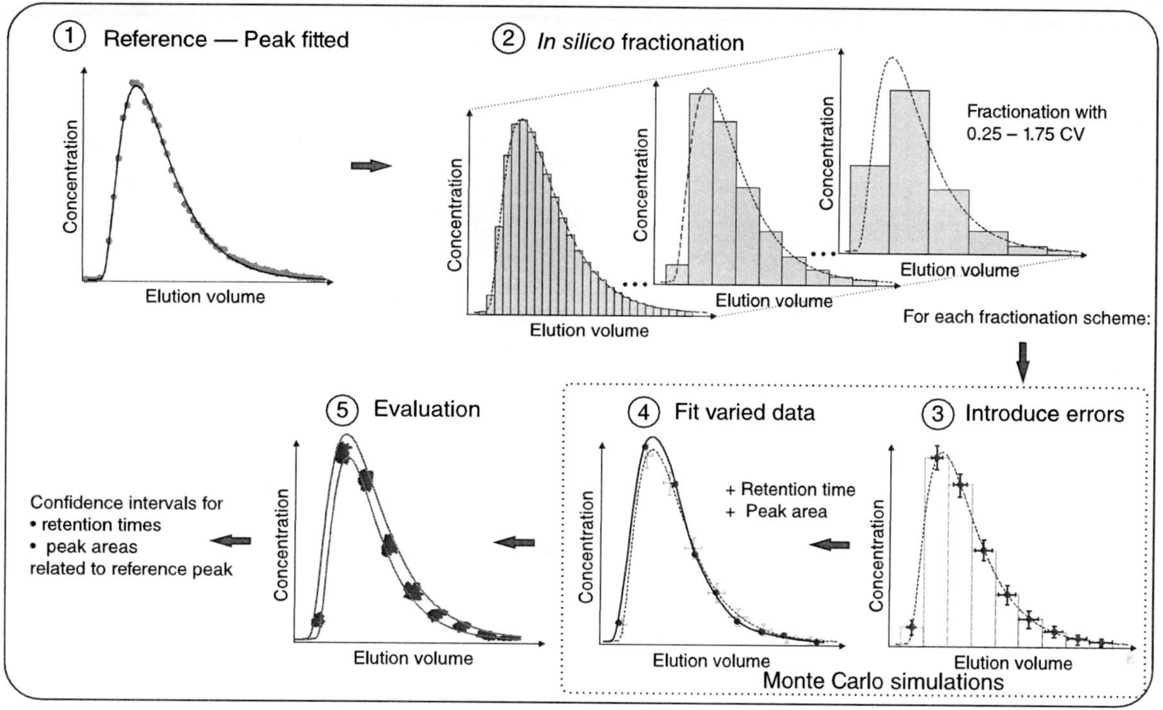

Figure 10.5 Illustration of the Monte Carlo simulation procedure. (1) Fit of experimentally derived elution (in this case the asymmetric peak). Characteristics of this peak serve as reference values. (2) *In silico* calculation of theoretical concentrations in fractions of varying fraction volume. (3) For each fractionation, randomly distributed errors of different levels are added on volume (*x*-coordinate) and concentration (*y*-coordinate) in 10 000 combinations. (4) Each combination of coordinates is fitted for the determination of peak retention time and peak area. (5) Retention time and area of each simulated peak is related to the reference peak. Finally, the 95% confidence intervals of the mean retention time and mean peak area of all simulated peaks per fraction volume are calculated.

$$\tilde{y}_V = y_{\text{ref,frac},V} + E_{\text{syst},V} + E_{\text{random},V}, \tag{10.2}$$

with

$$E_{\text{random},V} = y_{\text{ref,frac},V} \cdot s_{V,\text{frac}}, \tag{10.3}$$

and

$$\tilde{y}_c = y_{\text{ref,frac},c} + E_{\text{syst},c} + E_{\text{random},c}, \tag{10.4}$$

with

$$E_{\text{random},c} = y_{\text{ref,frac},c} \cdot \left(\text{randn} \cdot s_{c,\text{frac}} + \text{randn} \cdot s_{\text{pip,frac}}\right). \tag{10.5}$$

The systematic deviation was given by the absolute values E_{syst}. The statistical errors E_{random} were calculated by multiplying a random number, randn, normally distributed around the mean zero with functions $s_{V,\text{frac}}$, $s_{c,\text{frac}}$, or $s_{\text{pip,frac}}$ representing the RSTD. An overview of the used error terms is given in Table 10.1. The error functions for the determination of concentration were selected to cover a wide range of analytical error levels and were based on data reported for photometric analysis in 96-well plates [19, 27]. The functions incorporated the experimental observation that the common random error increases for very small and for large concentrations. Systematic errors were not considered.

For the determination of the volume by photometric analysis, a RSTD of 2% was experimentally derived and used in all simulations unless stated differently. This value included the pipetting error during calibration and photometric precision with a RSTD of ~0.5%. Additionally, a concentration-dependent systematic deviation of the determined volume $E_{\text{syst},V}$ was calculated for each fraction representing the error caused by meniscus formation in each fraction. This absolute error linearly depended on the liquid level and hence on the choice of half-area or full-area plates (Table 10.1).

The variance introduced by pipetting for dilution was added to fractions with concentrations >0.2 g/L. In this study, we chose to add an average variance for the dilution step on the concentration value rather than calculating the dilution error in dependence of the concentration, as the concentrations of the considered peak were relative low. Derived from the data presented in Figure 10.3, for a dilution using 10 μL protein solution, concentration values were varied with a RSTD of 3.8%, and a RSTD of 0.6% was used for a dilution using 50 μL protein solution.

The effect of a varying number of fractions per peak width on the peak area and peak retention time is displayed in Figures 10.6 and 10.7. For both investigated

Table 10.1 Error correlations used in Monte Carlo simulations.

Parameter	Error term	Error source	Notation
Volume (V)	$s_{V,\text{frac}} = \frac{a}{100}$	Calibration + detector	$V_{a\%}$
	$E_{syst,V} = 0.075 \cdot \exp(-11c) \cdot \frac{V}{k}$	Meniscus	$V_{\text{syst,meniscus}}$
Concentration (c)	$s_{c,\text{frac}} = 0.05 \cdot \exp(-8c) + 0.001 \cdot c^3 + p$	Analytical error	
	$p = 0.004$		$c_{0.5\%}$
	$p = 0.019$		$c_{2\%}$
	$p = 0.099$		$c_{10\%}$
	$E_{\text{syst,c}} = 0$		
Pipetting	$s_{\text{pip10µL,frac}} = 0.041$	Precision, 1 mL plunger	$\text{Pip}_{\text{dil:V}}$
	$s_{\text{pip50µL,frac}} = 0.0064$	Precision, 1 mL plunger	$\text{Pip}_{\text{dil:V}}$

The path length in full-area plates was assumed to be identical to the length in half-area MTPs with half of the volume. The constant k was 75 µL in the case of fractions <150 µL (half-area MTPs) and 150 µL for fractions ≥150 µL (full-area MTPs).

peak types, error scenarios are compared for errors introduced during determination of the fraction volume (V) (Figures 10.6b, c, and 10.7b, c), the concentration (c) (Figures 10.6d, e, and 10.7d, e), and dilution procedures (Pip) (Figures 10.6f, g, and 10.7f, g).

10.3.3.2 Effect of Peak Fitting

The error caused by peak fitting was relatively small (<0.2%) for the symmetrical peak (Figure 10.6b and c). For the asymmetrical peak, however, a low fraction number of <8 – 10 resulted in deviations up to 9 and 4% for the determined peak area and retention time, respectively (Figure 10.7b and c). It becomes clear that even if the applied analytics was free of errors, accurate results for asymmetrical peaks would only be achieved when using a large number of fractions per peak width (e.g., 20). The reason for a decreased accuracy when fitting asymmetrical peaks lies in the steep peak front of the asymmetrical peak. With changing fraction sizes, the fraction that covers the peak maximum is pointing either to the peak front or the peak tail. For which number of fractions the peak maximum shifts is furthermore influenced by the start of the fractionation. Nevertheless, the shown overall trends and the demonstrated sensitivity of the asymmetrical peak towards the number of fractions are expected to hold true regardless of a change in starting volumes or the elution peak width.

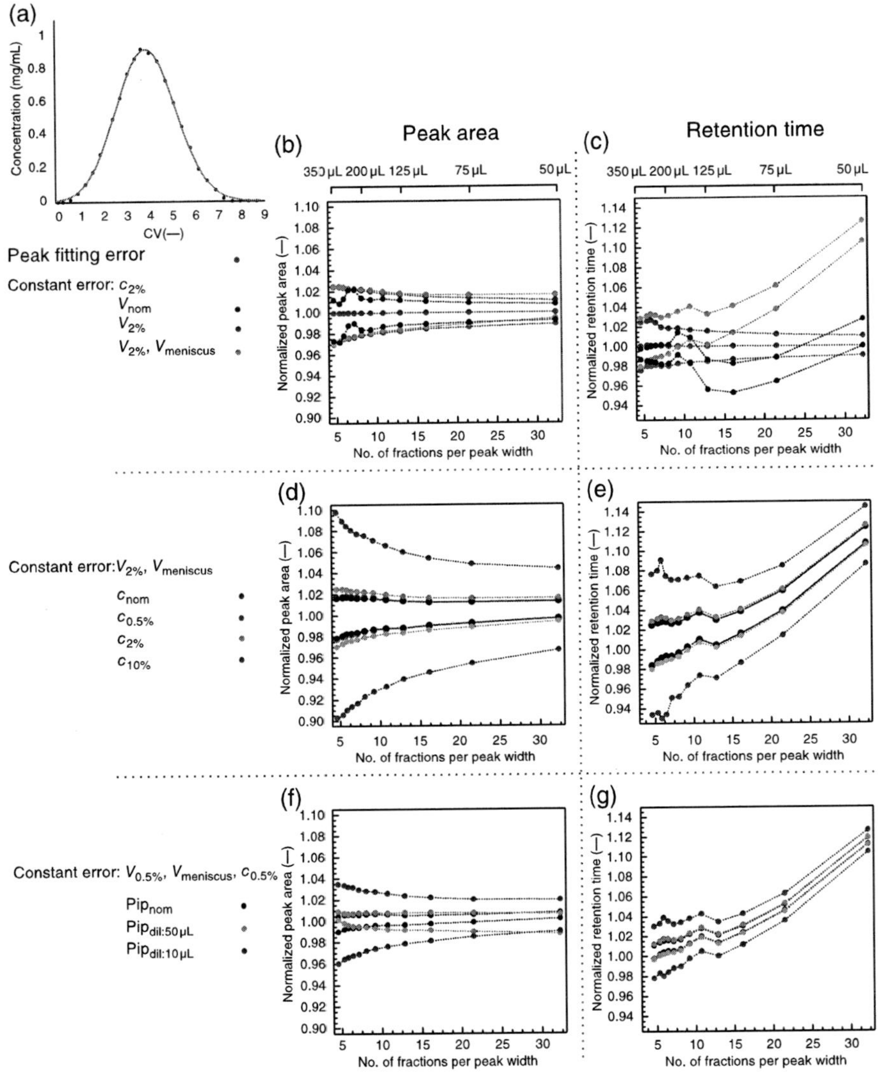

Figure 10.6 Effect of the number of fractions per peak width on the peak area (b, d, f) and peak retention time (c, e, g) determined after peak fitting for a symmetrical elution peak displayed by 95% confidence intervals. The confidence intervals were calculated for peak area and retention time of the simulated peaks related to the values of the reference peak (a). Lower and upper boundaries of the intervals are plotted in the same color. Introduced error categories were (b, c), determination of volumes; (d, e), determination of protein concentration; and (f, g), dilution procedures. The following errors introduced were set constant: (b, c), $c_{2\%}$; (d, e), $V_{2\%}$, $V_{meniscus}$; (f, g), $V_{0.5\%}$, $V_{meniscus}$, $c_{0.5\%}$. (*See insert for color representation of the figure.*)

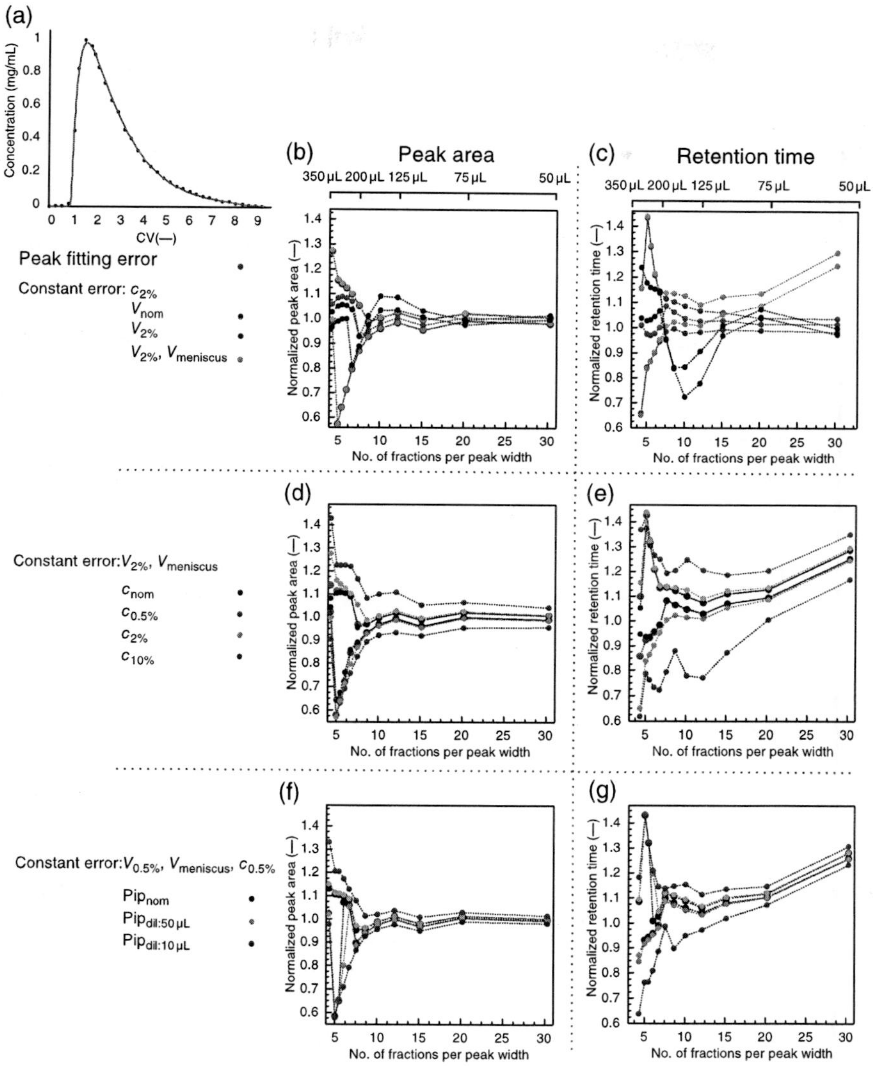

Figure 10.7 Effect of the number of fractions per peak width on the peak area (b, d, f) and peak retention time (c, e, g) determined after peak fitting for an asymmetrical elution peak displayed by 95% confidence intervals. The confidence intervals were calculated for peak area and retention time of the simulated peaks related to the values of the reference peak (a). Lower and upper boundaries of the intervals are plotted in the same color. Introduced error categories were (b, c), determination of volumes; (d, e), determination of protein concentration; and (f, g), dilution procedures. The following errors introduced were set constant: (b, c), $C_{2\%}$; (d, e), $V_{2\%}$, $V_{meniscus}$; (f, g), $V_{0.5\%}$, $V_{meniscus}$, $C_{0.5\%}$. (*See insert for color representation of the figure.*)

10.3.3.3 Effect of Fraction Number: General Trends

For all investigated error levels, the confidence intervals decreased with an increasing number of fractions per peak width. This is as expected, because a peak shape is obviously better represented by more data points as well as the errors are averaged and partly cancel each other out when fitting the peak. From Figures 10.6 and 10.7, it can be seen that the decrease of the confidence interval approached a width, which did not decrease further with increasing number of fractions. This limit was reached for fraction volumes of <100 μL and represented the approximately maximal feasible experimental precision at corresponding error levels. For a number of fractions ≥15, the width of the confidence intervals of asymmetric and symmetric peaks became similarly small, thus indicating that an effect of the peak shape can be neglected if a large number of data points per peak width are available to fit the peak. Precision of the determination of the retention time and peak area increased with increasing number of fractions. Accuracy was only increasing for the peak area, whereas for the retention time, an increasing discrepancy to the nominal retention time with increasing number of fractions was observed as explained in the following section.

While the overall trend of increasing precision with increasing number of fractions was the same for both peak shapes, huge differences in the absolute value of the confidence intervals as well as in the shape of the confidence interval bands were determined. The confidence intervals obtained for the asymmetrical peak were up to ± 45% when using a fraction number of five, whereas the intervals obtained for the symmetrical peak shape were up to 10% for five fractions. For the asymmetrical peak, large variations in the confidence intervals were obtained even if the number of fractions was changed by only one (e.g., changing the number of fractions from five to four, Figure 10.7). These spikes in the confidence intervals result from the peak fitting algorithm, by which for the altered number of data points a different peak maximum and hence retention time is identified.

10.3.3.4 Accuracy of Retention Times

In all simulations in which the meniscus formation of a liquid in a MTP was taken into account (V_{meniscus}), an increase in retention times with increasing number of fractions was obtained. This can be explained with the data given in Section 10.3.2 and Figure 10.4. For the simulations, we chose a calibration scenario for the determination of the fraction volumes using a protein solution with a concentration >0.2 g/L. From Figure 10.4 it becomes clear that the volume with concentrations ≳0.2 g/L will be determined correctly using this calibration, whereas the volume of fractions with lower concentrations will be determined higher than the actual values. As shown in Figure 10.4, the offset of an additional false volume for buffer fractions compared with protein fractions is up to about 6 and 16 μL for half-area MTPs and full-area MTPs, respectively.

Randomized volumes up to these values were therefore added as absolute values to the fraction volume, and hence, the larger the number of fractions, the larger the absolute error and shifting towards later retention times. The widths of the confidence intervals and hence the precision, however, was not influenced by V_{men}. Obviously, the retention times would become shorter when calibrating using a buffer solution, or they might barely be influenced when a different volume offset is present (protein specific). It is therefore essential to investigate the calibration for the volume determination when using a high number of fractions and aiming at an accurate determination of the retention times.

10.3.3.5 Effect of Volume Errors

The effect of errors introduced during the determination of fraction volumes is shown in Figure 10.6b and c for a symmetric peak and in Figure 10.7b and c for an asymmetric peak. Under the assumption of a constant error of 2% introduced by the determination of protein concentration, $c_{2\%}$, three different scenarios were compared: (a) assuming the preset constant nominal fraction volume, V_{nom}; (b) volumes with a constant RSTD of 2%, $V_{2\%}$; and (c) volumes with a constant RSTD of 2% combined with the effect caused by meniscus formation, $V_{meniscus}$.

As can be seen when comparing case (b) and (c), the addition of errors caused by meniscus formation had very little effect on the widths of the confidence intervals. However, the accuracy of the determination of retention times strongly decreased with an error of about +11 and +27% for the symmetric and asymmetric peak, respectively, due to the assumption of using a protein solution for the calibration of the volume determination as explained earlier. When omitting the determination of fraction volumes and using the nominal fraction volume (case (a)) instead, the precision was higher for a number of fractions $\lesssim$10, and the accuracy of the determined retention time was higher for a number of fractions $\gtrsim$20. This is an interesting result demonstrating that an additional effort of measuring and calculating the true volumes (V_{men}) in fact can result in less accurate retention times. Omitting the volume determination would furthermore be beneficial as the experimental and evaluation method could be simplified. However, when not measuring the fraction volumes, significant deviations to the actual retention times were seen for a number of fractions of 12–22 in Figure 10.6c and for a number of fractions of 8–13 in Figure 10.7c. The reason for such deviations lies in the collection of the eluate in droplets. The fractions often differ in the number of collected droplets, which was taken into account for the reference and simulated peaks by alternating the fraction size by half of a drop volume (13 μL). Hence, the x-coordinate values for the concentrations being fitted were different for the scenario using fixed nominal values compared with the scenario taking the droplets into account. For both considered peaks, fitting of the peak was thereby affected resulting in deviations of accuracy when using V_{nom}.

10.3.3.6 Effect of Concentration Errors

The effect of errors introduced during the determination of protein concentration is shown in Figure 10.6d and e for a symmetric peak and in Figure 10.7d and e for an asymmetric peak. The errors applied for the determination of protein concentrations were set to 0% (c_{nom}), 0.5, 2, and 10% (see Table 10.1) while keeping an error of the determination of fraction volumes ($V_{2\%} + V_{meniscus}$) constant for all systems.

A RSTD of 0.5% in the concentration did not result in a noticeable change on the precision when compared to simulations with c_{nom}. Such low error levels apply if precise analytical techniques are available, for example, HPLC. A decrease in analytical precision from 0.5 to 2% resulted in an absolute increase of the confidence interval of max. 0.1 for the symmetric peak (Figure 10.6d). For the asymmetric peak, the effect was larger for a number of fractions <8 with an increase of the confidence interval by maximum of 0.25 (Figure 10.7e). In the case of the symmetric peak and for a large analytical error of 10%, the determined confidence interval for the peak area and the retention time decreased with increasing number of fractions from ~0.2 to 0.08 and from ~0.15 to 0.06, respectively. In the case of the asymmetric peak and an analytical error of 10%, the determined confidence interval for the peak area and the retention time decreased with increasing number of fractions from ~0.55 to 0.1 and from ~0.75 to 0.18, respectively.

It becomes clear that the compensation of the analytical error by using a larger number of fractions is greater for a large error in the determination of the concentration compared to a rather high precision of the analytical technique (e.g., $c_{0.5\%}$). For an asymmetrical peak, using a larger number of fractions (e.g., >20) would therefore be more effective than optimizing the precision of the analytical method. In contrast, the precision for symmetrical peaks is mostly dominated by the precision of the analytical technique and the optimization thereof outweighs the benefit from using more fractions.

10.3.3.7 Effect of Dilution Errors

The effect of errors introduced during a dilution process is shown in Figure 10.6f and g for a symmetric peak and in Figure 10.7f and g for an asymmetric peak. Three different scenarios for the pipetting during dilution were compared: (i) no pipetting error (Pip_{nom}), (ii) a dilution by pipetting 50 µL of protein solution ($Pip_{dil:50\mu L}$), and (iii) a dilution by pipetting 10 µL of protein solution ($Pip_{dil:10\mu L}$). For all scenarios, additional errors for the determination of fraction volumes ($V_{0.5\%} + V_{meniscus}$) and for the determination of protein concentrations ($c_{2\%}$) were kept constant.

Dilution using a large volume did not result in a significant decrease in precision, because the pipetting error is small as discussed in Section 10.3.1. With decreasing sample volume used for dilution and hence an increasing pipetting error, the effect on precision for the determination of the peak area and retention time increased. This effect was seen from the simulation results for both the symmetric and asymmetric peaks. For the symmetric peak, the confidence intervals increased up to 0.055 (Figure 10.6f), whereas for the asymmetric peak, the confidence intervals increased by up to 0.26. An increase in precision at a low number of fractions for the low pipetting error as seen in Figure 10.6f is assumed to be caused by the errors canceling each other out. Furthermore, for a larger number of fractions, the mean of the confidence interval shifted from ~1.01 for (Pip_{nom}) to the expected value of 1.00 $Pip_{dil:50\mu L}$ (Figure 10.6f), demonstrating that in some cases accuracy might increase despite a reduction of the precision.

The fraction volume needs to be adequately large to enable dilution providing highest precision. This, however, contradicts the previously drawn conclusion, that a higher number of fractions (smaller fraction volumes) should be used to improve precision. Therefore, the pipetting accuracy and expected peak shape must be known before a sound decision on the fraction volume and dilution volume can be made aiming at the highest overall precision.

10.3.4 Influence of Flow Regime

During HTCC, the pipetting needles leave the columns multiple times as explained in Section 10.1.1, Figure 10.1. In contrast to typical laboratory LC systems, the fluid flow is thus interrupted for a period of time required for the uptake of new buffer solution. This leads to longer residence times and potential reflux caused during withdrawal of the needle. In order to investigate the quantitative influence on the chromatographic behavior in isocratic HTCC, the following fluid flow pattern for a step elution using a 200 µL RoboColumn were analyzed:

1) "Continuous" flow
2) "Flow interruptions" every 150 µL for 19 s, keeping needles in columns
3) "Needle" movement out and into the column every 150 µL[1]
4) "Combination" of flow interruption and needle movement.[2]

The linear velocity during fluid flow was 90 cm/h (5 µL/s), which equaled a residence time of 40 s. The identical eight columns were used throughout all experiments in order to evaluate whether the inter-column variability significantly differ from the variance caused by the different flow patterns. Two sets of four columns were used with loading either lysozyme (set 1) or avidin (set 2).

The retention times, peak widths, and asymmetries determined after peak fitting with the EGH function are compared for all scenarios in Figure 10.8.

No effect of the chosen flow pattern on either retention time, peak width, or peak asymmetry was observed when using the protein lysozyme. The differences between the flow patterns were statistically not larger than the differences

[1] This implied an interruption of the flow, ~14 s.

[2] Flow was stopped for 40 s corresponding to the total time to perform all program steps for buffer uptake.

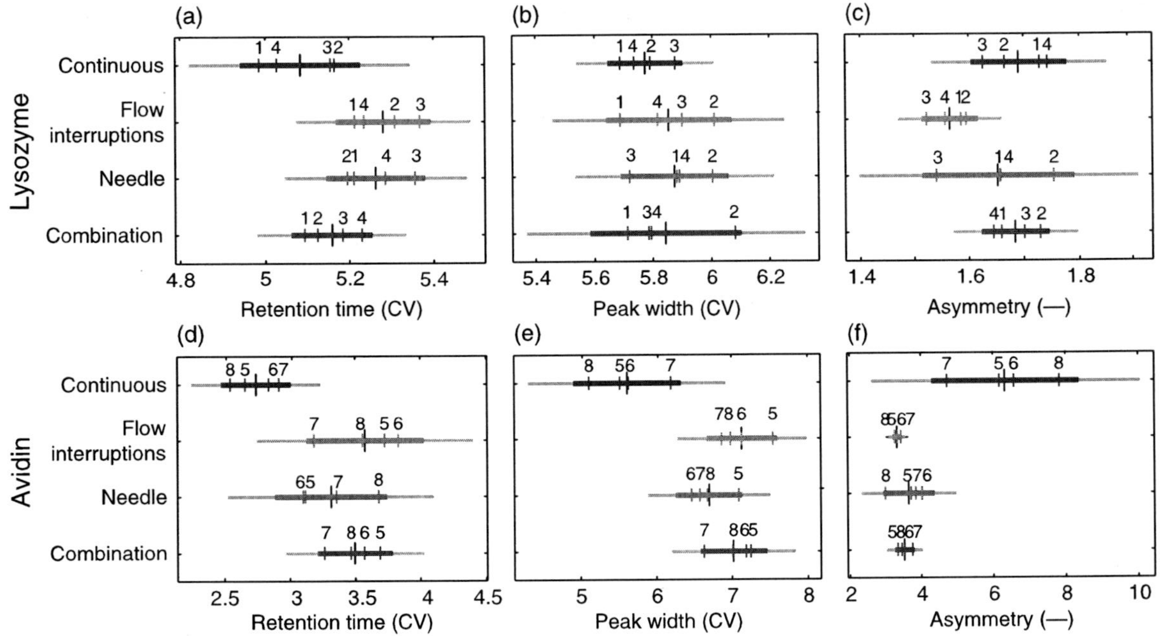

Figure 10.8 Influence of the used fluid flow pattern on peak retention time (a) and (d), peak width at 50% peak height (b) and (e), and peak asymmetry (c) and (f) for experiments using lysozyme and avidin, respectively. For each fluid flow pattern, the same four columns were used in parallel for each protein and are indicated with numbers. The determined values (|, numbers), the mean (|, long), and the 95 and 99% confidence intervals widths (horizontal bars, thick and light gray, respectively) are displayed. A trend was determined if the 95% confidence intervals did not overlay vertically. The effect was considered significant, if also the 99% confidence intervals for each flow pattern did not overlay vertically.

between the columns. The elution profiles of avidin, however, showed an effect when comparing the "continuous" flow with the other patterns, which all featured flow interruptions. In the case of avidin, the mean retention time and the peak width were both reduced by $\gtrsim$20% compared to the three other flow patterns. The asymmetry was increased by factor 1.8. Between the flow patterns' "flow interruptions," "needle," and "combination," no differences were determined.

The difference between the two proteins indicates that the influence of changing from a continuous flow to a flow pattern incorporating flow interruptions depends on the diffusion rates specific for the protein/resin system as well as on the flow rate applied. With changing residence times and flow rates, the ratio of external mass transfer rate to intra-particle diffusion, expressed by the Biot number, changes, eventually resulting in a change in the rate-limiting step [28, 29]. The diffusion coefficient of avidin is about three times smaller than the value for lysozyme explaining among others the larger difference of the peak width and asymmetry between continuous flow and flow with interruptions for avidin compared with lysozyme.

Concentration changes after flow interruptions were not detected in the elution profiles of any case (data not shown). This indicates a "smoothing" effect when collecting the eluate in fractions compared to the continuous signal from a flow cell. The effect of flow interruptions, however, will be detectable when performing breakthrough curves. Wiendahl et al. [9] reported that the concentration during protein breakthrough dropped significantly when interrupting the flow for aspirating new protein solution, hence affecting the shape of the breakthrough curve.

The frequent movements of the pipetting needles out of the column and into the columns did not lead to a detectable change in retention time and peak shape. There was no liquid volume detected by visual detection being soaked from the column outlet into the columns by an underpressure generated when the needles were withdrawn out of the columns.

When comparing the results for each column with each other, no statistically relevant differences were detected that would indicate deviations in packing quality or bed volume. Since identical buffers and protein solutions were used and each column was operated each time with the same pipetting channel, the systematic errors were considered being negligible compared to the overall variance. The mean mass and volume balances and their uncertainties for all 2×16 experiments were calculated in order to estimate the precision for yield calculations. For lysozyme, the average mass balance was 94 ± 1.5% (RSTD of 3.1%). The mean volume balance for lysozyme experiments was 101 ± 1.3% (1.7% RSTD). In the case of avidin, the mean mass balance was 98 ± 2.1% (RSTD 4.1%) and the mean volume balance was 102 ± 0.4% (RSTD 0.8%). The obtained deviations were therefore close to the theoretical value for the peak area confidence interval (~2%) determined by simulations of fractionation schemes and a corresponding 2% error level (see Section 10.3.3).

10.3.5 Gradient Elution Experiments

When performing isocratic elution in HTCC mode, flow interruptions can be reduced to a minimum required for changing fraction collection plates. When performing gradient elutions, however, flow interruptions for buffer uptake cannot be avoided when using current HTCC hardware setups. When mimicking a salt gradient by sequential pipetting of buffers (Eq. 10.1), it is obvious that the smaller the applied salt steps (and thus the more steps), the closer the approximation to a gradient applied on a laboratory LC system. In a study published by Welsh et al. [30], the effect of three different salt step heights on the elution profile was qualitatively shown demonstrating a significant influence of the salt steps on the elution profile. Furthermore, with each flow interruption required, the discrepancy between applied flow rate and average residence time increases. In Table 10.2, ratios between times of flow and interruptions due to buffer uptake were exemplarily calculated for 12 and 48 interruptions during elution. Significant differences in residence times occur depending on the chosen flow rate and frequency of flow interruptions.

In the following sections, investigations of gradient elutions in HTCC are presented addressing:

1) The effect of different salt step heights mimicking a gradient
2) The combined influence of salt steps, flow interruptions, and flow rate on elution performance
3) The comparability of HTCC, simulation, and common laboratory LC experiments.

As it is not possible to investigate separately the effect of different salt step heights and flow interruptions for salt gradient runs experimentally on a LHS, simulations were used for the comparison of various scenarios of salt step heights, flow interruptions, and fractionation. The employed solver, which is

Table 10.2 Relation of flow rate and residence times during elution with 12 and 48 interruptions in HTCC experiments.

Flow rate (μL/s)	No. of pauses	Average residence time (s)	Ratio to continuous flow
1	12	220	1.1
1	48	280	1.4
3	12	147	2.2
3	48	87	1.3
8	12	45	1.8
8	48	105	4.2

The duration of a flow interruption ("pause") for wash and buffer uptake was 30 s.

Table 10.3 SMA isotherm parameters for lysozyme and avidin using SP Sepharose FF at pH 4.5.

Parameter	Lysozyme	Avidin
ν	6.25	11.18
σ_{SMA}	10	45
k_{ad}	55	256
k_{des}	80	811

described by von Lieres and Andersson [31], uses the general rate model. The code was adapted for the simulation of HTCC, that is, incorporating a multiple step elution with variable step height and step volume and the corresponding flow interruptions for the uptake of each salt buffer step. The steric mass action (SMA) model introduced by Brooks and Cramer [32] was chosen as sorption model. The determination of the SMA isotherm parameters, characteristic charge ν, shielding factor σ, and ad- and desorption equilibrium constants k_{ad} and k_{des}, was performed using the inverse method presented by Osberghaus et al. [33]. For this approach, experimental data from elution on the LHS with salt gradient slopes of 525 mM NaCl over 8, 13, 18, 23, and 28 CV were used. The isotherm parameters for lysozyme and avidin resulting from the fitting procedure are given in Table 10.3. The ionic capacity Λ was set to 0.8 M [33]. The process parameters and column model parameters are summarized in Table 10.4.

10.3.5.1 Salt Step Height

In the first step, the separation of lysozyme from avidin was simulated using NaCl step heights between 11 and 66 mM (corresponding to buffer steps from 75 to 450 μL, respectively). The qualitative effect of the salt step height can be seen from the simulated chromatograms in Figure 10.9.

The application of buffer salt steps caused a deviation from the common peak profile. With increasing salt step height, the deformation of the peaks became more pronounced. When using salt step heights below 11 mM, an effect of buffer steps was not visually detected. For salt step heights of 21.9 mM, the effect of using salt steps still appeared to be negligible. For salt step heights of 54.7 and 65.6 mM, the avidin peak eventually split into two main peaks. It became clear that peak fitting is required in order to enable determination of peak retention times. However, despite the large peak deformation, a similar qualitative conclusion on the separation efficiency of the system of resin, solvent, and gradient would be drawn regardless of the salt step height used.

Simulation results would obviously be different for other protein systems, flow rates, and binding kinetics. However, the observation based on the

Table 10.4 Process model parameters for HTCC simulations.

Parameter	Value	Determination
Process and column parameters		
Column length, L	0.01 m	Manufacturer information
Column diameter, d_{col}	0.005 m	Manufacturer information
Linear flow rate, u	36–144 cm/h	—
Column porosity, ε_{col}	0.33	Experimental, moment analysis
Particle porosity, ε_p	0.82	Experimental, moment analysis
Particle diameter, d_p	$90 \cdot 10^{-6}$ m	Manufacturer information
Axial dispersion coefficient, D_{ax}	$6.8 \cdot 10^{-8}$ m^2/s	Estimated by correlation [34]
Mass transfer parameters		
Film mass transfer coefficient, k_f	NaCl: $1.9 \cdot 10^{-4}$ m/s	Literature [35]
	Lysozyme: $3.5 \cdot 10^{-5}$ m/s	Estimated by correlation [36, 37]
	Avidin: $2.4 \cdot 10^{-5}$ m/s	Estimated by correlation [36, 37]
Effective intra-particle diffusion coefficient, D_p	NaCl: $5.9 \cdot 10^{-10}$ m^2/s	Literature [35]
	Lysozyme: $5.75 \cdot 10^{-11}$m^2/s	Estimated from literature [37]
	Avidin: $2.33 \cdot 10^{-11}$m^2/s	Estimated from literature [37]

Porosities were obtained in pulse experiments using a penetrating (acetone) and non-penetrating tracer (2 MDa Dextran).

presented simulated elution profiles matches the experimental data using a monoclonal antibody reported by Welsh et al. [30]. As general conclusion one can state that the suspected "noise" often observed in data from robotic experiments is mostly caused by the application of salt steps. These peak deformation must therefore not be considered as noise but in fact reflect the elution profile of a multiple step elution. The reason why these unsmooth elution profiles are not always detected in the obtained pseudo-chromatograms lies in the chosen fractionation volumes (as will be shown in the next subsection), but it depends also on the step length and hence on the gradient slope.

10.3.5.2 Salt Steps and Flow Interruptions

In the next step, simulations were performed taking the flow interruptions caused by each buffer uptake in gradient HTCC into account by setting the flow rate to zero for 30 s before every salt step. Furthermore, we expanded the

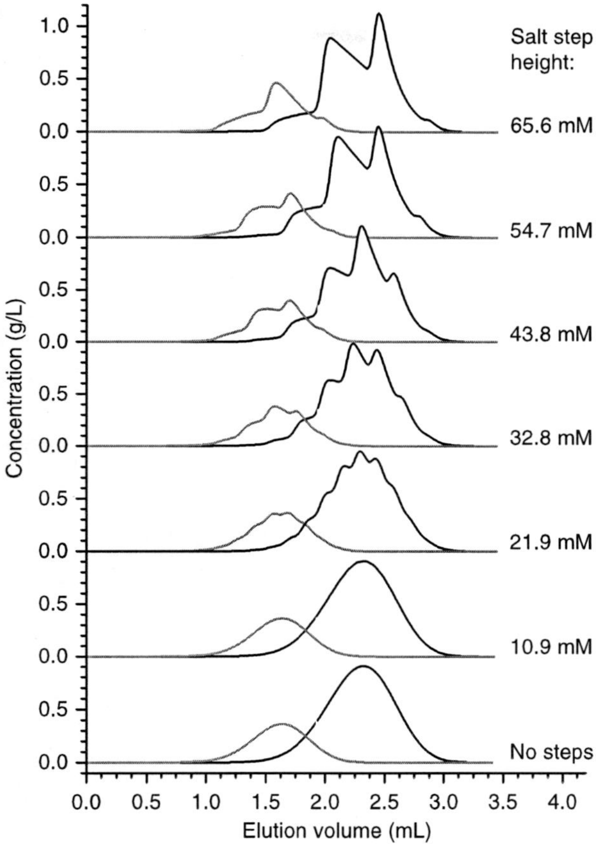

Figure 10.9 Separation of lysozyme (gray) from avidin (black) simulated using different NaCl concentration steps for elution. Flow rate: 54 cm/h (3 μL/s).

simulations introducing *in silico* fractionation with fractions of 75, 150, 225, 300, and 375 μL. In order to determine retention time, asymmetry, and peak width for each of these fractions, the theoretical average concentration was calculated, and a peak fit was performed using these average values. It was furthermore assumed that no additional pauses would be introduced during a salt step, and thus, the number of breaks corresponded to the number of applied salt steps.

10.3.5.2.1 Effect of Salt Steps and Flow Interruptions: Qualitative Evaluation

Simulated elution profiles obtained for NaCl concentration step sizes of 10.9 and 43.8 mM, each with fractionation in 75 or 300 μL, are displayed for flow rates of 54 and 144 cm/h in Figure 10.10.

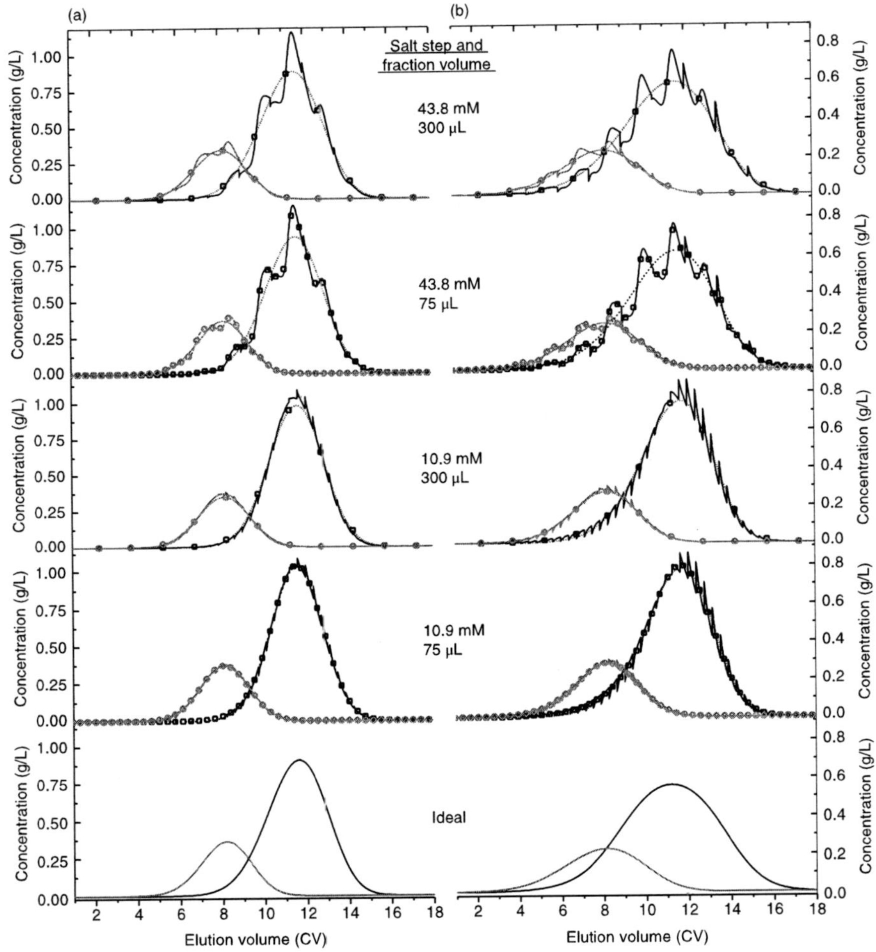

Figure 10.10 Effect of salt step height, fractionation, and flow rate on the elution profile for the separation of lysozyme (gray) from avidin (black). (a) 54 cm/h. (b) 144 cm/h. The gradient slope was +525 mM NaCl over 18 CV. Solid lines: simulated elution profiles for corresponding salt step height. Markers (circles: lysozyme, boxes: avidin): concentration values obtained in theoretical fractions of corresponding size (300 or 75 μL). Dashed lines: fit of the fraction concentrations using the EGH function. The step size of 11 and 43.8 mM corresponded to a step length of 75 and 300 μL, respectively. The "ideal" chromatogram displays the simulation result assuming a linear gradient without flow interruptions.

Similar to the result shown previously (Figure 10.9), the influence of the salt step was large for the 43.8 mM salt step, whereas when using 11 mM steps, the peak shapes were very similar to those of the reference ("ideal") gradient elution. The introduction of flow interruptions generated prompt changes in the concentration after each break.

The concentration in the bulk phase decreases at peak front after each stop in flow because protein predominantly diffuses into the beads. At peak tail, the protein concentration in the bulk phase increases after each stop in flow due to protein diffusion out of the beads. The magnitude of this effect is more pronounced at high flow rates at which the ratio of convection to diffusion is larger. Compared with the effect of different salt steps, the effect of flow interruptions on the overall peak shape appeared smaller.

The influence of the fraction volume becomes very clear from the chromatograms shown in Figure 10.10. When using small fraction volumes, the number of distinct data points is high enough to reflect the actual elution profile including the variations in concentration. That means, an "unsmooth" elution peak will be observed potentially leading to a false conclusion of having large data noise levels. When using large fraction volumes, the elution profile will be smoothed, and the resulting common peak shape intuitively gives the impression of a more precise result. However, as demonstrated in Section 10.3.3, a fraction number per peak lower than 8–10 may lead to a significant reduction in precision of the determined retention time and mass balance (area).

By performing a peak fit, the resulting peaks were very similar regardless of the chosen fractionation or salt step height indicating the importance of fitting the peaks prior to evaluation.

10.3.5.2.2 Effect of Salt Steps and Flow Interruptions: Quantitative Evaluation

The quantitative deviation of the determined peak characteristics for different gradient patterns and fractionation schemes related to results from a simulated "ideal" gradient experiment are presented in Figure 10.11 for lysozyme (Figure 10.11a and b) and avidin (Figure 10.11c,d, and e). The effect on the determined peak resolution is shown in Figure 10.11f. The relative deviations were calculated for different fraction volumes and two flow rates, each grouped by a certain salt step height (and hence number of flow interruptions). In addition to simulations with flow interruptions, the theoretical scenario of performing salt steps without having flow interruptions are given for the smallest and largest fraction volume.

A clear difference between the output parameter retention times on one hand and peak asymmetry, peak width, and peak resolution on the other hand can be seen in Figure 10.11 from the magnitude of deviations. The effect of different experimental combinations on the determined retention times was rather low (<4.5% for all cases), whereas the range of deviations in peak asymmetry and avidin peak widths was very large (−29 to +22%). The range of deviations in peak width for lysozyme was −21 to +11% (data not shown). In accordance with the large deviations in peak width, the determined resolution values increased significantly (+48%) compared to the "ideally" performed gradient when using the highest number of flow interruptions, that is, smallest salt step heights and smallest fractions.

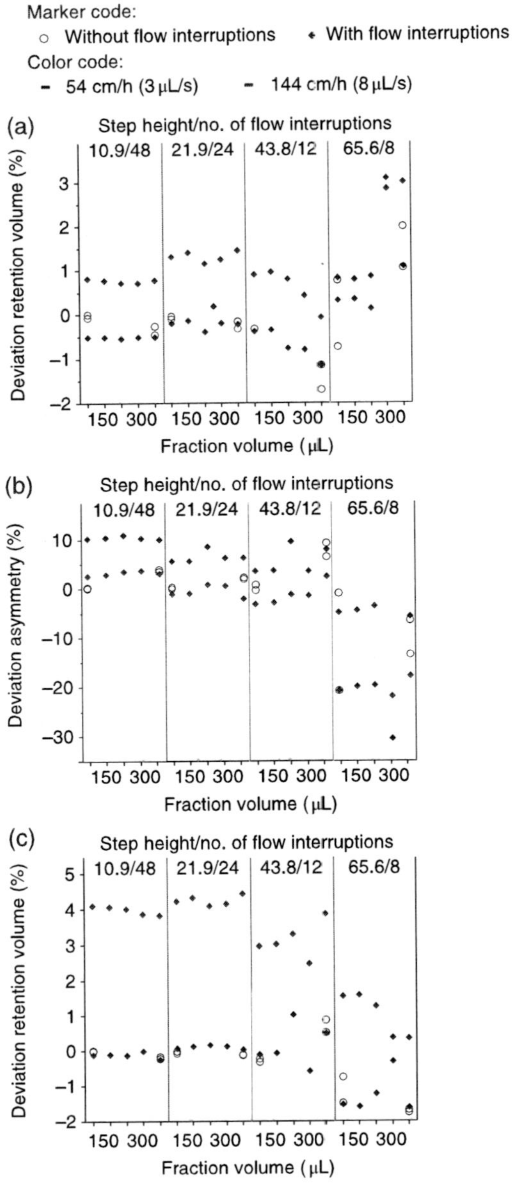

Figure 10.11 Relative deviations of the determined output parameters from the simulated "ideal" gradient experiment for varying fraction volumes and flow rates grouped by a certain salt step height. The effect on the determined retention time and peak asymmetry is given for lysozyme (a, b) and avidin (c, d). The effect on the peak width is only displayed for avidin (e). The effect on the resolution is displayed in subplot (f).

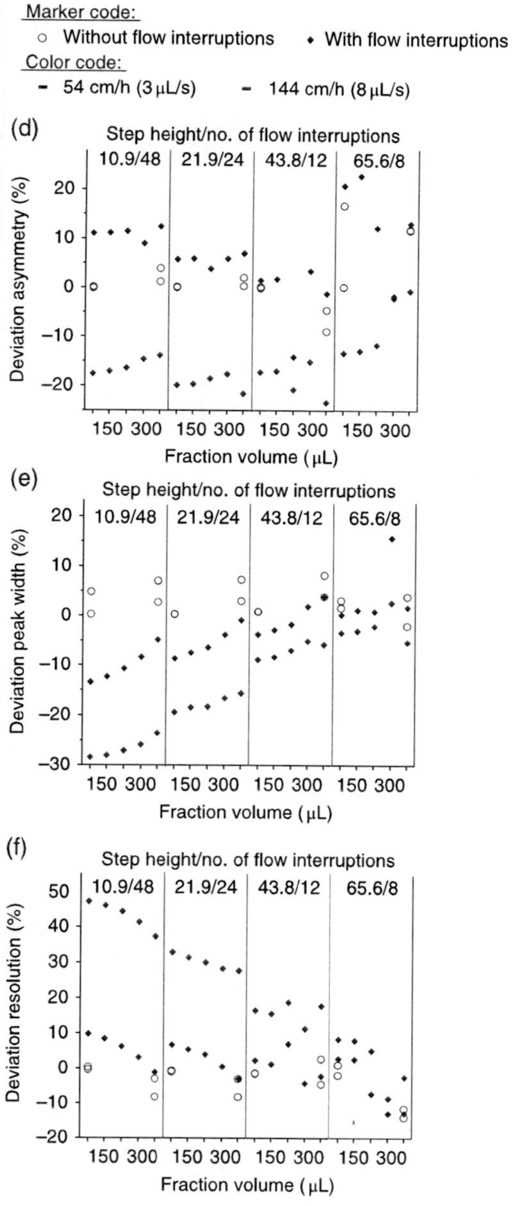

Figure 10.11 (Continued)

The effect of only introducing salt steps can be seen from the simulations without flow interruptions. The quantitative effect on the chromatographic output parameters was small (<8%) when applying the two small salt step heights. For a step height of 65.6 mM and a step height of 43.8 combined with the largest fraction volume, deviations of up to 20% were obtained for peak asymmetries (Figure 10.11b and d). For these larger salt step heights, the output parameters were more affected by the choice of the fraction volume compared to the salt step height.

The effect of the flow interruptions can be seen when comparing the simulation results of identical salt step heights with and without flow interruptions as displayed for each fraction volume of 75 and 375 μL. It becomes clear that the effect of flow interruptions on the quantitative results is significantly larger than the variation of the salt step height, although qualitatively the chromatograms indicated the opposite (Figure 10.10).

For a combination of salt steps and flow interruptions, the overall results were closer to the result from an "ideal" gradient experiment when using larger salt step heights compared with results using small salt steps. This is due to the fact that with increasing salt step heights, a lower number of buffers and hence a lower number of flow interruptions is required, which outweighs the peak deformation resulting from larger step heights.

For a variation of only the fraction volume at constant salt step height, peak widths and asymmetries for avidin increased with the chosen fraction volumes, and larger fraction volumes resulted in less deviation compared to the "ideal" case. This has several reasons: First, a lower peak height of the determined peaks was obtained when fitting the peak with less fractions. As the peak widths were calculated at 0.1 peak height, the peak widths increased with increasing fraction volume and thus are coming closer to the ideal case, in which peaks were broad. Second, the concentrations after a flow interruption at peak front were reduced, and the concentrations after a flow interruption at peak tail were increased (see Figure 10.10), leading to a different peak height and thus the determined peak width. Lastly, with an increasing fraction volume and hence using less fractions for fitting the peak, the influence of flow interruptions decreased due to the generated smoothing effect.

The trends described earlier were very similar for both flow rates. However, for a flow rate of 144 cm/h, the deviations in peak width and resolution were more than doubled compared to a flow rate of 54 cm/h. Flow interruptions have a larger effect on the concentration profile at higher flow rates as the discrepancy of mass transfer rates during flow and stopped flow increases. From this result, higher comparability of experiments in HTCC mode to traditional laboratory LC experiments would be expected using lower flow rates.

10.3.5.3 Comparability of Simulation, HTCC, and Laboratory LC Results

In the previous sections, it was shown in simulations that HTCC process parameters have both a quantitative and qualitative effect on the outcome of chromatography experiments. However, in order to investigate whether the determined effects and deviations from the "ideal" case can also be detected in experimental data, results from HTCC experiments were compared for two flow rates with data generated in traditional laboratory LC experiments and simulation data (Figure 10.12 and Table 10.5). For an evaluation of the variance between HTCC runs, four elution profiles from parallel HTCC experiments were compared.

From Figure 10.12 it can be seen that the experimental chromatographic profiles largely match the simulated elution profiles, hence confirming the use of simulations as a suitable tool for the presented investigation of HTCC process parameters. The concentration profiles of the elution peaks show a pattern of changing concentrations every second value. This pattern was seen for all four HTCC experiments indicating these variations to be reproducible. The concentration pattern is furthermore in agreement with the predictions from the simulated elution profiles using the same fractionation scheme consisting of step heights of 21.9 mM (step lengths = 150 μL) and a fraction size of 75 μL. The elution profile from laboratory LC experiments (continuous flow and ideal gradient) shows the expected Gaussian peak shape. Small variations to the ideal peak profile reflect the analytical error for the determination of the concentration. The experimental data confirmed the earlier findings that variations in concentration are more pronounced at higher flow rates (Figure 10.10).

The differences of retention times and peak widths obtained in experimental and simulated HTCC and laboratory LC experiments related to the reference of an simulated "ideal" robotic experiment (without salt steps and flow interruptions) are summarized in Table 10.5. In average, the deviations for the different experimental modes regarding the determined retention times were significantly lower (~2.1%) than the deviations determined for the peak widths (~11.3%). The RSTDs determined from four HTCC experiments for retention times and peak widths were ≤1.0 and ≤2.9%, respectively, with the exception of 7.5% RSTD for the avidin peak width. With respect to this variation, it can be concluded that there is no statistical effect of the experimental mode on the retention times. This conclusion is affirmed when taking potential deviations in the determined retention time due to different fractionation schemes into account as presented in Figures 10.6 and 10.7. Furthermore, the differences between results from HTCC and results from laboratory LC experiments were larger than the differences between the two simulated HTCC modes ("ideal" and "steps + interruptions").

A significant difference of HTCC to ideal gradient and laboratory LC experiments was however obtained for the peak widths. The peaks obtained from the

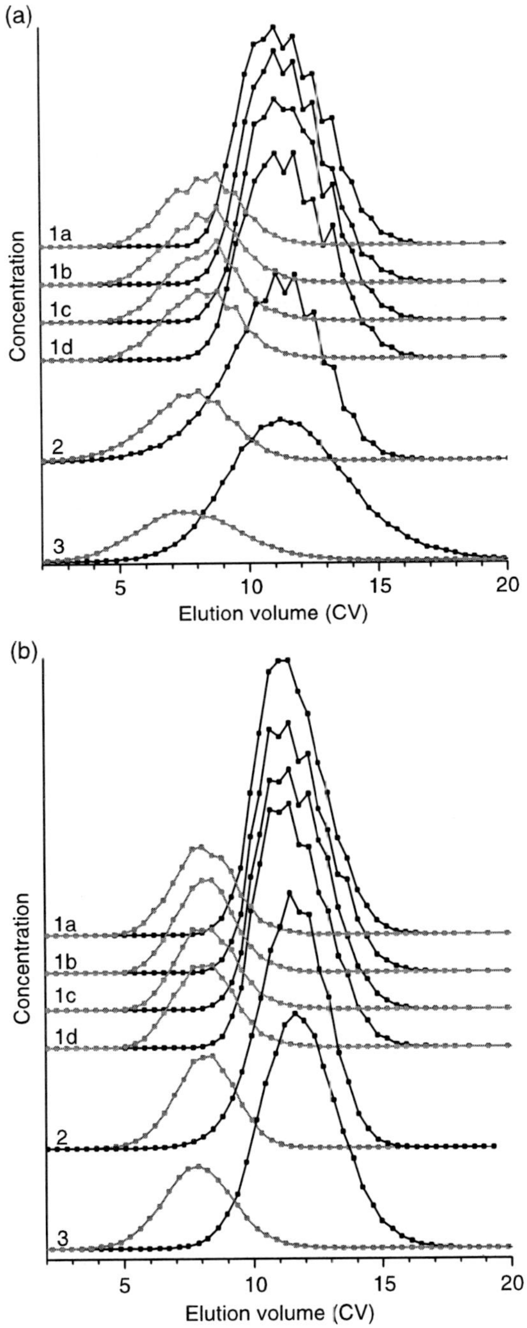

Figure 10.12 Chromatograms of the separation of avidin (black) from lysozyme (gray) at flow rates 144 cm/h (a) and 54 cm/h (b) on SP Sepharose FF (CV = 200 μL) using a salt step height of 21.9 mM and fractionation in 75 μL. Separations were performed on a LHS in four parallel experiments (1a–1d), simulated including flow interruptions and salt steps (2) and performed on a laboratory LC system (ideal gradient and without flow interruptions, 3).

Table 10.5 Deviations of retention time and peak width from the reference of a simulated "ideal" experiment without salt steps and flow interruptions ("pauses") for different experimental setups.

		Retention volume		Peak width		Resolution
Experimental mode	**Protein**[a]	**Deviation (%)**	**RSTD (%)**	**Deviation (%)**	**RSTD (%)**	**Mean (—)**
54 cm/h (3 μL/s)						
HTCC simulated "ideal"	Lys/avi	—	—	—	—	1.32
HTCC experimental "steps + interruptions"	Lys	−0.9	0.3	−1.5	2.7	—
HTCC simulated "steps + interruptions"	Lys	−0.2	—	−2.0	—	—
Laboratory LC experimental "ideal"	Lys	−4.3	—	14.4	—	—
HTCC experimental "steps + interruptions"	Avi	−3.1	0.5	−7.7	2.9	1.24
HTCC simulated "steps + interruptions"	Avi	0.1	—	−8.4	—	1.40
Laboratory LC experimental "ideal"	Avi	0.7	—	4.4	—	1.36
144 cm/h (8 μL/s)						
HTCC simulated "ideal"	Lys/avi	—	—	—	—	0.76
HTCC experimental "steps + interruptions"	Lys	2.5	0.8	−23.4	2.7	—
HTCC simulated "steps + interruptions"	Lys	1.5	—	−14.7	—	—
Laboratory LC experimental "ideal"	Lys	−7.4	—	6.0	—	—
HTCC experimental "steps + interruptions"	Avi	0.3	1.0	−31.3	7.5	0.98
HTCC simulated "steps + interruptions"	Avi	4.3	—	−19.3	—	1.00
Laboratory LC experimental "ideal"	Avi	0.3	—	2.3	—	0.86

The relative standard deviations (RSTD) were calculated from four parallel experiments on a LHS. The resolution values are given as absolute values. All separations were performed using the same gradient slope of 525 mM NaCl over 18 CV.

[a] Avi, avidin; Lys, lysozyme.

laboratory LC experiments were broader compared with the peaks from HTCC experiments. This result is in agreement with the findings discussed in Section 10.3.5.2, Figure 10.11. Furthermore, for a flow rate of 144 cm/h, the peak widths determined from experimental HTCC were significantly smaller than the simulated values (Table 10.5). This indicates that the peak width and hence peak resolution might in fact be even more affected by the HTCC experimental setup than concluded from preceding simulation results.

10.4 Summary and Conclusion

In this study, we systematically investigated in which way robot-specific experimental parameters contribute to the experimental error of HTCC. Furthermore, differences between elution profiles obtained in HTCC and profiles obtained with laboratory LC systems were explained.

It was shown that the precision of pipetting (e.g., for dilution prior to analysis) increases exponentially with increasing pipetting volumes. Furthermore, the use of MTPs was demonstrated to cause specific challenges. First, meniscus formation being different for solvent and varying protein concentrations affects the path length in absorption measurements and thus causes inaccuracies in the determination of concentration and volume. Calibration ranges need therefore to be chosen adequately and without forcing the calibration curve through zero (buffer blank). Secondly, when collecting eluate in MTPs, the elution signal comprises a series of distinct absorption values rather than a continuous absorption signal, and a chromatogram needs to be rebuilt by fitting discrete absorption/concentration data points. The importance of using peak fitting functions to obtain suitable information on retention times and peak resolution as well as to increase precision was highlighted.

It was shown that a low number of collected elution fractions significantly contribute to imprecision of the retention time and peak area (mass balance) obtained from HTCC data, particularly when elution peaks are asymmetric. Remarkably, using very precise analytical techniques cannot outweigh the increase of variance caused by a low number of fractions. If the analytical precision is known to be imprecise, however, a sufficiently precise determination of mass balance and retention time can still be reached by increasing the number of fractions. About 10 fractions collected per peak width appear sufficient in order to significantly reduce the variance, and the maximal possible precision is approached. A possible further gain of precision using more fractions might eventually not justify the increasing effort required for the analysis of more fractions. Furthermore, for low fraction volumes, the absolute error of the volume determination via absorption measurements increases. In this case, it becomes favorable to instead use the nominal fraction volumes.

It was explained that the "smoothness" of an elution peak in HTCC gradient experiments is mainly depending on the choice of the salt step height. By choosing small fractions, the actual, potentially uncommon, concentration profile will be reflected giving the impression of apparent noise in the data. In contrast, when collecting less fractions, the common peak shape is obtained. This is a fact which might be counterintuitive, as one would expect better representation of the common peak shape when collecting more fractions rather than less. It was quantitatively shown that when using peak fitting functions, the choice of the salt step height is not as important as the resulting number of flow interruptions. The effect of flow interruptions was furthermore demonstrated for isocratic elutions showing an effect only for the larger protein investigated. The peak widths were influenced most by the number of flow interruptions, whereas the fractionation scheme and analytical precision were the main factors affecting the determination of the retention time. Using larger salt steps (i.e., less flow interruptions) increased comparability to common laboratory LC experiment. However, larger deviations in retention times and asymmetry were seen in this case study for salt step heights >43.8 mM. Therefore, a suitable compromise for the salt step heights would be in a range of 22–44 mM. Future work could expand this study by taking other mass transfer scenarios into account, for example, by varying the diffusion rates (different proteins and resins), flow rate, and column dimensions.

There is no doubt that for screening purposes, the use of HTCC fulfills the requirements on accuracy and precision. However, the results of this study also demonstrate that robotic process parameters should be chosen thoughtful on a case by case basis with respect to the desired data quality. Variabilities in the determined retention times and peak widths need to be taken into account, if data are used to calibrate mechanistic models or if the technology is considered for detailed process development and robustness studies. The changes in residence times depending on the flow interruptions need to be considered when evaluating comparability to larger scales.

Acknowledgements

We thank Anna Osberghaus for the support regarding chromatography simulations as well as Sven Amrhein for his contributions to the MC simulations. Further, we thank Katharina Drechsel for carrying out parts of the isocratic elution experiments.

References

1. J.L. Coffman, J.F. Kramarczyk, B.D. Kelley, High-throughput screening of chromatographic separations: I. Method development and column modeling, *Biotechnol. Bioeng.* 100 (2008) 605–618.

2. B.D. Kelley, M. Switzer, P. Bastek, J.F. Kramarczyk, K. Molnar, T. Yu, J. Coffman, High-throughput screening of chromatographic separations: IV. Ion-exchange, *Biotechnol. Bioeng.* 100 (2008) 950–963.
3. J.F. Kramarczyk, B.D. Kelley, J.L. Coffman, High-throughput screening of chromatographic separations: II. Hydrophobic interaction, *Biotechnol. Bioeng.* 100 (2008) 707–720.
4. T. Bergander, K. Nilsson-Valimaa, K. Oberg, K.M. Lacki, High-throughput process development: determination of dynamic binding capacity using microtiter filter plates filled with chromatography resin, *Biotechnol. Prog.* 24 (2008) 632–639.
5. L. Connell-Crowley, E.A. Larimore, R. Gillespie, Using high throughput screening to define virus clearance by chromatography resins, *Biotechnol. Bioeng.* 110 (2013) 1984–1994.
6. J. Kittelmann, J. Hubbuch, Robust high-throughput batch screening method in 384-well format with optical in-line resin quantification, *J. Chromatogr. B* 988 (2015) 98–105.
7. P. Schulze Wierling, R. Bogumil, E.H.J. Knieps-Grünhagen, High-throughput screening of packed-bed chromatography coupled with SELDI-TOF MS analysis: monoclonal antibodies versus host cell protein, *Biotechnol. Bioeng.* 98 (2007) 440–450.
8. M.D. Wenger, P. DePhillips, C.E. Price, D.G. Bracewell, An automated microscale chromatographic purification of virus-like particles as a strategy for process development, *Biotechnol. Appl. Biochem.* 47 (2007) 131–139.
9. M. Wiendahl, P. Schulze Wierling, J. Nielsen, D. Fomsgaard-Christensen, J. Krarup, A. Staby, J. Hubbuch, High throughput screening for the design and optimization of chromatographic processes—miniaturization, automation and parallelization of breakthrough and elution studies, *Chem. Eng. Technol.* 31 (2008) 893–903.
10. S. Chhatre, D.G. Bracewell, N.J. Titchener-Hooker, A microscale approach for predicting the performance of chromatography columns used to recover therapeutic polyclonal antibodies, *J. Chromatogr. A* 1216 (2009) 7806–7815.
11. A. Susanto, K. Treier, E. Knieps-Grünhagen, E. von Lieres, J. Hubbuch, High-throughput screening for the design and optimization of chromatographic processes: automated optimization of chromatographic phase systems, *Chem. Eng. Technol.* 32 (2009) 140–154.
12. K. Treier, S. Hansen, C. Richter, P. Diederich, J. Hubbuch, P. Lester, High-throughput methods for miniaturization and automation of monoclonal antibody purification processes, *Biotechnol. Prog.* 28 (2012) 723–732.
13. K. Rege, N. Tugcu, S.M. Cramer, Predicting column performance in displacement chromatography from high-throughput screening batch experiments, *Sep. Sci. Technol.* 38 (2003) 1499–1517.
14. A. Susanto, E. Knieps-Grünhagen, E. von Lieres, J. Hubbuch, High-throughput screening for the design and optimization of chromatographic

processes: assessment of model parameter determination from high throughput compatible data, *Chem. Eng. Technol.* 31 (2008) 1846–1855.

15. A. Osberghaus, K. Drechsel, S. Hansen, S. Hepbildikler, S. Nath, M. Haindl, E. von Lieres, J. Hubbuch, Model-integrated process development demonstrated on the optimization of a robotic cation exchange step, *Chem. Eng. Sci.* 76 (2012) 129–139.
16. W.R. Keller, S.T. Evans, G. Ferreira, D. Robbins, S.M. Cramer, Use of minicolumns for linear isotherm parameter estimation and prediction of bench top column performance, *J. Chromatogr. A* 1418 (2015) 94–102.
17. A. Osberghaus, P. Baumann, S. Hepbildikler, S. Nath, M. Haindl, E. von Lieres, J. Hubbuch, Detection, quantification, and propagation of uncertainty in high-throughput experimentation by Monte Carlo methods, *Chem. Eng. Technol.* 35 (2012) 1456–1464.
18. L. Britsch, T. Schroeder, J. Friedle, Automated high-throughput chromatographic separation of biological compounds, *Am. Biotechnol. Lab.* 26 (2008) 20–23.
19. S.K. Hansen, E. Skibsted, A. Staby, J. Hubbuch, A label-free methodology for selective protein quantification by means of absorption measurements, *Biotechnol. Bioeng.* 108 (2011) 2661–2669.
20. P. Diederich, S. Amrhein, F. Hämmerling, J. Hubbuch, Evaluation of PEG/phosphate aqueous two-phase systems for the purification of the chicken egg white protein avidin by using high-throughput techniques, *Chem. Eng. Sci.* 104 (2013) 945–956.
21. I.H. Xie, M.H. Wang, R. Carpenter, H.Y. Wu, Automated calibration of TECAN genesis liquid handling workstation utilizing an online balance and density meter, *Assay Drug Dev. Technol.* 2 (2004) 71–80.
22. S.A. Oelmeier, F. Dismer, J. Hubbuch, Application of an aqueous two-phase systems high-throughput screening method to evaluate mAb HCP separation, *Biotechnol. Bioeng.* 108 (2011) 69–81.
23. E.L. McGown, D.G. Hafeman, Multichannel pipettor performance verified by measuring path length of reagent dispensed into a microplate, *Anal. Biochem.* 258 (1998) 155–157.
24. M.G. Cottingham, C.D. Bain, D.J.T. Vaux, Rapid method for measurement of surface tension in multiwell plates, *Lab. Investig.* 84 (2004) 523–529.
25. K. Lan, J.W. Jorgenson, A hybrid of exponential and Gaussian functions as a simple model of asymmetric chromatographic peaks, *J. Chromatogr. A* 915 (2001) 1–13.
26. J. Li, Comparison of the capability of peak functions in describing real chromatographic peaks, *J. Chromatogr. A* 952 (2002) 63–70.
27. S. Kreusch, S. Schwedler, B. Tautkus, G.A. Cumme, A. Horn, UV measurements in microplates suitable for high-throughput protein determination, *Anal. Biochem.* 313 (2003) 208–215.
28. Y. Tao, G. Carta, Rapid monoclonal antibody adsorption on dextran-grafted agarose media for ion-exchange chromatography, *J. Chromatogr. A* 1211 (2008) 70–79.

29. K.M. Lacki, High-throughput process development of chromatography steps: advantages and limitations of different formats used, *Biotechnol. J.* 7 (2012) 1192–1202.
30. J.P. Welsh, M.G. Petroff, P. Rowicki, H. Bao, T. Linden, D.J. Roush, J.M. Pollard, A practical strategy for using miniature chromatography columns in a standardized high-throughput workflow for purification development of monoclonal antibodies, *Biotechnol. Prog.* 30 (2014) 626–635.
31. E. von Lieres, J. Andersson, A fast and accurate solver for the general rate model of column liquid chromatography, *Comput. Chem. Eng.* 34 (2010) 1180–1191.
32. C.A. Brooks, S.M. Cramer, Steric mass-action ion exchange: displacement profiles and induced salt gradients, *AIChE J.* 38 (1992) 1969–1978.
33. A. Osberghaus, S. Hepbildikler, S. Nath, M. Haindl, E. von Lieres, J. Hubbuch, Determination of parameters for the steric mass action model-A comparison between two approaches, *J. Chromatogr. A* 1233 (2012) 54–65.
34. H. Schmidt-Traub (Ed.), *Preparative Chromatography*, 1st Edition, Wiley-VCH, Weinheim, 2005.
35. C.A. Orellana, C. Shene, J.A. Asenjo, Mathematical modeling of elution curves for a protein mixture in ion exchange chromatography applied to high protein concentration, *Biotechnol. Bioeng.* 104 (2009) 572–581.
36. T. Kataoka, H. Yoshida, K. Ueyama, Mass transfer in laminar region between liquid and packing material surface in the packed bed, *J. Chem. Eng. Jpn.* 5 (1972) 132–136.
37. M. Schröder, E. von Lieres, J. Hubbuch, Direct quantification of intraparticle protein diffusion in chromatographic media, *J. Phys. Chem. B* 110 (2006) 1429–1436.

11

Lab-Scale Development of Chromatography Processes

Hong Li, Jennifer Pollard, and Nihal Tugcu

Process Development and Engineering, Merck, Kenilworth, NJ, USA

11.1 Introduction

Many techniques have been used for the purification of complex protein and peptide mixtures including precipitation, extraction, filtration, and chromatography. Among these different techniques, chromatography has continued to play an essential role owing to its selectivity, power of resolution, and well-established methods for separation, optimization, and scale-up. Over the years, chromatography has continued to advance due to stationary phase development to establish better resins with higher capacity, unique selectivities, and mechanical stability. These advancements have led to higher productivity and purity, and so, chromatography remains an important tool for protein purification [1, 2].

There are many components that contribute to a purification process development strategy. Development of chromatographic processes requires a good understanding of the molecular properties of the protein/peptide to be purified as well as those of the impurities to ensure selection of the correct chromatographic technique (e.g., ion exchange, hydrophobic interaction, etc.) leading to selectivity and resolution. Additionally, the purity requirements for the intended use of the protein/peptide need to be set in order to evaluate the need for multiple purification steps (e.g., multiple chromatography steps). The ability to assess these properties and set specifications early is the key to establishing a purification strategy. A product developability assessment can offer the first opportunity to provide biochemical and biophysical properties of the product and establish analytical support for chromatography process development [3].

Depending on the molecular properties of the protein, some rules of thumb are used to aid in selecting the appropriate mode of chromatography. If net charge and charge distribution are differentiators between the protein of interest and the impurities, then ion exchange chromatography can be utilized, and initial conditions can be chosen based on the isoelectric point (p*I*) of the protein. Similarly, size-based separations can be carried out using size-exclusion chromatography (SEC), which generally can resolve globular proteins differing by at least twofold in molecular weight. Hydrophobicity differences may be exploited via the use of hydrophobic interaction chromatography (HIC). Chromatography based on mixed-mode interactions (e.g., hydroxyapatite, cation, and anion exchange mixed mode) may be beneficial where secondary interactions exist, such as hydrophobic interaction and hydrogen bonding, in addition to electrostatic interactions. There are cases where recognition (e.g., protein A (ProA) binding by monoclonal antibodies (mAbs)) exists and for which affinity chromatography can be utilized for the best selectivity [4]. There are also cases where displacement chromatography can be utilized to purify very closely related proteins or trace amount of impurities [5]. It is also important to include the feasibility for scale-up as a selection criterion. That is to say, the particle, pore size, and pressure drop across different flow rates should be considered if the resin is to be used at large scale [6–9]. It is important to note that some modes of chromatography, for example, SEC, can potentially present challenges during scale-up and might require additional process development.

To design a robust chromatography process for the purification of proteins from complex mixtures, many factors must be considered. Preparative chromatography is widely employed in linear and step gradient modes for the purification of proteins [10, 11]. The factors to be evaluated vary depending on the type of chromatography, but load concentration of the protein of interest, impurity profile (e.g., other constituents of the load), stationary phase, pH, eluting salt, solubility, and stability would be likely candidates [12].

Development of chromatography processes has embraced a new paradigm where high-throughput techniques are becoming a staple for the screening of chromatography resins and purification conditions. High-throughput scale-down methods have become a critical component in process development, as these strategies allow for screening a wide process design space, in support of quality by design (QbD) principles, while minimizing time and material requirements [13, 14]. QbD is a risk and science-based approach that requires a thorough understanding of a product and its process of manufacture. For QbD, the product and process knowledge base must include an understanding of variability in raw materials, the relationship between a process and the product's critical quality attributes (CQAs), and the association between CQAs and a product's clinical properties [15]. The wider design space achieved through the use of high-throughput scale-down methods ensures a better process understanding and more optimal conditions, in support of QbD. The ability to build a knowledge base around the design space for the manufacturing processes is critically important. Hence utilizing a combination of high-throughput process development (HTPD) and column experiments while developing chromatography steps greatly aids in understanding important parameters and their ranges in relation to CQAs [16].

With the use of automatic liquid handlers, 96-well filter plates and miniature columns, high-throughput screening of resins for capacity and selectivity, and the identification and optimization of operating conditions for chromatography at microliter scale are an integral part of process development [17]. For this, initial high-throughput applications focused on using resin slurry plates to determine capacity [18, 19] and partitioning [20] under different process conditions and determining the correlations that could be appropriately applied to packed columns [21]. More recently, miniature columns (0.2–0.6 mL) have been successfully implemented with automated liquid handling to predict the performance of larger-scale columns [22, 23].

Downstream purification typically consists of several chromatography steps including an initial capture step followed by one or more polishing steps operated in either flow-through or bind and elute mode. High-throughput methods have been applied to each of these chromatography steps. Slurry plates (96 well) have been used to develop purification process strategies with different resins [24], and miniature columns have been used to demonstrate each step of a typical three-step mAb purification protocol [25]. Discussions on the appropriate use of each technique have also been presented [26]. Additionally, strategies for coupling slurry plates and miniature columns have been demonstrated [27]. While the literature cited in the preceding text is primarily focused on mAb purification, the techniques and strategies are applicable to other therapeutic proteins. In this chapter, a methodology for lab-scale chromatography process development utilizing high-throughput tools in conjunction with traditional column-based methodologies is summarized. A workflow is presented to elaborate on the different requirements considered in order to ensure the developed process is robust and scalable. The proposed experimental plan for process development relies heavily on a design of experiments (DOE) approach supplemented with one-factor-at-a-time (OFAT) experiments. It fully utilizes HTPD and transitions into lab-scale column experiments where additional confirmation is required for defining parameter ranges and scale-up. This chapter also presents examples to demonstrate the chromatography process development methodology and workflow for the purification of mAbs. However, we believe the approach summarized in these examples can be widely utilized for the development of a range of therapeutic proteins.

11.2 Methodology and Proposed Workflow

There are two strategies that can be employed for determining the use of chromatography in developing a purification process. The first is a platform approach, where the unit operations are set and the process is reverse engineered to take the molecular properties into account. This is a typical strategy for mAbs, as the separation challenges tend to be conserved across molecules. The second strategy is to develop a custom purification process based on an assessment of the molecular properties while utilizing a modular toolbox. These properties (e.g., charge profile, hydrophobicity) typically can be studied as early as the product developability assessment stage, which is a crucial step to ensure the selection of candidates that guarantee development of robust manufacturing processes while achieving critical product attributes. This assessment of the candidate can either be performed using early, transiently produced material or with material obtained from the final stable cell line and process that is more representative of clinical manufacturing.

High-throughput screening can be used to support either strategies. Much of the resin selection or resin suitability testing can be performed using high-throughput experimentation as illustrated in the flowchart shown in Figure 11.1. With the platform approach, a range of loading and the buffer conditions for a given set of resins are investigated to evaluate the molecule's suitability for the existing platform technologies. For the custom development

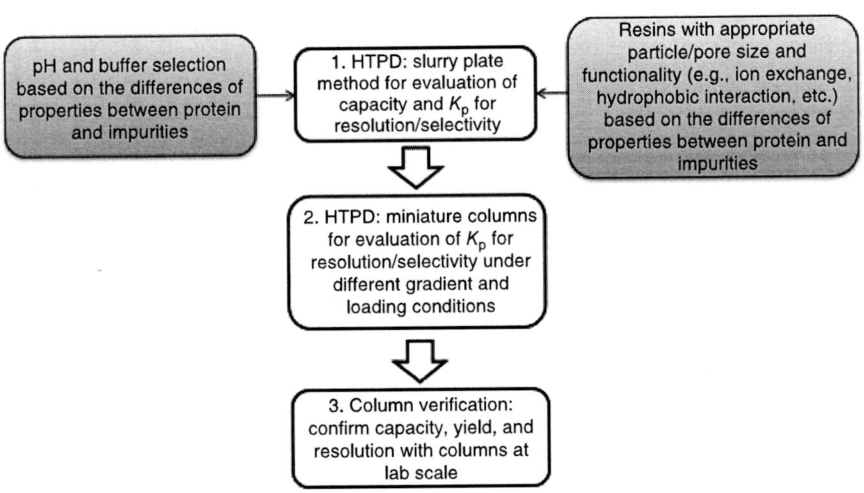

Figure 11.1 Resin screening methodology via use of high-throughput process development (HTPD) and lab-scale column runs. The partition coefficient (K_p) is defined as the amount of product bound to the resin in equilibrium with the amount remaining in the liquid phase.

approach, a selection of purification resins, in addition to loading and buffer conditions, is screened to establish possible alternatives for further evaluation. A two-stage process via use of slurry plates and miniature columns can be considered. Following completion of these stages, confirmation with lab-scale columns is recommended to ensure the validity of the HTPD findings as well as the investigation of parameters, such as linear velocity and temperature, which are not amenable to high-throughput formats. It is common practice to confirm capacity under different flow rates and load protein concentrations (selected based on expected process fit and equipment capability at large scale) via utilization of breakthrough curves [27, 28].

Once the resins are identified or suitability is confirmed, an experimental plan to establish and optimize processing conditions will be executed (examples are provided in the following sections). DOE methodologies are combined to evaluate operating conditions and their ranges and the sensitivity of yield, purity, and other predefined quality attributes within these ranges [22]. Statistical analysis is performed using standard software packages, such as JMP (SAS, North Carolina) or Design-Expert (Stat-Ease, Minneapolis, MN). The scope and number of parameters being evaluated determines the design used. For wide-range screening of multiple parameters, a response surface design, such as a central composite design, is typically employed. This limits the number of experimental runs while capturing centerpoint (CP) and axial conditions within a fractional factorial design. For more extensive investigation, a full factorial design is used, which encompasses all possible combinations of parameters and ranges. Due to the number of experiments this requires, HTPD is key to ensuring this design is practical. To determine the significance of the models resulting from these DOE designs, the p value is used as a threshold to determine the probability that the data are significant and not due to random error, with a cutoff of <0.05 showing significance. The data are also analyzed using various tools, such as predictor plots or interaction plots. These plots show the relationship between one categorical factor and a continuous response, which depends on a second factor.

The solubility and stability of the protein of interest is also an important factor in determining the operating conditions. If the recombinant proteins are expressed in mammalian cell lines, the evaluation of viral clearance under the proposed operating conditions will be necessary to ensure the minimum viral clearance expectations set by regulatory agencies are met. This would only be necessary for steps claimed to contribute to viral clearance. Another important consideration would be resin reuse and its impact on yield and purity. An overall chromatography process development strategy combining all these considerations is presented in Figure 11.2.

As discussed in the previous paragraphs, chromatography process development is performed at different scales depending on the type of information to be collected. A summary of the scales used for chromatography process development

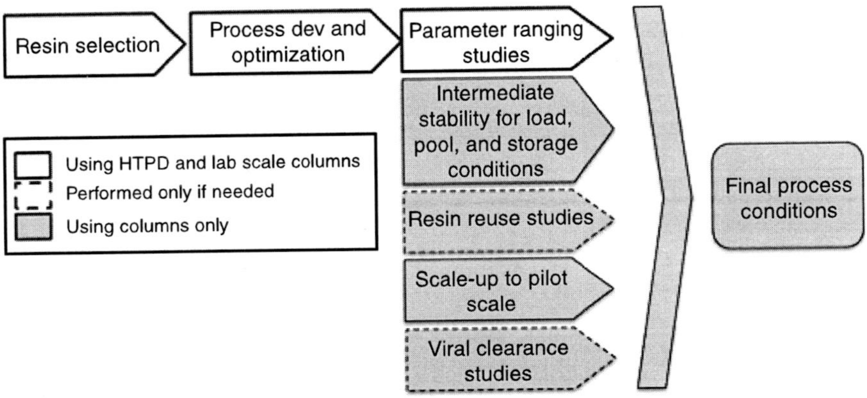

Figure 11.2 Workflow for lab-scale chromatography process development.

Table 11.1 Scales of chromatography used during process development.

Scale	Diameter (cm)	Height (cm)	Volume (mL)
Slurry plate	N/A	N/A	0.02 (0.004–0.05)
Miniature columns	0.5	3	0.2 or 0.6
Lab scale	0.66–5	15–25	5–490
Pilot scale	10–30	15–25	1200–18000
Manufacturing scale	45–140	15–25	24000–390000

and verification in relation to pilot and manufacturing scale, as used in the examples in this chapter, is shown in Table 11.1. In addition to column diameter, bed height could also be varied to maintain a certain residence time leading to a desired capacity and resolution [29]. A height range of 15–25 cm is commonly utilized for protein purification. For each project, qualification of the small-scale model is recommended to ensure that the selected scale is representative of pilot and manufacturing scale.

It should also be noted that process development for biologic clinical candidates very commonly goes through different phases in conjunction with the different demands in support of clinical trials. For early clinical trials, Phase I through II, an early-phase process is utilized. In accordance with the increased clinical demand and scale, processes are generally further optimized to increase productivity and robustness and defined as late-phase processes. This nomenclature will be utilized in this chapter to differentiate early- and late-phase process development.

All examples presented in this chapter pertain to mAbs. Production of high purity mAbs to support clinical development usually requires more than a single-step purification process. Purification schemes consisting of ProA affinity chromatography used for primary capture followed by ion exchange chromatography are commonly used in industry as platform processes to accelerate timelines [30–33]. The common usage of ProA resins is mainly driven by the fact that it selectively and efficiently binds antibodies in complex solutions such as harvested cell culture fluid (HCCF) and removes >99.5% of product impurities in a single step with high step yields and high throughput. Following ProA affinity chromatography, ion exchange and/or HIC are used as polishing steps. Cation exchange chromatography (CEX), which is generally operated under bind and elute mode, removes host- and product-related impurities such as host cell proteins (HCP), DNA, and aggregates (high molecular weight (HMW) species) and process-related impurities such as endotoxin and leached ProA ligand. Anion exchange chromatography (AEX) is also utilized for the removal of DNA, HCP, and endotoxin in flow-through mode and additional removal of aggregates and leached ProA when operated in weak partitioning and capture mode [20]. HIC has also been used for mAb purification as a polishing step. The HIC step, relying on differences of hydrophobicity between mAbs and other impurities, has been shown effective in removing aggregates, DNA, HCP, and endotoxin [34, 35]. In addition to these methodologies, mixed-mode resins have been developed for the purification of proteins delivering unique selectivities and solutions to challenging problems. Hydroxyapatite chromatography has been utilized for mAb purification and has been demonstrated to effectively clear aggregates, leached ProA ligand, and HCP [36]. In recent years, additional mixed-mode resins were developed to combine hydrophobic and electrostatic interactions through advanced ligand design. These resins, such as Capto adhere (GE Healthcare), have also been used for mAb purification and have shown effective clearance of aggregates and HCP [37].

11.2.1 High-Throughput Process Development

HTPD allows for earlier engagement with discovery to guide candidate selection and upstream development, due to lower material needs and the ability to evaluate larger design spaces, resulting in better process understanding. A multiple step approach utilizing various high-throughput techniques provides the flexibility and throughput of batch incubations with the increased separation capabilities of the packed bed format of miniature columns. Initially, resin slurry plates are used to scout feasible operating conditions; miniature columns are then used to determine the potential separation of impurities. Finally, a limited number of lab-scale column experiments are performed to confirm the conditions identified using high-throughput techniques and to provide a path forward toward large-scale processing [27]. The strategy for initial resin and condition screening is to employ resin slurry plates to investigate a very wide range of parameters. Resin slurry plates can either be purchased or customized. For our customized 96-well plates, we typically use a resin slurry of 30% for a target volume range of 2–50 μL. The amount of resin is varied to achieve different loadings for the same feed concentration. For example, for an AEX flow-through step in a mAb purification, four pH levels from 6.5 to 8.5 and four conductivity levels (3–10 mS/cm) are chosen at three different loadings (100–200 mg/mL). When considering a CEX bind and elute step, a pH range of 4.0–6.0 and a conductivity range from 0 to 1 M sodium chloride are typically evaluated. These ranges, as well as the addition of modifiers, can increase for mixed-mode resins as the optimal operating space may be more complicated to determine theoretically. Slurry plate screens are performed at low loading (5 mg/mL) to determine partition coefficients and are performed at high loading (>100 mg/mL) to determine capacity [17]. Once the range is narrowed down, miniature columns can be employed for elution development, to evaluate separation capacity using simulated gradients, where a linear gradient is approximated by employing multiple step gradients [23]. Miniature columns are purchased from Atoll GmbH in volumes of either 200 or 600 μL for use with a liquid handler. Depending on the purification challenge and the phase of development, this strategy can be modified. In early-phase development, when qualitative trends are sufficient, a heavier reliance on resin slurry plate screens is seen due to the much higher throughput and reduced material requirements. Later in development, miniature columns might be applied more as there is a stronger prediction of scale-up behavior. The ability to run parallel miniature columns using a liquid-handling system allows for higher resolution DOE experiments to be executed, which leads to better modeling of the chromatography system.

The following are five examples to demonstrate how the strategy can be applied to various purification challenges for mAbs: (i) early developability assessment for a mAb utilizing the resin slurry plate method, (ii) HIC polishing step resin screening on miniature columns, (iii) mixed-mode polishing step development using the two-step strategy employing resin slurry plates for initial ranges and confirmation with miniature columns, (iv) CEX polishing step development utilizing resin slurry plates for capacity and miniature columns for elution optimization, and (v) AEX optimization utilizing resin slurry plates.

11.2.1.1 Case 1: Utilizing HTPD for Early Developability Assessment

This example shows that an initial HTPD study on lead mAb candidates can quickly provide valuable information to determine the molecule's fitness for the existing platform technologies or whether an alternative is needed. As well as the amount of resources required for development. Among the high impact attributes used during this candidate developability profiling, isoelectric point (p*I*) was identified as a property that would impact the downstream process and the potential success of turning the candidate to a viable drug. In this case, with a limited amount of transiently expressed material available, high-throughput resin screenings were performed focusing on the polishing steps in both bind and elute and flow-through modes. Platform anion, cation, and mixed-mode resins were screened in slurry plates under different pH (6.0–7.5), conductivity (0–30 mS/cm), and loading conditions (50–175 mg/mL resin). Results showed that there were no significant differences in terms of partitioning behavior and aggregate clearance among the three candidates under the screening conditions with the three resins. Additionally, the screens revealed that all candidates behaved differently from a typical mAb under standard processing conditions. Lower yield and reduced or no aggregate clearance were observed under the flow-through conditions with anion and mixed-mode resins, whereas the cation exchange resin binding study showed characteristic binding behavior under the screened conditions, however with no clearance of the aggregates. Figure 11.3 shows an example of the yield comparison among the low p*I* (<7.0) candidates for the flow-through AEX screenings. Data for the other two resin screenings are not shown. Under the conditions screened, there was no significant yield difference among all three candidates, with a range from 40 up to 70%. The results narrowed down the process conditions needed for further development when sufficient process relevant material was available.

Overall, these screens were able to evaluate the behavior of all three candidates on three resins (Poros HQ50, Poros HS50, Capto adhere) with only 300 mg required for each molecule. The initial HTPD study did not differentiate the three candidates but did provide the information that these candidates would not readily fit the platform. The screening results from the two alternative resins, Poros HS50 and Capto adhere, indicated that additional development effort was needed in order to identify a suitable chromatography modality for aggregate removal.

11.2.1.2 Case 2: Polishing Resin Screening with Hydrophobic Interaction Chromatography Using Miniature Columns

This example demonstrates the use of miniature columns to approximate the separation capability of traditional lab-scale columns. To support early-phase studies, a poorly expressed mAb, with a p*I* in the normal range of 7.0–9.0, needed to be purified in a short amount of time, with very little material available for development. The biggest purification challenge was controlling the level of fragments resulting from clipping in the heavy chain constant region. Given the early phase of the project, the fragments were not characterized, and little information could be used to narrow down the modality. An initial resin screen was performed in slurry plates to narrow the focus to two resins, a cation exchanger (Poros HS50) and a HIC resin (Butyl Sepharose 4 Fast Flow). These two resins were evaluated with miniature columns (600 μL) using a liquid handler to run four conditions for each resin in parallel. For Poros HS50, the pH range evaluated was 4.5–6.0 with an approximate linear gradient for elution

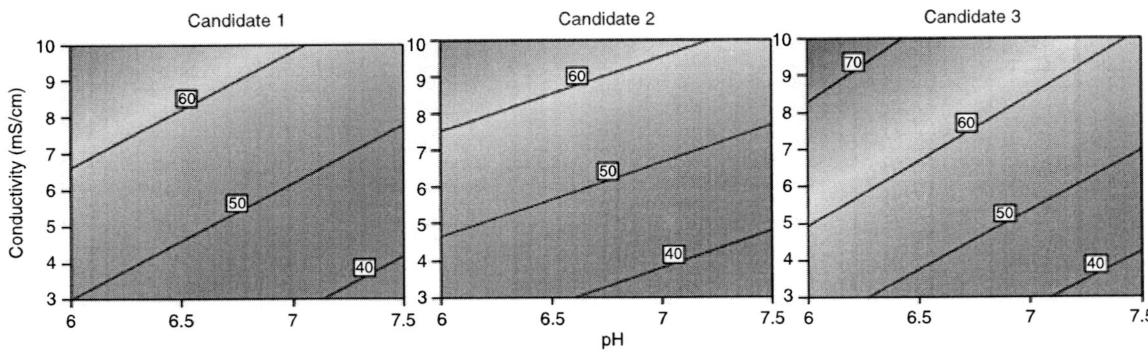

Figure 11.3 Flow-through AEX screenings for step yield (AEX load = 100 mg/mL resin).

from 0 to 500 mM sodium chloride over 36 column volumes (CV). For Butyl Sepharose, the pH range evaluated was 6.5–8.0 with an approximate linear gradient for elution from 1000 to 0 mM ammonium sulfate over 36 CV. Fractions (200 μL) were collected every 1/3rd CV and were analyzed for protein concentration by measuring UV absorbance in 96-well plates at 254, 280, and 300 nm. The chromatograms were then used to select fractions to be analyzed for purity by CE-SDS using a high-throughput Caliper LabChip system. Figure 11.4 shows the chromatograms for the Poros HS50 screens at pH 5.0–6.0. The pH 4.5 condition is not included due to an operational failure for that column.

As can be seen in Figure 11.4, there is no apparent separation of the fragments based on the chromatogram. This is confirmed with fractional analysis by nonreduced CE-SDS as shown in Figure 11.5. Across the peak, the fragment at ~120 kDa co-elutes with the product. There is some apparent purification as the fragments seen at 68 and 48 kDa are not observed. The feed purity was 82.7% and the best purity achieved by pooling the Poros HS50 fraction CV 17–23 at pH 5.0 was 94.5% with an estimated yield of 72%.

Figure 11.6 shows the chromatograms for the Butyl Sepharose 4 FF columns at the four pH conditions.

Unlike the Poros HS50 resin, there is some apparent separation based on the chromatogram for the majority of the pH conditions tested with two peaks

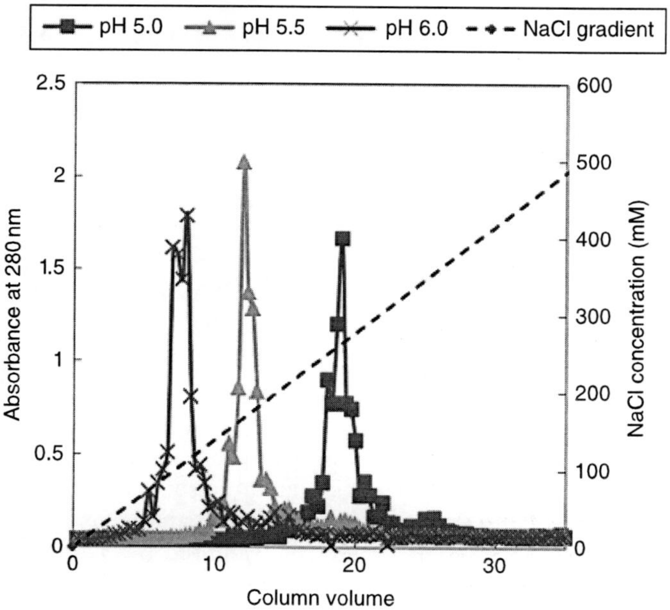

Figure 11.4 Poros HS50 chromatograms at pH 5.0–6.0 with an approximate NaCl linear gradient.

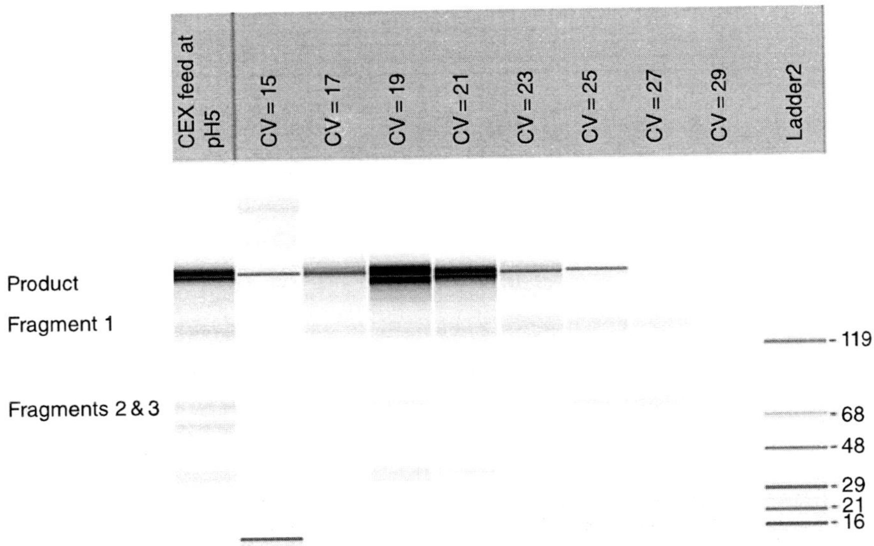

Figure 11.5 Caliper Labchip CE-SDS (nonreduced) results for Poros HS50 fractions at pH 5.0. The simulated gel shows the product at ~150 kDa with fragments at ~120, 68, and 48 kDa.

seen eluting prior to the product peak. Based on the chromatograms, fractions from the pH 7.0 and 8.0 experiments were analyzed by nonreduced CE-SDS and shown in Figure 11.7.

Based on the chromatograms, it appeared that both pH 7.0 and 8.0 provided adequate separation. But the fractional analyses show that at pH 7.0 there is practically baseline separation between the two fragments and the product. The fragment at 68 kDa elutes first at the start of the gradient. The second main fragment at 120 kDa elutes just prior to the product peak. The product purity at pH 7.0 is 100% with an estimated yield of 80%. At pH 8.0, there is some co-elution of the second fragment and the product. The purity at pH 8.0 is ~97% with an estimated yield of 70%. Figure 11.8 shows the HP-SEC results for pH 7.0, where peak 1 is the first fragment, peak 2 is the second fragment, and peak 3 is the product.

From the HP-SEC data, it appears that the first eluting peak in Figure 11.6b contained both the 68 kDa fragment and potential aggregates. Based on these results, the Butyl Sepharose 4 FF at pH 7.0 was scaled up to successfully provide rapid purification of product from fragments and aggregates. The development of this polishing chromatography step only required 3 days from receipt of material to scale up. These aggressive timelines were achieved by utilizing miniature columns in parallel and the high-throughput Caliper Labchip CE-SDS system, which allowed for all fractional analysis to occur in only a few hours.

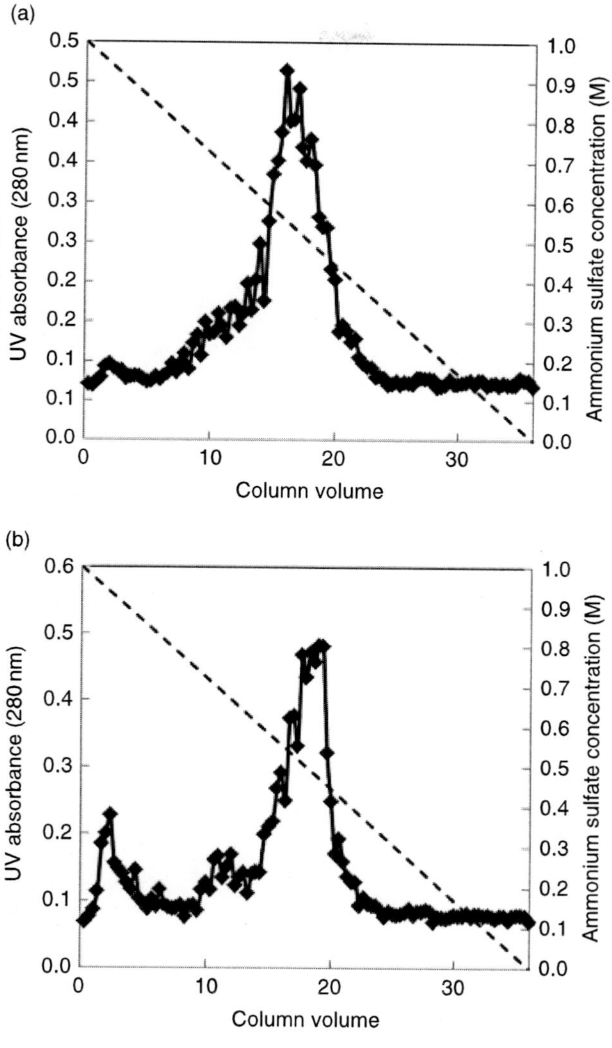

Figure 11.6 Butyl Sepharose 4 FF chromatograms with an approximated ammonium sulfate gradient. (a) pH 6.5, (b) pH 7.0, (c) pH 7.5, and (d) 8.0 with the dashed line representing the gradient.

11.2.1.3 Case 3: Flow-through Chromatography Step Optimization Using Resin Slurry Plates and Miniature Columns

This example demonstrates the strategy of using resin slurry plates to determine process behavior over a wide range of operating conditions with confirmation of a few points with miniature columns for a mAb with a p*I* in the

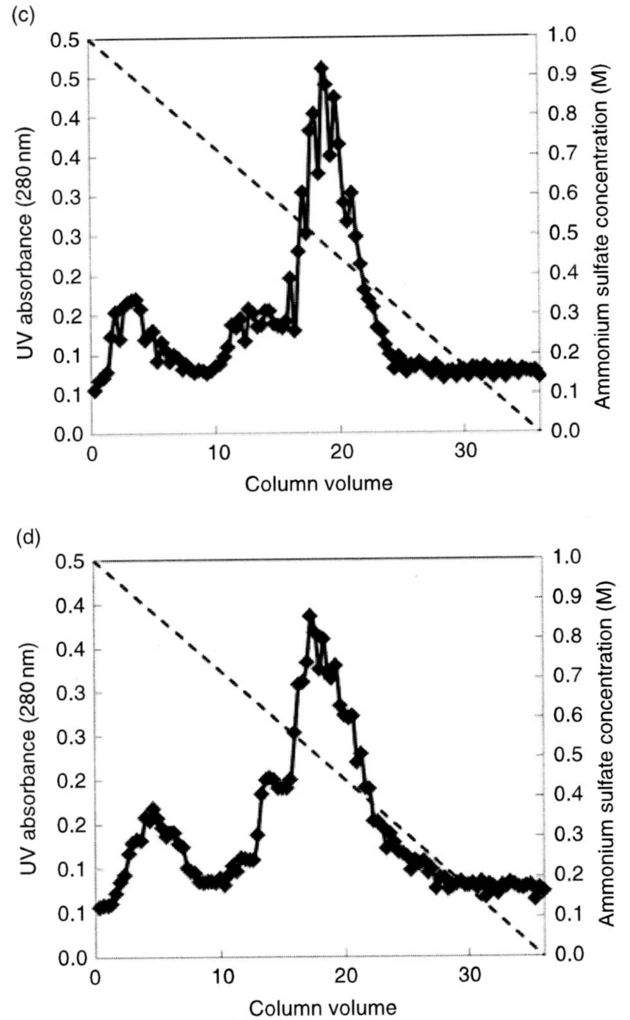

Figure 11.6 (Continued)

normal range. The goal of this first polishing step was to reduce aggregates and fragments for a mAb using GE Healthcare Capto adhere, an anion exchange mixed-mode resin, in flow-through mode. The early-phase process included a ProA chromatography step as the capture step prior to the polishing step. As part of late stage process development, that affinity step was replaced with a mixed-mode chromatography step, leading to increased conductivity for the flow-through polishing feed. The early-phase process polishing step CP condition

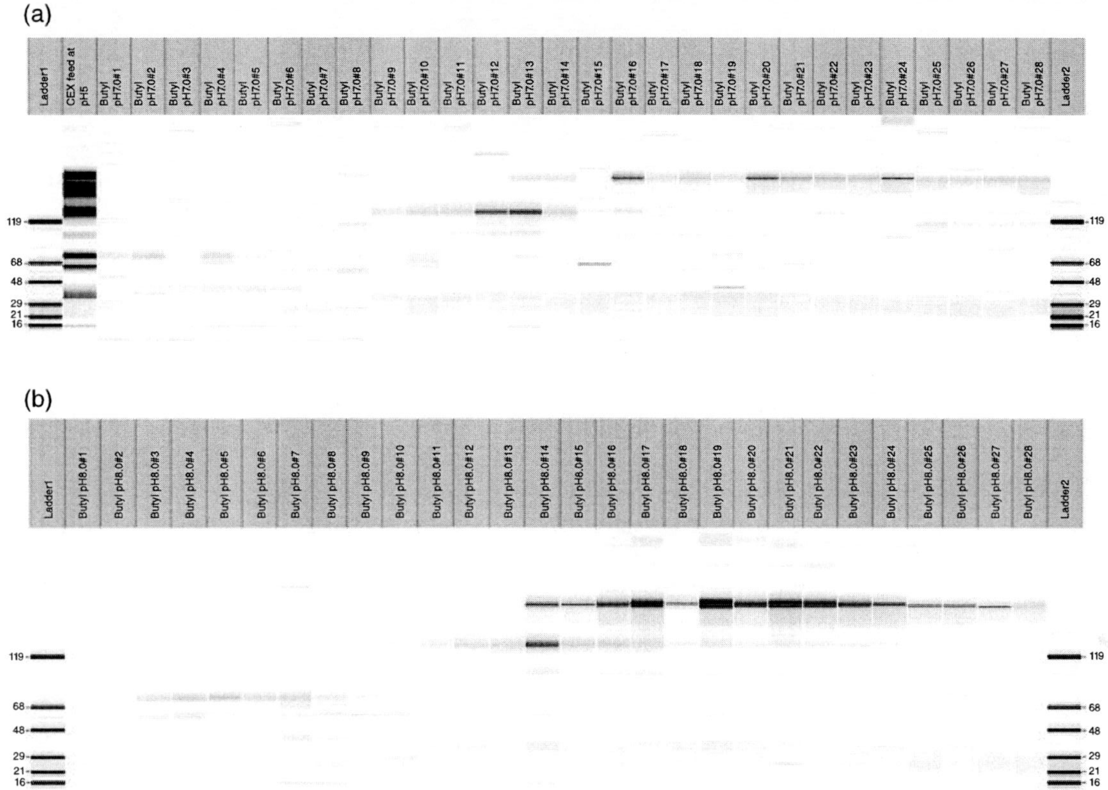

Figure 11.7 Caliper Labchip CE-SDS (nonreduced) results for Butyl Sepharose 4 FF fractions at (a) pH 7.0 and (b) pH 8.0. The simulated gel shows the product at ~150 kDa with fragments at ~120, 68, and 48 kDa.

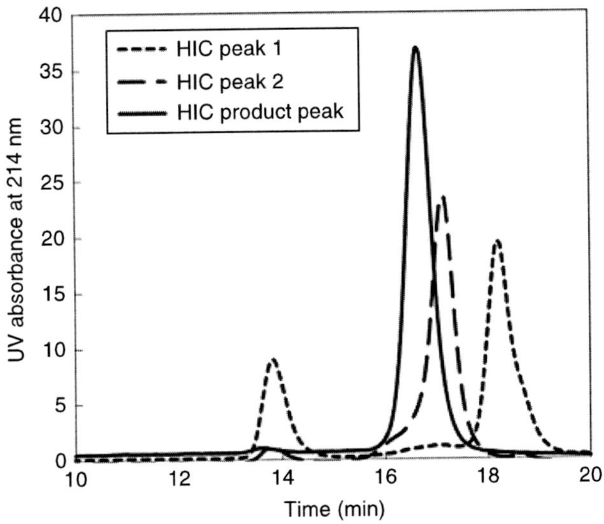

Figure 11.8 HP-SEC retention times for the Butyl Sepharose 4 FF pH 7.0 pools. Peak 1 is the first fragment, peak 2 is the second fragment, and peak 3 is the product.

was at a conductivity of ~4.5 mS/cm and a pH of 8.0. Due to this upstream process change, the conductivity of the late stage feed to the polishing step was higher than typically seen, so the initial slurry screen investigated a range from 3 to 10 mS/cm. The pH range was from 7.0 to 8.5 with loadings of 100, 150, and 200 mg/mL resin. The loadings were achieved by varying the resin volumes from 4 to 8 μL and keeping the feed volume constant. Feed preparation was performed by titration and dilution to achieve the targeted pH and conductivity values. The feed used for these experiments contained an aggregate level of 3.3%. Figure 11.9a and b shows the response surfaces for yield and aggregate levels as a function of conductivity and pH, respectively.

Based on the response surface data, higher yields are expected at high conductivity in a pH range of 7.0–7.5. The early-phase CP yield is predicted to be ~80% with the new feed conditions, which is in line with the early-phase process experience. Lower yields are seen at low conductivity in the high pH range. For aggregate levels, the trends are opposite those of yield. The best aggregate clearance is seen at conditions that give the lowest yield, producing a product with only 1.1% aggregate. Figure 11.10 shows the desirability plot showing the preferred area of operation, to achieve at least an 80% yield and aggregate levels below 2%, while maintaining or minimizing residual HCP and DNA, as this was not the purification focus of this step.

From the initial screen, the most robust operating space appears to be in a pH range of 7.5–8.0, with a conductivity of 3.0–6.0 mS/cm. To confirm the

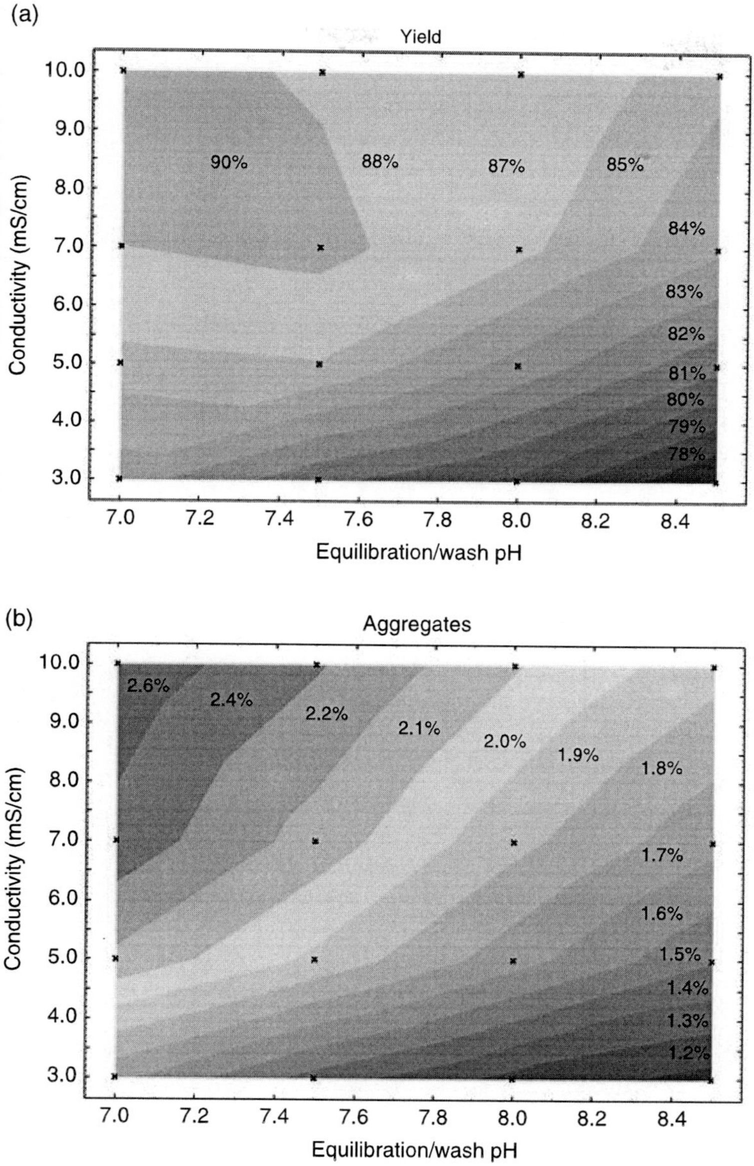

Figure 11.9 Response surfaces for (a) yield and (b) aggregate content as a function of pH and conductivity at all loadings. A total of 16 points, as indicated by the black stars, were used to determine the response surface.

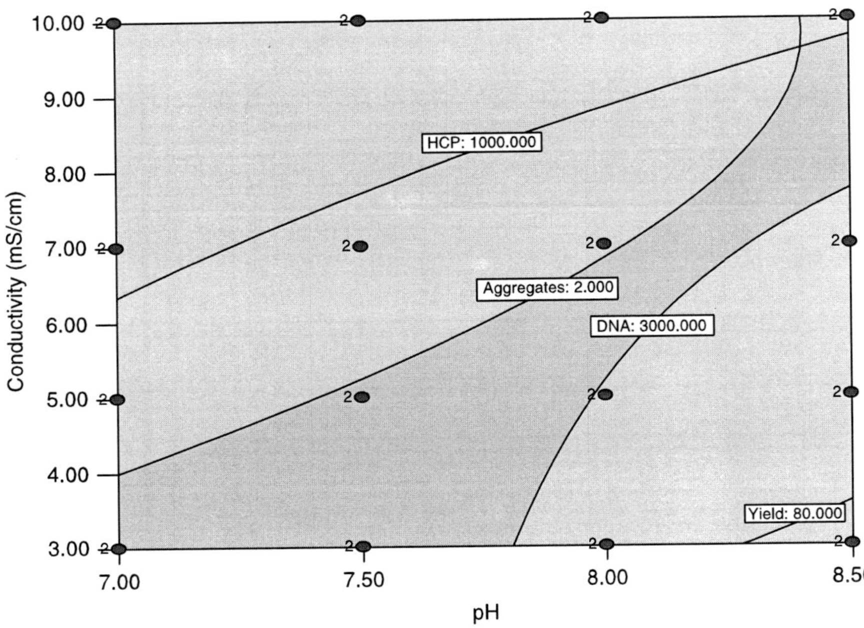

Figure 11.10 Desirability plot showing the target operating region (dark area) to achieve a yield of >80% and <2% aggregates, minimizing residual HCP (ppm) and DNA (ppb). Dots represent experimental conditions.

Table 11.2 Experimental design for miniature column DOE.

Run	Feed pH	Feed conductivity (mS/cm)
1	7.5	3.3
2	7.5	8.8
3 (CP)	8.0	4.8
4 (CP)	8.0	4.8
5	8.5	3.3
6	8.5	8.8

operating region, a full factorial DOE including all possible combinations of feed pH and conductivity was performed on miniature columns (600 μL), with pH and conductivity as the parameters, with two additional CPs. Loading was maintained at 100 mg/mL. The experimental design is shown in Table 11.2.

The miniature column chromatograms are shown in Figure 11.11a, with Table 11.3 providing the yield, aggregate, and fragment levels for all conditions

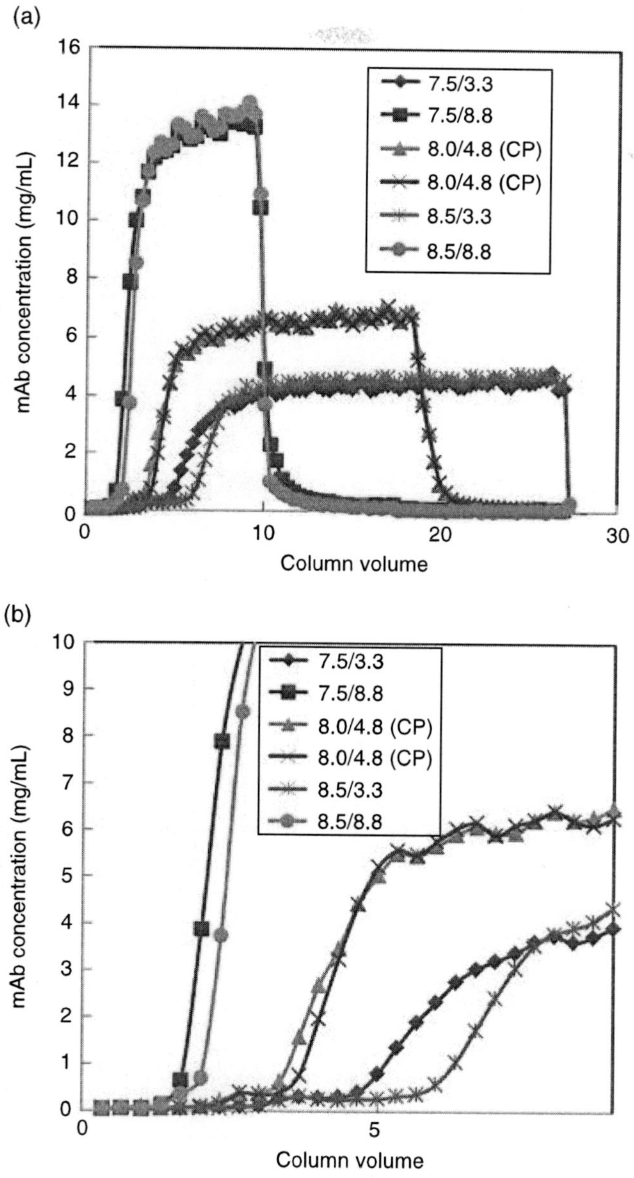

Figure 11.11 Miniature column chromatograms showing (a) the full chromatogram and (b) a focus on the initial breakthrough. (*See insert for color representation of the figure.*)

Table 11.3 Experimental design for miniature column DOE.

Run	Yield (%)	Aggregate (%)	Fragment (%)
1	72.1	1.2	1.2
2	78.6	2.7	1.1
3 (CP)	73.8	1.4	0.9
4 (CP)	73.5	1.3	0.8
5	70.0	1.2	0.9
6	72.5	2.5	0.9

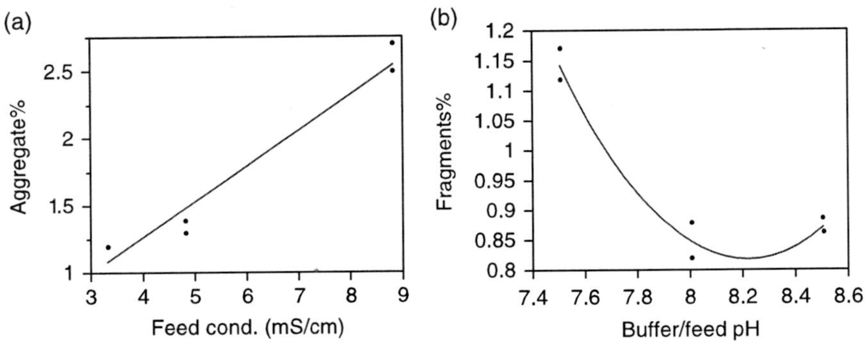

Figure 11.12 Factor interaction plots for (a) aggregate levels as a function of feed conductivity and (b) fragment levels as a function of feed pH.

tested. The mini-column data confirmed the trends observed with the slurry plate data for the mixed-mode anion exchange flow-through step. The absolute yields are ~10–15% lower than that predicted with the slurry plates, which is in line with historical experience. It is clear from the chromatograms that feed conductivity is correlated to the initial breakthrough, as seen in Figure 11.11b. The highest conductivity points gave the highest yields with minimal aggregate clearance. For conductivities lower than 5 mS/cm, the aggregate level was approximately the same. The trend is different in terms of fragment clearance, which depended on the feed pH, with better clearance at higher pH.

Using DOE software, the data was modeled and showed that the feed conductivity was the most significant parameter for aggregate clearance, with a p value of 0.0006. For fragment clearance, feed pH was the most significant parameter with a p value of 0.0038. Both factor interaction plots are shown in Figure 11.12.

Based on both resin slurry plate and miniature column data, the CP conditions of pH 8.0 and feed conductivity of 4.5 mS/cm were kept as the process conditions and successfully scaled up to lab scale.

11.2.1.4 Case 4: Bind and Elute CEX Polishing Chromatography Step Optimization Using Resin Slurry Plates and Miniature Columns

This example demonstrates the strategy of using resin slurry plates to determine capacity and miniature columns to optimize the elution conditions of a bind and elute polishing step using Poros HS50 resin. The objective of these experiments was to compare the static binding capacities and elution conditions to those previously developed for the initial early-phase process. For the slurry resin plate screen, conductivities between 3 and 20 mS/cm (corresponding to ~0–200 mM NaCl) and a pH range of 4.5–5.5 were tested. The miniature column experiment consisted of six linear gradients at the same pH values tested for the slurry plate screen. The miniature column experiments used a 30 mg/mL loading that was lower than the lowest static capacity determined in the slurry screen to ensure no product breakthrough in the loading step.

A response surface for the resin slurry plate screen is shown in Figure 11.13. The highest capacities of 42–45 mg/mL were seen at low pH at conductivities

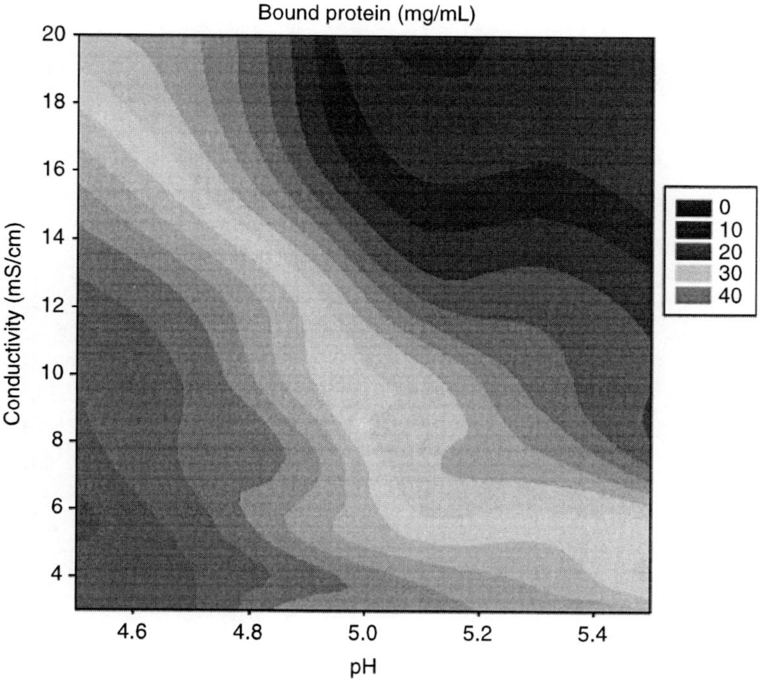

Figure 11.13 Contour plot of capacity with respect to pH and salt concentration. (*See insert for color representation of the figure.*)

below 10 mS/cm. Capacities below 20 mg/mL, indicating unfavorable binding conditions and potential elution conditions, were seen at high pH and conductivities above 12 mS/cm.

Based on the capacity screen results, all pH conditions could give >30 mg/mL capacities at low conductivity (<7 mS/cm). Capacities >40 mg/mL might be utilized at pH values below 4.9, but further testing at higher conductivity values would be needed for estimating elution conditions.

Given this data, the entire pH range of 4.5–5.5 could potentially be utilized for binding. Thus, the same buffer conditions were further screened using miniature columns to determine elution conditions (k') for both the monomer and aggregate peaks, as well as determining peak resolution between the two species. A loading of 30 mg/mL was used to ensure no product breakthrough even at the higher pH conditions tested, and the elution gradient was run from 0 to 500 mM NaCl followed by a strip of the column performed at 1 M NaCl. Fractions at 1/3 CV were collected. Figure 11.14 shows absorbance elution chromatograms for the elution at each pH condition. Negligible protein was seen in the flow-through, wash, and strip fractions, and a mass balance approximation indicated that product recovery was ~100% in the collected fractions.

Fluctuations in the concentration as indicated by UV absorbance is an artifact of the miniature column method used. The lower concentration is correlated to some back mixing with the liquid-handling system water. This did not impact the overall trends in the chromatogram. A clear trend of earlier elution and narrower peak shape can be seen with increasing pH, as would be expected for a typical CEX and mAb combination, as at higher pH values, the

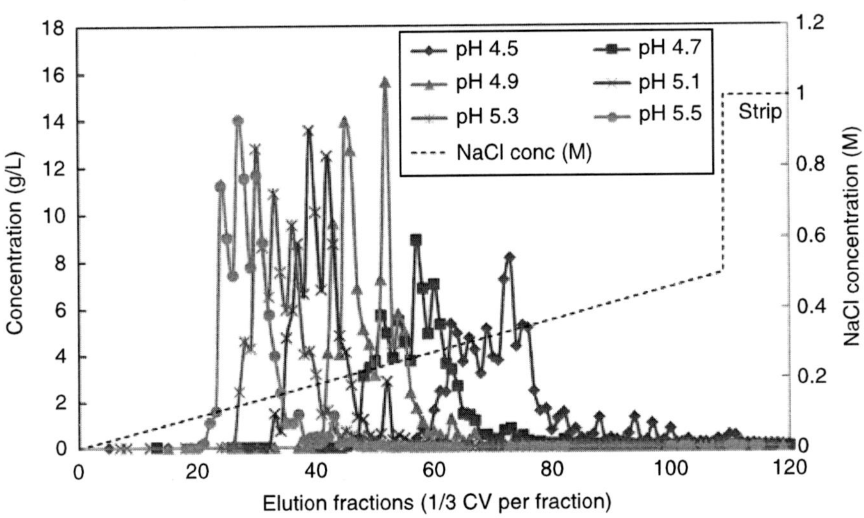

Figure 11.14 Poros HS50 chromatograms for all six pHs.

net surface charge is less and requires a reduced amount of salt for elution. For the purposes of this analysis, only every third fraction was sampled further for more quantitative concentration estimate and aggregate measurements, equating to 1 CV fractional analysis. Figure 11.15 and Table 11.4 depict chromatograms and peak characteristics for monomer and aggregate peaks. While peaks for both species are clearly narrower at higher pH, separation is reduced, and the overall resolution numbers are relatively constant.

$$R_S = \frac{1.18 \times \left(k_{\text{aggregates}}' - k_{\text{monomer}}'\right)}{\left(w_{1/2\text{peakaggregates}} + w_{1/2\text{peakmonomer}}\right)}$$

The monomer elution at pH 5.5 and 116 mM NaCl is in good agreement with the previously developed early-phase clinical process that utilized a linear gradient from ~75 to 150 mM NaCl for elution. The resolution values (R_S) ranged from 0.54 to 0.79 and are likely within the experimental error of this screen.

The resin static capacity screening and subsequent miniature column experiments demonstrated that 30 mg/mL loading does not exceed capacity in the pH 4.5–5.5 range. It is possible that higher capacities could be supported at the lower end of this range, but more column tests would need to be run for confirmation. None of the conditions showed a significant advantage in terms of resolution, though the higher pH conditions did show narrower peaks with less separation. In summary, these high-throughput screens provide supporting data for 30 mg/mL loadings with good gradient separation of aggregates in the pH 4.5–5.5 range.

11.2.1.5 Case 5: AEX Chromatography Optimization Utilizing Resin Slurry Plates

This example demonstrates the strategy of using resin slurry plates to evaluate the robustness of an AEX flow-through step for a mAb with a low p*I*, <7.0. Data was generated at process conditions (loadings from 50 to 175 mg/mL resin) and low loading (5 mg/mL) to determine both the platform performance and the partitioning behavior. These studies focused on the impact of loading, feed and equilibration pH and conductivity (Table 11.5) on HCP, aggregate clearance, and yield. The feed utilized for these studies was the product from ProA chromatography spiked with 1% ProA flow-through to increase residual levels, producing an aggregate level of 1.1% and HCP level of 1000 ppm.

Figure 11.16 shows the yield as a function of pH and conductivity. Overall yields were fairly robust with all points tested falling between 89 and 95%, demonstrating acceptable yields at even the least favorable conditions tested (high pH and low conductivity).

The partition coefficient (K_p) is defined as the amount of product bound to the resin in equilibrium with the amount remaining in the liquid phase. The partitioning behavior is determined at low loading (<5 mg/mL) to ensure operation

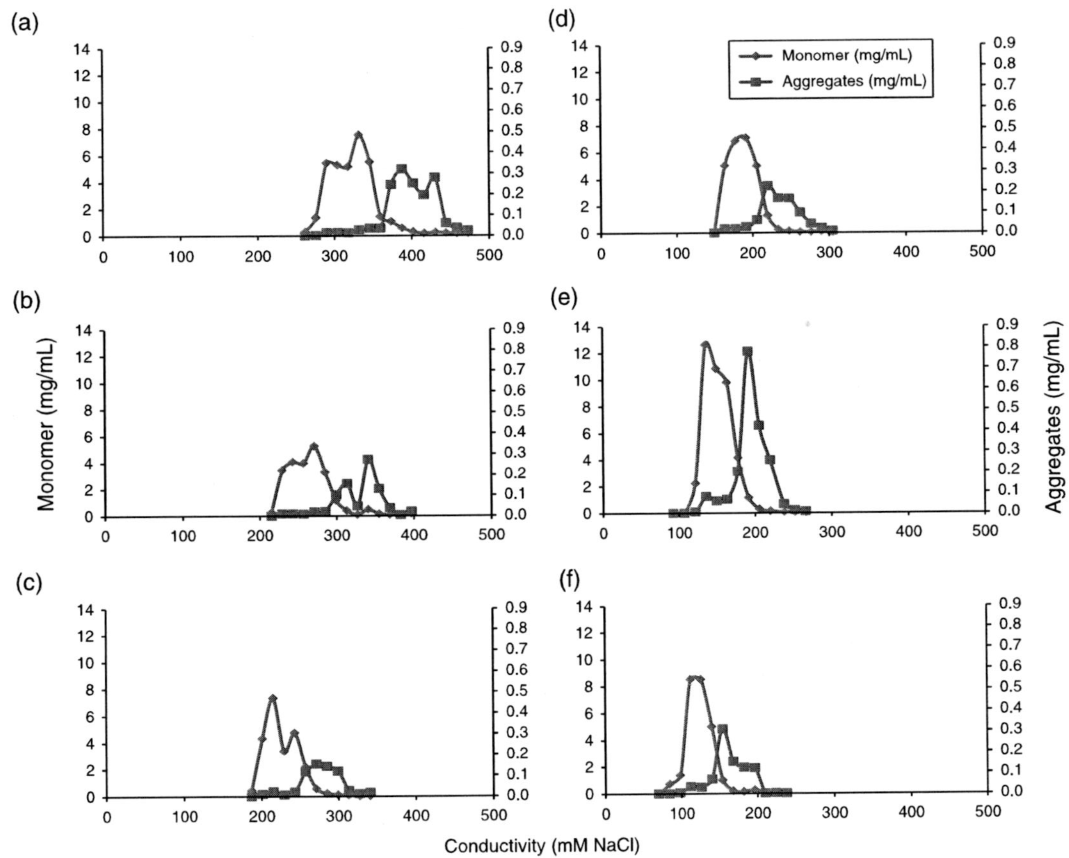

Figure 11.15 Resolution chromatograms of monomer and aggregates at pH (a) 4.5, (b), 4.7, (c) 4.9, (d) 5.1, (e) 5.3, and (f) 5.5.

Table 11.4 Peak characteristics and resolution values for Poros HS50 elutions.

pH	Monomer *k′* (mM NaCl)	Peak width at half max (mM NaCl)	Aggregate *k′* (mM NaCl)	Peak width at half max (mM NaCl)	R_S[a]
4.5	318	68	392	70	0.637
4.7	256	64	329	46	0.787
4.9	214	50	270	55	0.632
5.1	180	52	226	49	0.543
5.3	143	45	188	26	0.735
5.5	116	39	153	23	0.714

[a] The resolution number, R_S, is calculated as a function of salt concentration elution moments and the peak width at half max for both monomer and impurity species [38].

Table 11.5 Experimental conditions for slurry-based screen.

Resin	Poros HQ50
Load (mg/mL resin)	50, 125, 175
Feed/equilibration buffer pH	6.0, 6.5, 7.0, 7.5
Conductivity (mS/cm)	3.4, 5, 7, 10
Feed concentration (mg/mL)	1.14

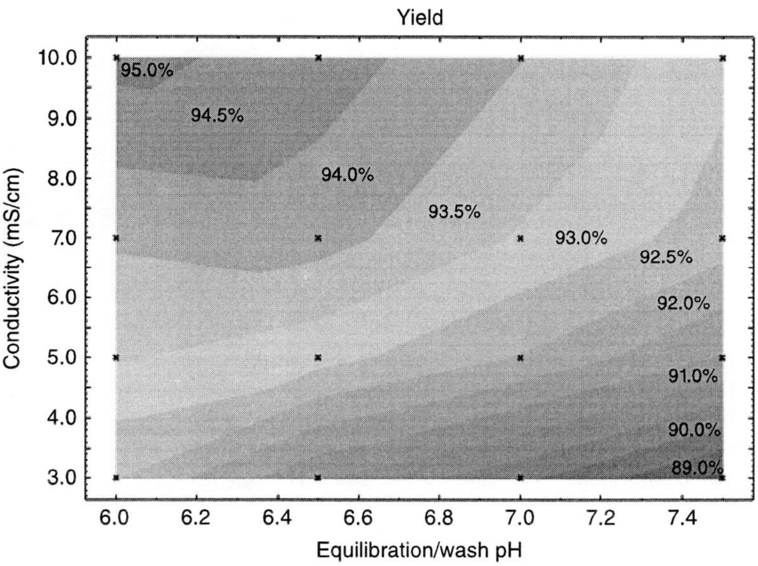

Figure 11.16 Yield as a function of pH and conductivity. The black stars indicate the experimental points used to construct the response surface.

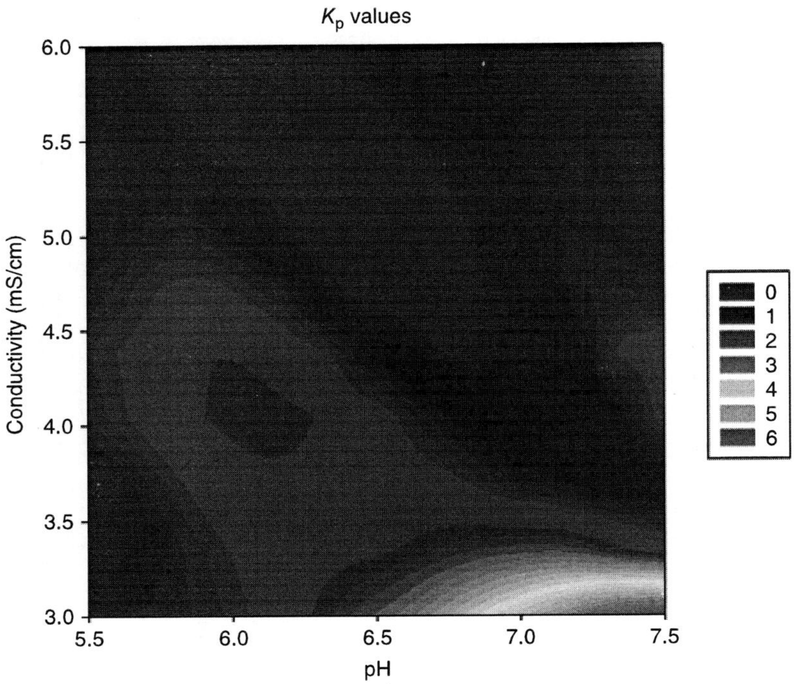

Figure 11.17 Partitioning coefficient values calculated over the range of screening conditions. (*See insert for color representation of the figure.*)

in the linear range of the isotherm. For traditional flow-through conditions, the K_p value is near zero. For bind and elute operation, the typical K_p range would be 10–15. Any K_p value >20 would be very difficult to recover from the resin. A range for pH and conductivity was found where weak partitioning would be observed, a K_p between 1 and 10 [20] as shown in Figure 11.17. Calculated K_p values showed weak partitioning to the resin at conditions above pH 6.5 and below 4 mS/cm. The highest K_p was 6.1 at pH 7.5 and 3 mS/cm.

Figure 11.18a and b summarizes the impact of pH and conductivity on the amount of aggregates and HCP in the resulting product. For all conditions tested, loading was shown to not have an effect on the outputs and is not shown here.

From the cumulative data observed with this screen, conductivities between 3 and 5 mS/cm and the entire pH range of 6.0–7.5 were recommended to minimize HCP levels and maximize the yield. Finally, loadings did not have a strong effect on any of the factors tested, so higher loadings could be pursued with this resin and molecule if desired. Based on these conditions a target load range of 50–150 mg/mL resin with AEX equilibration/wash conditions at pH 7.2 ± 0.2 and conductivity 5.0 ± 1.0 mS/cm were carried forward for lab-scale column evaluations as described in Section 11.2.2.2.2.

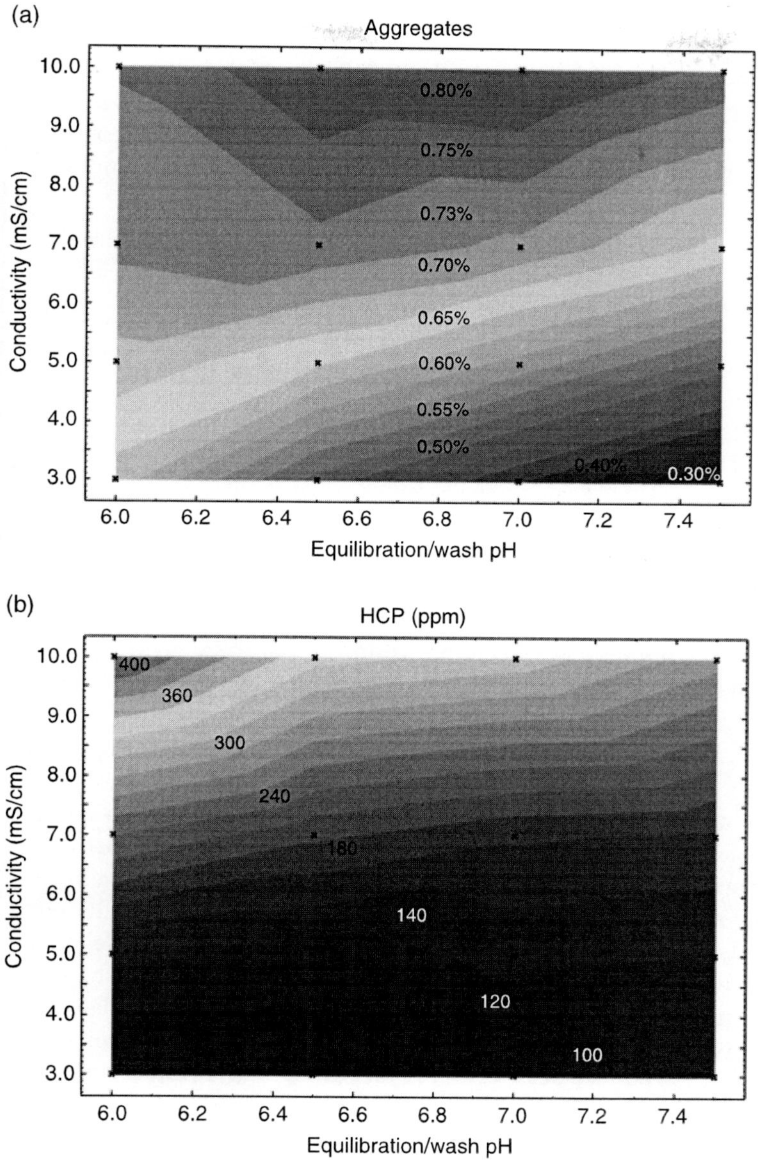

Figure 11.18 (a) Aggregates as a function of pH and conductivity. (b) HCP as a function of pH and conductivity. The black stars indicate the experimental conditions used to construct the response surface.

11.2.2 Column Verification and Final Process Definition

As summarized in the previous section, if HTPD capabilities are available, subsequent process development using columns would be simplified. If not, a similar DOE and OFAT approach could be utilized but using lab-scale columns. Due to limited material availability, the DOEs used in this case would likely need to be simplified, utilizing lower resolution designs.

Once the HTPD results are available, some key experiments are recommended to be performed at lab scale to verify the results and ensure the design space is confirmed and does not show much variation (±15%). It is also recommended to perform the step at pilot scale to effectively troubleshoot any potential changes based on scale-up.

11.2.2.1 Verification of Dynamic Binding Capacity

As mentioned in the preceding text, once a range of loadings have been tested with HTPD, simple breakthrough experiments could be utilized to verify the results and range by using lab-scale columns. An example is shared in this section using a mAb where breakthrough curves and dynamic binding capacity at 10% breakthrough were evaluated for a ProA affinity resin in miniaturized columns as well as lab-scale columns. Figure 11.19 shows the breakthrough curves as a function of column loading (mg/mL resin). When the dynamic binding capacity is measured at 10% breakthrough, very comparable values were observed as summarized in Table 11.6.

11.2.2.2 Verification of Operating Conditions and Ranges

An experimental design using lab-scale columns is recommended to be executed upon availability of data from HTPD studies. Verification of the CP operating conditions qualifies the lab-scale column model as a foundation for further scale-up and troubleshooting as needed. Confirmation of the edge points established with the HTPD work via the lab-scale column model is also recommended. Three different cases based on mAb purification are shared in this section: (i) ProA ranging study performed using miniature and lab-scale columns, (ii) AEX polishing step ranging study on miniature columns with limited verification runs on lab-scale columns, and (iii) CEX polishing step development mainly performed with lab-scale columns.

11.2.2.2.1 Case 1: Protein A Affinity Chromatography Step Development

An experimental design was initially utilized using miniature columns to better understand the impact of the loading (mg/mL resin) and elution pH on step performance (yield) and some selected quality attributes (levels of HCP (ppm) and aggregates (HMW)). The criteria for acceptable operating conditions were set as yield >80% and aggregates <1.9%. A loading range of 20–40 mg/mL resin and an elution pH range of 3.30–3.70 were evaluated using a central composite

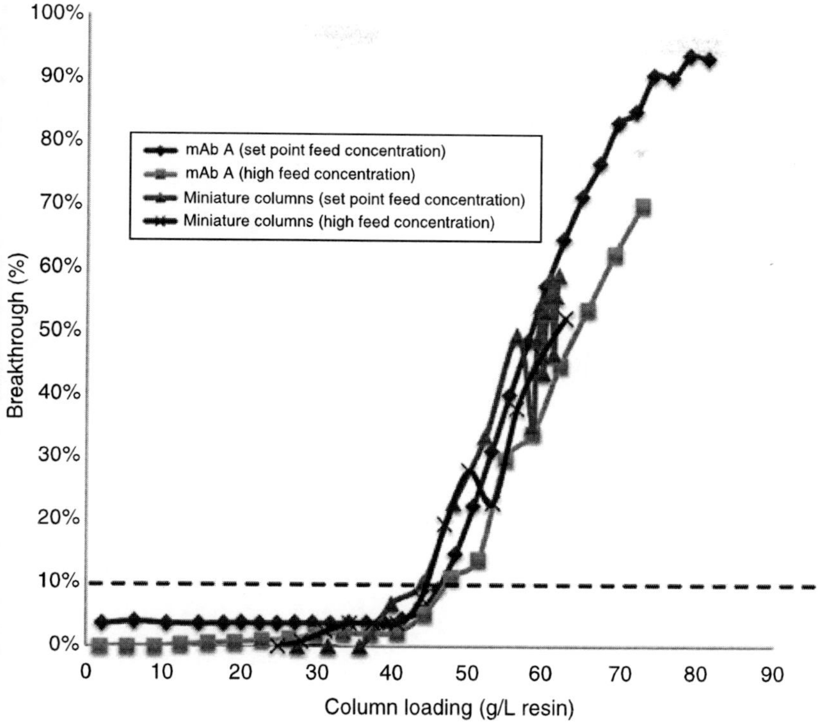

Figure 11.19 Comparison of breakthrough curves as observed at miniature and lab-scale columns (operated at 300 cm/h) using monoclonal antibody loads at two different concentrations, set point, and 10× higher concentration.

Table 11.6 Comparison of DBC values as evaluated with miniature and lab-scale columns.

	DBC at 10% breakthrough (mg/mL resin)	
Concentration/scale	**Miniature columns**	**Lab-scale columns**
Set point feed concentration	43.9	46.2
10× feed concentration	44.5	47.2

design with four axial points and eight CPs to better understand the variation. In this particular experiment, more CP runs than necessary were included to better understand the robustness of the operating parameters. As a follow-on, a subset of these experiments (seven runs), mainly the edge points and CPs, were repeated using a lab-scale column. Table 11.7 summarizes the experiments

Table 11.7 Summary of runs performed using miniature and lab-scale columns to evaluate impact of column loading and elution pH.

Runs	Loading (mg/mL resin)	Elution pH
1	**16**	**3.50**
2	20	3.30
3	20	3.70
4	**30**	**3.22**
5	30	3.50
6	30	3.50
7	**30**	**3.50**
8	**30**	**3.50**
9	**30**	**3.50**
10	30	3.50
11	30	3.50
12	30	3.50
13	**30**	**3.78**
14	40	3.30
15	40	3.70
16	**44**	**3.50**

Experiments shown in bold were executed with both miniature and lab-scale columns.

performed using miniature and lab-scale (0.66 cm diameter × 20 cm height, 6.8 mL) columns.

Neither column loading nor elution pH was found to be significant for step yield, with both parameters having *p* values of 0.1. As seen in Figure 11.20, elution pH (*p* value = 0.0007) and column loading (*p* value = 0.0079) had significant impacts on the amount of aggregates in the product pool. Decreasing elution pH as well as increased loading would cause an increase in aggregate levels. As seen in the desirability plots shown in Figure 11.21, similar behavior and operating ranges were observed with both lab- and miniature-scale columns. In summary, based on these results, column loadings in the range of 20–40 mg/mL resin and an elution pH in the range of 3.40–3.60 are recommended for the step operation leading to the yield >85% and aggregates <1.9%.

The step was successfully scaled up to pilot and manufacturing scale with the established operating parameters. As can be seen in Table 11.8, consistent results were obtained upon scale-up.

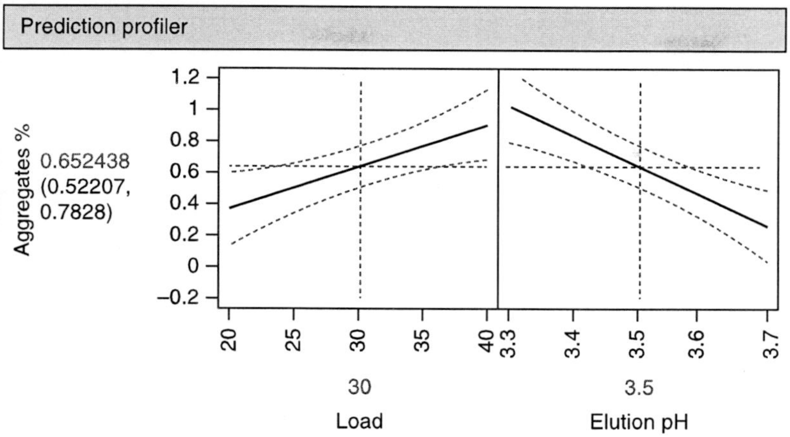

Figure 11.20 Interaction plots showing the impact of column load and elution pH on aggregate levels.

11.2.2.2.2 Case 2: Anion Exchange Chromatography Step Development

In mAb purification, the AEX step is widely utilized in flow-through mode (due to a common isoelectric point, p*I*, of mAbs in the range of 7–9) where mAb elutes in the flow-through and the impurities are bound to the column for removal of HCP, DNA, aggregates, and leached ProA ligand. It is also an effective step in removal of viruses as will be discussed in the next section.

In this example, additional experiments were performed to address the robustness of the AEX step operating parameter ranges (loading of 50–150 mg/mL resin, pH 7.2 ± 0.2, and conductivity 5.0 ± 1.0 mS/cm). The robustness was originally established based on the HTPD slurry plate method as discussed in Section 11.2.1.5. This section will focus on verifying the ranges via miniature and lab-scale columns. To evaluate the impact the end points of these ranges would have on impurity clearance and yield, lab-scale experiments (0.66 cm diameter × 20 cm height, 6.8 mL volume) at set point (pH 7.2, conductivity of 5.0 mS/cm) and at the worst-case end point (pH 7.0, conductivity of 6.0 mS/cm) were fractionated to further characterize the HCP and aggregate breakthrough. Figure 11.22 shows the chromatogram and the result of fractionation under worst-case conditions. In this experiment, the load contained 196 ppm HCP and 1.85% HMW species. The data suggests that the impurities begin breaking through from the start of the load and slowly increase throughout the load and then decrease during the start of the column wash until co-eluting with the back portion of the "washout peak." The CP run had the same trend at lower percentages, and the HCP trend closely follows HMW. The worst-case fractionation run "pool" had a step yield of 90.3% with HMW reduction of 62% in comparison to a yield of 84% with HMW reduction of 81% under CP conditions.

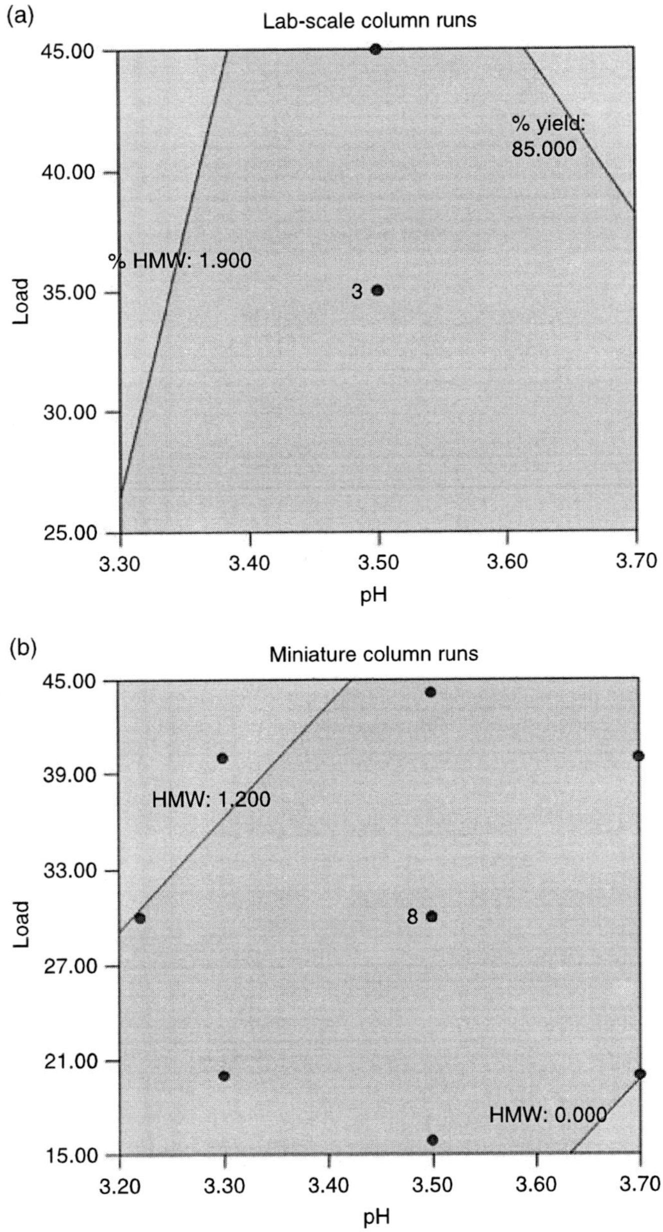

Figure 11.21 Comparison of operating window as obtained by lab- and miniature-scale experiments (dark section depicts the operating window).

Table 11.8 Scale-up of protein A chromatography step.

Scale	Load (mg/mL resin)	Column scale Diameter (cm)	Height (cm)	Volume (L)	Aggregates (%)	Yield (%)
Lab	35	0.66	20	0.0068	1.5	89
Pilot	35	25	20	9.9	1.8	86
Manufacturing	38	30	24	17.2	1.8	86.8

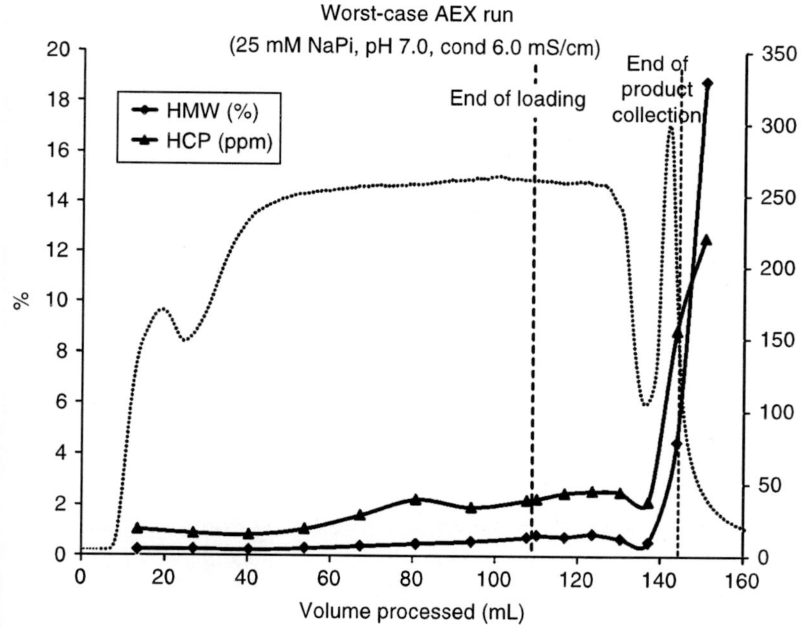

Figure 11.22 Worst-case run with HMW and HCP breakthrough.

To further evaluate the operating parameter ranges and the robustness of the process, a study was performed using miniature columns with verification of CP and axial points using lab-scale columns. The experiments performed with miniature and lab-scale columns are summarized in Table 11.9.

Figure 11.23 shows the summary of the impact of pH, conductivity, and loading on yield, aggregates, and the HCP levels as observed for both miniature (a) and lab-scale columns (b). While the absolute values are different in certain cases, the trends such as increasing yield with decreasing pH and increasing conductivity are similar with both sets of experiments. It is also important to note

Table 11.9 Summary of runs performed using miniature and lab-scale columns to evaluate impact of pH, conductivity, and column loading.

Run #	pH	Conductivity (mS/cm)	Loading (mg/mL)
1	**7.2**	**5.0**	**50**
2	**6.8**	**5.0**	**125**
3	**7.2**	**3.5**	**125**
4	7.2	5.0	125
5	7.2	5.0	125
6	**7.2**	**6.5**	**125**
7	**7.6**	**5.0**	**125**
8	**7.2**	**5.0**	**200**
9	7.0	4.0	80
10	7.0	6.0	80
11	7.4	4.0	80
12	7.4	6.0	80
13	7.0	4.0	170
14	7.0	6.0	170
15	7.4	4.0	170
16	7.4	6.0	170

Experiments shown in bold were executed with both miniature and lab-scale columns.

that because only a subset of the experiments were performed with lab-scale columns, the resolution of the lab-scale models were reduced leading to differences observed between Figure 11.23a and b. In the case of HTPD with miniature columns, where more experimental conditions were possible, a higher resolution model was achieved leading to the presence of curvature (two-factor interactions), which was absent in the models generated with lab-scale columns. Figure 11.24 shows the desirability plots as obtained for both (a) miniature and (b) lab-scale columns where yield has been maximized while decreasing aggregate and HCP levels (the lower the values shown on the figure, the more optimum the operation is). These desirability plots also lead to a very similar operating range for elution pH and conductivity as obtained by both miniature and lab-scale columns. Based on these results, the final operating conditions with ranges were determined as pH of 7.0–7.4, conductivity of 4.0–6.0 mS/cm with a loading range of 50–200 mg/mL while maintaining the yields >88%.

The step was successfully scaled up to pilot and manufacturing scale with the established operating parameters. As can be seen in Table 11.10, consistent results were obtained upon scale-up.

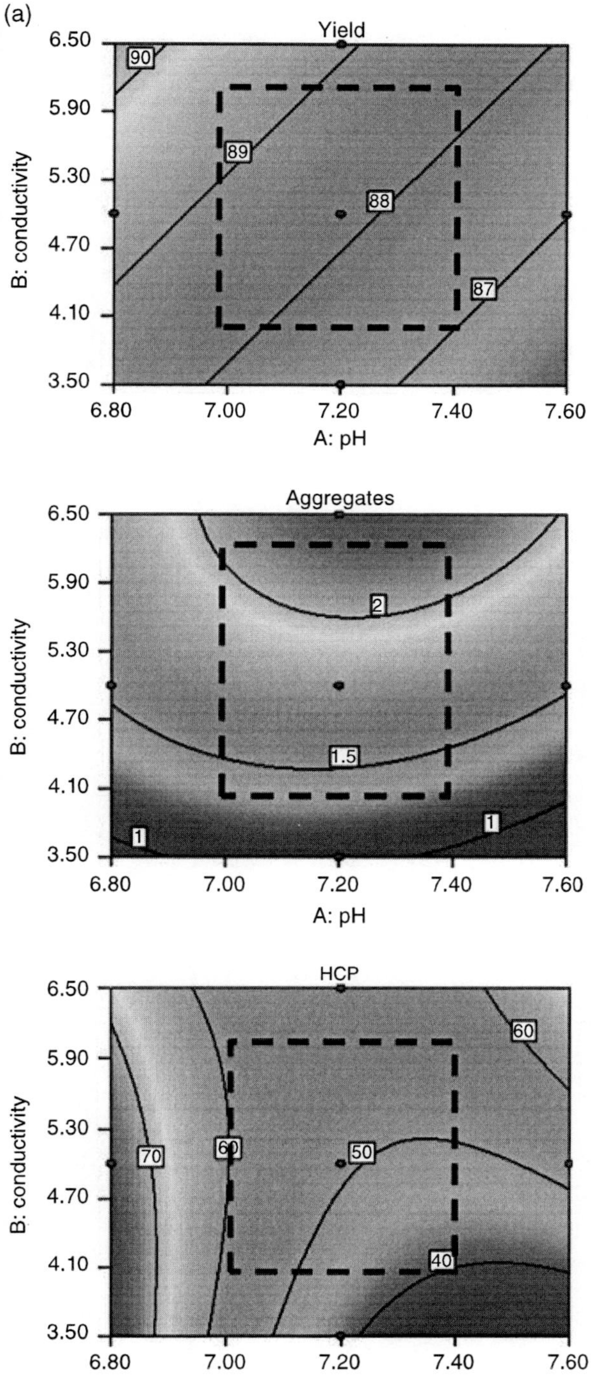

Figure 11.23 Yield, aggregates, and HCP as a function of elution pH and conductivity as obtained by (a) miniature and (b) lab-scale columns (dashed box shows the ranges for operating parameters set for large-scale experiments). (*See insert for color representation of the figure.*)

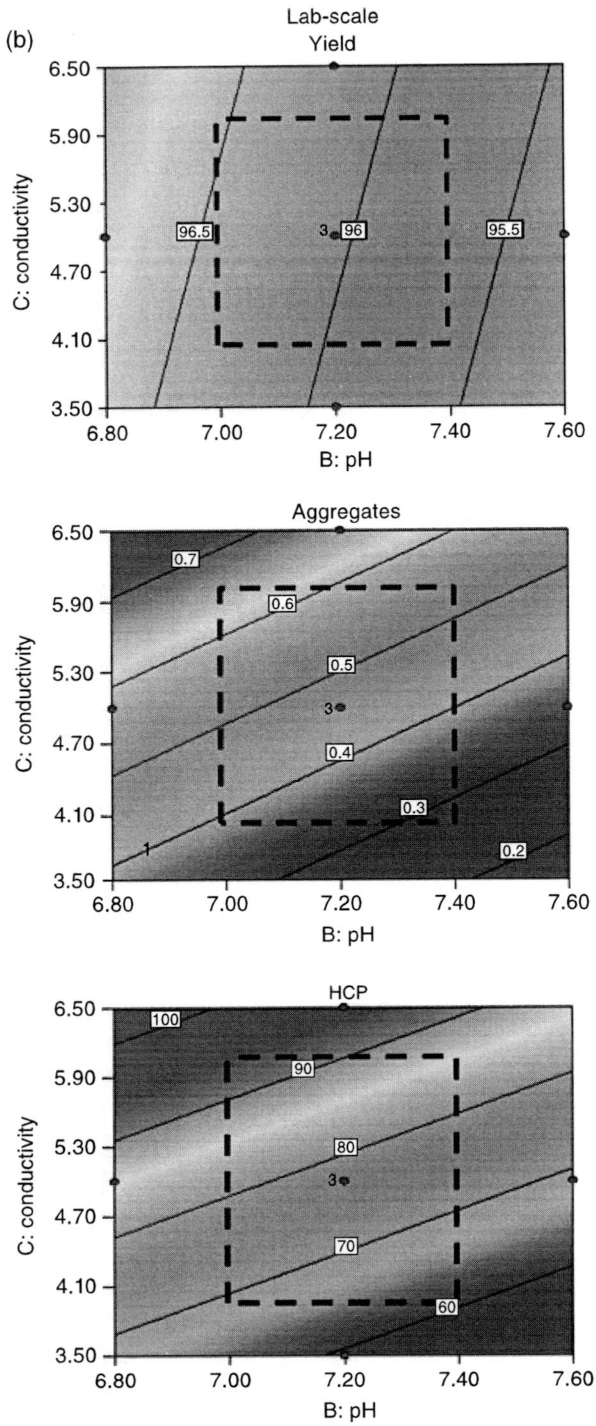

Figure 11.23 (Continued)

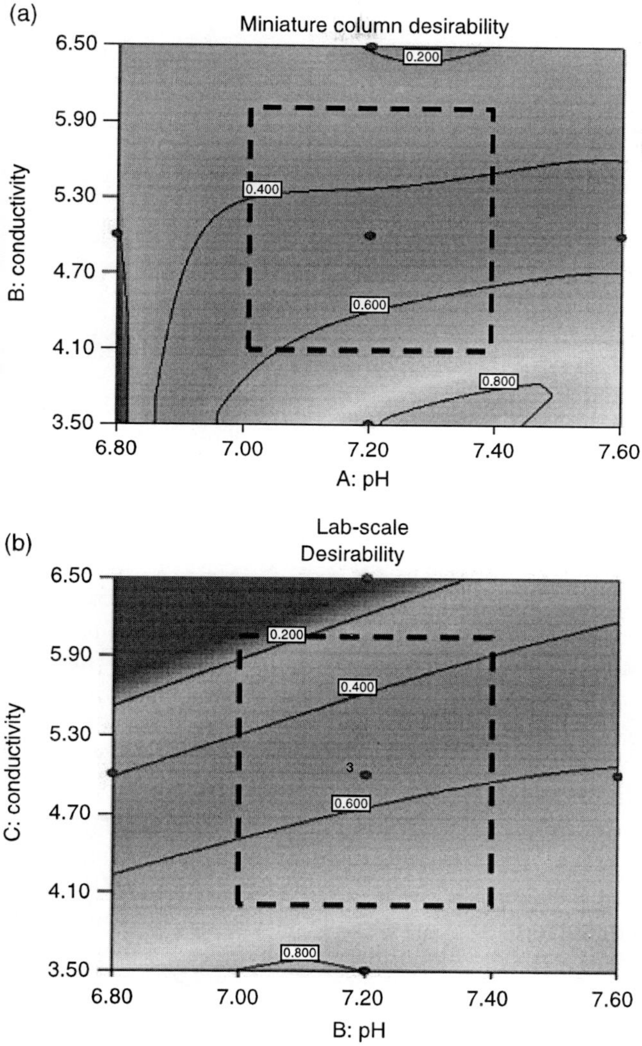

Figure 11.24 Desirability plots for maximizing yield while decreasing aggregate and HCP levels as obtained by (a) miniature and (b) lab-scale columns (dashed box shows the ranges for operating parameters set for large-scale experiments). (*See insert for color representation of the figure.*)

11.2.2.2.3 Case 3: Cation Exchange Chromatography Step Development

In this example, the development of a CEX step as a polishing step for purification of a mAb operated in bind and elute mode at ≤35 mg/mL resin (due to common p*I* range of 7–9 for mAbs) is shared. This example demonstrates the development of the CEX step without the benefit of HTPD. At the end of late

Table 11.10 Scale-up of AEX chromatography step.

Scale	Load (mg/mL resin)	Column scale: Diameter (cm)	Height (cm)	Volume (L)	Aggregates (%)	HCP (ppm)	Yield (%)
Lab	125	0.66	20	0.0068	0.4	81	94
Pilot	102	20	20	6.3	0.6	61	94
Manufacturing	138	30	20	13.8	0.5	88	98.5

stage process development, clipped species were identified in the drug substance upon further characterization. Following the identification of this new impurity based on SDS-PAGE results, additional development needs were identified for the CEX step to keep the levels of clipped species ≤1%. The process first involved resin screening to evaluate capacity via slurry plate experiments. Among the three resins tested, Poros HS50 was carried forward for process development due to higher capacity. This was followed by additional experiments to understand the resolution of the clipped species from antibody using different salt gradients at pH values of 4.5, 5.0, and 6.0. Based on the data summarized in Table 11.11, CEX chromatography on Poros HS50 resin was effective in removing a significant amount of the impurity (40–60%) at acceptable yield (>75%). Clearance appears to be better at lower pH (4.5) and higher pH (6.0) with a minimum near pH 5.0; therefore for later experiments a pH in between 5.0 and 6.0 was studied. Given these results, additional lab-scale column (0.66 cm diameter × 20 cm height, 6.8 mL) experiments were performed using gradients and step elutions at pH 5.6 and 4.5 for comparison. It is important to note that as the process step was being developed, the analysis of clipped species was mainly based on quantification of protein bands on SDS-PAGE gels. Later in development, an analytical method with better quantification, RP-HPLC, was developed and utilized for fraction analysis. Hence, some changes were observed in the reported reduction values shown in Tables 11.11 and 11.12. A fractionated step elution with 15 mS/cm conductivity at pH 5.6 (Figure 11.25) suggests that a cut of the tailing of the product pool peak could remove a significant fraction of the major impurities. Yield under these conditions was ~85% with 30–50% reduction of the clipped species. Two different conditions were repeated using a 40 mL column (1.6 cm diameter × 20 cm height), elution with 35 mS/cm conductivity at pH 4.5 and elution with 15 mS/cm conductivity at pH 5.6. As summarized in Table 11.12, better resolution of clipped species is obtained at pH 5.6, 15 mS/cm step elution conditions.

Once the CP operating conditions were identified, the robustness of the step was tested using a DOE approach. The key operating parameters identified

Table 11.11 Summary of CEX experimental conditions tested.

Experiments	Elution conditions	Reduction of clipped species (%) based on SDS-PAGE analysis	Yield (%)
pH 4.5 gradient elution	0.30–0.60 M in 20 CV	53	95.3
pH 4.5 step elution 1	0.33 M	30	80.6
pH 4.5 step elution 2	0.36 M	37	79.8
pH 4.5 step elution 3	0.40 M	43	99.1
pH 5.0 gradient elution	0.10–0.45 M in 20 CV	14	97.5
pH 5.0 step elution	0.25 M	5	97.5
pH 6.0 gradient elution	0–0.15 M in 20 CV	65	78.1
pH 6.0 step elution 1	0.08 M	23	64.8
pH 6.0 step elution 2	0.06 M	19	69.7

Table 11.12 Experimental results for pH comparison.

Experiment	Yield (%)	Reduction of clipped species based on RP-HPLC analysis (%)	Aggregate reduction (%)
pH 4.5, 35 mS/cm step elution	85.4	18	50
pH 5.6, 15 mS/cm step elution	87.8	32	45

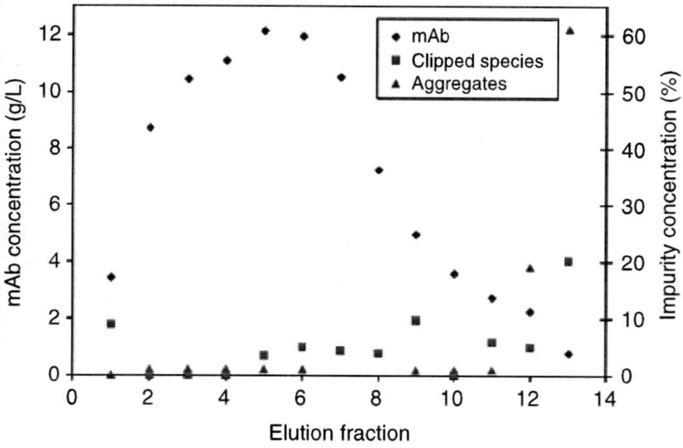

Figure 11.25 Resolution of clipped species and aggregates at pH 5.6 with step elution at 15 mS/cm.

Table 11.13 Summary of conditions tested via DOE using lab-scale column.

Parameters	Low range	Target	High range
Feed pH	5.4	5.6	5.8
Loading (mg/mL resin)	15	40	50
Wash pH	5.4	5.6	5.8
Wash conductivity (mS/cm)	8	9	10
Elution pH	5.4	5.6	5.8
Elution conductivity (mS/cm)	14	15	17

(feed pH, feed loading, wash pH and conductivity, and elution pH and conductivity) were evaluated using a 26-run minimum run resolution V (that captures all main effects and two-factor interactions) DOE study (four CPs). Table 11.13 provides a summary of the CEX robustness study. Outputs evaluated were yield, reduction of clipped species, and aggregates (Figure 11.26). Elution conductivity was not symmetrical around the CP as 13 mS/cm was unable to elute product from the resin. All experiments were performed on a Poros HS50 column at the 18.6 mL scale (1.1 cm diameter × 20 cm bed height).

Based on these results, the elution conductivity range was established as 14.5–16.0 mS/cm and the elution pH range was established as 5.5–5.8 at ≤40 mg/mL resin loading while reducing the levels of clipped species to ≤1.0% and aggregates to ≤0.7% with step yields ≥60%. The CEX step was then successfully scaled up to 307 L (140 cm diameter × 20 cm height) under the conditions given in the preceding text, and the step reduced levels of clipped species and aggregates to 0.8 and 0.2%, respectively, with 86% yield across this step.

11.2.3 Additional Considerations

11.2.3.1 Intermediate Stability

Because of the molecular complexity and relative fragility of protein therapeutics, intermediate hold times are critical not only for their manufacture but also for understanding process robustness. Intermediates are typically obtained from representative batches and held at set temperatures for predetermined amount of time. The intermediates are sampled periodically and assayed for product quality attributes (e.g., aggregates, fragments, oxidation, and acidic species) that may be affected by the hold and for microbial control (e.g., bioburden and endotoxin). Data to justify the hold time can be collected during development on lab-scale and pilot-scale batches or during process validation [39]. During the process development stage, hold time studies typically applied to process intermediates (e.g., column load and pool) can further inform the process

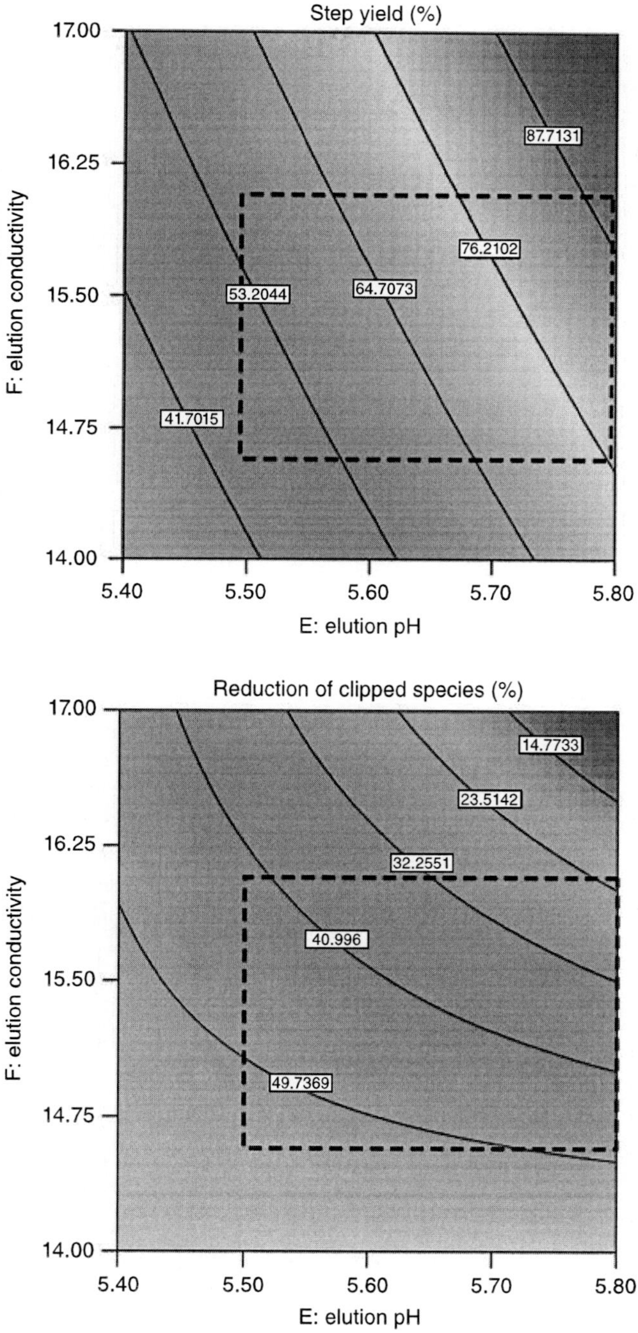

Figure 11.26 Yield, reduction of clipped species, and aggregates as a function of elution pH and conductivity.

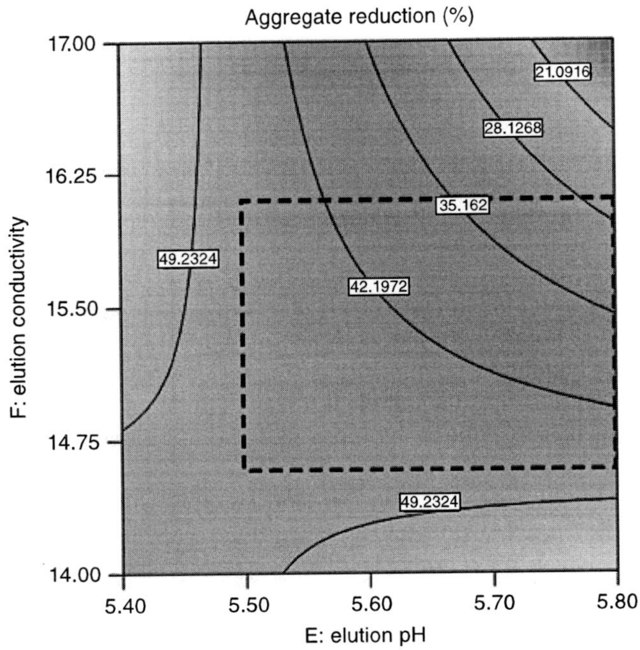

Figure 11.26 (Continued)

operating parameters. Evaluation of selected steps (e.g., steps for which high impurity removal or viral clearance are claimed) operating in worst-case and/or challenging conditions (e.g., maximum hold times) could be performed to demonstrate the robustness of the process as well.

An example shown here is the ProA pool hold time study that informs low pH viral inactivation conditions. Low pH inactivation is commonly used in mAb purification processes to inactivate large enveloped viruses. The inactivation comprises a number of steps including off-line sampling, pH measurement, and the accurate addition of the required amounts of acid and/or base—all to ensure that the pH is maintained in a predetermined range. During process development, optimal low pH holding conditions that allows sufficient time for virus inactivation while minimizing mAb quality issues usually aggregation are identified. Although a typical viral inactivation pH is 3.5, pH overshoot, time-to-process deviations, and operator deviations could all impact the robustness of the step. In this example, a mAb ProA product was titrated to five different pH values (pH 3.0, 3.3, 3.5, 3.7, 4.0) and held for up to 24h. Protein quality at hold time points of 1, 4, and 24h was compared, and an optimal pH value was determined for the low pH viral inactivation step based on the quality impact, as represented by product aggregation and fragmentation measured by HP-SEC (Figure 11.27). At lower pH values, 3.0 and 3.3, this mAb showed significant formation of HMW

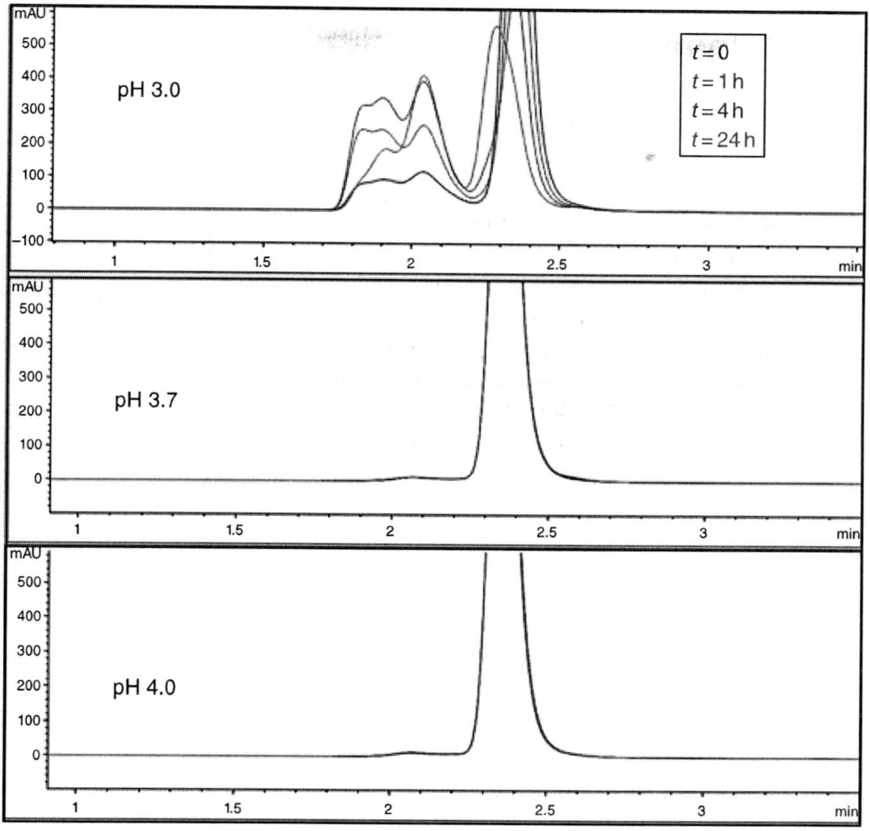

Figure 11.27 HP-SEC results of the protein A product low pH hold stability.

aggregates at hold time of 1 h or longer, whereas, at pH 3.5 and above, no product aggregation was seen for up to 24 h of holding. The process conditions chosen for this mAb viral inactivation step was pH 3.6 ± 0.1. The holding study of the ProA product clearly defined an operating pH range for the viral inactivation step. In some cases, manufacturing strategies are also leveraged, for example, when there are multiple cycles of ProA runs and different product hold is needed.

11.2.3.2 Viral Clearance Studies

Viral clearance studies are an important part of the purification process development for proteins expressed in mammalian cell cultures. The implementation of biological products in human clinical trials and as marketed products requires the use of effective viral clearance steps. Regulatory agencies have established laws and guidelines for safety testing at multiple stages throughout the drug development process, starting at the investigational new drug (IND) [40, 41]. Demonstration of viral clearance is a critical step in assuring the safety of biological products [42]. Some common process steps are introduced into downstream purification schemes that provide a physical means to separate and/or remove viruses from the therapeutic protein. The physical steps often include virus-removing filters and chromatographic resins in column or membrane configurations but can also include the introduction of irradiation, high heat steps, or other means for destroying the infectivity of a virus. Chemical treatment steps are often utilized as a means to inactivate a wide variety of virus types [43].

During the demonstration studies, model viruses are used to ensure that the purification processes have two or more orthogonal steps with sufficient viral clearance capacity. The number and type of viruses vary depending on the phase of the clinical trials, generally two model viruses prior to entering Phase I trials and four model viruses prior to marketing. If a chromatography step is one of these orthogonal steps to be claimed for viral clearance, it is essential to confirm that the proposed operating ranges deliver a robust viral clearance [40, 41].

An example is shared here where the preliminary operating conditions used for an AEX chromatography step that was developed as summarized in the example presented in Section 11.2.2.2.2 were evaluated using viral clearance studies prior to finalizing the operating conditions. It should be noted here that the viral clearance studies utilized in this case was designed to be preliminary and for information only, not supporting any filing or registration. Hence, it included only one experiment with each virus and preliminary operating conditions. The study included testing the log virus reduction (LRV) of minute virus of mice (MMV) and xenotropic murine leukemia virus (XMuLV) at worst-case scenarios: (i) high loading, target pH, and high conductivity and (ii) high loading, low pH, and high conductivity. The AEX flow-through was fractionated during the AEX load, and the column wash (which is a part of the product pool) was collected separately. The results summarized in Table 11.14

Table 11.14 Summary of LRV as tested for worst-case scenarios.

			LRV				
Run #	**Condition**	**Virus**	**0–50 mg/mL**	**50–100 mg/mL**	**100–165 mg/mL**	**Wash**	**Pool**[a]
1	pH 7.2 Cond 6.0 mS/cm	MMV	>4.20	>4.19	>4.19	>4.06	5.79
		XMuLV	>4.68	>4.68	>4.68	>4.58	>6.25
2	pH 7.0 Cond 6.0 mS/cm	MMV	>4.44	>4.44	4.44	>4.32	4.15
		XMuLV	>4.57	>4.57	>4.57	>4.48	>6.15

[a] Large-volume sampling results in lower limit of quantification (LOQ).

demonstrate effective clearance for the range tested (including the worst case), which provided sufficient data to ensure that the conditions being pursued for the AEX chromatography step would indeed lead to acceptable LRV.

11.3 Conclusions

In this chapter, we attempted to summarize a workflow for developing robust chromatography processes using both high-throughput and lab-scale column-based techniques to shorten development timelines and material needs. The workflow presented in this section ensures not only the ranges for operating parameters but also the stability of load/feed used for, and the product generated through the chromatography steps are evaluated. Within the QbD paradigm, we believe following such a workflow will not only benefit the early developability assessment but also the further scale-up of the processes developed in later stages.

Acknowledgments

We would like to acknowledge those who have contributed to this chapter via sharing data and valuable reviews provided during preparation: William Rayfield, Svetlana Schussler, Haiying Bao, John Welsh, Michael Rauscher, Darshini Shah, Sunitha Kandula, David Roush, and Thomas Linden.

References

1. Cramer, S. M. & Holstein, M. A. (2011). Downstream bioprocessing: recent advances and future promise. *Current Opinion in Chemical Engineering,* 1, 27–37.
2. Wheelwright, S. (1991). *Protein Purification: Design and Scale-Up of Downstream Processes.* New York: Oxford Press.
3. Yang, X., Xu, W., Dukleska, S., Benchaar, S., Mengisen, S., Antochshuk, V., et al. (2013). Developability studies before initiation of process development. *MAbs,* 5, 787.
4. GE Healthcare. (1999). *Protein Purification Handbook.* Uppsala: GE Healthcare.
5. Tugcu, N., Deshmukh, R., Sanghvi, Y., & Cramer, S. (2003). Displacement chromatography of anti-sense oligonucleotide and proteins by using saccharin as a non-toxic displacer. *Reactive and Functional Polymers,* 54, 37–47.
6. Deen, W. (2004). Hindered transport of large molecules in liquid-filled pores. *AIChE Journal,* 33, 1409–1425.

7. DePhillips, P. & Lenhoff, A. (2001). Determinants of protein retention characteristics on cation-exchange adsorbents. *Journal of Chromatography A*, 933, 57–72.
8. Kopaciewicz, W., Fulton, S., & Lee, S. (1987). Influence of pore and particle size on the frontal uptake of the proteins: implications for preparative anion-exchange chromatography. *Journal of Chromatography A*, 409, 111–124.
9. Walters, R. R. (1982). High-performance affinity chromatography: pore size effects. *Journal of Chromatography A*, 249, 19–28.
10. Cramer, S. M. & Subramanian, G. (1989). Preparative liquid chromatography of biomolecules—New directions. In G. Keller & R. Yang, *New Directions in Sorption Technology* (pp. 187–226). Stoneham: Butterworth.
11. Regnier, F. (1983). High-performance liquid chromatography of biopolymers. *Science*, 222, 245–252.
12. Gallant, S. R., Kundu, A., & Cramer, S. M. (1995). Modeling non-linear elution of proteins in ion-exchange chromatography. *Journal of Chromatography A*, 702, 125–142.
13. Rege, K., Pepsin, M., Falcon, B., Steele, L., & Heng, M. (2006). High-throughput process development for recombinant protein purification. *Biotechnology and Bioengineering*, 93, 618–630.
14. Welsh, J. P., Rauscher, M. A., Bao, H., Meissner, S., Han, I., Linden, T. O., et al. (2016). High throughput strategy and analytical methods for optimizing a domain antibody downstream process. *Engineering in Life Sciences*, 16, 133–142.
15. Rathore, A. S. & Winkle, H. (2009). Quality by design for biopharmaceuticals. *Nature Biotechnology*, 27, 26–34.
16. Bhambure, R., Kumar, K., & Rathore, A. (2011). High-throughput process development for biopharmaceutical drug substances. *Trends in Biotechnology*, 29, 127–135.
17. Coffman, J. L., Kramarczyk, J. F., & Kelley, B. D. (2008). High-throughput screening of chromatographic separations: 1. Method development and column modeling. *Biotechnology and Bioengineering*, 100, 605–618.
18. Carta, G. (2012). Predicting protein dynamic binding capacity from batch adsorption tests. *Biotechnology Journal*, 7, 1216–1220.
19. Linden, T. (2001). Untersuchungen zum inneren Transport bei der Proteinadsorption an poröse Medien mittels konfokaler Laser-Raster-Mikroskopie, Doctoral thesis. Düsseldorf: Heinrich-Heine-Universität Düsseldorf.
20. Kelley, B., Tobler, S., Brown, P., Coffman, J., Godavarti, R., Iskra, T., et al. (2008). Weak partitioning chromatography for anion exchange purification of monoclonal antibodies. *Biotechnology and Bioengineering*, 101, 553–566.
21. Bergander, T., Nilsson-Valimaa, K., Oberg, K., & Lacki, K. M. (2008). High-throughput process development: determination of dynamic binding capacity using microtiter filter plates filled with chromatography resin. *Biotechnology Progress*, 24, 632–639.

22. Pirrung, S. M. & Ottens, M. (2017). High throughput process development. In A. Staby, S. Ahuja, & A. Rathore, *Preparative Chromatography for Separation of Proteins* (pp. 269–291). Hoboken, NJ: John Wiley & Sons, Inc.
23. Wiendahl, M., Wierling, O. S., Nielsen, J., Christensen, D. F., Kraup, J., Staby, A., et al. (2008). High throughput screening for the design and optimization of chromatographic processes—miniaturization, automation, and parallelization of breakthrough and elution studies. *Chemical Engineering and Technology*, 31 (6), 950–963.
24. Toueille, M., Uzel, A., Depoisier, J. F., & Gantier, R. (2011). Designing new monoclonal antibody purification processes using mixed-mode chromatography sorbents. *Journal of Chromatography B, Analytical Technologies in the Biomedical and Life Sciences*, 879, 836–843.
25. Treier, K., Hansen, S., Richter, C., Diederich, P., Hubbuch, J., & Lester, P. (2012). High-throughput methods for miniaturization and automation of monoclonal antibody purification processes. *Biotechnology Progress*, 28, 723–732.
26. Lacki, K. M. (2012). High-throughput process development of chromatography steps: advantages and limitations of different formats used. *Biotechnology Journal*, 7, 1192–1202.
27. Welsh, J. P., Petroff, M. G., Rowicki, P., Bao, H., Linden, T., Roush, D. J., et al. (2014). A practical strategy for using miniature chromatography columns in a standardized high-throughput workflow for purification development of monoclonal antibodies. *Biotechnology Progress*, 30, 626–635.
28. Tugcu, N., Roush, D. J., & Goklen, K. E. (2008). Maximizing productivity of chromatography steps for purification of monoclonal antibodies. *Biotechnology and Bioengineering*, 99, 599–613.
29. Hahn, R., Deinhofer, K., Machold, C., & Jungbauer, A. (2003). Hydrophobic interaction chromatography of proteins II. Binding capacity, recovery and mass transfer properties. *Journal of Chromatography B*, 790, 99–114.
30. Fahrner, R. L., Knudsen, H., Basey, C., Galan, W., FeuerHelm, D., Vanderlaan, M., et al. (2001). Industrial purification of pharmaceutical antibodies: development, operation, and validation of chromatography processes. *Biotechnology and Genetic Engineering Reviews*, 18, 301–327.
31. Fahrner, R. L., Whitney, D. H., Vanderlaan, M., & Blank, G. (1999). Performance comparison of protein A affinity-chromatography sorbents for purifying recombinant monoclonal antibodies. *Biotechnology and Applied Biochemistry*, 30, 121–128.
32. Iyer, H., Henderson, F., Cunningham, E., Webb, J., Hanson, J., Bork, C., et al. (2002). Considerations during development of a protein A based antibody purification process. *BioPharm*, 1, 14–20.
33. Shukla, A. A., Hubbard, B., Tressel, T., Guhan, S., & Low, D. (2007). Downstream processing of monoclonal antibodies—application of platform technologies. *Journal of Chromatography B, Analytical Technologies in the Biomedical and Life Sciences*, 848, 28–39.

34. Follman, D. K. & Fahrner, R. L. (2004). Factorial screening of antibody purification processes using three chromatography steps without protein A. *Journal of Chromatography A*, 1024, 79–85.
35. Gagnon, P. (1996). *Purification Tools for Monoclonal Antibodies*. Tucson: Validated Biosystems, Inc.
36. Wensel, D. L., Kelley, B. D., & Coffman, J. L. (2008). High throughput screening of chromatographic separations: III. Monoclonal antibodies on ceramic hydroxyapatite. *Biotechnology and Bioengineering*, 100, 839–854.
37. Shukla, A. A. & Thommes, J. (2010). Recent advances in large-scale production of monoclonal antibodies and related proteins. *Trends in Biotechnology*, 28, 253–261.
38. Harrison, R. G., Todd, P., Rudge, S. R., & Petrides, D. P. (2003). Liquid chromatography and adsorption. In R. G. Harrison, *Bioseparations Science and Engineering* (pp. 203–204). New York: Oxford University Press.
39. Thillaivinayagalingam, P. & Newcombe, A. (2011). Validation of intermediate hold times. *BioProcess International*, 9, 52–57.
40. U.S. Department of Health and Human Services. (1997). *Points to Consider in the Manufacture and Testing of Monoclonal Antibody Products for Human Use*. Rockville, MD: U.S. Department of Health and Human Services.
41. U.S. Department of Health and Human Services. (1998). *Guidance for Industry: Q5A Viral Safety Evaluation of Biotechnology Products Derived from Cell Lines of Human or Animal Origin*. Rockville, MD: U.S. Department of Health and Human Services.
42. Miesegaes, G., Lute, S., & Brorson, K. (2010). Analysis of viral clearance unit operations for monoclonal antibodies. *Biotechnology and Bioengineering*, 106, 238–246.
43. Cipriano, D., Burnham, M., & Hughes, J. (2012). Effectiveness of various processing steps for viral clearance of therapeutic proteins: Database analyses of commonly used steps. In V. Voynov & J. Caravella, *Therapeutic Proteins: Methods and Protocols* (pp. 277–292). New York: Humana Press.

12

Problem Solving by Using Modeling

Martin P. Breil[1], Søren S. Frederiksen[2], Steffen Kidal[2], and Thomas B. Hansen[2]

[1] *DAPI Modelling & Optimization, Novo Nordisk A/S, Bagsværd, Denmark*
[2] *Mathematical Modelling, Novo Nordisk A/S, Bagsværd, Denmark*

12.1 Introduction

In this chapter we present a number of real-life cases relating to an anion exchange column in the purification of a GLP-1 compound. The modeled system comprises of four components: the product, denoted C; the closest eluting impurity peak before the product peak, denoted B; the closest impurity peak eluting after the product peak, denoted D; and the fourth component eluting very early in the chromatogram, denoted A. Normally, it is adequate to include the closest eluting impurities. Component A is only included because its concentration has a large variation in the previous process step, as seen in one of the cases.

The problem solving—described in this chapter—involves a number of different types of issues: what is the influence of the volume in buffer mixer, what is the effect of a high salt concentration in buffer, what is the effect of an increased level of impurities, and what is the influence of the simulation origin from a salt pulse in a dead volume? When the process step can be described reasonably well by a first-principle model, then it is possible to mimic the real-life observations *in silico* and thus be able to assert what the core of the problem is and how to resolve the situation. Additionally, the model can be used to optimize the process *in silico*. That is essentially what a mathematical model of a process step is aimed at: understanding the physics involved and proposing a valid solution to the problem.

12.2 Theory

As previously mentioned, this chapter describes a number of problems that have been encountered in real life on an ion exchange process step. The first step is to develop a mathematical model, which is based on physics that makes it possible to make sensible explanations and predictions of the process step. The model consists of a mathematical description of the column (a partial differential equation (PDE)) and an algebraic equation for the isotherm. Additionally, a model for a column mixer with a linear gradient in the inlet is given.

12.2.1 Column Model

The model comprises of two parts: a time-dependent 1D-model for the flow and mass balance in the column and an algebraic equation for the adsorption isotherm (steric mass action (SMA)). The concentration of each component (the negatively charged proteins and the salt anion) in the mobile phase is described by a mass balance (mathematically by a PDE). The integrated mass balance governs the distribution of salt and protein in both mobile and stationary phase throughout the column.

The mass balance for component i reads

$$\frac{\partial c_i}{\partial x} + \varepsilon \frac{\partial c_i}{\partial \tau} + (1-\varepsilon)\varepsilon_p K_d \frac{\partial}{\partial \tau}(c_i + q_i) - \frac{1}{\text{Pe}} \frac{\partial^2 c_i}{\partial x^2} = 0$$

where x is the dimensionless axial position, $x = z/L$, and τ is the dimensionless time that originates from real time as $\tau = v_s \cdot t/L$, where v_s is the superficial velocity (velocity on an empty column), whereas ε is the column porosity and ε_p is the porosity of the particle. The Péclet number, $\text{Pe} = (v_s/\varepsilon) \cdot L/D_{\text{eff}}$, describes the ratio between the convective forces and the diffusive forces. When the Péclet number is large, then convection is the predominate effect compared with diffusion. This is usually the case for a chromatographic column since the diffusion coefficient is $\sim 10^{-9}\,\text{m}^2/\text{s}$, Ref. [1], and the flow is roughly 1 cm/min. Additionally, all mass transfer is pooled into the Péclet number for convenience.

All zone spreading mechanisms have been pooled into the Péclet number. The boundary conditions are

$$c_{i,\text{in}}(\tau) - c_i(\tau, x=0) = -\frac{1}{\text{Pe}} \frac{\partial c_i}{\partial x}\bigg|_{x=0}$$

and the concentration is finite for $x \rightarrow \infty$.

All the mass balances are coupled by an adsorption isotherm (in this case the SMA equation). The equilibrium is given by the SMA isotherm, Ref. [2]:

$$c_i = \frac{q_i}{K_i}\left[\frac{c_s}{\Lambda - \sum_j q_j(\nu_j + \sigma_j)}\right]^{\nu_i}$$

where c_s is the salt concentration, K_i is the equilibrium constant, ν_j is the binding charge, and σ_j is the shielding parameter. The SMA isotherm describes the concentrations of protein in the mobile and stationary phase at equilibrium as a function of salt concentration.

Despite being a relatively simple model, it is striking how much process insight can be obtained by using this model.

The PDEs were solved using the commercially available software COMSOL Multiphysics™. Combining the two types of equations previously mentioned, the retention volume as a function of the salt concentration for a linear isotherm can be expressed by

$$V_{R,i} = V_{NR,i} + (1-\varepsilon)\varepsilon_p K_d K_i \left(\frac{\Lambda}{z_s c_s}\right)^{\nu_i}$$

where $V_{NR,i}$ is the non-retained volume of component i, corresponding to $V_{NR,i} = \varepsilon + (1-\varepsilon)\varepsilon_p \cdot K_{d,i}$.

12.2.2 Gradient Mixer

Only a fraction of all preparative purification processes run at isocratic conditions; the majority of the processes involve some of sort of a gradient. The reason for this is simply that isocratic elution is (too) time consuming and not necessarily the optimal solution because very long retention times also give very broad peaks. Therefore it is desirable to apply gradients during elution. This is achieved by inserting a mixer in which the buffers (at least two—otherwise mixing would be redundant) are mixed to form a homogeneous solution.

The applied buffers usually only differ in one property, that is, salt concentration, pH, or organic solvent concentration. There is, of course, no reason why the number of changing properties should not exceed one, but a multi-property variance in the buffer compositions is more complicated and could easily cause additional challenges.

The impact of the mixer volume cannot always be neglected. When the equipment reaches production scale, it is imperative that the mixer volume (the dead volume in general) is well defined. Otherwise the applied gradient changes (ramps and steps) are muffed or dampened. If the mixer volume is large, it would be virtually impossible to make any step changes to the gradient.

The mathematical description of the mixer is that of a perfectly stirred tank defined by its residence time (i.e., the ratio between the volumetric flow rate,

Q, and the volume of the mixer, V_{mixer}). In the present derivation the flow rate and the volume are considered constant. The differential equation reads

$$V_{\text{mixer}} \cdot \frac{dc}{dt} = Q \cdot (c_{\text{in}} - c)$$

where Q is volumetric flow rate, c is the concentration of the property that changes during the mixing, and c_{in} is the inlet concentration to the mixer (mixing chamber). Normally, we operate in terms of column volumes (CV) rather than time. Therefore the differential equation is rewritten as

$$\frac{dc}{dV} = \frac{1}{V_{\text{mixer}}} \cdot (c_{\text{in}} - c)$$

where V (=$Q{\cdot}t$) is the accumulated volume that has entered into the mixer. The inlet concentration is a function of time (and therefore a function of the accumulated volume), and here we will limit ourselves to linear changes only. However, any nonlinear change in the inlet concentration can eventually be represented by a set of (small) linear changes. The inlet concentration is made up of a number of linear segments; each segment is assigned a linear expression

$$c_{\text{in}}(V) = a \cdot V + b, \quad V_{\text{start}} < V \leq V_{\text{end}}$$

where the coefficients a and b are constants of that particular segment. The differential equation is solved by Laplace transformation

$$c(s) = \frac{c_0}{s+1} + \frac{1}{s+1} c_{\text{in}}(s)$$

where c_0 is the initial concentration, and the inlet concentration is transformed to

$$c_{\text{in}}(s) = \frac{a}{s^2} + \frac{b}{s}$$

The solution is

$$c(V) = c_0 \cdot e^{-w} + \left(1 - e^{-w}\right) \cdot c_{\text{in}}(V) - a \cdot V_{\text{mixer}} \cdot \left(1 - (1 + w) \cdot e^{-w}\right)$$

where w is the accumulated volume that has entered mixer relative to the volume of the mixer and $w = V / V_{\text{mixer}}$.

Table 12.1 Used column parameters for Source 30Q column.

	Λ (mol/L pore)	ε	ε_p
Source 30Q	0.30	0.45	0.57

12.3 Materials and Methods

All experiments necessary to determine the model parameters were performed on an ÄKTAexplorer 100 using Source 30Q resin and an NaCl/TRIS pH 7.5 (20°C) buffer system. A 2.75 mL (length 3.5 cm) column was used for parameter experiments. In the pulse experiments the absorbance was measured at both 214 and 280 nm using a 10 mm light path to ensure an adequate response. In the preparative experiments a 2 mm light path was used, and the absorbance was only measured at 280 nm due to the higher concentrations. The concentration of the impurities might be orders of magnitude lower than the concentration of the product. To make an application solution where the concentrations of the impurities have more comparable concentrations, the pulse experiments were conducted with an application solution originating from a fraction collected before the main peak in a previous run. The same was done with a fraction collected after the main peak. This ensures a high concentration of the impurity of interest. A further description of the experiments can be found in Ref. [3].

The column parameters are the capacity of the gel (matrix) and the porosity of the packed column, and the porosity column is subjected to a pulse of pure water in a nitrate saturated column. The retention time of the water pulse is

$$V_R = V_{col} \cdot \left(\varepsilon + (1-\varepsilon) \cdot \varepsilon_p\right)$$

where V_{col} is the CV. As seen there is one equation with two unknowns. Only the total porosity can be determined, and therefore one of the porosities is set, and the other one is calculated. In Table 12.1 the two porosities are given. The particle porosity and the capacity of the gel are those of Ref. [4], and interstitial porosity is modified to match the non-retained volume. Any inaccuracy in these column parameters will later on be compensated for by the equilibrium parameters.

12.4 Determination of Model Parameters

All equilibrium parameters except the maximum binding capacity can be found from isocratic pulse experiments. These are relatively easy to perform and only require a small amount of material.

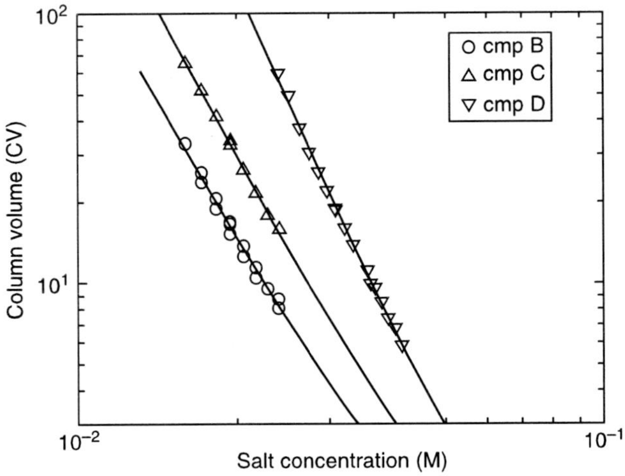

Figure 12.1 Retention volume as a function of salt concentration of the components B, C, and D.

In the simplest case the separation problem can be reduced to a three-component system consisting of the product and closest eluting impurities before and after the main component. The parameters for these three components were determined in two different series: the first series used an application solution, which was spiked with a pool fraction from the front of the main peak (as described earlier), and the second series used a spiking fraction obtained from the rear of the main peak. Both fractions contain the product of interest, and the two series therefore also serve as a repetition to determine the retention time as a function of the salt concentration of the main peak. From the chromatograms of the pulse experiments, the retention volume as well as the peak width can be fitted. The retention volumes from all experiments are given in Figure 12.1.

The parameters for the early eluting component A are not determined from isocratic pulse experiments but estimated from preparative chromatograms.

The shielding factor is determined for the product, and due to the similarity among the four components, it is readily assumed that the shielding factor is the invariant. The value of the shielding factor is determined by a breakthrough experiment where the column is loaded until a breakthrough is observed. The resulting isotherm is given in Figure 12.2. Alternatively the shielding parameter can be determined from fitting it to the chromatogram at preparative load.

Finally, it is assumed that the distribution coefficient, K_d, is unity for all four components inasmuch as it is assumed that the entire pore volume is accessible

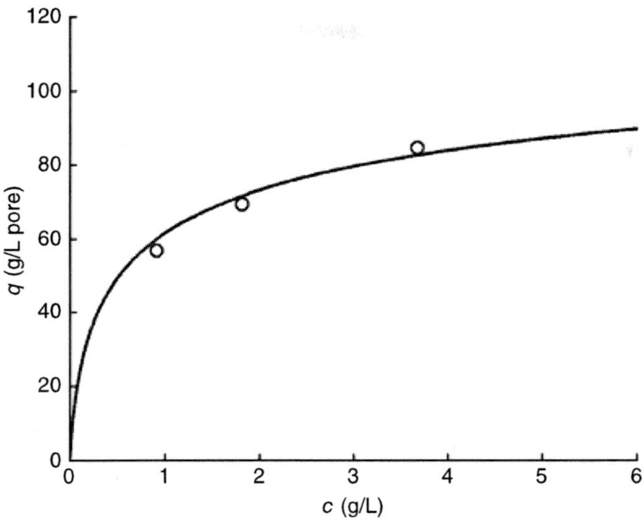

Figure 12.2 Measured capacity from breakthrough experiments and fitted isotherm for compound C.

Table 12.2 Protein parameters.

Component	ν	K	σ	K_d	Pe
A	3.20	0.00185	4	*1*	*75*
B	3.47	0.00368	4	*1*	*75*
C	3.66	0.00459	4	*1*	*75*
D	4.43	0.00251	*4*	*1*	*75*

The shielding parameter, steric exclusion factor, and the Péclet number are set to the same for all components. The parameters for component A are estimated from preparative experiments.

for these components. The Péclet number is adjusted to give the same peak width in the modeled data as in the pulse experiments.

All the determined model parameters are given in Table 12.2.

Based on the charge curves of the product and the predominant by-products, see Figure 12.3, the optimal pH range for the anion exchange separation is predicted. An optimal pH range has to fulfill two criteria: the range should be robust to (small) perturbations in the acidity of the application solution (i.e., the charge curve is flat), and there should be a charge difference between the components (otherwise ion exchange is nearly futile).

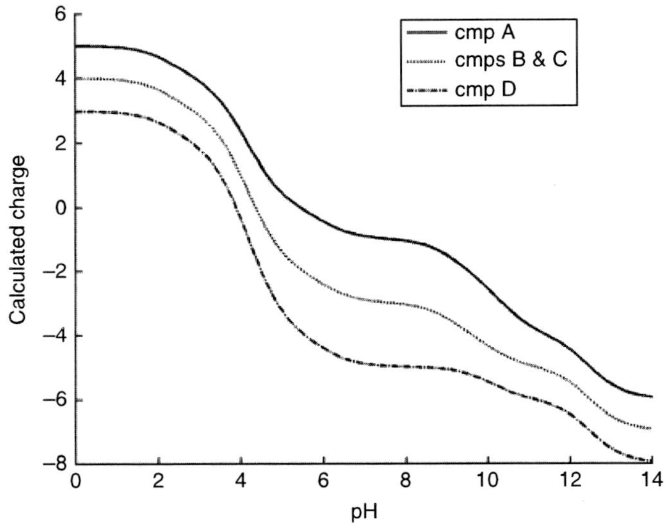

Figure 12.3 Charge curves for components A, B, C, and D. The process step is run at pH = 7.6. The charge curve is calculated based on the pK_a values of Ref. [5] and the corresponding dissociation reactions. *Source:* Campbell [5]. Reproduced with permission of Elsevier.

12.5 Optimization *In Silico*

Once the model parameters have been determined, it is time to apply them to a chromatographic simulator so we can perform optimizations *in silico*—it costs far less than ordinary experiments in terms of the chemicals and technician. Another obvious advantage of a mathematical model of a chromatographic process step is that it is possible to make a million simulations (e.g., Monte Carlo simulations) and then just pick the optimal one. The same brute-force approach is not a realistic solution in the lab.

Usually the chromatographic process steps are developed based on previous experience. This also means that the shortcomings of the past are sometimes passed on to the next generation. One way to redeem this situation is to do optimizations *in silico* because it is faster and cheaper—but still based on sound physics. After the *in silico* optimization, the suggested process is tested in laboratory scale to confirm the results.

One of the first applications of the mathematical model, which we have derived in the first sections of this chapter, was to increase the production capacity by increasing the load on the column at a constant flow (and thus a constant production cycle). Figure 12.4 shows how the reference method performs, and Figure 12.5 shows the chromatogram from the *in silico* optimized process. It is noted that the retention time in the optimized method is

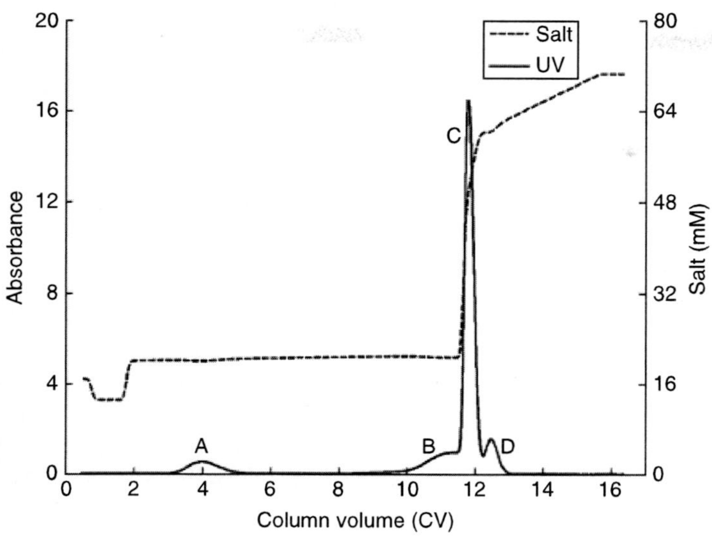

Figure 12.4 Reference chromatogram for process before the *in silico* optimization.

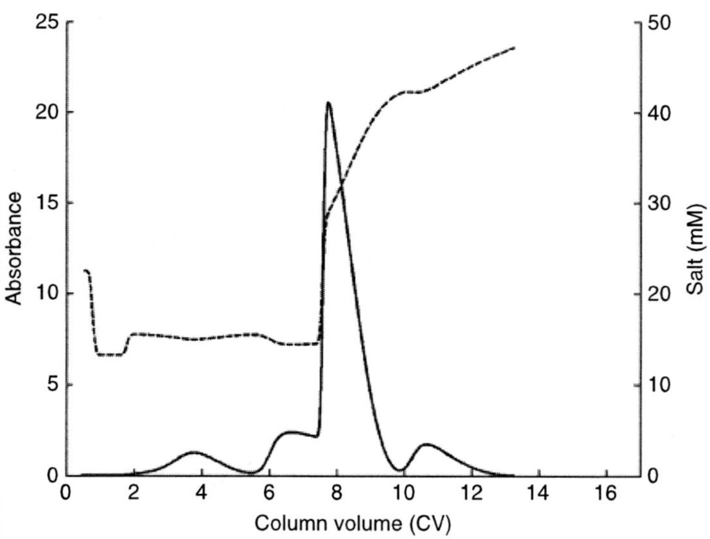

Figure 12.5 Chromatogram for process after the *in silico* optimization. The full line is the absorbance response (referring to the ordinate axis on the left), and the dashed line is the salt concentration (referring to the ordinate axis on the right).

reduced by 20%, whereas the resulting yield and purity are still quite satisfactory. It is worth to notice that the load in the optimized process is almost three times higher.

12.6 Extra-Column Effects

A gradient mixer might be applied in a chromatographic separation process to reduce fluctuations from the strong and weak solvent pumps. In addition to this, other dead volumes are present in the system. These can originate from, for example, piping or air traps. In a pilot facility the column sizes might vary significantly, but often the system with piping, air traps, etc. remains the same. A relatively large dead volume in comparison with the CV might therefore occur. These volumes must be taken into account when modeling the process since these will act as mixing zones in the separation process. These extra-column effects can be estimated and accounted for before running the process.

When running an isocratic gradient, these dead volumes will have no effect, whereas they might have a large effect when the gradient is changed stepwise. As seen in the current process, it is designed with a step-up in salt concentration at the start of the pool collection. This is done to reduce the volume of the collected pool.

It is worth noting in Figure 12.6 that the impurity eluting before the main peak, cmp B, only changes slightly in shape. This is due to the relatively small step-up in salt concentration early in the chromatogram (around 2 CV). After the front peak is eluted, the pool collection starts, and a significant step-up is made in the salt concentration to reduce the volume of the collected pool. The effect of an increased mixing volume results in a shallower salt gradient (at the

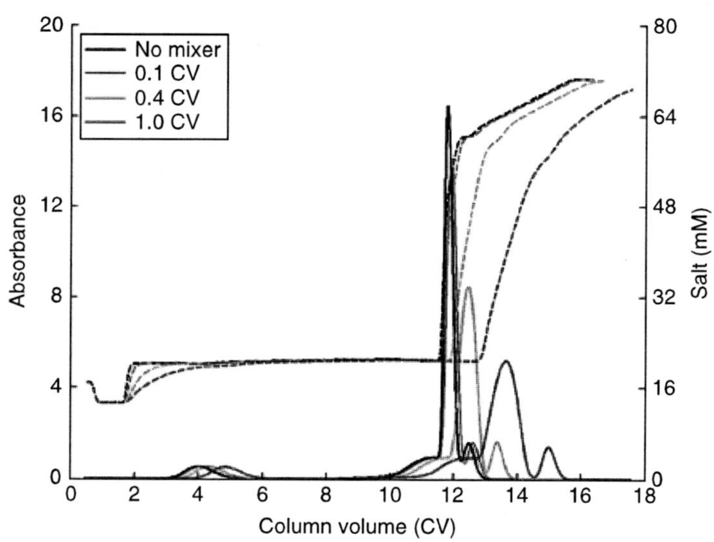

Figure 12.6 Resulting salt concentrations and chromatograms for systems with differences in the mixing volume (0.1, 0.4, and 1 CV). (*See insert for color representation of the figure.*)

large step-up in salt around 10 CV in Figure 12.6). This leads to a broader peak and to a better separation of the late eluting impurity.

These simulations were calculated to demonstrate the effect of an increased dead volume upstream to a column during the design of a pilot facility. Based on the results from these calculations, the dead volume was reduced in the pilot facility.

The dead volumes from the column outlet to the pool collection are normally relatively small and would not lead to the same variation in the chromatogram. It is therefore worth noting *where* the dead volumes are located in the process.

Case I Trouble Shooting of Experiment

As mentioned earlier, a first-principle model also gives the possibility to explain the (few) batches that behave unexpectedly. During the development of the process, a run in the pilot facility resulted in a premature elution (see Figure 12.7). Normally, when these oddities are observed, a workshop is held where the problem is presented, and potential reasons for the unexpected behavior are discussed. Among these were the following reasons:

- High salt in application solution (observed)
- High salt in buffer 1 (observed)
- High salt in buffer 2 (checked and the salt concentration turned out to be correct)
- Wrong pH (checked and pH turned out to be correct)

Buffers 1 and 2 are used during the elution phase. The difference between buffers 1 and 2 is the salt concentration; buffer 1 has the lower salt concentration.

In this case the hypotheses were easily tested simply by checking the conductivity and the pH of the solvents. As seen earlier two salt concentrations deviated from what was normally seen. The mathematical model of the process allowed for an *in silico* investigation of the effect of an increased conductivity/salt concentration in the application solution and buffer 1. The simulations indicated that an increased salt concentration in the application solution could not lead to an early elution of the product.

From the observed increased conductivity in buffer 1, the salt concentration in this solvent could be estimated, and this could be used in the simulation of the process. The resulting chromatogram is given in Figure 12.8. A very good agreement between the simulated and the observed chromatogram is seen. Based on this it was decided to discard buffer 1, and the campaign was continued. The following run led to the expected chromatogram, as seen in Figure 12.4, which confirmed that early elution was caused by an increased conductivity in buffer 1.

The strength of the model is put to the test: does model predict to the observed behavior if the ionic strength in the elution gradient is corrected? Figures 12.7 and 12.8 show that the model does an excellent job in describing the process at increased buffer conditions. But then again, we would expect the model to do well because its parameters have been determined in a broader interval than just at the reference run conditions. Furthermore, the model is based on sound physics, PDEs, and mass balances, which must be fulfilled even at a slightly increased ionic strength.

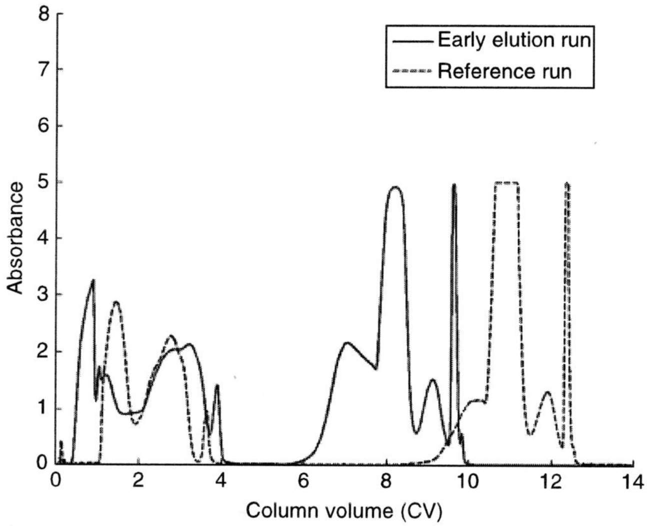

Figure 12.7 Overlay of experimental reference chromatogram, dashed line, and deviation chromatogram where the main peak is eluting early, full line.

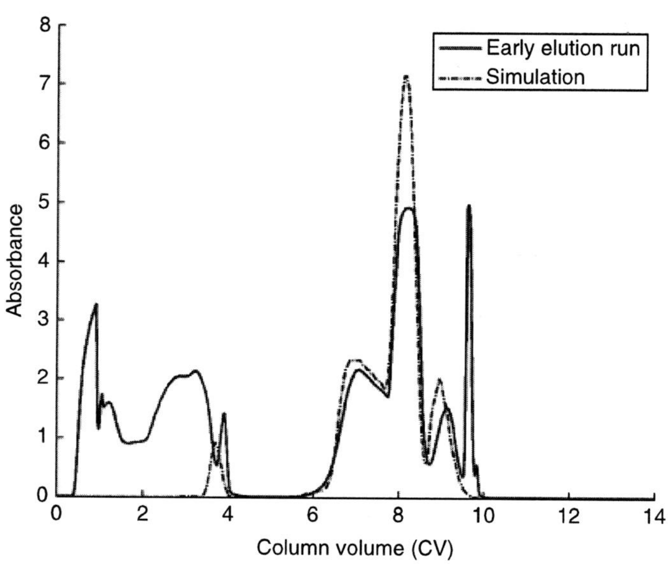

Figure 12.8 Overlay of experimental early elution run and *in silico* simulation of process with higher salt concentration in buffer 1. Only the four components named in Table 12.2 are simulated as described earlier.

Case II Salt Pulse in Wash

During the scale-up of the process, the purification step was moved from a pilot facility to a larger facility. The transfer of a process, from one facility to another, can result in some initial problems due to differences among the facilities.

The first run in the production plant gave the chromatogram shown in Figure 12.9 that deviates significantly from the expected chromatogram from the laboratory-scale experiments, Figure 12.4.

A workshop was held, and a number of possible reasons for this nonconformity were given:

- Bad column packing
- Unexpected composition of the feed
- Wrongly applied gradient

A further investigation of the chromatograms from the new production facility revealed a high conductivity in the wash phase, Figure 12.9, significantly higher than normally observed in laboratory scale.

It was identified that this high conductivity originated from a small volume of a regeneration solvent used in the end of each run. A small part of a regeneration solvent was standing in the piping upstream the column and eluted in the next run after the application phase. Even though the volume was very small, the salt concentration in this regeneration solvent was significantly higher than the salt concentration in the elution buffers. The salt pulse will elute at a volume corresponding to the total porosity in the column plus the dead volume of the system, ~1 CV in Figure 12.9. At the inlet of the column, the salt concentration will be very high. The high salt concentration will as described by the SMA isotherm elute all protein bound to the column in the previous loading/application phase, and a part of the protein will elute with the salt peak around 1 CV. Due to zone spreading mechanisms, a part of the protein will "escape" this high salt concentration traveling through the column. Thereby the protein "sees" a lower salt concentration allowing the protein to bind to the column. The result is that a part of the protein will elute with the high salt pulse, whereas the remaining protein bound to the column is spread all over the column and not be bound to the first part of the column as intended (see Figure 12.10). The protein would normally elute after ~10 CV, which corresponds to the volume it would take for the protein to elute from the start of column to the column outlet. With the protein spread across the column, the protein bound at the end of the column will only have a very short distance to elute. With the protein spread across the column, a diffuse/shallow UV-peak from 2 to 12 CV is obtained as seen in Figure 12.9. The remaining part of the protein will elute at the normal retention volume, at the step-up in salt concentration (at 12 CV) as shown in Figure 12.9. Figure 12.9 also reveals that there is the good agreement between the actual run and the simulation.

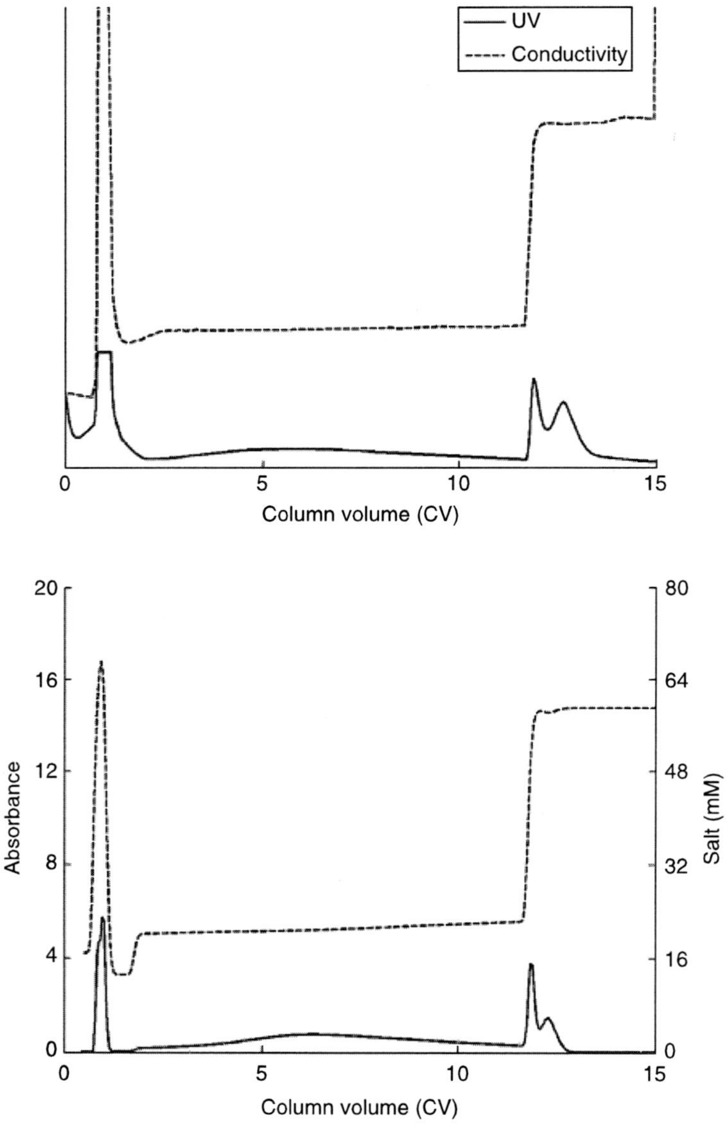

Figure 12.9 *Top*: experimental chromatogram from new pilot facility with pulse of regeneration solvent. *Bottom*: modeled chromatogram of process with high salt concentration in wash phase.

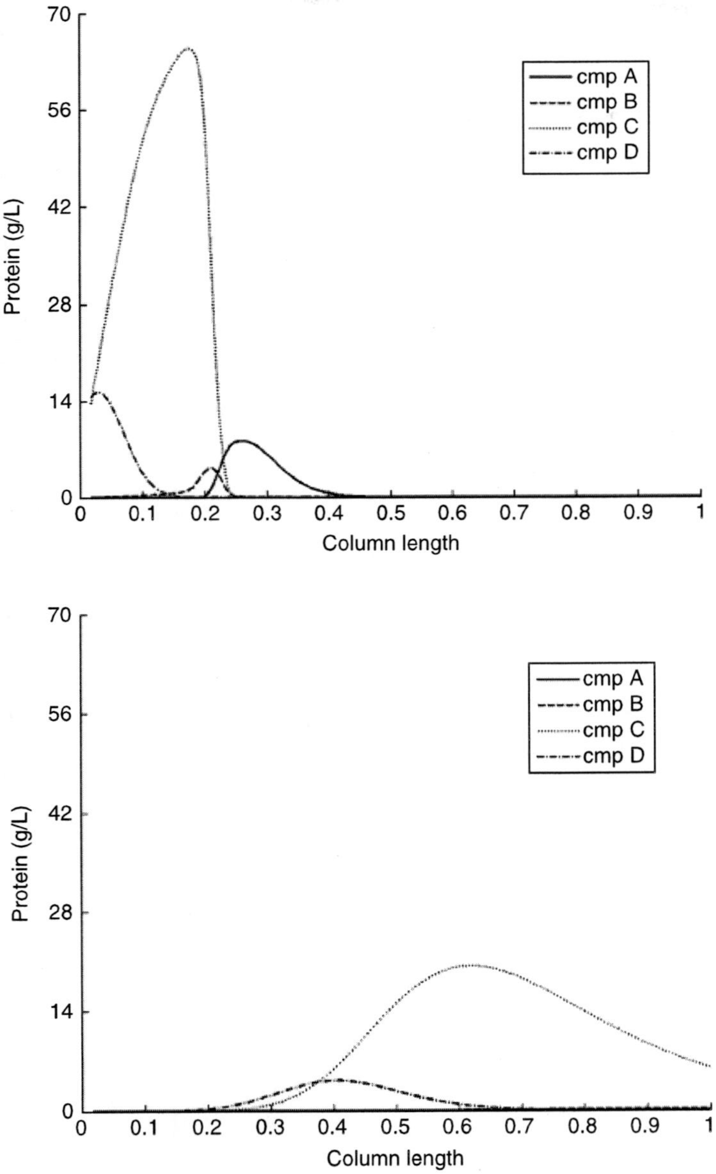

Figure 12.10 *Top*: bound protein concentrations in column after a normal wash phase. *Bottom*: bound protein concentrations after a wash phase containing a small pulse of regeneration solvent.

Case III High Impurity Level

Prior to the purification step (which we have described in details in this chapter), there is a reactor in which a number of new impurities are inheritably formed due to the nature of the main reaction. However, these levels can be reduced to

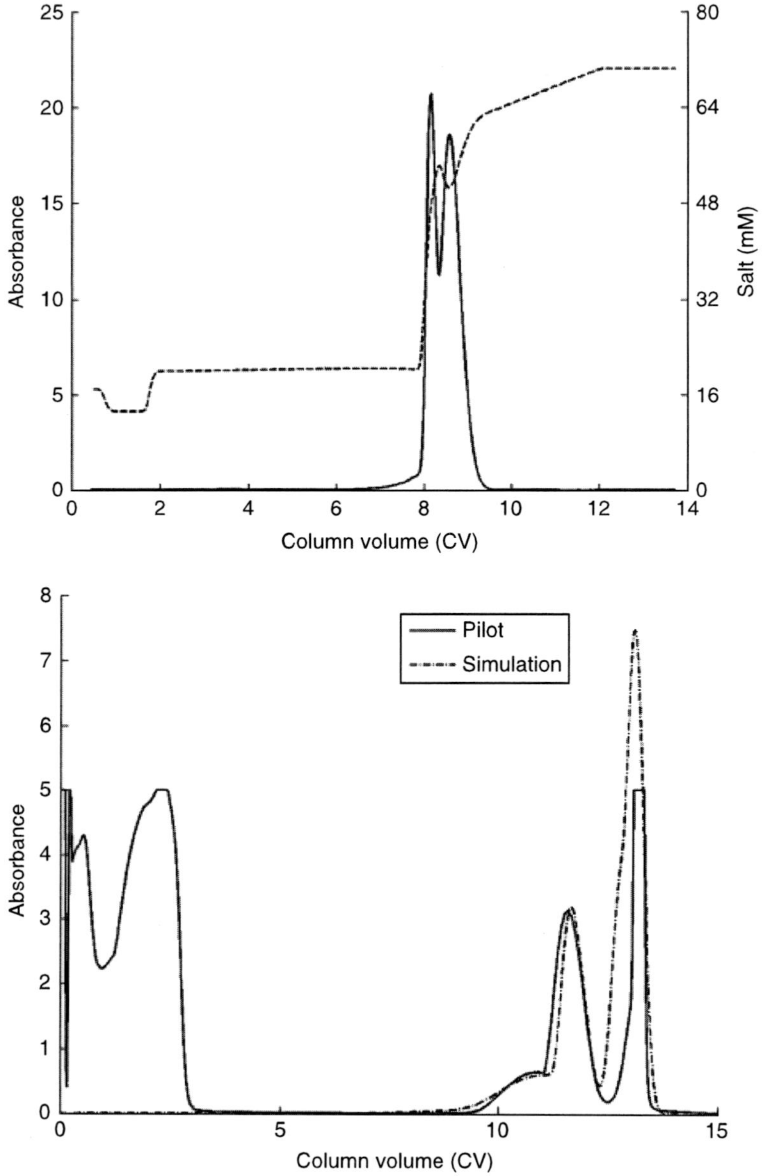

Figure 12.11 *Top*: resulting chromatogram with reference gradient and high-impurity batch. *Bottom*: resulting chromatogram with modified gradient.

a minimum by controlling the reactor settings; this is possible because all the reactions (concerning the product and by-products) are well known.

Because the mathematical model of the purification step exists, it is possible to design a modified set of elution gradients to handle unusual levels of impurities. In the present case, the composition of the reaction mixture was 1% A, 1% B, 39% C, and 59% D—instead of the set point composition 8% A, 2% B, 81% C, and 9% D. The chromatogram of the purification step at reference conditions is shown in Figure 12.5. It is noted that the reference program provides a good resolution of the product, cmp C.

When compared with the chromatogram of the high-level impurity batch, Figure 12.11 top, it is clear that the reference set point of the elution gradient is inadequate for handling this particular reaction mixture; the resolution of the main peak is unacceptably poor. Therefore a tailor-made elution gradient was used for this batch based on process simulations. The new elution gradient was applied for the purification of the batch, and the resulting experimental chromatograms and simulated chromatograms are compared in Figure 12.11 bottom; it is noted how incredibly well the predicted chromatogram concurs with the experimental purification of the batch. The performance of the purification step is given in Table 12.3.

Table 12.3 Normal composition for the loading solution in the reference process and the modified process.

Component	Inlet composition (%)		ΔModified—reference process
(—)	Reference	Modified	Outlet composition (%) or yield (%)
A	8	1	0.0
B	2	1	−0.2
C	81	39	0.6
D	9	59	0.0
Yield	—	—	−7.0

The last column shows the difference in purity and yield between the modified process and the reference process.

Abbreviations

c (mol/L)	mobile phase concentration
c_s (mol/L)	salt concentration
CV	number of column volumes, dimensionless
i, j	component index
K	equilibrium constant

K_d	steric exclusion factor
L (m)	column length
Pe	Péclet number, dimensionless
q (mol/L pore volume)	solid-phase concentration
Q (m^3/s)	volumetric flow
t (s)	time
v_s (m/s)	superficial velocity
V (CV)	dimensionless volume
V_{NR} (CV)	retention volume, non-retained
V_R (CV)	retention volume in column volumes
x	dimensionless axial column position
z (m)	axial column position
ε	interstitial porosity
ε_p	particle porosity
Λ (mol/L)	capacity or ligand concentration
ν	characteristic charge of the protein
σ	steric factor
τ	dimensionless time
PDE	partial differential equation

References

1. Jones J.R., D.L.G. Rowlands, and C.B. Monk, Diffusion coefficient of water in water and in some alkaline earth chloride solutions at 25°C. *Transactions of the Faraday Society*, 61 (1965) 1389–1398.
2. Brooks C.A. and S.M. Cramer, Steric mass-action ion exchange: displacement profiles and induced salt gradients. *AIChE Journal*, 38 (1992) 1969–1978.
3. Kidal S., T.B. Hansen, and A. Staby, "Methods of optimizing chromatographic separation of polypeptides" WO/2008/028974, 2008.
4. Pedersen L., J. Mollerup, E. Hansen, and A. Jungbauer, Whey proteins as a model system for chromatographic separation of proteins. *Journal of Chromatography B*, 790 (2003) 161–173.
5. Campbell M.K., *Biochemistry*, 2nd Edition, Saunders College Publishing, Wiley, New York, 1995.

13

Modeling Preparative Cation Exchange Chromatography of Monoclonal Antibodies

Stephen Hunt, Trent Larsen, and Robert J. Todd

KBI Biopharma, Boulder, CO, USA

13.1 Introduction

Mechanistic modeling has been effectively applied to process or product development in many industries including aerospace, automotive, and petrochemical. In these cases, the application of mechanistic models has significantly improved the cost and timing of process or product development and the quality of the products being developed. Biotechnology has been slow to adopt mechanistic modeling approaches for a number of reasons. The unit operations are complex, and mathematical descriptions of the processes are often not available (e.g., cell culture metabolism). Furthermore, the product and impurity profiles are complex, and there can be insufficient analytical resolution to perform effective modeling (e.g., host cell proteins (HCPs)).

Recent advances in computational power, scientific understanding, and analytical capabilities are enabling more mechanistic modeling in biotechnology [1–4]. One area where mechanistic modeling has been successfully demonstrated is protein process chromatography [5–8]. This chapter describes the development and application of a mechanistic model for industrially relevant monoclonal antibody–cation exchange (CEX) chromatography processes.

Monoclonal antibodies have become the primary modality for biotherapeutic products, and chromatography remains the primary separation method used for their purification. Typical monoclonal antibody purification processes utilize a protein-A chromatography capture step followed by two polishing chromatography steps [9]. CEX chromatography is being used extensively in monoclonal antibody purification as one of the polishing steps and has been shown to be effective at removing high molecular weight species and providing some removal of HCPs and DNAs [10].

For the processes we will be modeling, the primary function of the CEX step is the clearance of dimer from the monomeric product species. Therefore, prediction of dimer clearance will be the focus of the modeling effort. The CEX chromatography process is operated in bind and elute mode with the following process steps: (i) equilibrate column with solution at low salt concentration, (ii) load column to a target product concentration, (iii) wash with equilibration solution, and (iv) elute product with a linearly increasing salt gradient. Figure 13.1 is a typical CEX chromatogram showing the product eluting as a single peak with dimer contained in the trailing edge of the peak. Product is typically collected from a fixed absorbance on the front of the peak (start collect) to a percent of the peak maximum on the back of the peak (stop collect).

The mechanistic modeling approach we chose utilizes the general rate model to describe transport behavior in the column and beads and the steric mass action (SMA) binding model to describe protein binding to the resin matrix. This approach has been used successfully to describe the primary mechanisms involved in CEX chromatography of proteins [5, 11]. An open source chromatography solver was used to estimate model parameters and evaluate the impact of operating parameters on process performance [12]. Model parameters were estimated by performing a set of specific model calibration experiments. Pulse injection experiments were used to estimate the general rate model transport

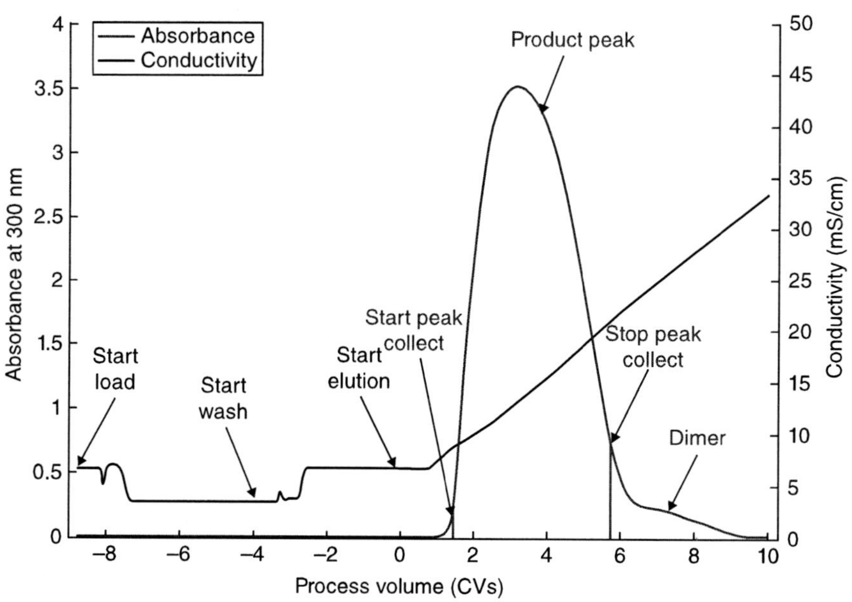

Figure 13.1 Process chromatogram for a typical monoclonal antibody–cation exchange step.

parameters, while SMA binding parameters were estimated by backfitting the model to a set of fractionated gradient elution runs [13].

13.2 Theory

Downstream purification of monoclonal antibodies involves large molecular compounds that have complex mass transfer mechanisms. These mechanisms are often slower than mass transfer of small molecules and strongly influence the chromatographic profiles.

There are a number of chromatography models described in the literature [6, 12, 14–17]. The simplest is the ideal model of chromatography. The ideal model assumes infinite column efficiency, no axial dispersion, and constant equilibrium between the liquid and solid phases. The equilibrium–dispersive model builds on the ideal model by including an axial dispersion term to the column differential mass balance. The lumped kinetic model combines the equilibrium–dispersive model with a kinetic equation that describes the sorption of the solutes, relating the solute concentration in the liquid phase to that in the stationary phase as it varies down the column. The general rate model attempts to describe all the fundamental mechanisms occurring in a chromatography process (Figure 13.2). These include the convection of species by flow within the packed bed, axial dispersion due to the tortuous path that liquid takes as it flows through the packed bed, transport of species into and out of the chromatography beads, diffusion of species through the pores of the beads, and binding of the protein species to the chromatographic matrix. The general rate model was used in this case study due to the slow mass transfer kinetics of monoclonal antibodies (relative to small molecules) and its influence on the resulting chromatogram.

13.2.1 General Rate Model

The general rate model for chromatography is composed of several coupled transport and sorption mechanisms at different scales. For a mixture of species $i = 1,\ldots,N_c$, the convective flow through the packed bed is described by

$$\frac{\partial c_i}{\partial t} = -\mu\frac{\partial c_i}{\partial z} + D_{\mathrm{ax}}\frac{\partial^2 c_i}{\partial z^2} - \frac{1-\varepsilon_{\mathrm{c}}}{\varepsilon_{\mathrm{c}}}\frac{3}{r_p}k_{f,i}\left(c_i - c_{\mathrm{p},i}\left(r_{\mathrm{p}}\right)\right) \tag{13.1}$$

where c_i is the mobile-phase molar concentration of solute i in [mol/L], t is time, μ is the mobile-phase interstitial velocity in [m/s], D_{ax} is the axial dispersion coefficient in [m^2/s], z is the axial distance down the column in [m], ε_{c} is the column porosity, r_{p} is the resin bead radius in [m], $k_{f,i}$ is the external mass transfer coefficient in [m/s], and $c_{\mathrm{p},i}(r_{\mathrm{p}})$ is the concentration of the solute at the

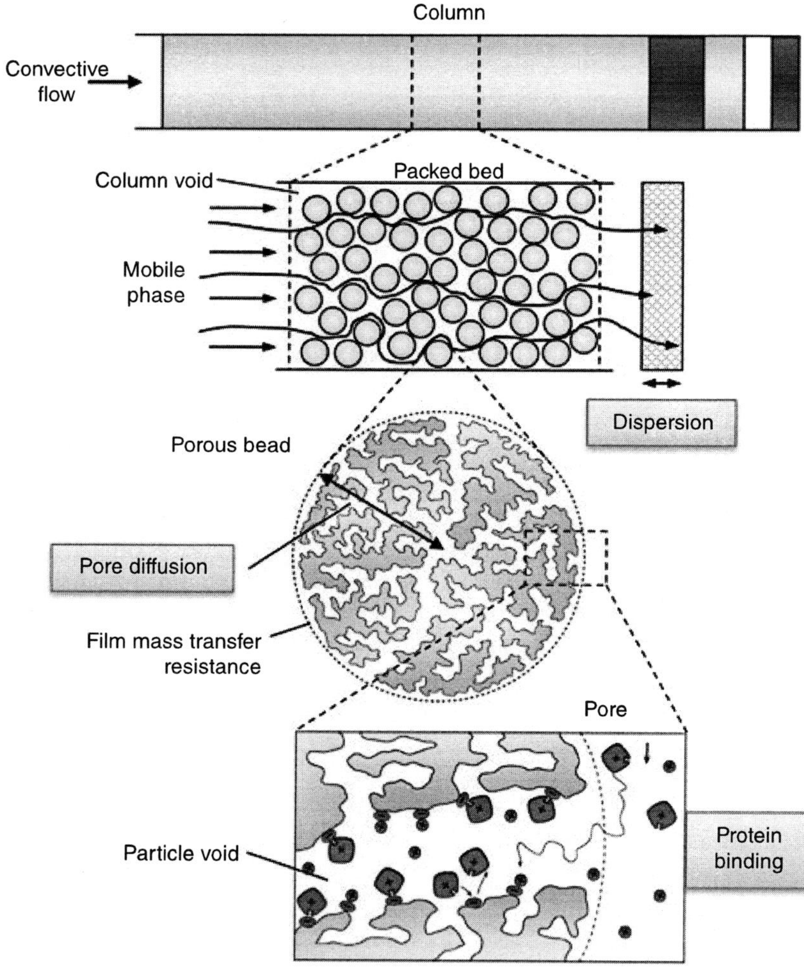

Figure 13.2 Fundamental mechanisms of chromatography.

surface of the resin bead in [mol/L]. The interstitial velocity, μ, is defined as the superficial velocity divided by the column porosity, ε_c. The three terms on the right hand side of Equation 13.1 represent convection, dispersion, and film mass transfer, respectively.

Once a solute reaches the bead surface, mass transfer into the resin bead is governed by

$$\frac{\partial c_{p,i}}{\partial t} + \frac{1-\varepsilon_p}{\varepsilon_p}\frac{\partial q_i}{\partial t} = D_{p,i}\left(\frac{\partial^2 c_{p,i}}{\partial r^2} + \frac{2}{r}\frac{\partial c_{p,i}}{\partial r}\right) \tag{13.2}$$

where $c_{p,i}$ is the liquid-phase concentration of solute i within the resin bead, ε_p is the bead porosity, q_i is the bound concentration of solute i, $D_{p,i}$ is the liquid-phase diffusion coefficient in the pores of the resin bead, and r is the average resin bead radius.

13.2.2 Steric Mass Action Binding Isotherm

The SMA isotherm describes the equilibrium binding behavior of proteins as a function of salt concentration and small ion capacity of the resin, Λ, as shown in Equation 13.3,

$$q_i = k_{eq} c_{p,i} \left(\frac{\Lambda - \sum_{j=1}^{n} (\nu_j + \sigma_j) q_j}{c_{p,salt}} \right)^{\nu_i} \tag{13.3}$$

where q_i is the amount of component i bound to the resin, k_{eq} is the equilibrium binding constant, $c_{p,i}$ is the bead pore liquid-phase component concentration, ν_i is the characteristic charge, σ_i is the steric factor, and $c_{p,salt}$ is the bead pore salt concentration [18].

The SMA model in Equation 13.3 neglects the effect of pH on adsorption. To overcome this limitation, k_{eq} and ν were modified to be functions of pH as follows:

$$k_{eq}(\mathrm{pH}) = k_0 10^{k_e \cdot \mathrm{pH} + k_{ee} \cdot \mathrm{pH}^2} \tag{13.4}$$

$$\nu(\mathrm{pH}) = \nu_0 * (\nu_p \cdot \mathrm{pH} + 1) \tag{13.5}$$

where k_0, k_e, and k_{ee} are regression coefficients that result from fitting k_{eq} versus pH, and ν_0 and ν_p are regression coefficients that result from fitting ν versus pH.

13.3 Model Development

13.3.1 Model Solution

The solution of the general rate model had been discussed extensively in the scientific literature [12, 19, 20]. Finite difference schemes are commonly used when convection in the column dominates and the contribution from axial dispersion is very small, as is the case with ideal nonlinear chromatography. When the contribution from axial dispersion is considered significant, it is often preferable to use segment approximations like finite element methods to obtain smooth continuous solutions for the convective–diffusion equations. More recently, finite volume methods have been implemented due to their mass conservation property and uncomplicated implementation, particularly at the boundaries.

In this case study, the Chromatography Analysis and Design Toolkit (CADET) was utilized to solve the general rate model as described in Equations 13.1–13.5 for CEX chromatography. The CADET source code is freely available and can be downloaded at github.com [12]. CADET implements several state-of-the-art scientific computing techniques combined for maximal solver performance: (i) the model equations are spatially discretized with finite volumes and the weighted essentially non-oscillatory (WENO) method, (ii) a solver with variable step width and order is applied for time integration, and (iii) the internal linear solver module is replaced by customized code that is based on domain decomposition and can be executed on parallel computers.

Discretization of the model within CADET consists of choosing an appropriate number of points on which to numerically solve the general rate model. The number of discretization points is influenced by both operating conditions and the system under study. Typically 50 points in the Z-direction (direction of flow) and 5 points in the radial direction (inside the porous resin bead) were sufficient to provide grid-independent results for a typical monoclonal antibody CEX chromatography model.

The solution provided by CADET includes the liquid-phase and bound-phase solute concentrations at each discretization point within the column as a function of time. This data was used to construct a process chromatogram showing the elution behavior of each component in the model as a function of process time or volume, which in turn was used to calculate relevant process performance parameters: pool concentration, step yield, and pool purity.

13.3.2 Determination of Transport Parameters

Table 13.1 provides typical transport parameter values obtained for Fractogel SO_3 (M) resin based on experience evaluating four different monoclonal antibodies.

Fractogel SO_3 (M) resin has a range of particle sizes from 40–90 μm based on the manufacturers' literature. An average bead size of 65 μm was utilized for all modeling.

Column porosities were estimated from the elution volume of a blue dextran pulse injection, V_{elution}, which is excluded from the pores of the resin beads, using Equation 13.6

$$\varepsilon_{\text{c}} = \frac{\left(V_{\text{elution,BD}} - V_{\text{hold-up}}\right)}{V_{\text{column}}} \tag{13.6}$$

where $V_{\text{hold-up}}$ represents the volume of all system components external to the column. For compressible resins, such as Fractogel, column porosity is a function of how much the resin is compressed during packing. The compression factor,

Table 13.1 Typical physical and mass transport parameters for cation exchange chromatography model.

Transport parameter	Typical values	Units
Resin bead diameter	65	μm
Column porosity (ε_c)	0.30–0.35	
Resin bead porosity (ε_p)	0.35–0.39 (monomer)	
Axial dispersion coefficient	$6–30 \times 10^{-8}$	m^2/s
Film mass transfer coefficients	2.1×10^{-5} (monomer)	m/s
	1.8×10^{-5} (dimer)	
Pore diffusion coefficients	$8–10 \times 10^{-12}$ (monomer)	m^2/s
	3×10^{-12} (dimer)	

CF, is defined as the volume of uncompressed resin used to pack the column divided by the volume of the compressed column bed, V_{column}. For commercially relevant compression factors (<1.33), column porosity decreases linearly with increasing compression factor [21]. The model adjusts the column porosity based on the known compression factor using Equation 13.7, where $\varepsilon_{c,0}$ is the column porosity for uncompressed resin (CF = 1).

$$\varepsilon_c = 1 - \left(1 - \varepsilon_{c,0}\right)\mathrm{CF} \tag{13.7}$$

An $\varepsilon_{c,0}$ of 0.45 was used for Fractogel (M) resin, which is consistent with the published data [21].

Resin bead porosities were determined from the elution volume of monoclonal antibody monomer under nonbinding conditions using Equation 13.8:

$$\varepsilon_p = \frac{\dfrac{\left(V_{elution,mAb} - V_{hold-up}\right)}{V_{column}} - \varepsilon_c}{1 - \varepsilon_c} \tag{13.8}$$

Although accessibility to the pores is dependent on solute size, the general rate model considers only a single bead porosity. We chose to use the monomeric monoclonal antibody bead porosity for our modeling since it represents the primary component we were modeling. The main impact of this assumption is that the salt is transported through the column more quickly in the model than is observed experimentally due to the higher bead porosity of salt.

Axial dispersion coefficients were determined from the dispersion of a salt (NaCl) pulse injection by fitting the general rate model or using Equation 13.9, where u_o is the superficial velocity, L is the column length, and s^2 is the second central moment or variance of the peak [22]:

$$D_{ax} = \frac{u_0 L\, s^2}{2\left(V_{\text{elution,NaCl}} - V_{\text{hold-up}}\right)^2} \qquad (13.9)$$

This approach assumes that pore diffusion is fast and that axial dispersion is the dominant mechanism driving peak broadening, a reasonable assumption for salt at commercially relevant flow rates (100–150 cm/h). Axial dispersion results from a combination of molecular diffusion in the interparticle pores (between the beads) and eddy diffusion. For the flow rates used in our studies, molecular diffusion of proteins is slow relative to eddy diffusion, and therefore axial dispersion is caused primarily by eddy diffusion. Given that eddy diffusion is independent of solute size and a weak function of flow rate [23], the axial dispersion coefficient was held constant for each column modeled.

Film mass transfer coefficients were estimated using the Wilson and Geankoplis correlation [24] calculated at 150 cm/h. Although the film mass transfer coefficient is a weak function of flow rate, film mass transfer coefficients were held constant to simplify the model. Because mass transfer into the beads is fast relative to pore diffusion for proteins, the impact of this simplification is minimal.

Once all other parameters were available, pore diffusion coefficients for product monomer were determined by fitting the general rate model to the product monomer pulse injection elution peak under nonbinding conditions. Pore diffusion coefficients for the four commercial mAbs studied were very similar. This is expected given that mAbs are very similar in size. Although it has been reported that the pore diffusion rate in Fractogel (M) resin can be impacted by changes in the pore structure and by the amount of product bound in the pores [25], these effects are not considered in the models that were developed.

The pore diffusion coefficient for dimer was estimated based on the width of the dimer elution peak from a set of gradient fractionation runs, given a known set of dimer SMA parameters that had been determined independently through pulse gradient experiments.

To confirm the validity of the mass transport model that was used, the accuracy of model predictions was evaluated. Figure 13.3 compares model predictions to experimental results for mAb pulse injections under nonbinding conditions over a range of flow rates. Axial dispersion and pore diffusion coefficients were obtained at a single flow rate of 150 cm/h for this study. The model accurately predicted the peak shape over a range of flow rates from 50 to 200 cm/h. When the flow rate is reduced to 25 cm/h, the experimental peak was sharper than the model predictions, which is likely due to a reduction in axial dispersion at lower

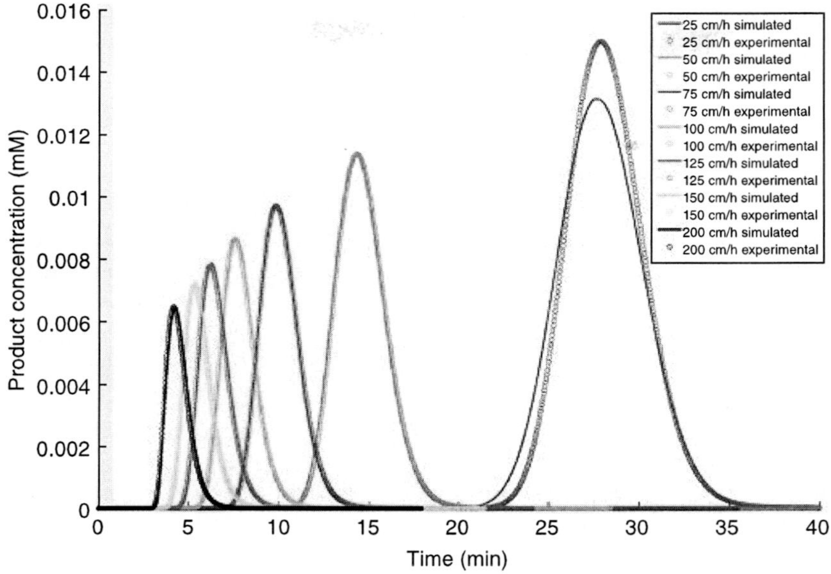

Figure 13.3 Prediction of nonbinding monoclonal antibody transport behavior as a function of flow rate.

flow rates. These results confirmed that our methodology for determining the physical and transport parameters for monomeric mAb was sufficiently robust and that our model simplifications did not have a significant impact on the quality of the model.

13.3.3 Determination of SMA Parameters

The primary function of the CEX step in the monoclonal antibody process is the removal of high molecular weight species. As such, the modeling efforts focused on product-related species (i.e., monomeric variants and dimer). Because the CEX step follows protein-A chromatography in the purification process, host-related impurities (HCPs and DNAs) are at low enough levels that they do not have a significant impact on binding capacity or step performance and were therefore not modeled. Furthermore, the level of low molecular weight (partial antibodies) and higher molecular weight species (larger than dimer) are typically negligible in the load and were therefore not modeled.

There are numerous methods reported for the determination of protein SMA parameters. Gadam et al. and Gallant et al. used a combination of isocratic or pulse gradient elution experiments and breakthrough experiments [26, 27]. Chilamkurthi et al. has more recently used high-throughput techniques to measure binding isotherms directly via batch adsorption or small column breakthrough experiments [28]. Osberghaus et al. described the use of backfitting of experimental data under typical processing conditions to the full model [13].

We attempted to use high-throughput batch binding experiments to directly determine binding isotherm parameters. Unfortunately, we were not able to get sufficiently reproducible data from our high-throughput platform. More traditional methods using pulse gradient, breakthrough experiments, and model backfitting approaches were then evaluated to identify the most efficient method for obtaining SMA parameters. Backfitting the model to experimental data under typical process conditions proved to be the most effective method as it allowed SMA parameters for all species to be estimated from the same experiment and avoided the need to purify individual species. To minimize experimental and analytical burden, a single fractionation run was typically used to estimate SMA parameters. In some cases, supplemental breakthrough or pulse gradient experiments were required to gain sufficient confidence in the parameter estimates.

Figure 13.4 shows a typical CEX chromatogram that has been fractionated (sampled over time) and the fractions analyzed by cation exchange HPLC (CEX-HPLC) and size-exclusion HPLC (SE-HPLC). CEX-HPLC was used to determine the levels of charge variants, which are reported as acidic, neutral, and basic species. SE-HPLC was used to determine the levels of size variants, which are reported as low molecular weight, monomeric, and high molecular

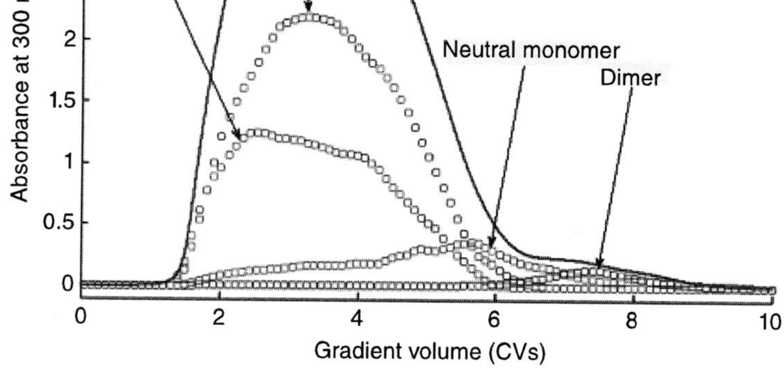

Figure 13.4 Fractionation of typical cation exchange chromatogram showing product-related species.

weight species (primarily dimer). As one might expect, some level of separation between monomeric charge variants was observed on the process column. The acidic and neutral monomer occasionally co-elute and can be modeled as a single species, while the basic monomer was fairly well resolved and has an impact on the back side peak shape, so it was modeled as a separate species. The dimer elutes last and was modeled as a single species. The CEX model for monoclonal antibody purification therefore has four relevant product species: acidic monomer, neutral monomer, basic monomer, and dimer.

Although it is understood that detailed product characterization can identify a multitude of subspecies within each of these categories, it was only possible to calibrate the model to the level of detail available from the analytical data. This is also the same level of detail used to monitor and control process performance.

Fitting of the SMA model parameters to the fractionation data for each species was performed both manually and using various fitting algorithms. Manual fitting was successful in many cases, although there is a fair level of subjectivity to this approach. Because the parameter space is large, we simplified the fitting problem by making some reasonable assumptions based on the mechanistic description of the SMA model. It was typically assumed that the characteristic charge and the steric factor was the same for all of the monomeric species. It was also assumed that the characteristic charge for the dimer was larger than the monomer due to its size and strength of interaction. The level of dimer was typically low enough (<5%) that the fit was independent of the steric factor, effectively behaving as a linear binding isotherm.

Figure 13.5 shows a typical model fit to a set of fractionation data that was performed manually. The SMA model does a good job of fitting the peak shape, although the model peak tended to be sharper on the front of the peak than the experimental data, especially when the product binds tightly to the resin (i.e., high ν and k_{eq} values). Because in many cases, it was not possible to obtain an exact fit to the monomer data, the fitting focused on matching specific aspects of the main peak shape. Although having the model fit the fractionation data for each component was important, it tended to be more important to ensure the overall peak shape was as close as possible to the experimental data. This was especially true for the backside of the peak, which was critical for accurately predicting the stop collection criteria and therefore the resulting product pool dimer levels.

Figure 13.6 shows an enlargement of the fit to the dimer peak. For all monoclonal antibodies studied, the dimer had a low level of co-elution with the monomeric peak. The mechanism for this behavior is not yet fully understood and is not well modeled using the SMA isotherm. Therefore, when manually fitting, focus was placed on matching the dimer peak location, essentially ignoring the constant level of dimer co-eluting during the monomer peak. This ultimately resulted in underprediction of dimer levels in the pool and required minor correction; typically a small offset in the model predicted pool dimer levels.

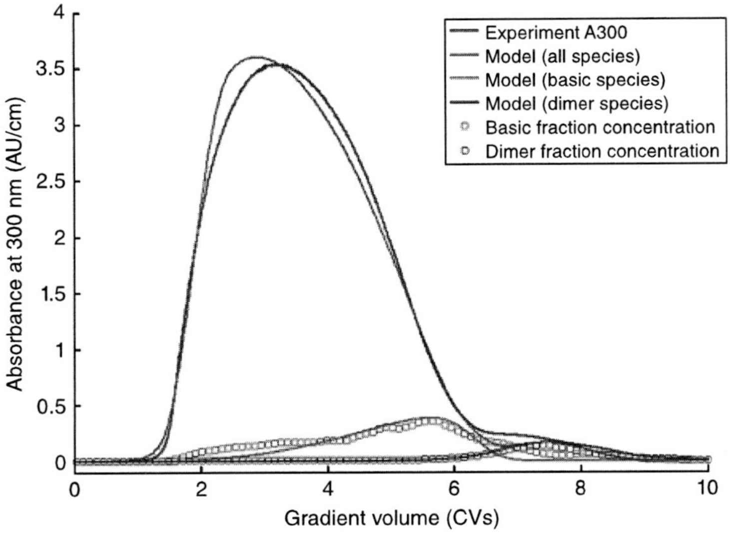

Figure 13.5 Model fit to typical cation exchange fractionation study. (*See insert for color representation of the figure.*)

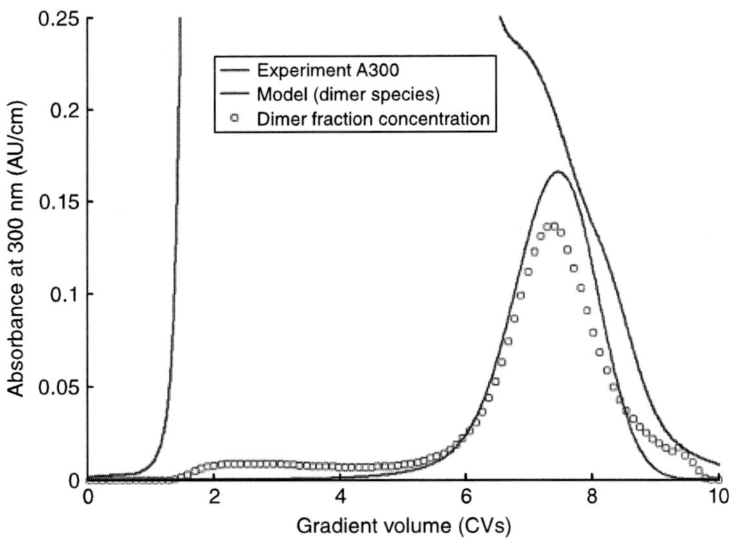

Figure 13.6 Model fit to typical CEX dimer profile.

Table 13.2 Typical monoclonal antibody SMA parameters.

SMA parameter	Typical values	Units
Small ion capacity (Λ)	74–87 (Fractogel SO_3 (M))	mmol/L
	83–94 (Fractogel COO (M))	
Characteristic charge (ν)	5–20 (monomer)	
	12–20 (dimer)	
Steric factor (σ)	20–60 (monomer)	
	80 (dimer)	

Given the complexity of fitting the peak shapes as described earlier, fitting with computer algorithms was less successful. Although reasonably good fits were obtained using methods that minimize the sum of squared errors (SSE), it was challenging to have the computer recognize the appropriate compromises reached with manual fitting. As a result, the parameters obtained by computer fitting often did not provide accurate predictions in the most critical areas of the chromatogram (e.g., back side peak shape).

Table 13.2 provides the typical ranges observed for SMA parameters of monoclonal antibody monomer and dimer species. Small ion capacity values were obtained from the resin manufacturer. Characteristic charge values for monomer species have ranged from 5 to 20, depending on the product and pH studied. Characteristic charge values for the dimer typically range from 12 to 20. Steric factors for monomer species have ranged from 20 to 60, a portion of which represents ligands that are contained in the inaccessible pore volume and therefore unavailable to the monoclonal antibody. The dimer fits were typically insensitive to changes in the steric factor due to the low level of dimer in the load and so the steric factor was fixed at 80. Equilibrium binding constant values were wide ranging depending on the product, pH, and characteristic charge required for successful fitting.

Protein binding in CEX chromatography is significantly impacted by pH. To accommodate this in the modeling, the SMA parameters need to change appropriately as a function of pH. Borg et al. assumed that the steric factor and characteristic charge remained constant over a small range of pH and that k_{eq} changed exponentially with pH [5]. We utilized a similar approach, where the steric factor was held constant as a function of pH, the characteristic charge, ν, either remained constant or decreased linearly with pH (Eq. 13.5), and k_{eq} decreased exponentially as a function of pH (Eq. 13.4).

To obtain the specific pH dependence for the SMA parameters for each species in the model, the SMA parameters obtained from fractionation studies (minimum of three different pHs) were fit to Equations 13.4 and 13.5. The linear correlation

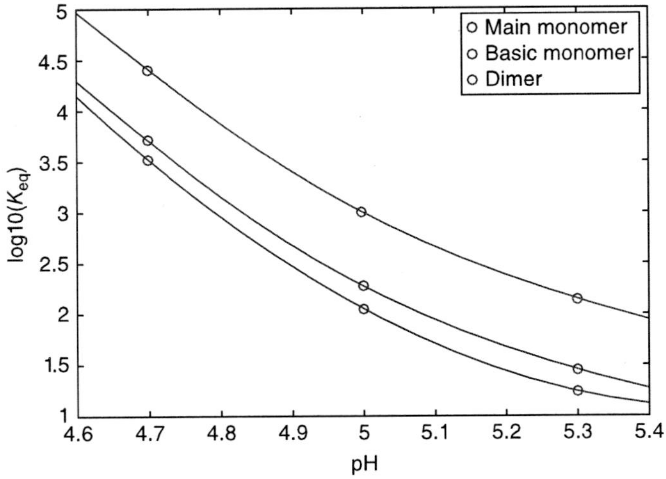

Figure 13.7 Equilibrium binding constants for monoclonal antibody species as a function of pH (dimer: upper curve; main monomer: lower curve).

between the characteristic charge and pH, when required to obtain acceptable peak shapes, was included in the fractionation fitting exercise. An example plot of the log of the binding constants versus pH is provided in Figure 13.7. Some curvature was typically seen in this relationship, which is why curvature was included in the fitting of the data.

13.3.4 Model Qualification

The level of rigor used in qualifying the model was dependent on the development phase in which the model has been used. Early in development, minimal model qualification was typically required, while more rigorous qualification was performed when the model was used to support process characterization.

Early in development, confidence in the accuracy of the model parameters was sufficient to utilize the model. This confidence was typically obtained by confirming parameters using alternate parameter estimation techniques. For example, pulse gradient experiments were used to confirm the characteristic charge and binding constants that were obtained through model backfitting.

Prior to process characterization, a more comprehensive model qualification was performed. Model qualification involved comparing model predictions against experimental data that was independent of the model development data in two studies. The first study utilized data from a small-scale design of experiment (DOE), and the second study used data from commercial-scale production (when available).

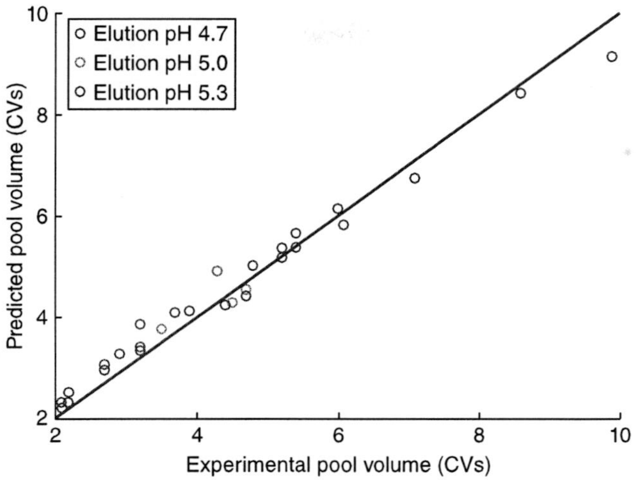

Figure 13.8 Model predicted versus experimental pool volumes. (*See insert for color representation of the figure.*)

In the first study, the ability of the model to accurately predict chromatographic peak shape and performance parameters over a range of operating parameter conditions was evaluated. Figure 13.8 is a qualification example comparing experimental data to model predictions for pool volume on a 28 run DOE study. In this example, the model did an excellent job predicting experimental pool volumes over a wide range of conditions (fivefold variation in pool volume). This type of study provides confidence that the model is responding correctly to changes in operating conditions.

In the second study, model predictions can be compared to data from commercial-scale runs (or largest scale available). Adjustments to the axial dispersion coefficient and column porosity can be made based on commercial-scale HETP or transition analysis data. Adjustments to the resin small ion capacity can be made based on resin manufacturer data. Successful prediction of commercial-scale data, typically only performed at process center point conditions, provides confidence in the predictive capabilities of the model in a commercially relevant setting.

13.4 Model Application

Once the CEX model was developed and the appropriate level of qualification performed, the opportunities for its application to process development were essentially unlimited. Provided in the following text are examples of how a mechanistic model was used to support the development of a commercial monoclonal antibody–CEX process.

13.4.1 Resin Selection and Process Optimization

During process development, two resins were identified that provided good separation of dimer from monomeric species, Fractogel SO_3 (M) and Fractogel COO (M). A mechanistic model was developed using three fractionation runs spanning the expected pH range and a number of breakthrough experiments at the pH center point for each resin. Figure 13.9 shows the model fit for both the Fractogel SO_3 and Fractogel COO resins at pH 5.0. The behavior of the monomeric and dimeric species was reasonably well explained by the model, providing a good representation of the trailing edge of the monomer peak and enabling accurate calculation of stop collection.

The primary goal of the CEX process development effort was to reduce the dimer levels to < 0.4% (target of 0.25%) with dimer levels in the load potentially

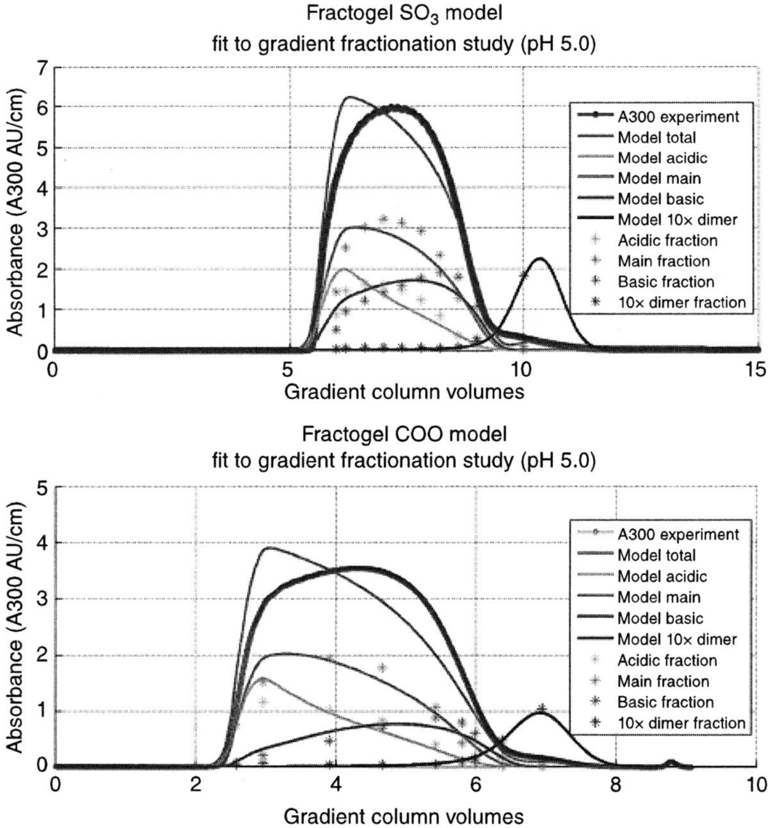

Figure 13.9 Model fit to cation exchange fractionation data at pH 5.0 for Fractogel SO_3 (top) and Fractogel COO resins (bottom). (*See insert for color representation of the figure.*)

ranging from 1 to 8%. Secondary process goals included maximizing process yield, maximizing process loading, and maintaining pool concentrations of at least 15 g/L to ensure the CEX pool would fit into the CEX pool tank in an existing facility. In order to choose the preferred resin, an initial process map was created from which the optimal conditions for each resin were determined. The primary operating parameters that typically impact performance of the CEX step are pH, gradient slope, column loading, and stop collect point.

As a first step, the optimal pH and gradient slope for each resin was determined using the model. As shown in Figure 13.10, the impact of pH and gradient slope on dimer levels and pool concentration was calculated at a loading of 40 g/L and assuming 5% dimer in the load. The best dimer clearance for both resins was

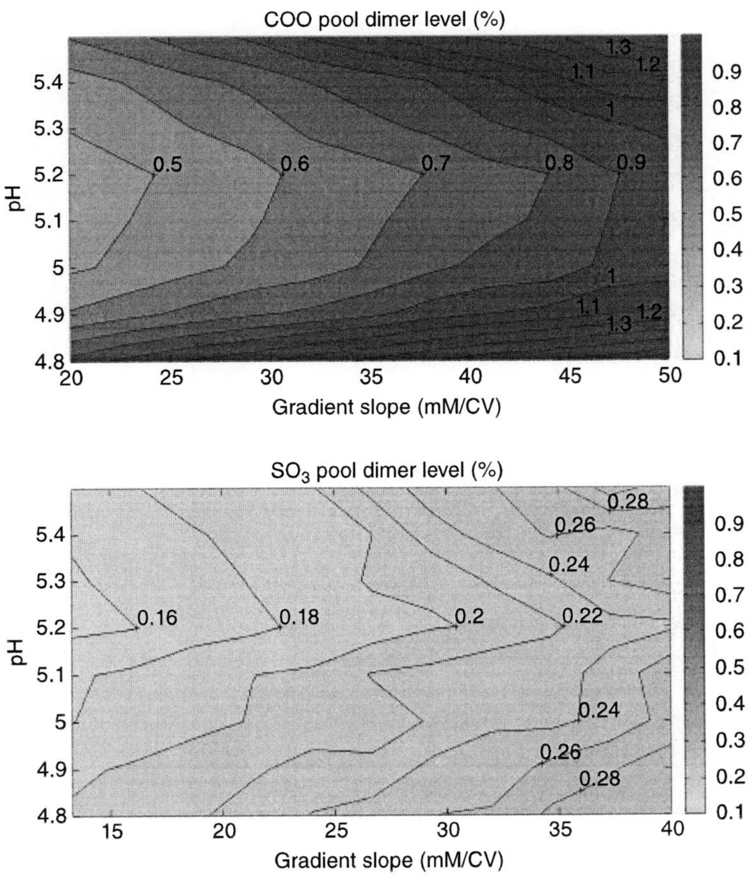

Figure 13.10 Impact of pH and gradient slope on cation exchange pool dimer levels and pool concentration.

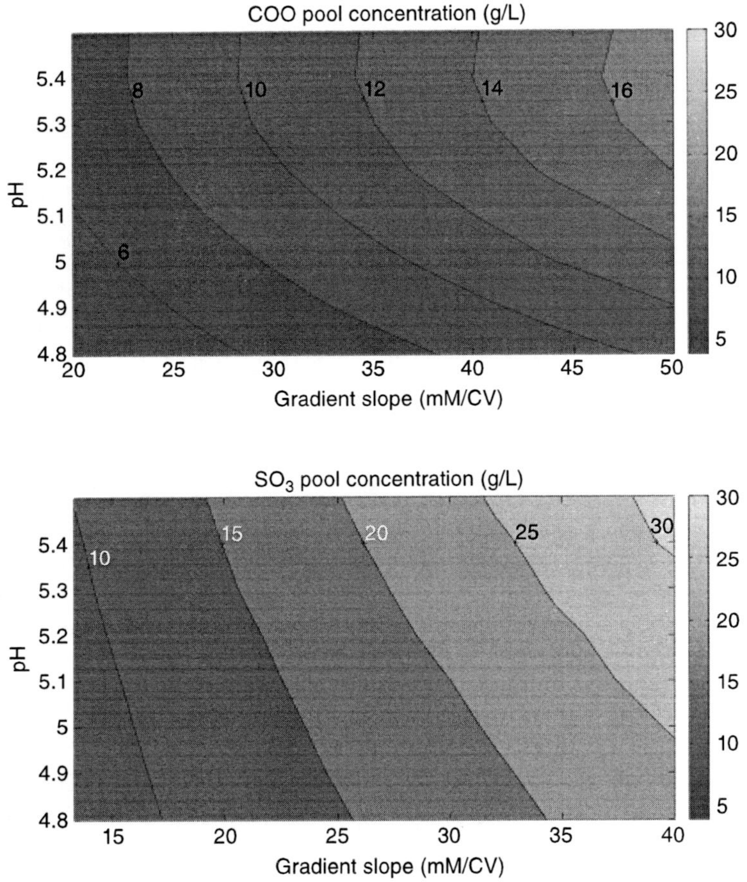

Figure 13.10 (Continued)

obtained at approximately pH 5.2. Step yield was acceptable over the entire range tested (data not shown). Acceptable pool concentration at pH 5.2 was obtained at 45 and 22.5 mS/CV for Fractogel COO and Fractogel SO_3 resins, respectively.

Once the optimal pH and gradient slope for each resin was determined, the model was used to generate selectivity curves (dimer level vs step yield) over a range of load rates with 5% dimer in the load (Figure 13.11). For Fractogel COO resin, step yields were reduced above 45 g/L due to product breakthrough that occurred during loading. Therefore, the maximum acceptable load rate was set at 40 g/L. At this load rate, it was necessary to set the stop collect point at 60% of peak maximum to achieve the target purity of 0.25%. This resulted in a process that met all the targets, but with relatively low loading (40 g/L) and step

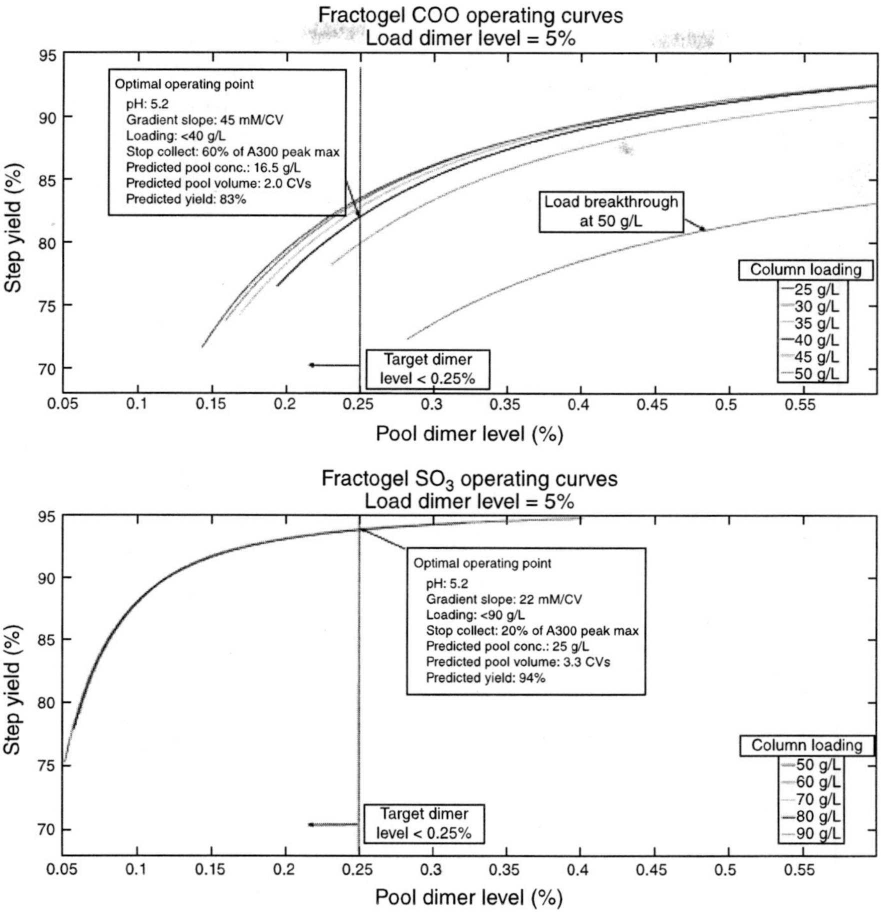

Figure 13.11 Selectivity curves for Fractogel COO (top) and Fractogel SO_3 (bottom).

yield (83%). In contrast, the Fractogel SO_3 resin did not show any product breakthrough up to 90 g/L. At 90 g/L loading, the target purity could be met with a stop collect of 20%, giving high product yield and pool concentrations near 25 g/L.

Based on these modeling results, it is clear that the Fractogel SO_3 resin was the preferred choice, providing the same level of purification at higher column loading and with increased product yield and pool concentration. Although it is likely that a traditional experimental development program would have come to the same conclusion as the mechanistic modeling, the ability to quickly evaluate and visualize the impact of the operating space on the performance for each resin made the decision fast and clear.

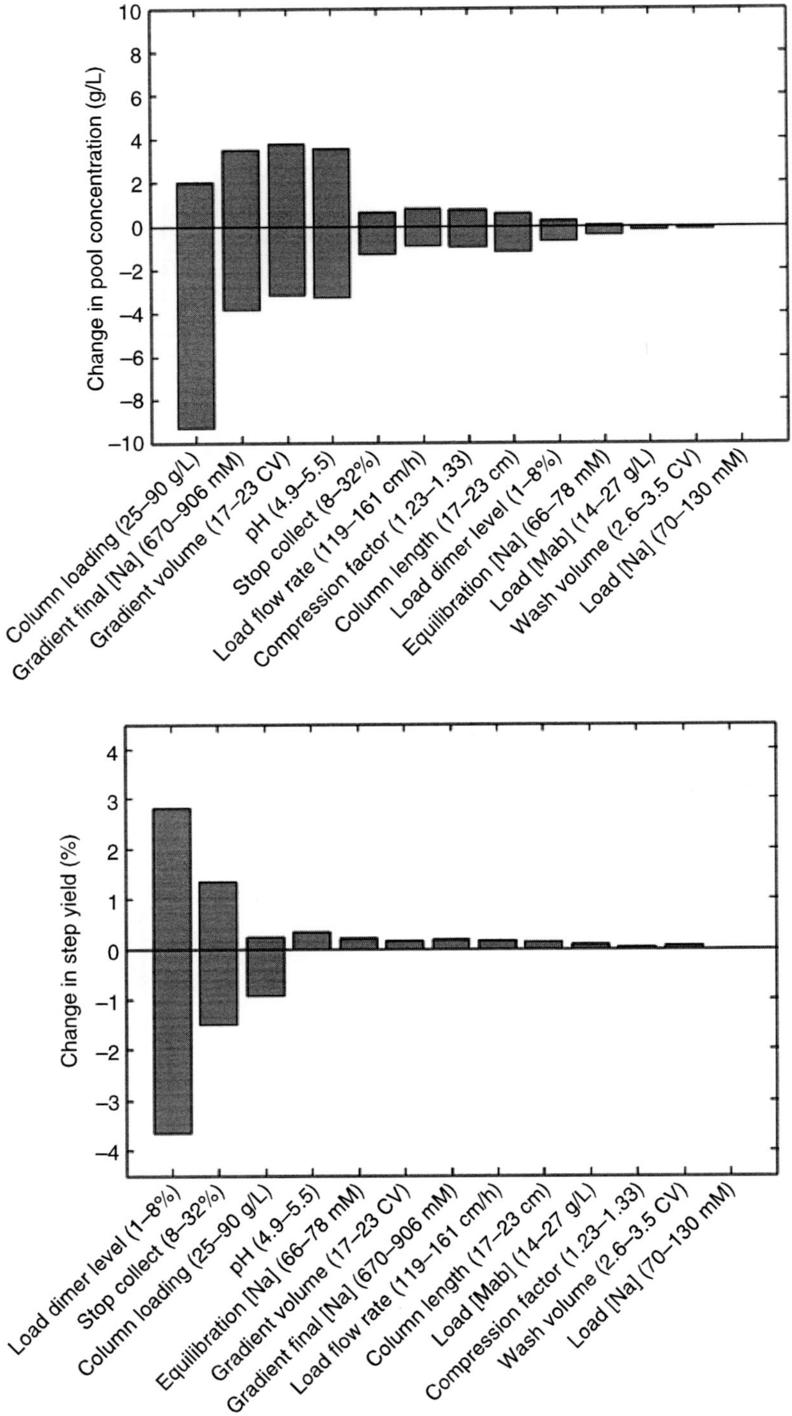

Figure 13.12 Impact of operating parameters on cation exchange pool concentration (top), step yield (middle), and pool dimer levels (bottom).

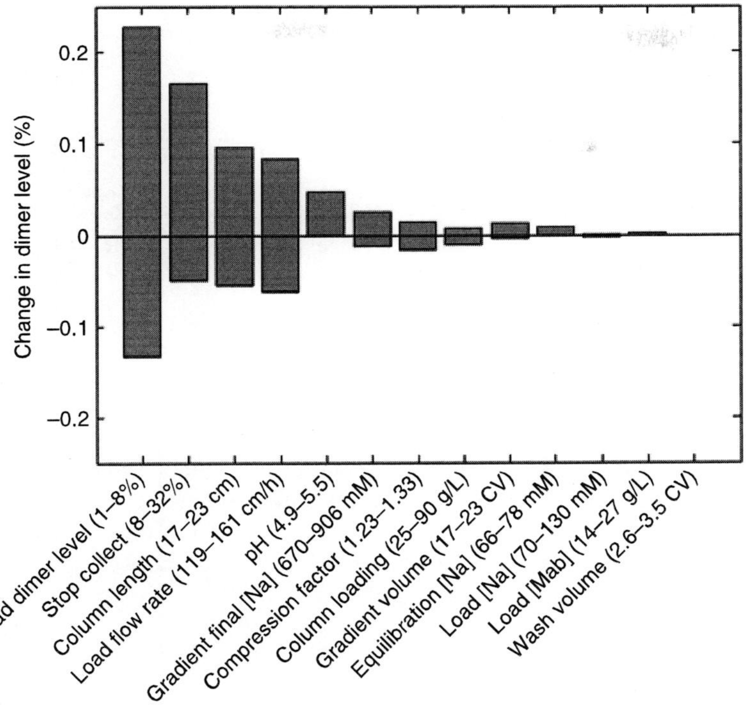

Figure 13.12 (Continued)

13.4.2 Process Robustness and Control Strategy

Once the resin was selected and the operating set points determined, the mechanistic model was used to evaluate process robustness. This evaluation included assessing the impact of varying operating parameters on process performance, independently at first and then in combination.

Operating parameters were first varied independently over a commercially relevant range around their set points and their impact on process performance observed. Results are summarized in Figure 13.12, which illustrated the magnitude of impact associated with varying each operating parameter. The operating parameters with the greatest impact on the pool volume were the column loading, the gradient slope, and the pH. Step yield was primarily impacted by the amount of dimer in the load and the stop collect point. Pool dimer level was primarily impacted by the level of dimer in the load and the stop collect point. Column length and flow rate also had an affect.

For highly impactful operating parameters, the model was used to understand the functionality of the response. As an example, Figure 13.13 shows how

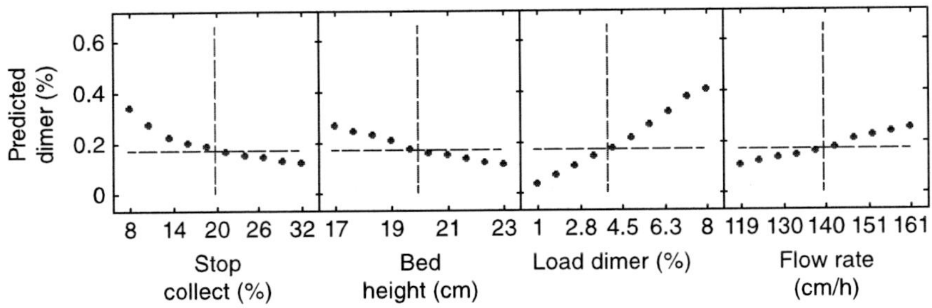

Figure 13.13 Impact of operating parameters on dimer levels.

the pool dimer levels responded to independent changes in load dimer level, stop collect point, column length, and flow rate. Although most parameters impacted dimer level in a fairly linear manner, the stop collect point impacted dimer level nonlinearly with dimer level increasing rapidly near the low end of the stop collect range.

This study enabled rapid understanding of which operating parameters had the largest impact on process performance and the magnitude of the impact and whether the impact was linear or nonlinear in nature.

Monte Carlo approaches were then used to evaluate the effect of operating parameter interactions on process robustness. 10 000 simulations were performed with operating parameters varying randomly within their normal operating ranges. The results are summarized in Figure 13.14, which shows the expected distribution of step yields and dimer levels over the entire range of load dimer levels (1–8%). Also shown is the expected distribution of step yields and dimer levels when the load dimer is within the most probable range based on upstream process development (2–3.5%). Based on these results, the process will consistently deliver dimer levels below the target and high product yields as long as the dimer level in the load is between 2 and 3.5%. If the load dimer level is allowed to span the full range of 1–8%, then approximately 10% of the results would have dimer levels above the process goal (<0.4%).

In an effort to improve control of pool dimer levels over the full range of load dimer levels, the model was used to develop an alternate stop collect strategy. To accomplish this, the impact of stop collect and load dimer levels on CEX pool dimer levels was fully mapped out using the model as shown in Figure 13.15. Rather than having a fixed stop collect point of 20%, which is typically employed, the stop collect point can be adjusted based on the load dimer level to optimize performance of the step. This approach maximizes yield by having a low stop collect point when the load dimer level is low and then increasing the stop collect point along the 0.25% contour line as the load dimer level increases to maintain product purity.

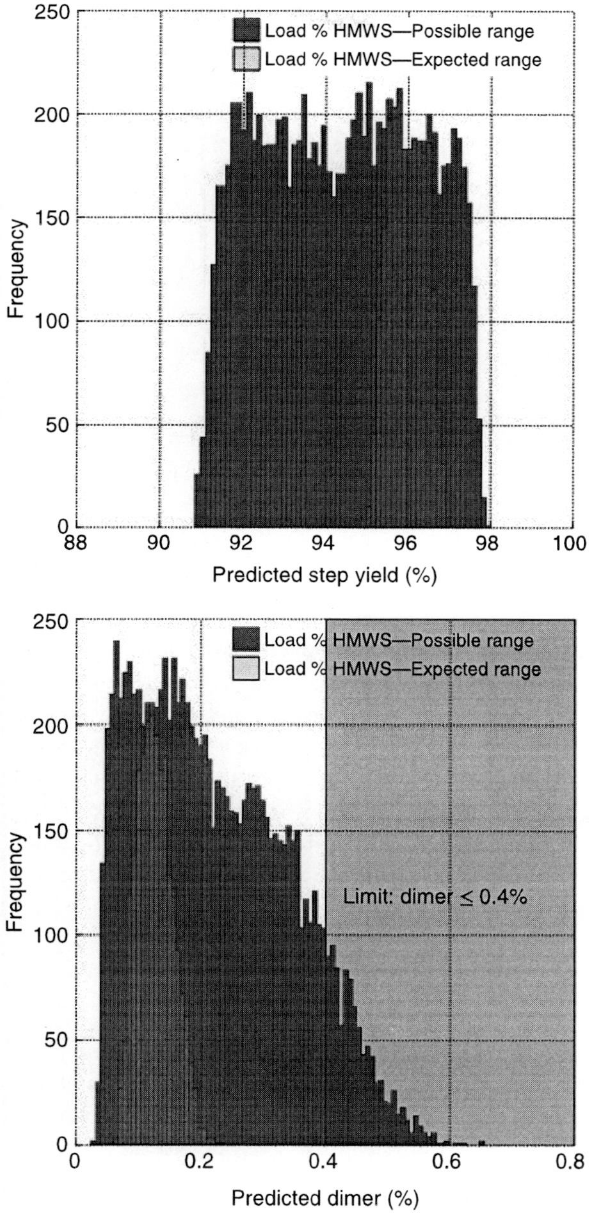

Figure 13.14 Expected distribution of step yield (top) and dimer levels (bottom) by Monte Carlo simulation.

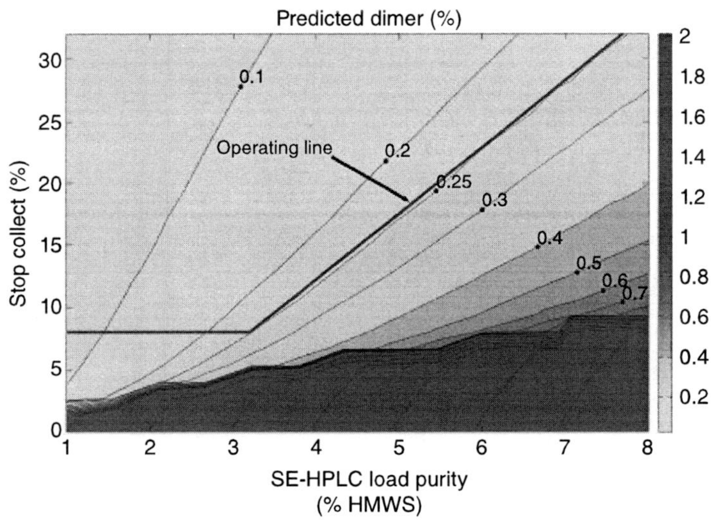

Figure 13.15 Impact of load dimer level and stop collect on pool dimer levels.

The proposed control strategy was evaluated by performing another Monte Carlo study that utilized the adjustable stop collect strategy described earlier. As shown in Figure 13.16, the alternate stop collect strategy provided acceptable dimer clearance over the entire range of load dimer levels (dimer <0.4% for all runs). Furthermore, although the step yield is slightly reduced for those runs with high dimer levels, the yield was improved by approximately 1% on average for the load dimer levels expected for the process (2–3.5%). The mechanistic model, in this example, was a valuable tool for understanding sources of process variability and developing and testing strategies to control process variability.

13.4.3 Raw Material Variability

Another application of the mechanistic model is to understand the impact of raw material variability on process performance. A number of raw material properties are directly correlated to model parameters and can therefore be studied with the mechanistic model. Examples include average bead diameter, average pore size (impacts bead porosity and pore diffusion rate), and ligand density.

Figure 13.17 demonstrates the impact of raw material-related model parameters on dimer clearance. This data suggests that bead pore size, which would correlate positively with bead porosity and pore diffusion rate, plays an important role in dimer clearance. Exact interpretation of this data is

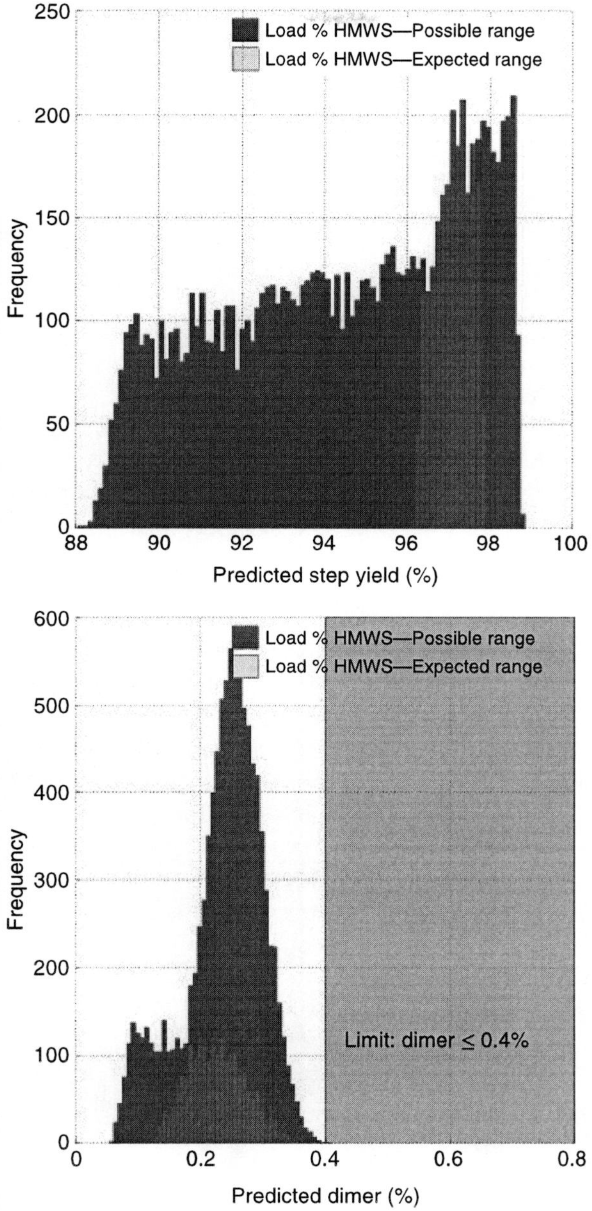

Figure 13.16 Expected distribution of step yield (top) and dimer levels (bottom) with alternative stop collect strategy.

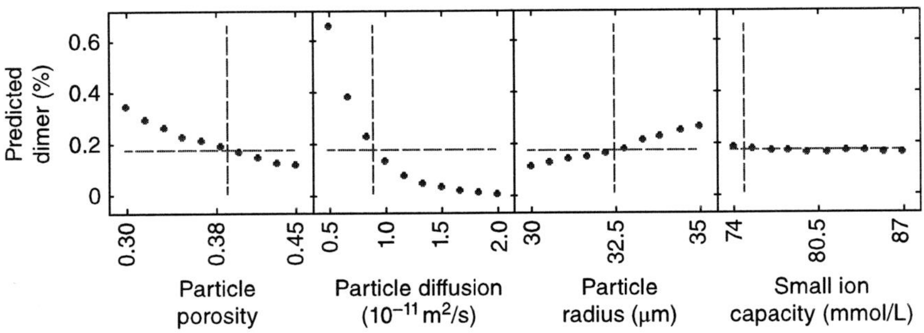

Figure 13.17 Impact of raw material-related parameters on dimer clearance.

difficult because the impact of changing bead porosity and ligand density on the monoclonal antibody SMA parameters has not been well studied. Nonetheless, the potential for understanding process performance as it relates to raw material properties is certainly possible using mechanistic modeling.

13.5 Conclusions

Developing and applying mechanistic models for monoclonal antibody–CEX processes has generated significantly increased process understanding. The models have been used to quickly identify optimal operating conditions during process development and evaluate process robustness and control strategies during process characterization. Although some effort is required to develop a predictive CEX model, it has the potential to significantly increase the speed of process development while at the same time reducing the overall experimentation required to fully understand process performance.

Truly leveraging a mechanistic modeling approach requires changing the way process development is performed. Rather than depending on a causal understanding of processes based on statistical analysis of experiments, fundamental process mechanisms must be identified, investigated, and ultimately quantified. Although this transformation takes time and investment, the benefits are significant, and it can ultimately change the way biotechnology processes are developed and controlled.

As scientific understanding and analytical techniques continue to improve, mechanistic modeling will become more broadly applicable to a range of products and resin types. The use of high-throughput methods for determining binding isotherm parameters should reduce the effort required to develop models and provide the opportunity to develop models earlier in the process development life cycle. Molecular simulation could eventually be used to determine binding parameters, dramatically reducing the experimental effort required to develop models. Software will become available that will enable scientists and engineers to more easily develop and apply mechanistic models, and computer speed will continue to increase, enabling advanced process optimization algorithms to be utilized. Eventually, mechanistic modeling will become a mainstream tool in every chromatography development toolbox.

13.6 Acknowledgments

This material is based upon work performed at Amgen Inc.

Nomenclature

c_i	Concentration of component i in the flowing mobile-phase mol/L
$c_{\mathrm{p},i}$	Liquid-phase pore concentration of solute i within the resin bead mol/L
$c_{\mathrm{p},i}(r_\mathrm{p})$	Liquid-phase concentration of solute i at the surface of the resin bead mol/L
$c_{\mathrm{p,salt}}$	Liquid-phase pore salt concentration within the resin bead mol/L
CF	Column compression factor, dimensionless
D_{ax}	Axial dispersion coefficient, m^2/s
i	Component
k_{eq}	Equilibrium binding constant
$k_{f,i}$	External mass transfer coefficient of component i, m/s
L	Column length, m
q_i	Bound concentration of solute i within the resin bead, mole/L
r	Bead radial dimension, m
r_p	Resin bead radius, m
s^2	Second central moment or variance, m^2
t	Time, s
u_0	Superficial velocity, m/s
V	Volume, m^3
V_{column}	Empty column volume, m^3
V_{elution}	Elution volume, m^3
$V_{\mathrm{hold-up}}$	System holdup volume, m^3
Z	Column axial dimension, m

Greek letters

ε_c	Column porosity, dimensionless
ε_p	Resin bead porosity, dimensionless
Λ	Small ion capacity of resin
μ	Interstitial velocity, m/s
ν_i	SMA model's characteristic charge for component i
σ_i	SMA model's steric factor for component i

References

1. Ghosh P, Vahedipour K, Leuthold M, Von Lieres E. 2014. Model-based analysis and quantitative prediction of membrane chromatography: Extreme scale-up from 0.08 ml to 1200 ml. *J. Chromatogr. A* 1332:8–13.
2. Johnson C, Natarajan V, Antoniou C. 2014. Evaluating two process scale chromatography column header designs using CFD. *Biotechnol. Prog.* 30:837–844.
3. Kempe H, Axelsson A, Nilsson B, Zacchi G. 1999. Simulation of chromatographic processes applied to separation of proteins. *J. Chromatogr. A* 846:1–12.
4. Kremer DM, Hancock BC. 2006. Process simulation in the pharmaceutical industry: A review of some basic physical models *J. Pharm. Sci.* 95:517–529.
5. Borg N, Brodsky Y, Moscariello J, Vunnum S, Vedantham G, Westerberg K, Nilsson B. 2014. Modeling and robust pooling design of a preparative cation-exchange chromatography step for purification of monoclonal antibody monomer from aggregates. *J. Chromatogr. A* 1359:170–181.
6. Degerman M, Westerberg K, Nilsson B. 2009. A model-based approach to determine the design space of preparative chromatography. *Chem. Eng. Technol.* 32:1195–1202.
7. Jakobsson N, Degerman M, Stenborg E, Nilsson B. 2007. Model based robustness analysis of an ion-exchange chromatography step. *J. Chromatogr. A* 1138:109–119.
8. Nfor BK, Ripić J, van der Padt A, Jacobs M, Ottens M. 2012. Model-based high-throughput process development for chromatographic whey proteins separation. *Biotechnol. J.* 7:1221–1232.
9. Shukla AA, Hubbard B, Tressel T, Guhan S, Low D. 2007. Downstream processing of monoclonal antibodies: Application of platform approaches. *J. Chromatogr. B Analyt. Technol. Biomed. Life Sci.* 848:28–39.
10. Liu HF, McCooey B, Duarte T, Myers DE, Hudson T, Amanullah A, van Reis R, Kelley BD. 2011. Exploration of overloaded cation exchange chromatography for monoclonal antibody purification. *J. Chromatogr. A* 1218:6943–6952.
11. Shukla AA, Barnthouse KA, Bae SS, Moore J, Cramer SM. 1998. Structural characteristics of low-molecular-mass displacers for cation-exchange chromatography. *J. Chromatogr. A* 814:83–95.
12. Von Lieres E, Andersson J. 2010. A fast and accurate solver for the general rate model of column liquid chromatography. *Comput. Chem. Eng.* 34:1180–1191.
13. Osberghaus A, Hepbildikler S, Nath S, Haindl M, von Lieres E, Hubbuch J. 2012. Determination of parameters for the steric mass action model: A comparison between two approaches. *J. Chromatogr. A* 1233:54–65.
14. Felinger A, Cavazzini A, Guiochon G. 2003. Numerical determination of the competitive isotherm of enantiomers. *J. Chromatogr. A* 986:207–225.

15. Gu T, Tsai G-J, Tsao GT. 1990. New approach to a general nonlinear multicomponent chromatography model. *AIChE J.* 36:784–788.
16. Hahn T, Huuk T, Osberghaus A, Doninger K, Nath S, Hepbildikler S, Heuveline V, Hubbuch J. 2016. Calibration-free inverse modeling of ion-exchange chromatography in industrial antibody purification. *Eng. Life Sci.* 16: 107–113.
17. Müller-Späth T, Ströhlein G, Aumann L, Kornmann H, Valax P, Delegrange L, Charbaut E, Baer G, Lamproye A, Jöhnck M, Schulte M, Morbidelli M. 2011. Model simulation and experimental verification of a cation-exchange IgG capture step in batch and continuous chromatography. *J. Chromatogr. A* 1218:5195–5204.
18. Brooks CA, Cramer SM. 1992. Steric mass-action ion exchange: Displacement profiles and induced salt gradients. *AIChE J.* 38:1969–1978.
19. Alhumaizi K. 2004. Comparison of finite difference methods for the numerical simulation of reacting flow. *Comput. Chem. Eng.* 28:1759–1769.
20. Guiochon G, Lin B. 2003. *Modeling for Preparative Chromatography First.* San Diego: Academic Press, 342 p.
21. Keener RN, Fernandez EJ, Maneval JE, Hart RA. 2008. Advancement in the modeling of pressure-flow for the guidance of development and scale-up of commercial-scale biopharmaceutical chromatography. *J. Chromatogr. A* 1190:127–140.
22. Guiochon G, Felinger A, Shirazi DG, Katti AM. 2006a. *Fundamentals of Preparative and Nonlinear Chromatography* 2nd ed. San Diego: Academic Press, 975 p.
23. Guiochon G, Felinger A, Shirazi DG, Katti AM. 2006b. *Fundamentals of Preparative and Nonlinear Chromatography First.* San Diego: Academic Press, 975 p.
24. Wilson EJ, Geankoplis CJ. 1966. Liquid mass transfer at very low Reynolds numbers in packed beds. *Ind. Eng. Chem. Fundam.* 5:9–14.
25. Thomas H, Coquebert de Neuville B, Storti G, Morbidelli M, Joehnck M, Schulte M. 2013. Role of tentacles and protein loading on pore accessibility and mass transfer in cation exchange materials for proteins. *J. Chromatogr. A* 1285:48–56.
26. Gadam SD, Jayaraman G, Cramer SM. 1993. Characterization of non-linear adsorption properties of dextran-based polyelectrolyte displacers in ion-exchange systems. *J. Chromatogr. A* 630:37–52.
27. Gallant SR, Kundu A, Cramer SM. 1995. Modeling non-linear elution of proteins in ion-exchange chromatography. *J. Chromatogr. A* 702:125–142.
28. Chilamkurthi S, Willemsen J-H, van der Wielen LAM, Poiesz E, Ottens M. 2012. High-throughput determination of adsorption equilibria for chromatographic oligosaccharide separations. *J. Chromatogr. A* 1239:22–34.

14

Model-Based Process Development in the Biopharmaceutical Industry

Lars Sejergaard[1], Haleh Ahmadian[2], Thomas B. Hansen[1], Arne Staby[3], and Ernst B. Hansen[1]

[1] *Mathematical Modelling, Novo Nordisk A/S, Bagsværd, Denmark*
[2] *Protein Purification Technology, Novo Nordisk A/S, Måløv, Denmark*
[3] *CMC Project Planning & Management, Novo Nordisk A/S, Bagsværd, Denmark*

14.1 Introduction

The aim of this chapter is to give an example of how mechanistic modeling may be applied in the biopharmaceutical industry. The specific application is a size exclusion step with the main purpose of reducing aggregated product forms for the commercial production of turoctocog alfa (NovoEight®). The example shows how the different quality by design (QbD) elements of risk assessment and process knowledge are linked through identification of key critical quality attributes (CQAs), which may be affected by the step and the different process parameters responsible for such influence on the CQAs (see also Chapter 1). The approach presents how the true critical process parameters are identified and how modeling is used to optimize the step and secure process robustness.

Furthermore this chapter contains examples of MATLAB programming code, which can be directly implemented for modeling of size exclusion chromatography (SEC). Multiple academic papers have been published where models are established and examples are shown, for example [1, 2]; however, setting up the code can often be an entry barrier to the field. This chapter should be regarded as an aid for upcoming practitioners to get started. In the appendix mathematical expressions and derivations for the partial differential equations are provided to help the understanding of these models.

14.2 Molecule—FVIII

FVIII is one of the most complex recombinant proteins that has been produced, and as such it may be regarded as a highly complicated molecule to be subject to process modeling. The human FVIII gene encodes a mature protein consisting of 2332 amino acid residues [3, 4]. This sequence is synthesized in cells as a single chain composed of six distinct domains: A1 and A2 with 740 amino acid residues (1–740); a B-domain of 908 residues (741–1648); and then three smaller domains, A3, C1, and C2, with 684 residues (1649–2332). During secretion, the single chain is cleaved by furin transmembrane proteases into a heavy chain containing the A1, A2, and B-domains and a light chain comprising the A3, C1, and C2 domains. The two chains are bound by non-covalent divalent metal ion linkage between the A1- and A3-domains [5, 6]. FVIII is activated in plasma by thrombin cleavage at three sites, resulting in a three-chain protein without the B-domain. Figure 14.1 presents the domain structure and activation process.

The expression in mammalian cells of a full-length FVIII molecule including the intact B-domain of 908 amino acid residues has been considerably more difficult than B-domain-truncated variants mainly because of the 19 potential N-glycosylation sites in this domain. Furthermore, the B-domain is cleaved in the activation process, and it does not seem to be required for factor VIII function [7]. The B-domain is subjected to cleavage at several sites by proteases, resulting in FVIII species with various lengths of B-domain.

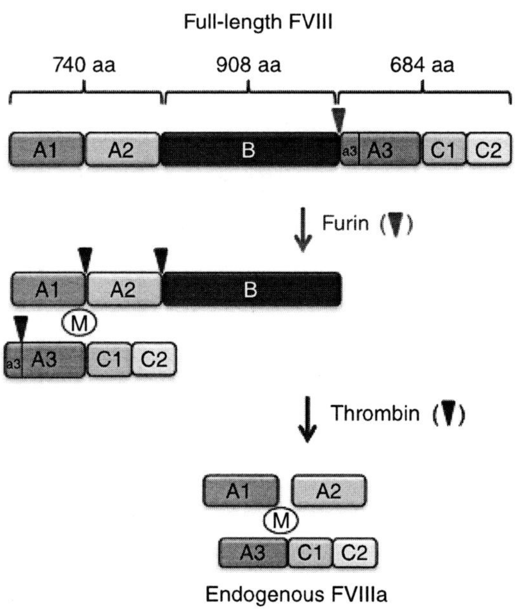

Figure 14.1 Domain structure and *in vivo* activation of endogenous FVIII and the rearrangement of the different domains. M denotes a metal.

Generally, truncation of the B-domain has been shown to increase the homogeneity of the final product.

Consequently when developing turoctocog alfa, a FVIII encoding sequence consisting of an intact human FVIII heavy chain sequence (A1 and A2 domains) of 740 amino acid residues followed by a 21-amino acid residue truncated B-domain was chosen. The light chain sequence of 684 amino acid residues was kept intact. The 21-amino acid residue truncated B-domain comprises 10 amino acids from the N-terminal of the naturally occurring B-domain linked to 11 amino acids from the C-terminal of the B-domain. This truncated B-domain is removed following thrombin cleavage of turoctocog alfa in plasma, generating an rFVIIIa molecule identical to endogenous FVIIIa.

14.3 Overall Process Design

Given the fragile nature of FVIII molecules, a considerate purification process is needed after serum-free (SF) cultivation. A five-step purification process is applied for turoctocog alfa (Figure 14.2) starting with capture on a mixed-mode resin and detergent treatment followed by immunoaffinity and anion exchange chromatography (AIEC), virus filtration, and SEC:

1) Capture is performed on a mixed-mode resin, enabling direct loading of the harvest to the column without the need for further conditioning such as adjustment of pH and conductivity. The main rationale for this step is reduction of water and virus.

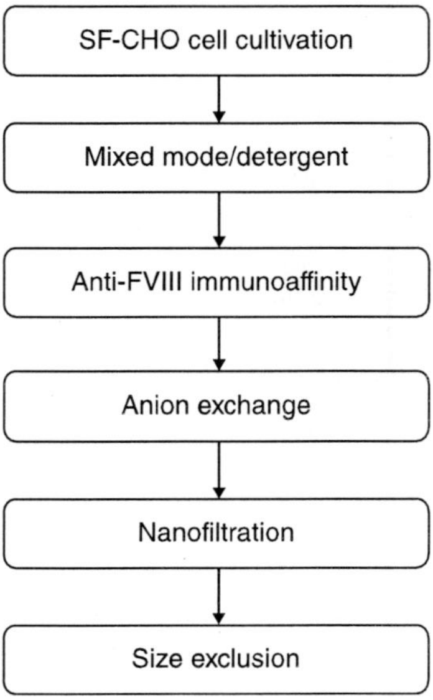

Figure 14.2 Overview of downstream process steps.

2) The second step is an immunoaffinity step using an in-house-developed antibody proved to be specific against FVIII. The aim of this step is host cell proteins (HCP) reduction.
3) The purpose of the AIEC step is to concentrate the product, in order to obtain the target drug substance concentration after SEC. Furthermore, this step reduces HCP and antibody leakage from the preceding step.
4) Virus reduction is performed using a virus filter to ensure a high level of viral clearance.
5) The purpose of the last step, SEC, is to reduce HMWP, HCP, and mAb leakage. Furthermore, the buffer is exchanged to the final drug substance buffer.

14.4 Use of Mathematical Models to Ensure Process Robustness

During process development of turoctocog alfa, a mechanistic model was used to evaluate process robustness and ensure a predictable outcome of the single steps. An example from the step 5 (SEC) is presented. In the following sections the use and outcome of the exercise is described. Subsequently, a general guide in using MATLAB is provided in Appendix A including an example for general testing of the models and improving them.

An initial risk assessment and evaluation identified the potency and impurities to be the main CQAs that could be affected by the SEC, and these CQAs are directly linked to the pool concentration and to the dimer and HMWP content, respectively. Various variables or process parameters that may affect the CQAs were studied experimentally and optimized during the development of the process. The following process parameters were identified and selected due to their potential importance:

- Load volume
- Load (g/L resin)
- Load concentration
- Peak collection criteria
- Flow rate
- Bed height

Initially the loading set point for the SEC step was to load 0.05 column volumes (CV) of feed solution. As it is the normal practice for SEC, loading is measured as volume of the loading solution per liter resin. Applying this approach, variation in feed concentration obviously results in varying amounts of protein loaded on to the SEC column with a direct impact on pool concentration. Using computer simulations (see also Appendices A and B), pool concentration was investigated as a function of feed concentration and feed volume. The procedure for column loading was changed from volume based (fixed CV) to mass based (load in g/L resin). This means that for dilute feed solutions, a higher volume is loaded, while for concentrated solutions a smaller volume is loaded. This significantly reduced variation in pool concentration. It is important to obtain a sufficient pool concentration to facilitate proper process integration with formulation and at the same time avoid excessively high concentrations, which could lead to inclusion of HMWP impurities in the final product. Furthermore, the loading volume was limited to 0.10 CV to obtain the required separation.

To set up the model, basic principles/theories need to be applied, various assumptions need to be taken (see also Chapter 2), and proper data needs to be generated. The basic theories applied were that convection and dispersion occur in the packed column, as well as mass transfer into the particles and diffusion within the pores, while no adsorption/desorption occurs in the SEC column. Assumptions taken included that the column is homogenous, that mass transfer kinetics can be lumped, that no aggregate formation occurs during processing, and that the polymers/aggregates can be split into two forms, dimers and octamers, representing dimers and higher order oligopolymer/polymers/aggregates, respectively.

Data is needed for development of the model, and the load volume, dimer and HMWP content, and salt concentration in the buffer were used as input parameters. In this case k_d (distribution coefficient) and Stanton number (mass transfer coefficient) were determined based on calibration experiments, whereas other parameters were estimated based on correlations found in the literature (see also Appendix B and Chapter 2). Four calibration experiments were conducted, where load, flow rate, and column length were varied according to Table 14.1.

In each experiment, the pool was analyzed for dimer content, HMWP content, and FVIII concentration and then compared with values obtained from simulations to calibrate the model. The chromatograms from calibration experiments are compared to the resulting simulated chromatograms in Figure 14.3.

Table 14.1 Lab scale calibration experiments used for model development.

Experiment number	Load (CV)	Flow rate (CV/h)	Column length (cm)	Column diameter (cm)	Column volume (mL)
1	0.1	1	9.1	2.5	44.6
2	0.05	1	9.1	2.5	44.6
3	0.05	0.2	38.6	2.5	189.3
4	0.05	1	38.6	2.5	189.3

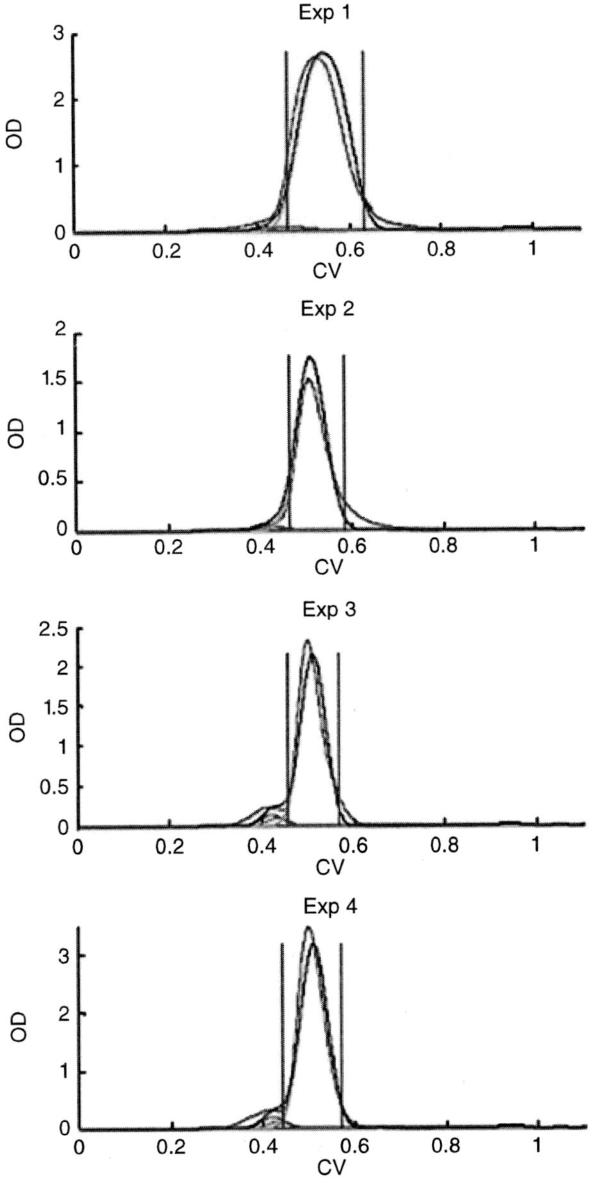

Figure 14.3 Simulated chromatograms compared to calibration experiments. Blue lines are experimental chromatograms; other colors are simulations. Vertical lines represent suggested peak collection criteria to obtain required purity and concentration. (*See insert for color representation of the figure.*)

14.5 Experimental Design of Verification Experiments

A series of computer simulations of the process using the calibrated model was conducted. The experimental design of this study was chosen in a way where some experiments were expected to be close to the specification limit, while other experiments were spread out in the parameter ranges to verify that the model was capable of predicting realistic results. A full factorial simulation of the process parameters was performed with 216 simulations in the parameter ranges given in the following text (dimer and aggregate content set at the highest possible levels in the feed):

- Column length (cm), $L = 25, 40$
- Load (CV) = 0.07
- Flow rate (CV/h) = 0.2, 1.1
- OD collection, front = 0.3, 0.4, 0.5
- OD collection, trailing edge = 0.1, 0.2, 0.3
- Feed concentration (g/L) = 0.5, 0.75, 1, 2, 3, 4
- Feed: aggregate/HMWP (%) = 8
- Feed: dimer (%) = 4

A visual sensitivity analysis exploring trends of CQAs as a function of process parameters was set up. The sensitivity analysis output for step yield, pool content of dimers and HMWP, and pool concentration revealed that flow rate and column length had no significant effect on the CQAs within the experimental ranges [8]. Further, pool concentration was not significantly affected by peak collection criteria, and dimer and HMWP content were not affected by the trailing edge pooling criterion. The most significant result of the sensitivity analysis was the effect of feed/load concentration on all the steps of CQAs. The results are depicted in Figure 14.4 where outputs are plotted as function of load concentration.

Figure 14.4 presents the expected trend of decreasing pool concentration and yield with decreasing load concentration in (B) and (D), while a nonobvious trend for HMWP and thus for purity is observed in (A) and (C), respectively. The highest HMWP content, and thus the lowest purity, is in fact found inside the operating range at a load concentration of 0.75 g/L. As discussed in Chapter 2 of this book, such a trend should be easily captured using a mechanistic model with proper assumptions fit for the purpose of the model, while a statistical-based model would require a very high density of experiments to capture this trend.

To evaluate the validity of the model for prediction, a number of verification experiments were performed varying influential parameters identified by the sensitivity analysis. Verification results are compared to the simulations in

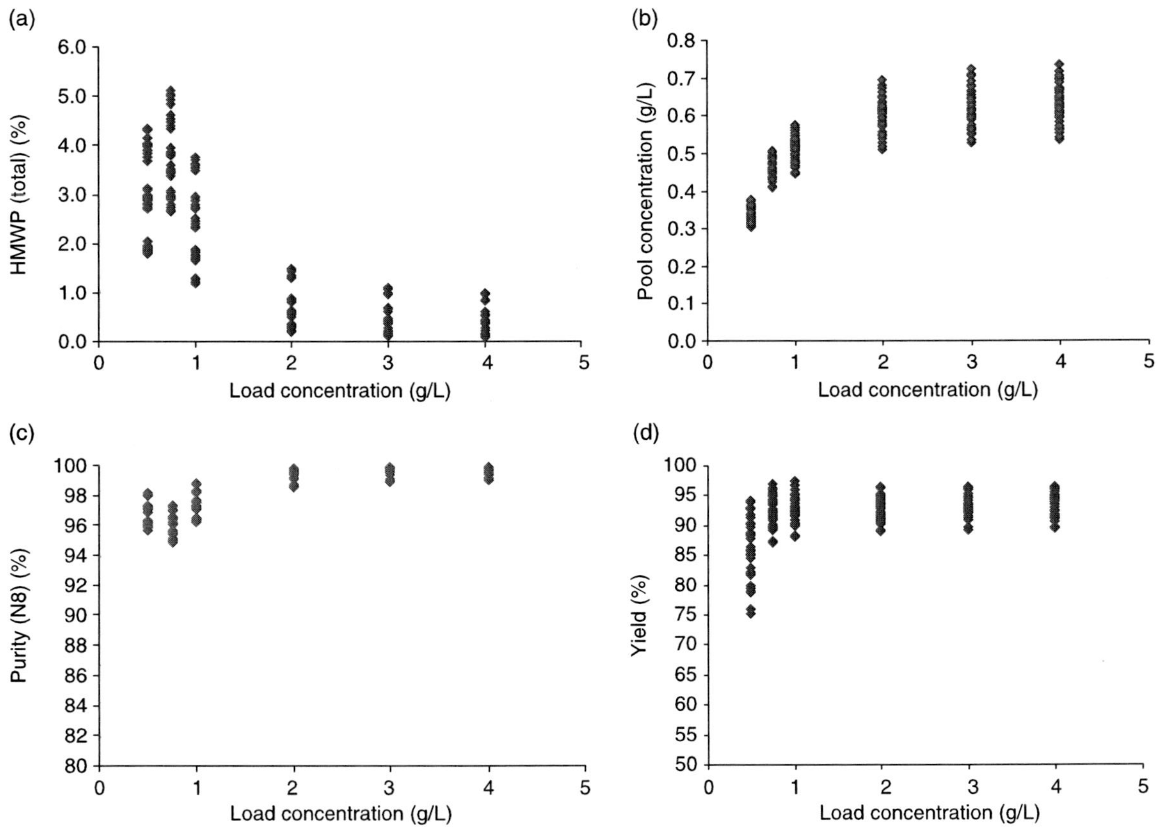

Figure 14.4 Results for the 216 computer simulations in terms of total HMWP (a), pool concentration (b), purity (c), and yield (d) as function of the load concentration.

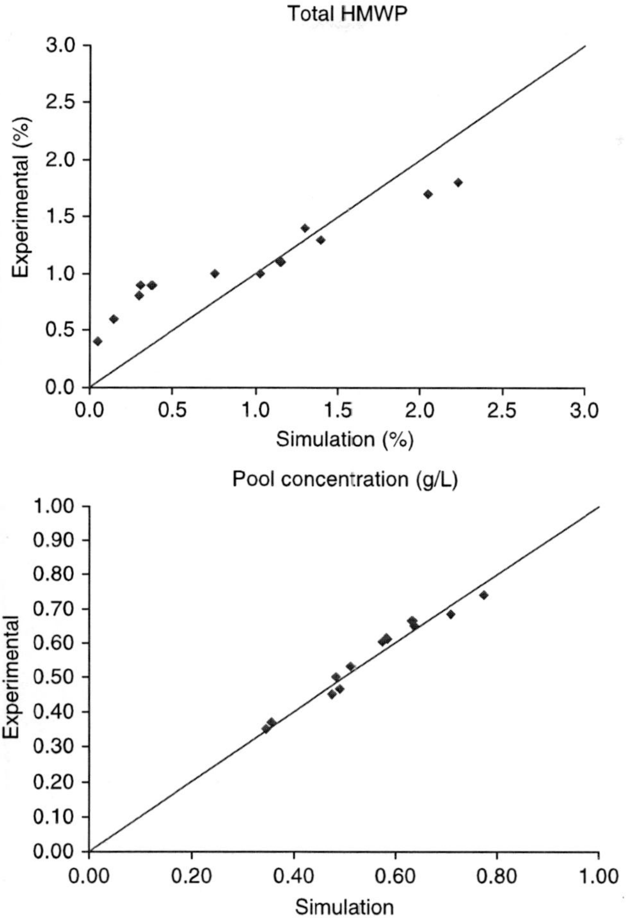

Figure 14.5 Verification results compared to simulations for total HMWP (top) and pool concentration (bottom).

Figure 14.5 for total HMWP and pool concentration. There is a general good agreement between simulation and experiments, especially for pool concentration. Of particular importance is that the pool concentration predictions are accurate in the low end near the specification limit. For the verification experiments of the dimer/HMWP analytical method, data results below the limit of quantification (LOQ—not shown) were included in order to correlate simulations and experimental observations at the low end of the range. At low levels of total HMWP, the simulations predict a slightly lower value than experimental results, while at high levels of total HMWP, the simulations predict a slightly higher level. Consequently, the model will overestimate the total HMWP level closer to the specification limit (which is still at a much higher value) and thus it is safe to use the model for prediction near the specification limit, which is the most important capability of the model. The differences between a model and experiments are likely due to the model assumptions of aggregate presence represented by only dimers and octamers, poor peak integration in the low end of total HMWP content, and the uncertainty of using analytical results below the LOQ.

14.6 Discussion

The current SEC example shows that with limited experimental effort and proper assumptions, quite good models may be established for *in silico* optimization and robustness analysis of a chromatographic step. Figure 14.3 shows a good agreement, but not a perfect overlap between experiments and a model; however, this is not required to obtain sufficiently good models for most purposes as discussed in Chapter 1. It is evident that the experimental setup required for generating mechanistic models is very different from that used in statistical modeling. Further, the models can guide you to the worst case conditions for the different CQAs focusing experimental verification and challenge efforts at these conditions instead of undue experimentation at conditions that are unlikely to ever occur.

Mechanistic modeling may also show derived, beneficial outcomes. The SEC step is designed for reduction of HMWP; however, the HCP level was found to be high in the HMWP fraction. A plot of HCP versus HMWP shows a direct correlation between HCP reduction and HMWP reduction as seen in Figure 14.6. Thus HCP reduction may also indirectly be correlated by the process model despite that it is not part of the model.

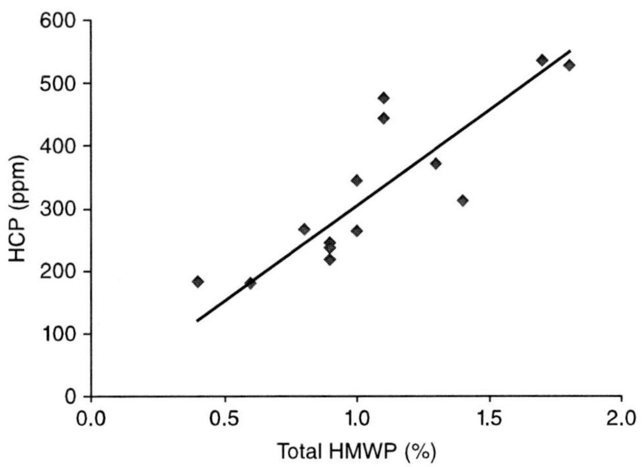

Figure 14.6 Correlation between HCP and HMWP reduction.

As shown earlier, modeling may be very beneficial for numerous purposes; however, it is not straightforward to implement for the biopharmaceutical industry as it requires not only skilled people but also access to computer code, for example, for MATLAB programming. Appendix A presents a practical MATLAB guide to SEC that may be applied—the code can simply be copied into a MATLAB file and be executed directly. Appendix A also discusses the discretization of partial differential equations—a topic and methodology that is difficult to find concrete examples of in the literature. The aim of Appendix A is thus to guide practitioners through the steps to solve partial differential equations for column simulations, starting with very simple examples and subsequently building up to a full column simulator using MATLAB.

Appendix B presents the mass balance equations applying various assumptions as well as the derivation of these models to be used for column simulations with increasing degree of complexity. Definitions and expressions of dimensionless variables and parameters needed are also included as well as equations for scale-up. The aim of Appendix B is to help practitioners with the different solutions required for modeling of chromatography depending on the level of complexity required to model their specific tasks.

14.7 Conclusion

The turoctocog alfa process case study in this chapter demonstrates that by only a few calibration experiments, a useful model can be established. Furthermore these types of models can be very helpful in the understanding and implementation of certain chromatographic unit operations. The data provided in this chapter serves as part of the background knowledge for the process as presented in the regulatory filing application for turoctocog alfa in the United States and European Union.

Acknowledgements

The MATLAB programming that has been performed at Novo Nordisk A/S has been facilitated by Bernt Nilsson's group at the University of Lund, Sweden [9, 10]. We are very thankful for all their advice. A large part of the tips and tricks presented in this chapter have their origin in that group.

Appendix 14.A Practical MATLAB Guideline to SEC

Use of computer simulations for chromatography has been published for decades. However, the number of practitioners using computer simulation has been limited mainly due to the entry barrier in solving and calibrating the models. There might also have been a mental or organizational distance between people doing simulations and people who plan and execute laboratory experiments. The aim of this appendix is to guide practitioners through the steps to solve partial differential equations for column simulations, starting with very simple examples and thereby building up to a full column simulator using MATLAB. The code examples shown in this chapter are designed to be compact, and the code can obviously be improved. However it is the intention to give experimentalists a smooth start. The code can be directly copied and will be ready for execution. In the end of this chapter, an example of using computer simulations to evaluate a size exclusion step is demonstrated. Derivation of used models can be found in Appendix B.

14.A.1 Discretization of Partial Differential Equations

Partial differential equations can rarely be solved analytically. Thus solving these equations requires a numerical solution. Numerical solution of partial differential equations is usually done in two steps. First a so-called discretization to a discrete number of ordinary differential equations (ODEs) is performed and followed by a numerical solution of this ODE system, which very often is straightforward using modern software like MATLAB.

Discretization is transformation of a partial differential equation of, for example, two variables (could be t and x where t is time and x is a position) to a discrete number of ODEs each located at different positions in the system.

14.A.2 Example

A very simple model that describes ideal convection in a tube can be expressed as (for derivation see Appendix B)

$$\frac{\partial c(\theta,x)}{\partial \theta} = -\frac{\partial c(\theta,x)}{\partial x} \tag{14.A1}$$

where c is a concentration, for example, salt, c is a function of two variables θ and x, θ is CVs (dimensionless time), and x is the position in the range 0–1, where $x = 0$ is the column inlet and $x = 1$ is the outlet. Furthermore at time = 0 ($\theta = 0$), the concentration in the entire column is 0 so $c(0,x) = 0$. The inlet concentration is 1, which is denoted $c(\theta,0) = 1$.

To make the discretization the following approximations can be used. Higher order finite difference method (FDM) schemes provide better precision. Using the fourth- or sixth-order methods actually decrease computation time since they are more stable and need fewer grid points (lower N), and this is easier for the ODE solver:

$$\frac{\partial c(\theta,x)}{\partial x}=\frac{c_i-c_{i-1}}{1\,\mathrm{h}}\text{(backward difference—tank in series model)} \quad (14.A2)$$

$$\frac{\partial c(\theta,x)}{\partial x}=\frac{c_{i+1}-c_{i-1}}{2\,\mathrm{h}} \quad (14.A3)$$
(central difference of second order—often unstable)

$$\frac{\partial c(\theta,x)}{\partial x}=\frac{-c_{i+2}+8\cdot c_{i+1}-8\cdot c_{i-1}+c_{i-2}}{12\,\mathrm{h}} \quad (14.A4)$$
(central difference of fourth order)

$$\frac{\partial c(\theta,x)}{\partial x}=\frac{-c_{i-3}+c_{i+3}-9\cdot c_{i+2}+9\cdot c_{i-2}-45\cdot c_{i-1}+45\cdot c_{i+1}}{60\,\mathrm{h}} \quad (14.A5)$$
(central difference of sixth order)

In Figure 14.A.1, a column is divided into six nodes, where the first node is the boundary condition (inlet concentration). The simplest form of discretization is chosen, that is, Equation 14.A2.

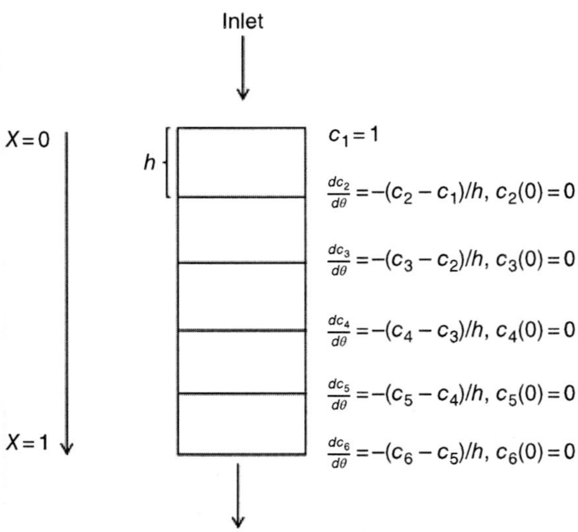

Figure 14.A.1 Example of a discretized PDE.

To solve this system the MATLAB code in the following text can be used. Copy the code into an ".m" file, name the file "solver1.m," and execute the file. As solver for this type of systems, ode15s is always a good choice.

Solver code for $N = 6$:

```
%%%%%%%%%%%%%%%%%%%%%%%%%%%%%%%%%%%%%%%%%%%%%%%%%%%%%
function solver1
N=6;
%boundary condition
bc=1;
c=zeros((N),1);
%Initial conditions
c(1:N)=0;
[t, c] = ode15s(@dsys,[0 5], c, [], N,bc);
plot(t,c(:,N),'b');
ylabel('c');xlabel('cv');title('Break through')
end

function cdot=dsys(t, c, N,bc)
h=1/(N-1);
%Zeros dot vektors
cdot =zeros(N,1);
%Boundary condition
c(1)=bc    ;
cdot(2)=-(c(2)-c(1))./h;
cdot(3)=-(c(3)-c(2))./h;
cdot(4)=-(c(4)-c(3))./h;
cdot(5)=-(c(5)-c(4))./h;
cdot(6)=-(c(6)-c(5))./h;
end
%%%%%%%%%%%%%%%%%%%%%%%%%%%%%%%%%%%%%%%%%%%%%%%%%%%%%%
```

The output of the aforementioned code is depicted in Figure 14.A.2. A breakthrough curve is obtained, and it does not look like a curve for an ideal tube. If $N \to \infty$ the curve will look more and more ideal.

To obtain a more ideal and real solution to the description, the ability to vary the number of nodes is required. The code for a variable number of nodes, N, is provided later for $N = 1000$:

```
%Code starts here%%%%%%%%%%%%%%%%%%%%%%%%%%%%%%%%%%%%
function solver2
N=1000;
%boundary condition
```

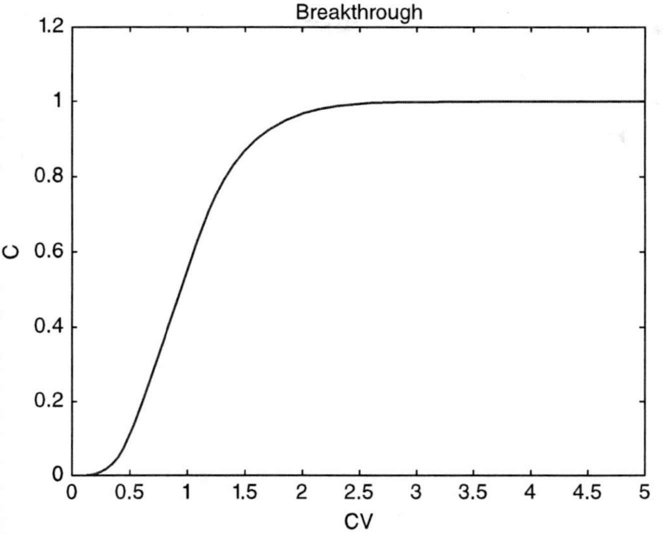

Figure 14.A.2 Breakthrough curve for $N=6$.

```
bc=1;
c=zeros((N),1);
%Initial conditions
c(1:N)=0;
[t, c] = ode15s(@dsys,[0 5], c, [], N,bc);
plot(t,c(:,N),'b');
ylabel('c');xlabel('cv');title('Break through')
end

function cdot=dsys(t, c, N,bc)
h=1/(N-1);
%Zeros dot vektors
cdot =zeros(N,1);
%Boundary condition
c(1)=bc    ;
for i=2:N
cdot(i)=-(c(i)-c(i-1))./h;
end
end
%Code end here%%%%%%%%%%%%%%%%%%%%%%%%%%%%%%%%%%%%
```

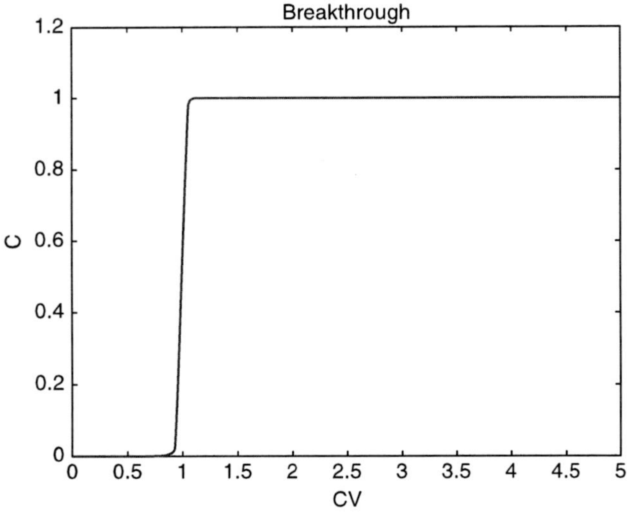

Figure 14.A.3 Breakthrough curve for $N = 1000$.

The outcome of the calculation is depicted in Figure 14.A.3. For $N = 1000$ the curve appears almost ideal. Note that high values of N require longer computation times.

To establish a very efficient and easy to program solution, MATLAB has to use matrix operations. To make the first order finite difference approximation (Equation 14.A2), the following matrix multiplication can be conducted:

$$\text{Gradient matrix FDM1} \times c = \partial c$$

$$\begin{bmatrix} -1 & 1 & 0 & 0 & 0 & 0 \\ -1 & 1 & 0 & 0 & 0 & 0 \\ 0 & -1 & 1 & 0 & 0 & 0 \\ 0 & 0 & -1 & 1 & 0 & 0 \\ 0 & 0 & 0 & -1 & 1 & 0 \\ 0 & 0 & 0 & 0 & -1 & 1 \end{bmatrix} \times \begin{bmatrix} c_1 \\ c_2 \\ c_3 \\ c_4 \\ c_5 \\ c_6 \end{bmatrix} = \begin{bmatrix} c_2 - c_1 \\ c_2 - c_1 \\ c_3 - c_2 \\ c_4 - c_3 \\ c_5 - c_4 \\ c_6 - c_5 \end{bmatrix}$$

where

$$\frac{\partial c}{\partial x} \approx \frac{\partial c}{h} \tag{14.A6}$$

The aforementioned matrix notation is a mathematical way of combining a list of different operations ($c_2 - c_1$, $c_3 - c_2$, …). The MATLAB program is built for these operations, and it is able to perform them almost as fast as a corresponding "for loop" operation in, for example, C++ or FORTRAN languages.

The MATLAB code for the simulation with $N = 1000$ using matrix multiplication is shown as follows:

```
%Code starts here%%%%%%%%%%%%%%%%%%%%%%%%%%%%%%%%%%%%%
function solver3
N=1000;
GM=GradientMatrixFDM1(N);
%boundary condition
bc=1;
c=zeros((N),1);
%Initial conditions
c(:,1)=0;
[t, c] = ode15s(@dsys,[0 5], c, [], N,bc,GM);
plot(t,c(:,N),'b');
ylabel('c');xlabel('cv');title('Break through')
end

function cdot=dsys(t, c, N,bc,GM)
h=1/(N-1);
cdot =zeros(N,1);
%Boundary condition
c(1)=bc    ;
cdot=-GM*c./h;
cdot(1)=0;
end

function GM=GradientMatrixFDM1(N)
GM=zeros(N,N);C1=zeros(N,N);C2=zeros(N,N);

C1(1:end,1:end)=eye(N,N);C2(2:end,1:end-1)=eye(N-
1,N-1).*-1;
GM=C1+C2;GM(1,1)=-1;GM(1,2)= 1;
end
%Code ends here%%%%%%%%%%%%%%%%%%%%%%%%%%%%%%%%%%%%%%%
```

The output for $N = 1000$ is depicted in Figure 14.A.4, which is identical to Figure 14.A.3.

For column simulations, where the columns are close to ideal, more sophisticated solutions are preferred to reduce the number of nodes and to decrease the time used for simulation. If a higher order FDM scheme is used, the calculations will be more accurate. It is obvious that only a FDM scheme of the first order may

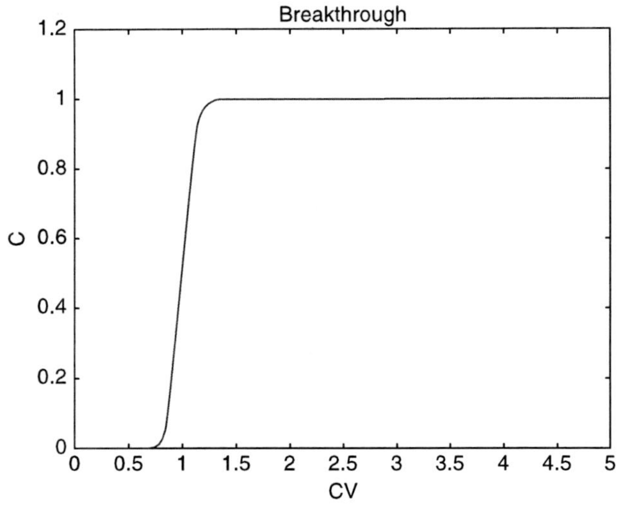

Figure 14.A.4 Breakthrough curve for $N = 1000$.

be used on the first and the last node. For the second node an FDM scheme of order 2 may be used, for the third node an FDM scheme of order 4 may be used, and the rest of the nodes can use an FDM scheme of order 6, and right before the outlet is reached a corresponding ramp down has to be used. A new gradient matrix has to be introduced to implement the scheme. An example for $N = 8$ is given as follows:

$$\text{Gradient matrix FDM6} = \begin{bmatrix} -1 & 1 & 0 & 0 & 0 & 0 & 0 & 0 \\ \frac{-1}{2} & 0 & \frac{1}{2} & 0 & 0 & 0 & 0 & 0 \\ \frac{1}{12} & \frac{-8}{12} & 0 & \frac{8}{12} & \frac{-1}{12} & 0 & 0 & 0 \\ \frac{-1}{60} & \frac{9}{60} & \frac{-45}{60} & 0 & \frac{45}{60} & \frac{-9}{60} & \frac{1}{60} & 0 \\ 0 & \frac{-1}{60} & \frac{9}{60} & \frac{-45}{60} & 0 & \frac{45}{60} & \frac{-9}{60} & \frac{1}{60} \\ 0 & 0 & 0 & \frac{1}{12} & \frac{-8}{12} & 0 & \frac{8}{12} & \frac{-1}{12} \\ 0 & 0 & 0 & 0 & 0 & \frac{-1}{2} & 0 & \frac{1}{2} \\ 0 & 0 & 0 & 0 & 0 & 0 & -1 & 1 \end{bmatrix}$$

When this approach is used in MATLAB programming, the second-order derivative may be obtained by multiplying the first derivative by gradient matrix FDM6 once again.

Thus, for SEC, models described in Equations 14.A7 and 14.A8 are used. Peak spreading is described by axial dispersion Peclet number (Pe) and film mass transfer by the dimensionless Stanton number (st_1). The phase ratio is ψ_1 and the column porosity is ε. The concentration c_1 is the concentration in the mobile phase, and c_{1p} is the concentration in the pore phase.

For the column

$$\varepsilon \cdot \left(\frac{\partial}{\partial t} c_1 \right) + \psi_1 \cdot \frac{\partial c_{1p}}{\partial t} + \frac{\partial c_1}{\partial x} - \frac{1}{Pe} \cdot \left(\frac{\partial^2}{\partial x^2} c_1 \right) = 0 \tag{14.A7}$$

For the pore

$$\frac{\partial c_{1p}}{\partial t} = st_1 \left(c_1 - c_{1p} \right) \tag{14.A8}$$

To solve the aforementioned model given by Equations 14.A7 and 14.A8, the following MATLAB code can be used:

```
%Code starts here%%%%%%%%%%%%%%%%%%%%%%%%%%%%%%%%%%
function solver3
%Parameters
vol=[0.042 1.2 0 0];Cp=2;Purity=[0.25 0.25 0.25 0.25];
epsilon=0.35;epsilonP=0.88;N=130;
Kd=[0.09 0.3 0.5 0.9];St=[1 1 1 1].*1e4;Pe=[1 1 1
1].*2000;
%solver
opts=odeset('initialstep', 1e-20, 'abstol', 1e-8);
y0=zeros((2*N),1);GM=GradientMatrixFDM6(N);hold on;
for CompNo=1:4
[t,y]=ode15s(@dsys,[0 sum(vol)],y0,...
opts,CompNo,Cp,Purity,vol,N,epsilon,epsilonP,Kd,GM,
St,Pe);
x=0:0.0001:sum(vol);
tout(:,CompNo)=x';
c(:,CompNo)=interp1(t,y(:,N),x');
end
plot(tout,c(:,:));plot(tout,sum(c,2),'k');
ylim([0 max(sum(c,2))]);xlabel('CV');ylabel('g/l');
end %function
```

```
function ydot=dsys(t,y,CompNo,Cp,Purity,vol,N,epsilon,
epsilonP,Kd,GM,St,Pe)
h=1/(N-1);ydot =zeros((2.*N),1);
%Boundary condition
if t>=vol(1)
y(1)=0;
else
    y(1)=Cp.*Purity(CompNo);
end
dydz=(GM*y(1:N))./h;
d2ydz2=(GM*dydz)./h;
ydot(1:N)=-dydz./epsilon + 1./Pe(CompNo)./epsilon.
*d2ydz2-St(CompNo).*...
(y(1:N)-y(N+1:2*N)).*Kd(CompNo).*(1-epsilon).
*epsilonP./epsilon;
ydot(N+1:2*N)=St(CompNo).*(y(1:N)-y(N+1:2*N));
ydot(1)=0;ydot(N+1)=0;
end %function

function UM=GradientMatrixFDM6(N)
UM=zeros(N,N);C1=zeros(N,N);C2=zeros(N,N);C3=zeros(N,N);
C4=zeros(N,N);
C5=zeros(N,N);C6=zeros(N,N);C1(4:end-3,1:N-6)=eye
(N-6,N-6).*-1/60;
C3(4:end-3,2:N-5)=eye(N-6,N-6).*3/20;C5(4:end-3,3:N-4)
=eye(N-6,N-6).*-3/4;
C2(4:end-3,end-(N-7):end-0)=eye(N-6,N-6).*1/60;
C4(4:end-3,end-(N-6):end-1)=eye(N-6,N-6).*-3/20;
C6(4:end-3,end-(N-5):end-2)=eye(N-6,N-6).*3/4;
UM=C1+C2+C3+C4+C5+C6;
UM(1,1)=-1;UM(1,2)= 1;UM(2,1)=-0.5;UM(2,3)=
0.5;UM(3,1)= 1/12;UM(3,2)=-2/3;
UM(3,4)=2/3;UM(3,5)=-1/12;UM(N-2,N-0)=-1/12;UM(N-2,
N-1)=2/3;
UM(N-2,N-3)=-2/3;UM(N-2,N-4)= 1/12;UM(end-1,end-2)
=-0.5;UM(end-1,end)= 0.5;
UM(N,N)=1;UM(N,N-1)=-1;
end %Function
%Code ends here%%%%%%%%%%%%%%%%%%%%%%%%%%%%%%%%%%%%%%
```

The output from the aforementioned MATLAB code is shown in the chromatogram in Figure 14.A.5.

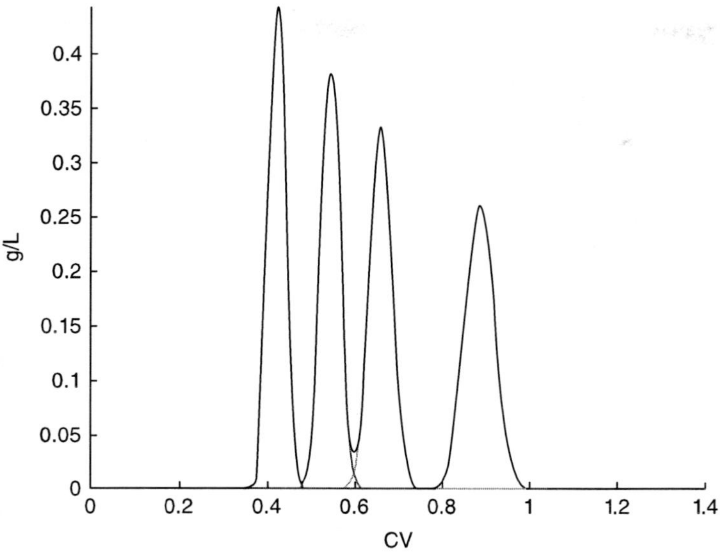

Figure 14.A.5 Size exclusion example with four parameters (fictional parameters).

Appendix 14.B Derivation of Models Used for Column Simulations

In this appendix the derivation of column models is shown. Derivation for convection, which is flow through a pipe, diffusion along a pipe, mass transfer from the mobile phase, which is the space between particles, and the mass transfer into the pores, is shown step by step. Adsorption is also shown, which is relevant for almost all chromatographic principles except size exclusion. Many different mathematical terms can be used to describe adsorption; however, they are not shown here, but they can be employed where the terms adsorption "ads" and desorption "des" are used in the following text.

Derivations are conducted based on a differential mass balance where

$$\text{In} - \text{out} = \text{accumulated}$$

The derivation is done in the time span from $t \rightarrow t + \Delta t$ and the distance from $x \rightarrow x + \Delta x$. Finally, Δx and Δt is $\rightarrow 0$ recalling that

$$\frac{\partial c(t,x)}{\partial t} = \frac{c(t+\Delta t,x) - c(t,x)}{\Delta t} \quad \text{for } \Delta t \rightarrow 0$$

14.B.1 Derivations

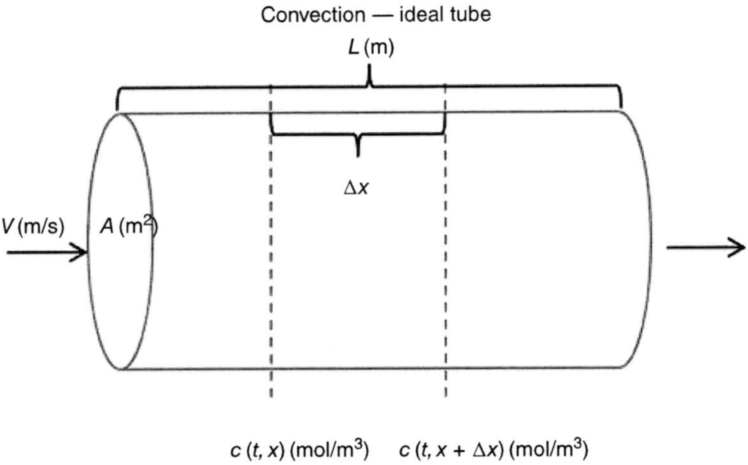

In − out = accumulated

$$c\,[t,x]\cdot v\cdot A\cdot\Delta t - c\,[t,x+\Delta x]\cdot v\cdot A\cdot\Delta t = c(t+\Delta t,x)\cdot A\cdot\Delta x - c(t,x)\cdot A\cdot\Delta x$$

$$v\cdot\frac{c\,[t,x+\Delta x]-c\,[t,x]}{\Delta x} = -\frac{c\,(t+\Delta t,x)-c\,(t,x)}{\Delta t}$$

$$\frac{\partial c\,(t,x)}{\partial t} + v\cdot\frac{\partial c\,(t,x)}{\partial x} = 0$$

Figure 14.B.1 Column model for convection in an ideal tube.

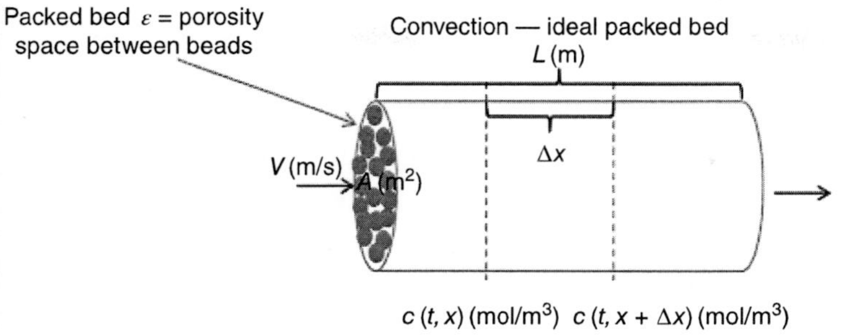

In − out = accumulated

$$c[t,x]\cdot v\cdot A\cdot\Delta t - c[t,x+\Delta x]\cdot v\cdot A\cdot\Delta t = c(t+\Delta t,x)\cdot A\cdot\Delta x\cdot\varepsilon - c(t,x)\cdot A\cdot\Delta x\cdot\varepsilon$$

$$v\cdot\frac{c[t,x+\Delta x]-c[t,x]}{\Delta x} = -\frac{c(t+\Delta t,x)-c(t,x)}{\Delta t}\cdot\varepsilon$$

$$\boxed{\varepsilon\cdot\frac{\partial c(t,x)}{\partial t} + v\cdot\frac{\partial c(t,x)}{\partial x} = 0}$$

Figure 14.B.2 Column model for convection in an ideal packed bed.

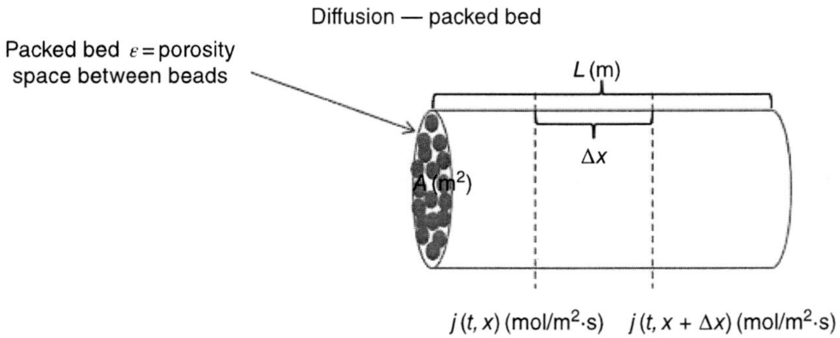

In − out = accumulated

$$j[t,x]\cdot A\cdot\Delta t - j[t,x+\Delta x]\cdot A\cdot\Delta t = c(t+\Delta t,x)\cdot A\cdot\Delta x\cdot\varepsilon - c(t,x)\cdot A\cdot\Delta x\cdot\varepsilon$$

Note: Ficks first law

$$j(t,x) = -k\cdot\frac{\partial c(t,x)}{\partial x}$$

$$\frac{j[t,x+\Delta x]-j[t,x]}{\Delta x} = -\frac{c(t+\Delta t,x)-c(t,x)}{\Delta t}\cdot\varepsilon$$

$$\varepsilon\cdot\frac{\partial c(t,x)}{\partial t} + \frac{\partial j(t,x)}{\partial x} = 0$$

$$\varepsilon\cdot\frac{\partial c(t,x)}{\partial t} - k\cdot\frac{\partial}{\partial x}\,\frac{\partial c(t,x)}{\partial x^2} = 0$$

$$\boxed{\varepsilon\cdot\frac{\partial c(t,x)}{\partial t} - k\cdot\frac{\partial^2 c(t,x)}{\partial x^2} = 0}$$

Figure 14.B.3 Column model for diffusion in a packed bed.

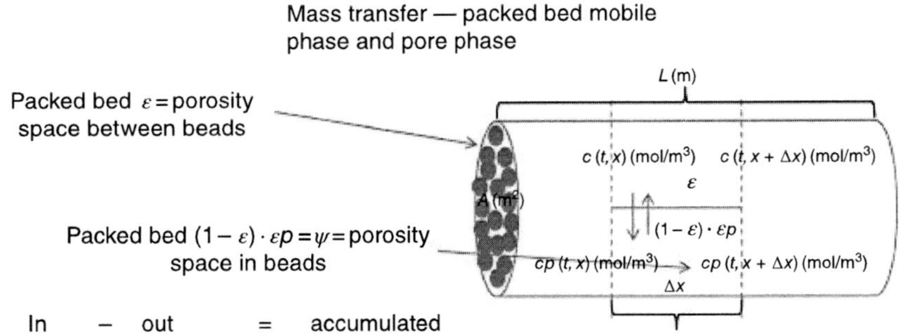

In − out = accumulated

$$\text{km} \cdot cp[t,x] \cdot A \cdot \Delta x \cdot \Delta t \cdot \psi - \text{km} \cdot c[t,x] \cdot A \cdot \Delta x \cdot \Delta t \cdot \psi = c(t+\Delta t, x) \cdot A \cdot \Delta x \cdot \varepsilon - c(t,x) \cdot A \cdot \Delta x \cdot \varepsilon$$

$$(c[t,x] - cp[t,x]) \cdot \text{km} \cdot \psi = \frac{c(t+\Delta t, x) - c(t,x)}{\Delta t} \cdot \varepsilon$$

$$\boxed{\varepsilon \cdot \frac{\partial c(t,x)}{\partial t} = (cp[t,x] - c[t,x]) \cdot \text{km} \cdot \psi}$$

In − out = accumulated

$$\text{kfilm} \cdot c[t,x] \cdot A \cdot \Delta x \cdot \Delta t \cdot \psi - \text{km} \cdot cp[t,x] \cdot A \cdot \Delta x \cdot \Delta t \cdot \psi = cp(t+\Delta t, x) \cdot A \cdot \Delta x \cdot \psi - c(t,x) \cdot A \cdot \Delta x \cdot \psi$$

$$(c[t,x] - cp[t,x]) \cdot \text{km} \cdot \psi = \frac{c(t+\Delta t, x) - c(t,x)}{\Delta t} \cdot \varepsilon$$

$$\boxed{\frac{\partial cp(t,x)}{\partial t} = -(cp[t,x] - c[t,x]) \cdot \text{km}}$$

Figure 14.B.4 Column model for mass transfer in the packed bed mobile phase and pore phase.

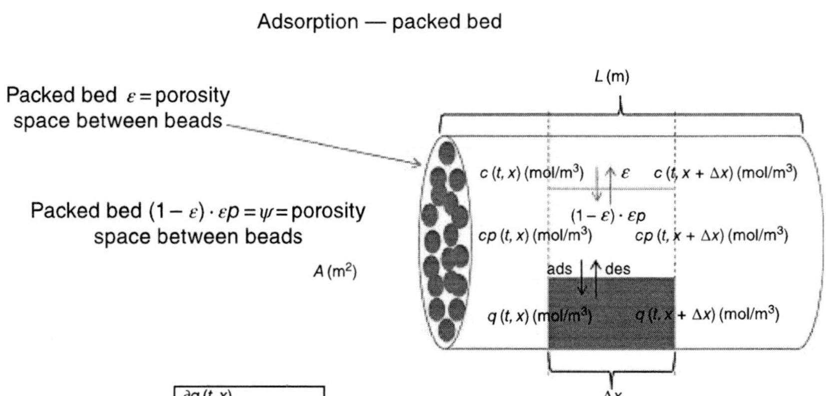

$$\boxed{\frac{\partial q(t,x)}{\partial t} = \text{ads} - \text{des}}$$

Figure 14.B.5 Column model for adsorption in a packed bed.

Collect all contributions to chromatography

$$\varepsilon \cdot \frac{\partial c\,(t,x)}{\partial t} + v \cdot \frac{\partial c\,(t,x)}{\partial x} - k \cdot \frac{\partial^2 c\,(t,x)}{\partial x^2} + (cp\,[t,x] - c\,[t,x]) \cdot \mathrm{km} \cdot \psi = 0$$

$$\frac{\partial cp\,(t,x)}{\partial t} = -(cp\,[t,x] - c\,[t,x]) \cdot \mathrm{km} - \mathrm{ads} + \mathrm{des}$$

$$\frac{\partial q\,(t,x)}{\partial t} = \mathrm{ads} - \mathrm{des}$$

Figure 14.B.6 Column model for all contributions.

Residence time

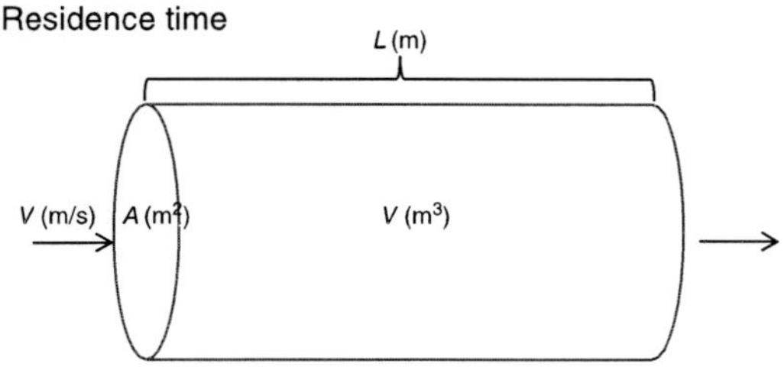

$$\text{Residence time} = \tau\,(1/\mathrm{s}) = \frac{L\,(\mathrm{m})}{v\,(\mathrm{m/s})} = \frac{v\,(\mathrm{m}^3)}{q\,(\mathrm{m}^3/\mathrm{s})}$$

Figure 14.B.7 Model for residence time.

Dimensionless time

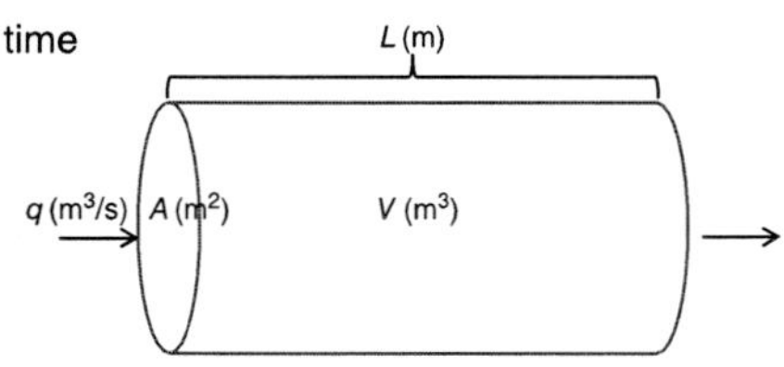

$$\text{Volume through reactor} = q\,(\mathrm{m}^3/\mathrm{s}) \cdot t\,(\mathrm{s})$$

$$\text{Column volumes through Column} = \frac{\text{volumes through Column}\,(\mathrm{m}^3)}{V\,(\mathrm{m}^3)} = \frac{q\,(\mathrm{m}^3/\mathrm{s})}{V\,(\mathrm{m}^3)} \cdot t\,(\mathrm{s}) = \frac{1}{\tau\,(\mathrm{s})} \cdot t\,(\mathrm{s}) = \frac{t\,(\mathrm{s})}{\tau\,(\mathrm{s})} = \text{dimensionless time}$$

Figure 14.B.8 Model for dimensionless time.

Space velocity — reciprocal residence time

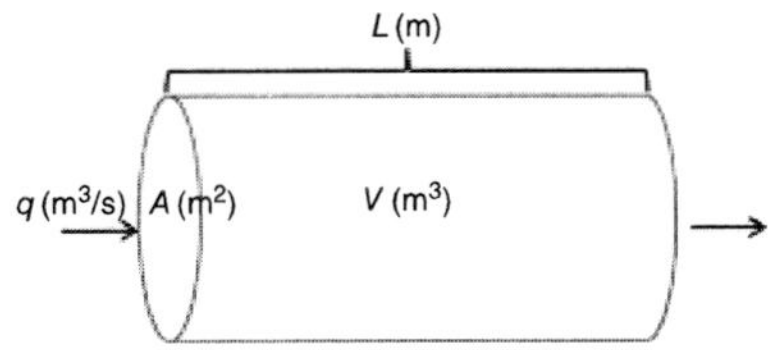

$$\frac{q\,(\mathrm{m^3/s})}{V\,(\mathrm{m^3})} = Q\,(1/\mathrm{s}) = \text{dimensionless flow } (1/\mathrm{s}) = \frac{1}{\tau\,(\mathrm{s})} = \frac{\text{column volumes}}{s}$$

Figure 14.B.9 Model for space velocity.

Make the model dimensionless

$$\varphi = \frac{t\,(\mathrm{s})}{\tau\,(\mathrm{s})} = \text{dimensionless time} \qquad z = \frac{x\,(\mathrm{m})}{L\,(\mathrm{m})} = \text{dimensionless length}$$

$$t\,(\mathrm{s}) = \varphi \cdot \tau\,(\mathrm{s}) \qquad x(\mathrm{m}) = z \cdot L\,(\mathrm{m})$$

$$\varepsilon \cdot \frac{\partial c\,(t,x)}{\partial\,(\tau)} + \frac{\partial c\,(t,x)}{\partial\,(z)} - \frac{k}{L^2 \cdot Q} \cdot \frac{\partial^2 c\,(t,x)}{\partial\,(z)^2} + (cp\,[t,x] - c\,[t,x]) \cdot \frac{\mathrm{km}}{Q} = 0$$

$$\frac{\partial cp\,(t,x)}{\partial\,(\varphi)} = -(cp\,[t,x] - c\,[t,x]) \cdot \frac{\mathrm{km}}{Q} - \frac{\mathrm{ads-des}}{Q}$$

$$\frac{\partial q\,(t,x)}{\partial\,(\varphi)} = \frac{\mathrm{ads-des}}{Q}$$

Figure 14.B.10 Equations for dimensionless model.

Scale-up — axial dispersion (Peclet number)

$$\frac{k}{L^2 \cdot Q} = \frac{1}{\text{Peclet}}$$

If the process is governed by Peclet number, the Peclet number is unaffected if Q is constant. If L is increased the column will be closer to ideal (better performance)

Figure 14.B.11 Model for scale-up (axial dispersion)—see also Chapter 7.

Scale-up — mass transfer and adsorption

$$\frac{\mathrm{km}}{Q} = \text{Stanton number}$$

$$\frac{\mathrm{ads-des}}{Q} = \text{dimensionless kinetics} = \text{"sort of damkohler number"}$$

Mass transfer and kinetics is obviously governed by space velocity, thus constant Q = constant performance

Figure 14.B.12 Model for scale-up (mass transfer and adsorption)—see also Chapter 7.

References

1. Zelic B, Nesek, B. *Eng. Life Sci.*, 2006; 6(2) 163–169.
2. Jakobsson N, Karlsson D, Axelsson JP, Zacchi G, Nilsson B. *J. Chromatogr. A*, 2005; 1063:99–109.
3. Gitschier J, Wood WI, Goralka TM, Wion KL, Chen EY, Eaton DH. *Nature*, 1984; 312:326–330.
4. Vehar GA, Keyt B, Eaton D, Rodriguez H, O'Brien DP, Rotblat F. *Nature*, 1984; 312:337–342.
5. Kaufman RJ, Wasley LC, Dorner AJ. *J. Biol. Chem.*, 1988; 263:6352–6362.
6. Svensson LA, Thim L, Olsen OH, Nicolaisen EM. *Biol. Chem.*, 2013; 394:761–765.
7. Burke RL, Pachl C, Quiroga M, Rosenberg S, Haigwood N, Nordfang O, Ezban M. *J. Biol. Chem.*, 1986; 261(27):12574–12578.
8. Degerman M, Hansen E, Sejergaard L, Hansen TB, Krarup J, Frederiksen S, Kidal S, Staby A. Model based process development in biopharmaceutical industry—case studies. In: Oral keynote presentation at BPP 2011, Puerto Vallarta, Mexico, September 18–22, 2011.
9. Nilsson B, Borgquist P, Axelsson A, Zacchi G. *Comput. Chem. Eng.*, 1999; 23: S715–S718.
10. Nilsson B, Borgquist P, Persson P. "A Simulator for Multi Component Chromatography." In: Preprints of SIMS 99 (editor P Fritzson), Scandinavian Simulation Society, Linköping, October 18–19, 1999.

15

Dynamic Simulations as a Predictive Model for a Multicolumn Chromatography Separation

Marc Bisschops[1] *and Mark Brower*[2]

[1] *Pall Corporation, Medemblik, the Netherlands*
[2] *Merck & Co., Inc., Kenilworth, NJ, USA*

15.1 Introduction

Over the past few years, multicolumn chromatography processes have been gaining momentum in the biopharmaceutical industry for the purification of recombinant proteins (see also Chapter 5). Heavy development investment in genetic engineering and cell culture processes has resulted in a significant increase in protein expression levels from an average of well below 1 mg/mL in the 1980s to 5 mg/mL and even higher in modern processes [1]. With this trend, the bottleneck in manufacturing biotherapeutics has shifted from the upstream bioreactor to the downstream process. In particular, capture chromatography steps, with low specific productivities, have become the primary bottleneck in the downstream process resulting from the fact that the column size is directly related to the total amount of protein to be purified. With increasing expression levels, the column sizes have surpassed the physical limit of what is practical in a clean room environment. The first obvious solution was to purify the amount of material in a limited number of injections using a standard column format, but this resulted in significantly longer processing times and had an undesired impact on the overall facility throughput. In addition, not all recombinant proteins are sufficiently stable to tolerate such long processing times in the early stage of the downstream process.

A very promising strategy to mitigate the bottleneck in chromatographic processes is the use of continuous multicolumn chromatography. This approach has been adopted by many other industries over the past decades, including small molecule active pharmaceutical ingredients, dairy, and vitamin production, resulting in various designs for multicolumn chromatography equipment [2].

The use of the existing equipment designs in a biopharmaceutical production environment, however, is limited due to the specific requirements related to the need for hygienic processing. Both the hardware and the control software need to comply with rigorous regulatory standards in order to be accepted by the industry. In earlier studies, the potential of multicolumn chromatography and simulated moving bed (SMB) technology for the capture of monoclonal antibodies on protein A chromatography was demonstrated [3]. This work was performed on a carrousel-type SMB system utilizing a sliding valve that connected 20 columns to the required pumps and collection containers. The design of the sliding valve in these systems is not conducive to sterilization procedures for hygienic processing, and hence alternative systems needed to be developed in order to materialize the promises of this technology. This has led to the development of suitable multicolumn chromatography systems for biopharmaceutical use, such as the BioSC from Novasep (Pompey, France) [4], the 3C PCC from GE Healthcare (Uppsala, Sweden) [5, 6], and the BioSMB system from Tarpon Biosystems (Worcester, MA) [7, 8]. The BioSMB also adheres to a greater trend in the biomanufacturing industry toward single-use processing by offering a disposable fluid contact flow path. Various studies that demonstrate the effectiveness of these technologies have been presented and/or published [5, 9–12].

During typical biopharmaceutical process development schedules, time and material are limited. Continuous processes are generally more efficient than batch processes, but they require multiple cycles to arrive at steady state conditions; hence they require significantly more products during process development. In addition to this, parallel processing strategies and/or high-throughput experimentation for continuous processes are not yet well established. As such, the development of a continuous chromatography process may negatively impact the overall time it takes for a biotherapeutic molecule to enter the clinic. A model-based assisted approach during continuous multicolumn chromatography process development will help to mitigate the risks of lengthening timelines and will reduce overall material demands thus enabling the use of the technology in the biopharmaceutical industry. The strategy described in this chapter relies on supplementing experimental data with simulation (or in silico) data to completely understand the design space for the process. The use of numerical simulations fits well within recent initiatives launched by the authorities, such as the European Medicines Agency, which are directed toward generating more process understanding during the process development and process characterization work. Probably the best example of this is the quality by design (QbD) initiative launched by the US Food and Drug Administration, which asserts that quality should be built into biotherapeutic molecules through process understanding and appropriate monitoring techniques. Within this framework, the use of mechanistic models, for instance, to support the definition of the design space is highly encouraged by the agency [13].

Within this chapter, the use of a numerical simulation tool to predict the performance of multicolumn chromatography processes for the capture of monoclonal antibodies is investigated. The in silico models described herein are easily applied to non-antibody recombinant protein chromatographic systems.

15.2 BioSMB Technology

The BioSMB technology is the first fully disposable multicolumn chromatography system in the market. It was designed to accommodate any chromatography process application that one would normally expect for the purification of biotherapeutics (packed bed, expanded bed, membrane chromatography, etc.). The disposable cassette can be pre-sterilized by irradiation to facilitate hygienic processing necessary for biopharmaceutical processing. The global architecture of the system is shown in Figure 15.1.

The valve system in the BioSMB system accommodates 240 individually actuated zero dead volume diaphragm valves. This provides the flexibility that is required to connect and disconnect the columns in virtually any configuration. This flexibility is key to using the system for both capture steps, such as protein A affinity chromatography, and polishing steps, such as size exclusion chromatography (SEC). The first mode resembles continuous ion exchange operations as practiced with various continuous ion exchange chromatography technologies in other industries, such as the purification of lysine [14] and vitamin C [15]. These applications require the process to be configured in

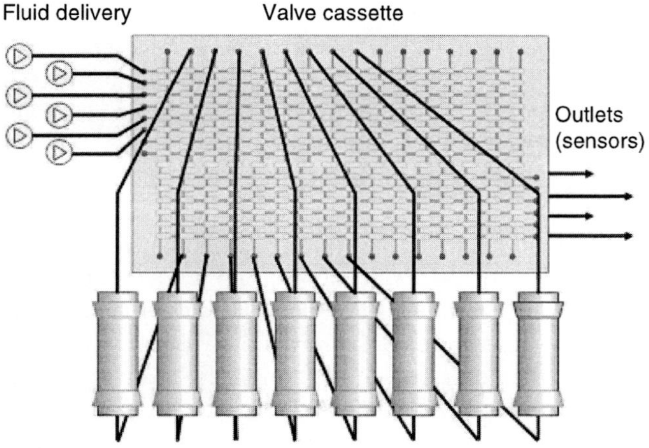

Figure 15.1 Global layout of the BioSMB system with multiple chromatography columns. The valves (shown in the center) are held in a single disposable cassette.

isolated steps in which the outlet of each column is not always connected to the next. SEC, on the other hand, can very effectively be used for binary fractionations. SEC applications require that the columns are configured according to a traditional SMB scheme where the outlet of each column is connected to the inlet of the next column.

The models developed in this chapter will specifically describe a protein A chromatographic capture for monoclonal antibodies, which is comprised of equilibration, loading, washing, elution, and CIP steps. Of primary concern in the model are the countercurrent loading zone and the elution zone. Intrinsically, multicolumn chromatographic processes are run for multiple cycles to reach steady state, and as such it is important that protein elution is properly characterized so that there is no accumulation and carry-over of product between consecutive cycles. Similar models, not addressed herein, can be developed within the in silico BioSMB software to simulate other chromatographic separations such as SEC and hydrophobic interaction chromatography, which are governed by a different set of physical parameters.

15.3 Protein A Model Description

Tarpon Biosystems has developed in silico BioSMB as a numerical simulation tool for any multicolumn chromatography process. Before being used in determining the design space for biopharmaceutical process development, the numerical simulation model needs to be qualified for an accurate representation of the physical system. As such, the models used must be based on a mechanistic description of the separation process. For this reason, the in silico BioSMB model was built upon the first principles that govern chromatography processes, which are (i) convective fluid flow, (ii) thermodynamic equilibrium, (iii) mass transfer kinetics, and (iv) axial dispersion in the columns.

The convective flows in the chromatography process are controlled by pumps and valve settings. These are programmed in the method that is executed on the actual multicolumn chromatography system. The numerical simulation model used in this study was developed to utilize the same valve switching schemes and column configurations as programmed in the BioSMB methods. With this, the in silico BioSMB software accommodates the same flexibility in configurations and options as offered in the physical system itself in addition to other third-party hardware configurations. By programming a method with a single column, the numerical simulation tool can also be used for batch chromatography processes. Because of this feature, traditional batch-based chromatography experiments such as frontal analysis can be used to extract model parameter sets necessary for the simulation of more complex multicolumn chromatography processes.

The adsorption equilibrium behavior for proteins on chromatographic media is most commonly described by either a linear isotherm or the multicomponent Langmuir isotherm. In this study, we will limit the work to the Langmuir isotherm:

$$q = \frac{Q_{\max}\ c}{1 + b\ c} \tag{15.1}$$

In this relation, $Q_{\max}$ describes the static binding capacity (SBC) of the chromatography media, and q and c represent the equilibrium solid-phase and liquid-phase concentrations. The coefficient b defines the interaction between the chromatography media and the protein. The Langmuir isotherm is the most common isotherm expression that supports single- and multicomponent behavior and covers a wide range of applications with reasonable predictive accuracy with a limited number of parameters.

The equilibrium parameters ($Q_{\max}$ and b) solely describe the thermodynamic equilibrium behavior of the process. As a result, the SBC and interaction parameters are independent of flow rates. In addition to this, it should be noted that this Langmuir equation given here expresses the concentration of the target molecule inside the beads. The target molecule in the chromatography column is commonly expressed per unit volume of the packed bed.

To characterize the mass transfer behavior, the classical two-film theory (or linear driving force model) has been adopted. This model was originally developed by Lewis and Whitman [16] for gas absorption but has been proven reliable in many two-phase mass transfer processes, including chromatographic separations [17–19]. In the linear driving force model, resistance to mass transfer is localized in two films at each side of the phase boundaries. The thickness of each film determines the steepness of the concentration gradient from the bulk concentration to the boundary between the two phases.

Several relations have been developed to estimate the film thickness associated with the liquid-phase mass transfer resistance relative to the particle size. For countercurrent processes, and based on the homogenous diffusion model assumption, the intraparticle film thickness is equivalent to one tenth the particle size. The Sherwood number for countercurrent chromatography therefore equals $Sh_S = 10$ [20]. The Sherwood number in the liquid phase, Sh_L, depends on the hydrodynamic behavior in the column and is determined according to the Snowdon–Turner correlation:

$$Sh_L = \frac{0.86}{\varepsilon_L}\ Re^{0.50}\ Sc^{0.33} \tag{15.2}$$

where ε_L is the void faction in the column. The Reynolds (Re) and Schmidt (Sc) numbers are dimensionless numbers that describe hydrodynamic and diffusivity characteristics of the solute in the fluid phase. They are defined as

$$\text{Re} = \frac{\rho_L \ v_L \ d_p}{\mu_L} \text{and Sc} = \frac{\mu_L}{\rho_L \ D_L} \tag{15.3}$$

where ρ_L is the fluid density, v_L is the fluid flow rate, d_p is the diameter of the particle, μ_L is the fluid viscosity, and D_L is the diffusion coefficient in the bulk liquid phase.

From the Sherwood numbers, the partial mass transfer coefficient in the liquid and solid phase (k_L and k_S, respectively) can be calculated:

$$k_L = \frac{\text{Sh}_L \ D_L}{d_p} \text{and} k_S = \frac{\text{Sh}_S \ D_S}{d_p} \tag{15.4}$$

where D_S represents the diffusion coefficient for the solute in the resin particle. The combined or lumped (overall) mass transfer coefficient then is

$$k_{oL} = \left(\frac{1}{k_L} + \frac{1}{K \ k_S} \right)^{-1} \tag{15.5}$$

In this relation, K represents the partition coefficient based on the local concentrations.

Deviations in ideal plug flow profiles are in the column commonly described by an axial dispersion coefficient, which, when not set to zero, complicates the governing set of partial differential equations used to describe the system. In order to address the flexibility that is required to support numerical simulations for all potential configuration options in the BioSMB technology, the model is based on the compartment model for axial dispersion. In the compartment model, each column is subdivided into small segments, each of which is modeled as an ideally mixed nonequilibrium stage. Dispersion in the column is more accurately simulated with a large number of smaller compartments in the column. For each of these compartments, the mass balance is written as a set of ordinary differential equations as

$$\varepsilon_L \frac{dc}{dt} = \frac{\phi}{V}\left(c_{in} - c\right) + k_{oL} a \left(\frac{q}{K} - c \right) \tag{15.6}$$

$$\left(1 - \varepsilon_L\right) \frac{dq}{dt} = -k_{oL} a \left(\frac{q}{K} - c \right) \tag{15.7}$$

In these mass balances, ϕ represents the liquid flow rate entering each compartment having volume V. The solution entering the compartment has a protein concentration c_{in}. Inside the compartment, the conditions are assumed to be homogeneous, represented by concentrations c in the liquid phase and q in the beads. The specific superficial area between the liquid and the beads is represented by a ($a = 6/d_p\ (1-\varepsilon_L)$, with d_p being the particle diameter). Instead of describing each column as a set of partial differential equations with very complex boundary conditions, this approach results in a large number of ordinary differential equations that have to be solved simultaneously. For this purpose, the in silico BioSMB uses the ode45 solver from MATLAB that uses the explicit Runge–Kutta method. The penalty for running simulations with a large number of compartments is the computational time needed to converge upon a solution. As a result, a moderate number of compartments are typically chosen (10–20) above which simulation results are minimally impacted.

15.4 Fitting the Model Parameters

For the numerical simulations to give representative results, the kinetic and thermodynamic input parameters must adequately describe the chromatographic system. Since continuous chromatography and batch chromatography rely on the same fundamental principles, the model parameters for both modes of operation will ideally be the same. As a result, the model parameters can be determined by fitting procedures to frontal analysis of batch chromatography. The model parameters for the adsorption zone can be fitted to the batch breakthrough curves, and desorption parameters can be fitted to the elution curve in the chromatogram.

For this purpose, the minimum of residual sum of squares is evaluated in two dimensions. If only the residuals in the vertical axis (concentration) would be taken into account, the deviations in the steep part of the breakthrough curve would dominate the residual sum of squares (SS_{res}). If the residual sum of squares would only be evaluated based on the residuals in the horizontal axis (volume, time, or loading), the deviations in the shallow parts of the breakthrough curve would dominate the residual sum of squares. For this reason, the residual sum of squares is evaluated by taking the combined residuals:

$$\delta = \Delta X \frac{\Delta Y}{\sqrt{\Delta X^2 + \Delta Y^2}} \tag{15.8}$$

where $\sqrt{\Delta X^2 + \Delta Y^2}$ represents the length of the hypotenuse of the right-angled triangle that is formed by ΔX and ΔY. Since the right-angled triangle that is formed by ΔX and δ has the same shape, the ratios of their hypotenuses ($\Delta Y/\sqrt{(\Delta X^2 + \Delta Y^2)}$) equal the ratios of their bases ($\delta/\Delta X$) (Figure 15.2).

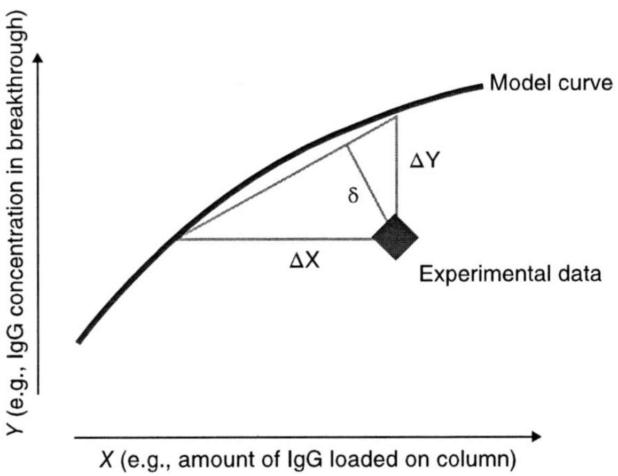

Figure 15.2 Combining the deviations in horizontal and vertical direction.

And hence,

$$\delta^2 = \frac{\Delta Y^2\ \Delta X^2}{\Delta X^2 + \Delta Y^2} \tag{15.9}$$

The residual sum of squares (SS_{res}) for the parameter fitting procedure thus becomes

$$SS_{res} = \sum \delta^2 = \sum \frac{\Delta X^2\ \Delta Y^2}{\Delta X^2 + \Delta Y^2} \tag{15.10}$$

By adopting this approach, neither small deviations in concentration during the start and end of the breakthrough curve nor small deviations in the amount loaded during the steeper part of the breakthrough curve contribute too heavily to the overall residual sum of squares.

15.5 Case Studies

The breakthrough curves for the capture of two-model monoclonal antibodies on two types of protein A chromatography media have been explored. The capture of the first monoclonal antibody (MAb1) was evaluated on MabSelect SuRe™ and on CaptivA™ PriMAB chromatography media. For the second monoclonal antibody (MAb2), the evaluation was performed on MabSelect SuRe only. The experimental conditions for the breakthrough experiments are summarized in Table 15.1.

Table 15.1 Experimental conditions for the batch breakthrough experiments.

Media	Product	Conc. (mg/mL)	Column dimensions	Flow rates (CV/h)
MabSelect SuRe (GE Healthcare)	MAb1	2.52	1.6 cm ID·2.5 cm H	10, 20, 30, and 40
CaptivA PriMAB (Repligen)	MAb1	2.07	1.2 cm ID·5.0 cm H	10, 20, 30, and 40
MabSelect SuRe (GE Healthcare)	MAb2	1.76	1.6 cm ID·2.5 cm H	10, 20, 30, and 40

Table 15.2 Model parameters for the best fit of capturing MAb1 on MabSelect SuRe chromatography media.

Flow rate (CV/h)	Q_{max} (mg/mL)	b (mL/mg)	D_L (m^2/s)	D_S (m^2/s)
10	108	45	$2.7 \cdot 10^{-11}$	$9.0 \cdot 10^{-14}$
20	107	9.2	$1.5 \cdot 10^{-11}$	$1.9 \cdot 10^{-13}$
30	107	50	$4.9 \cdot 10^{-11}$	$7.7 \cdot 10^{-14}$
40	106	150	$5.2 \cdot 10^{-11}$	$9.5 \cdot 10^{-14}$
10–40	107	44	$4.4.\ 10^{-11}$	$9.5\ 10^{-14}$

For each of the three cases, the parameter set was fitted to the data points from the experimentally derived batch breakthrough curves. The results of the fitting process are sets of model parameters that are fitted to each of the breakthrough curves at a given flow rate individually and also to the entire set of four breakthrough curves collectively.

The parameters that gave the best fit for the capture process for each matrix are summarized in Table 15.2 (MAb1 on MabSelect SuRe chromatography media), in Table 15.3 (MAb1 on CaptivA PriMAB chromatography media) and in Table 15.4 (MAb2 on MabSelect SuRe chromatography media). It should be noted that the maximum binding capacities reported in these tables refer to the capacity expressed in terms of bead volume. This is different from the SBC expressed in terms of the bed volume. The latter convention is typically reported from empirical data since the interparticle void volume is not taken into account. The SBC of a packed column is related to the maximum binding capacity of the beads using

$$SBC = (1 - \varepsilon_L)\ Q_{max} \tag{15.11}$$

The experimental data and model curves for the three case studies are plotted in Figure 15.3 (MAb1 on MabSelect SuRe chromatography media), Figure 15.4

Table 15.3 Model parameters for the best fit of capturing MAb1 on CaptivA PriMAB chromatography media.

Flow rate (CV/h)	Q_{max} (mg/mL)	b (mL/mg)	D_L (m²/s)	D_S (m²/s)
10	88	10	$1.8 \cdot 10^{-11}$	$1.2 \cdot 10^{-13}$
20	87	16	$7.5 \cdot 10^{-12}$	$1.1 \cdot 10^{-13}$
30	87	9.5	$1.9 \cdot 10^{-11}$	$1.0 \cdot 10^{-13}$
40	86	13	$7.9 \cdot 10^{-12}$	$9.3 \cdot 10^{-14}$
10–40	87	18	$1.4 \cdot 10^{-11}$	$9.2 \cdot 10^{-14}$

Table 15.4 Model parameters for the best fit of capturing MAb2 on MabSelect SuRe chromatography media.

Flow rate (CV/h)	Q_{max} (mg/mL)	b (mL/mg)	D_L (m²/s)	D_S (m²/s)
10	115	10	$2.5 \cdot 10^{-11}$	$1.2 \cdot 10^{-13}$
20	120	5.1	$2.9 \cdot 10^{-11}$	$1.2 \cdot 10^{-13}$
30	115	10	$2.1 \cdot 10^{-11}$	$9.6 \cdot 10^{-14}$
40	110	15	$2.6 \cdot 10^{-11}$	$1.2 \cdot 10^{-13}$
10–40	114	9.3	$2.2 \cdot 10^{-11}$	$1.2 \cdot 10^{-13}$

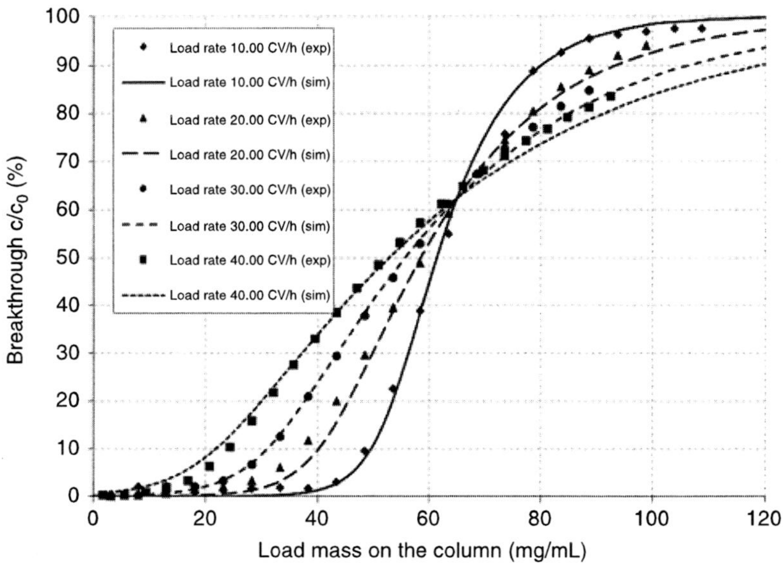

Figure 15.3 Breakthrough curves for MAb1 on MabSelect SuRe chromatography media. The markers represent the experimental data at various linear velocities, and the curves represent the model curve according to the simulation model, using one single parameter set that was optimized for the entire dataset (10–40 CV).

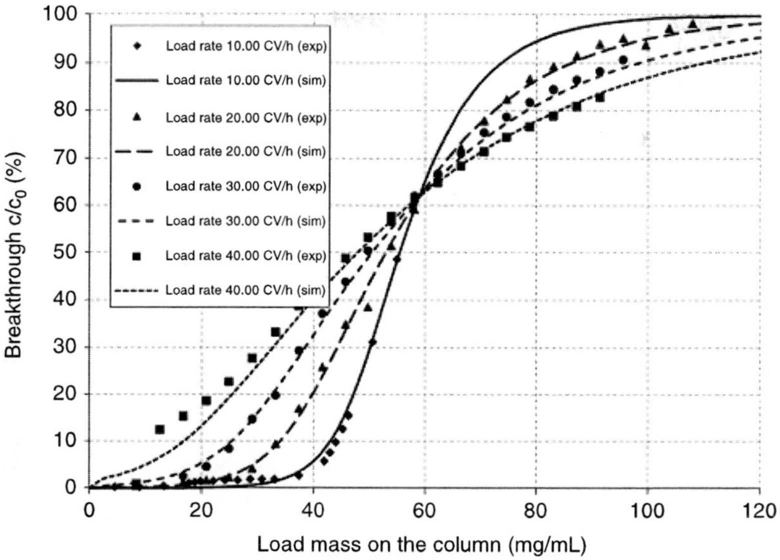

Figure 15.4 Breakthrough curves for MAb1 on CaptivA PriMAB chromatography media. The markers represent the experimental data at various linear velocities, and the curves represent the model curve according to the simulation model, using one single parameter set that was optimized for the entire dataset (10–40 CV).

(MAb1 on CaptivA PriMAB chromatography media), and Figure 15.5 (MAb2 on MabSelect SuRe chromatography media) using the parameter set that was optimized for the entire dataset (10–40 CV) collectively.

The model parameters that result from the fitting procedure are reasonably constant throughout the range of linear velocities. This indicates that the model is capable of describing the chromatographic capture throughout the process conditions evaluated in this study. At higher linear velocities, the model curves show some deviations from the experimental data. The breakthrough curves for MAb1 on MabSelect SuRe and on CaptivA PriMAB show a clear intersection for the breakthrough curves at different linear velocities.

The fitted parameters reported in Tables 15.2, 15.3, and 15.4 confirm that the equilibrium parameters (Q_{max} and b) as well as the diffusion coefficients (D_L and D_S) are independent of the flow rates. This suggests that the parameters found in the fitting procedure can be used to describe the load process over the entire range covered by the experimental data. The maximum (or equilibrium) binding capacities reported in these tables reflect the concentration of the target molecule expressed per unit volume of the beads. In order to translate this into a static capacity expressed in terms of bed volume, the void volume needs to be corrected for. The reported maximum capacities correspond to

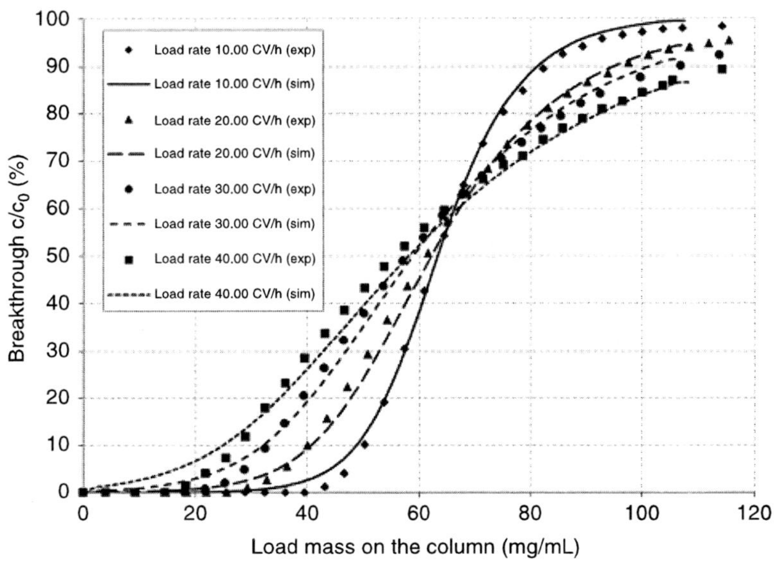

Figure 15.5 Breakthrough curves for MAb2 on MabSelect SuRe chromatography media. The markers represent the experimental data at various linear velocities, and the curves represent the model curve according to the simulation model, using one single parameter set that was optimized for the entire dataset (10–40 CV).

SBC expressed per unit volume of packed bed of 63 mg/mL for MAb1 on MabSelect SuRe, 51 mg/mL for MAb1 on CaptivaA PriMAB, and 65 mg/mL for MAb2 on MabSelect SuRe.

Tyn and Gusek reported a diffusion coefficient of $3.8 \cdot 10^{-11}$ m^2/s for γ-immunoglobulin (IgG) in water [21]. The protein A capture process involves diffusion of monoclonal antibodies in a complex clarified cell culture medium, which has a significantly higher viscosity than water. As a result, the diffusion coefficient will be lower and it will vary with the upstream process conditions (cell culture and clarification). For this reason, it is not a surprise that the diffusion coefficients found in the fitting procedure are lower than the diffusivity reported by Tyn and Gusek.

The diffusion coefficients of the proteins inside the chromatography beads are two orders of magnitude lower than the diffusivity in the surrounding solution. The linear driving force model that was used in this model assumes the homogeneous diffusion model. McCue et al. provided a correlation between the pore diffusion and homogeneous diffusion coefficient [22]:

$$\frac{D_S}{D_{pore}} \approx \left(\frac{\varepsilon_{pore} \ c_{Feed}}{Q_{max}} \right) \tag{15.12}$$

where D_S represents the diffusion coefficient in the resin bead according to the homogeneous diffusion model, D_{pore} represents the pore diffusion coefficient, ε_{pore} is the intraparticle void fraction, and c_{Feed} is the antibody concentration in the feed solution.

The pore diffusion coefficient for a polyclonal IgG in MabSelect SuRe was reported by Hahn et al. [23]. Depending on the IgG concentration in the liquid, they found pore diffusion coefficients in the range of $0.50–3.2{\cdot}10^{-11}\,m^2/s$ with a pore diffusion coefficient of $0.8{\cdot}10^{-11}\,m^2/s$ giving the best fit for the capture of IgG at 2 mg/mL feed concentration. McCue et al. reported similar data for porous glass-based protein A chromatography media [22, 24]. McCue et al. also compared the homogeneous diffusion model with the pore diffusion model. The diffusion coefficients for the homogenous diffusion model, which is the basis for the linear driving force model used in the numerical simulations described in this chapter, are 60–90 times lower than the pore diffusion coefficients A.

With this ratio applied on the diffusion coefficients reported by Hahn et al., the resin phase diffusion coefficient is expected to be $D_S = 1.3{\cdot}10^{-13}\,m^2/s$ (with $\varepsilon_{pore} = 0.88$). This corresponds reasonably well with the diffusivities found in this study, particularly if we bear in mind that the diffusion coefficients reported by Hahn et al. are evaluated for capture processes from a phosphate buffer rather than from a (more viscous) cell supernatant.

15.6 Results for Continuous Chromatography

The numerical simulation tool was used to predict the performance of the BioSMB process with multiple columns. In order to verify the validity of the model, the simulation results were compared to the experimental data described in a separate paper (M. Brower and M. Bisschops, Unpublished). The most relevant process conditions and results for these tests are summarized in Table 15.5 (for MAb1 on MabSelect SuRe), Table 15.6 (for MAb1 on CaptivA PriMAB™), and Table 15.7 (for MAb2 on MabSelect SuRe). All experiments were conducted with four columns in the load zone, connected in two passes in series with each two columns in parallel. The effluent of the first wash step (equal to 3 column volumes) was added to the effluent of the first pass and injected into the second pass of the load zone. This is needed in order to avoid loss of the unbound antibody present in the interstitial volume of the column after the loading has been completed. The concentrations in the effluent of the load zone (flow-through) are the averaged concentrations of the last two cycles of each BioSMB experiment.

The capture efficiencies as reported in Tables 15.5, 15.6, and 15.7 are based on the average concentration found in the flow-through. These concentrations are measured in the flow-through collected over various subsequent cycles

Table 15.5 Key experimental conditions and performance indicators for the BioSMB experiments used in this study: MAb1 on MabSelect SuRe.

Experiment	Feed rate (mL/min)	Feed conc. (mg/mL)	Sw. time (min)	Load cap. (mg/mL)	Flow-thr. (mg/mL)	Capture efficiency (%)
1	22.50	2.033	4.00	36.4	0.306	82.4
2	20.00	2.033	4.00	32.3	0.204	88.1
4	16.80	2.031	5.00	33.9	0.097	94.3
5	21.10	2.032	5.00	42.7	0.297	83.3
6	18.00	2.030	6.00	43.6	0.226	87.3
7	18.50	2.031	7.00	52.3	0.371	79.6
9	9.60	2.031	13.50	52.4	0.126	93.1
10	9.30	2.031	11.00	41.3	0.035	98.0
11	9.00	2.031	9.00	32.7	0.022	98.7
12	13.10	2.310	8.50	51.2	0.266	86.9
3	12.07	2.034	5.00	24.4	0.026	98.4
13	13.15	2.310	7.00	42.3	0.106	94.7
14	13.11	2.310	5.50	33.1	0.038	98.0
15	6.80	2.300	14.00	43.6	0.006	99.7
16	6.80	2.300	18.00	56.0	0.104	94.9
17	7.00	1.950	16.00	43.4	0.012	99.3
18	7.50	1.950	15.00	43.6	0.021	98.8
19	8.00	1.950	14.00	43.4	0.022	98.7

Table 15.6 Key experimental conditions and performance indicators for the BioSMB experiments used in this study: MAb1 on CaptivA PriMAB.

Experiment	Feed rate (mL/min)	Feed conc. (mg/mL)	Sw. time (min)	Load cap. (mg/mL)	Flow-thr. (mg/mL)	Capture efficiency (%)
1	9.30	2.504	12.25	50.4	0.344	84.2
2	9.30	2.523	9.50	39.4	0.098	95.4
3	9.30	2.523	8.00	33.2	0.038	98.2

during steady state operation. This was evaluated by collecting the flow-through over various cycles. In most experiments, steady state operation was obtained after the first or second cycle. The production concentration in subsequent cycles showed no significant fluctuations, and the overall mass

Table 15.7 Key experimental conditions and performance indicators for the BioSMB experiments used in this study: MAb2 on MabSelect SuRe.

Experiment	Feed rate (mL/min)	Feed conc. (mg/mL)	Sw. time (min)	Load cap. (mg/mL)	Flow-thr. (mg/mL)	Capture efficiency (%)
1	9.5	1.720	17.50	56.9	0.257	83.7
2	9.5	1.680	14.75	46.8	0.107	93.0
3	9.5	1.720	12.50	40.6	0.023	98.5
4	12.7	1.710	13.00	56.2	0.314	79.9
5	12.7	1.720	11.00	47.8	0.174	88.8
6	12.7	1.740	9.50	41.8	0.085	94.5

balance over such subsequent cycles was close to 100%. As a result, it is fair to consider the process performance over these cycles as representative for the steady state process performance.

The BioSMB process cycle covered all steps that are commonly found in a protein A chromatography process. After the load step, each column was subject to a first, second, and third wash step, followed by elution under acidic conditions. After elution, the column was cleaned and equilibrated before being subject to the load step again. The effluents of all outlets were analyzed for IgG content.

The numerical simulations, however, only focused on the capture step to predict the capture efficiency and process yield. The experiments showed some loss of product in the second wash step (in the range of 1–2%). This effect is not accounted for in the current simulation studies. In order to reduce simulation time, all other steps have been eliminated from the process cycle and replaced by a hypothetical elution step with very low equilibrium concentration and high diffusion coefficients in order to ensure that the protein concentration is zero once prior to that column reentering the load zone in the numerical simulations. The numerical simulations were done for three or four cycles. The steady state was verified by evaluating the capture efficiency of the last two cycles in the simulation. The results were deemed to have reached steady state performance if the relative difference in predicted breakthrough of the last two cycles was less than 0.5%.

With the parameter sets reported earlier, BioSMB experiments have been simulated, using the same load configuration as in the experiments. Simulations were performed at various feed flow rates and a range of load volumes according to the conditions listed earlier. The load volume can be translated into the protein load on the system, expressed in terms of a capacity utilization ratio or operating capacity:

$$Q^{\bullet}_{\text{op}} = \frac{\phi_{\text{F}}\ c_{\text{F}}}{\phi_{\text{bed}}} \tag{15.13}$$

where Q^*_{op} is the operating capacity, ϕ_F is the fluid flow rate, c_F is the concentration of the protein product in the feed, and ϕ_{bed} is the simulated bed transport rate. The utilized capacity thus corresponds to the amount of monoclonal antibody applied on the system per unit volume of the chromatography material. As soon as the operating binding capacity exceeds the SBC of the chromatography media, not all products can be captured even at infinite contact times. As a result, breakthrough of product will occur even in a countercurrent column configuration. The simulated bed transport rate (ϕ_{bed}) equals the total amount of chromatography media in the system divided by the process cycle time.

The capture efficiencies for the in silico BioSMB runs and for the experiments have been calculated using

$$\text{Eff} = 1 - \frac{(\phi_F + \phi_{W1})\ c_{FT}}{\phi_F\ c_F} \tag{15.14}$$

where ϕ_{W1} represents the time-averaged flow rate of wash 1 and c_{FT} represents the concentration of monoclonal antibody in the flow-through of the load zone.

In order to optimize the fit between the performances observed in the BioSMB experiments and the numerical simulations, the parameters can be fine-tuned to match the results better. This has been done for the dataset that involves capturing MAb1 on CaptivA PriMAB and for MAb2 on MabSelect SuRe. Any improvement of the match for the lower flow rates would have a negative impact on the match for the higher flow rates and vice versa. The BioSMB experiments for MAb1 on CaptivA PriMAB and for MAb2 on MabSelect SuRe did not cover such a wide range of flow rates, and hence a small adjustment to the design parameter resulted in a significantly better match between simulations and experimental data. The modified datasets are shown in Tables 15.8 and 15.9. The match between experimental data and numerical simulations for the continuous capture process of MAb1 on MabSelect SuRe did not justify modifications to the parameter set. Further the data collected for MAb1 on MabSelect SuRe were collected across a broad flow

Table 15.8 Summary of simulation parameters fitted for Batch and BioSMB experiments: MAb1 on CaptivA PriMAB.

Parameter	Batch	BioSMB
Max. capacity (Q_{max})	87 mg/mL	87 mg/mL
Interaction parameter (b)	18 mL/mg	18 mL/mg
Liquid diffusivity (D_L)	$1.4 \cdot 10^{-11}$ m^2/s	$1.4 \cdot 10^{-11}$ m^2/s
Resin diffusivity (D_S)	$9.2 \cdot 10^{-14}$ m^2/s	$1.1 \cdot 10^{-13}$ m^2/s

Table 15.9 Summary of simulation parameters fitted for Batch and BioSMB experiments: MAb2 on MabSelect SuRe.

Parameter	Batch	BioSMB
Max. capacity (Q_{max})	114 mg/mL	100 mg/mL
Interaction parameter (b)	9.3 mL/mg	5 mL/mg
Liquid diffusivity (D_L)	$2.2{\cdot}10^{-11}\,m^2/s$	$2.0{\cdot}10^{-11}\,m^2/s$
Resin diffusivity (D_S)	$1.2{\cdot}10^{-13}\,m^2/s$	$1.3{\cdot}10^{-13}\,m^2/s$

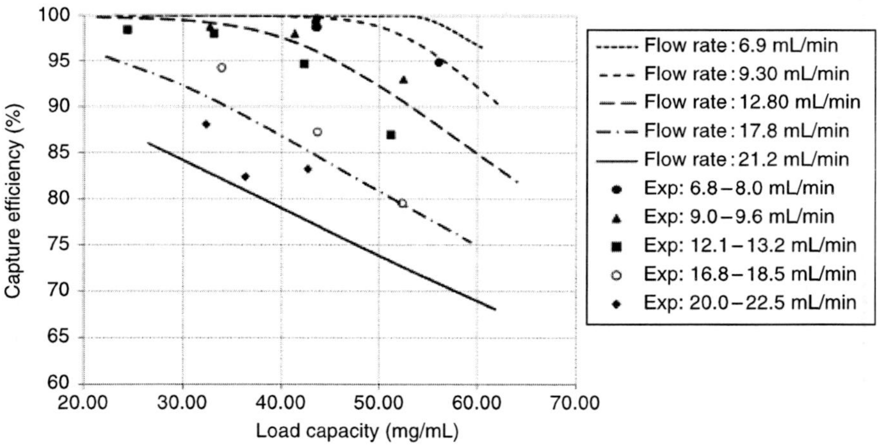

Figure 15.6 Capture efficiencies for MAb1 on MabSelect SuRe according to in silico BioSMB (curves) and derived from the experiments (data points). The capture efficiency is plotted against the operating capacity (or load capacity).

rate range due to difficulty in calibrating the pumps, not a repeated exact value about which the simulations are based. The capture efficiencies for the three separations evaluated in this study are plotted in Figures 15.6, 15.7, and 15.8.

In addition to the response curves, the numerical simulations have also been performed for exactly the same conditions as the experiments. The comparisons between the experimental and in silico results are plotted in the parity plot shown in Figure 15.9.

The differences in the best parameter fits to match Batch and BioSMB experiments can have various causes. The tests were not all performed with the same batch of cell supernatant. Any variation in upstream processing can have an impact on the properties of the feed solution. A difference in cell viability could, for instance, cause differences in viscosity and hence in diffusivities. Another difference between Batch and BioSMB experiments is the fact that the

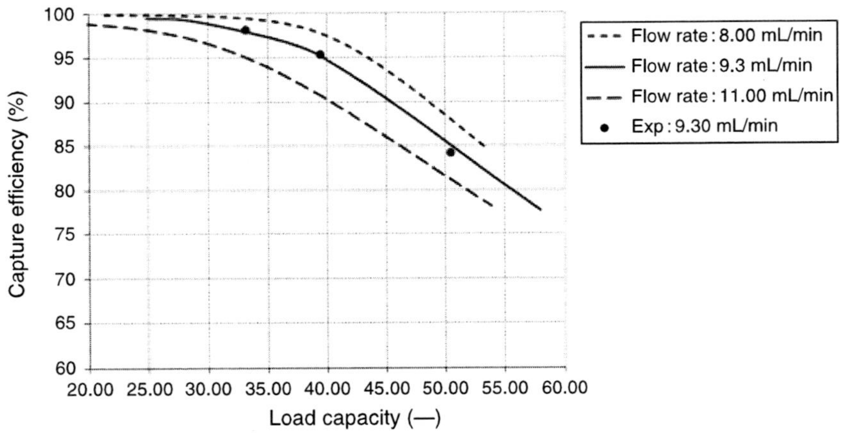

Figure 15.7 Capture efficiencies for MAb1 on CaptivA PriMAB according to in silico BioSMB (curves) and derived from the experiments (data points). The capture efficiency is plotted against the operating capacity (or load capacity).

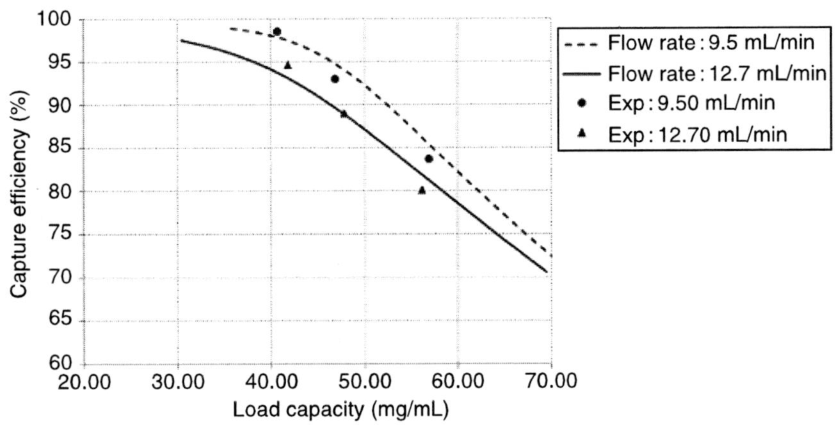

Figure 15.8 Capture efficiencies for MAb2 on MabSelect SuRe according to in silico BioSMB (curves) and derived from the experiments (data points). The capture efficiency is plotted against the operating capacity (or load capacity).

effluent of the first wash step is added to the effluent of the first pass in the load zone will result in slightly different conditions, such as viscosity and conductivity. In Batch experiments, the solution conditions are identical throughout the entire load step. Finally, it is possible that small variation in column packing density and pressure versus flow performance can impact ideal flow distribution in the loading zone, which would not be modeled correctly in the simulation, which assumes that all columns are packed and behave identically.

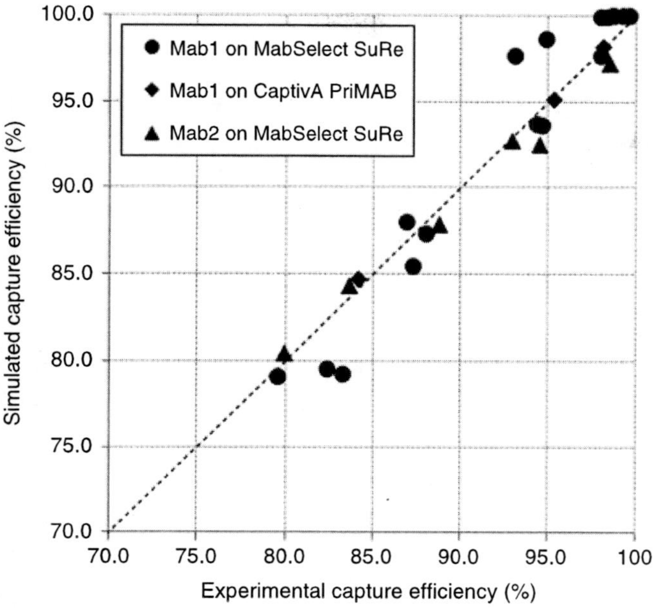

Figure 15.9 Parity plot for the capture efficiency of MAb1 on MabSelect SuRe as predicted by the numerical simulation model (in silico BioSMB) and from the experiments.

15.7 Conclusions

Numerical simulations are a valuable tool in developing and optimizing chromatography processes. Particularly for continuous chromatography process development, numerical simulations can supplement experimental data and expand the knowledge space within a short time frame. With this, numerical simulations as presented in this chapter address the issue of longer process development times that is often mentioned as a concern associated with continuous processing for biopharmaceuticals.

The numerical simulation model presented in this chapter is based on mechanistic models for thermodynamic equilibrium and mass transfer kinetics, with minimum simplifications to balance accuracy and reliability on one hand and computation speed on the other hand. The mechanistic model approach allows model predictions over a wide range of process conditions, using the same set of fundamental process parameters. As a result, a first reliable estimate for the simulation parameters for continuous chromatography can be derived from standard batch chromatography tests. The process parameters can be further fine-tuned once a first set of continuous chromatography experiments is available, thereby further improving the accuracy of the predictions. For a preliminary design, the model parameters derived from Batch experiments are good enough for basic predictions, whereas a complete process characterization or QbD study would require fine-tuning of the parameters as continuous chromatography data becomes available.

The parameters that were found during the fitting of the breakthrough curve and the BioSMB experiments match fairly well with earlier literature data reported for protein A chromatography of monoclonal antibodies. This indicates that the fundamental models, on which the numerical simulations are based, are correct and that parameters obtained from other sources can be used as an initial estimate for the design parameters in this tool. Such mechanistic approach in general provides a reliable platform for scale-up and troubleshooting.

The numerical simulations can be easily extended to include other steps in the process and to include other components of the crude cell culture fluid. This allows the use of the approach presented in this study to evaluate and optimize the product purity with a limited number of experiments. In addition to this, the models can be used to further optimize the process configuration and the buffer consumption in the continuous chromatography process.

References

1. B. Kelly (2009) Industrialization of mAb production technology: the bioprocessing industry at a crossroads, *MAbs*, 1 (5), 443–452.
2. C.Y. Chin and N.H.L. Wang (2004) Simulated moving bed designs, *Sep. Purif. Rev.*, 33 (2), 77–155.
3. J. Thommes, A. Sonnenfeld, L. Conley and J. Pieracci (2003) Protein A affinity simulated moving bed chromatography for continuous monoclonal antibody purification, Recovery of Biological Products XI, September 14–19, Banff.
4. M. Holzer, H. Osuna-Sanchez and L. David (2008) Multicolumn chromatography: A new approach to relieving capacity bottlenecks for downstream processing efficiency, *Bioprocess Int.*, 6 (8), 74–84.
5. V. Warikoo et al. (2012) Integrated continuous production of recombinant therapeutic proteins, *Biotechnol. Bioeng.*, 109, 3018–3029.
6. R. Godawat, K. Brower, S. Jain, K. Konstantinov, F. Riske and V. Warikoo (2012) Periodic counter-current chromatography—design and operational considerations for integrated and continuous purification of proteins, *Biotechnol. J.*, 7 (12), 1496–1508.
7. M. Bisschops, L. Frick, S. Fulton and T. Ransohoff (2009) Single-use, continuous countercurrent multicolumn chromatography, *Bioprocess Int.*, 9 (Suppl.5), 18–23.
8. M. Bisschops and M. Brower (2013) The impact of continuous multicolumn chromatography on biomanufacturing efficiency, *Pharm. Bioprocess.*, 1 (4), 361–372.

9. L. Allen (2011) Developing Purification Unit Operations for High Titre Monoclonal Antibody Processes, presented at 26th Antibody Development and Production Conference, March 16–18, Bellevue, WA.
10. A. Noyes, J. Coffman, R. Godavarti and M. Bisschops (2010) Development of a Protein A SMB Step for a mAb With Up to 10 g/L Titers, presented at Biopharmaceutical Manufacturing and Development Summit, Boston.
11. J. Pieracci, N. Mao, J. Thömmes, M. Pennings, M. Bisschops and L. Frick (2010) Using Simulated Moving Bed Chromatography to Enhance Hydrophobic Interaction Chromatography Performance, presented at Recovery of Biological Products Conference XIV, August 1–6, Lake Tahoe.
12. H. Jiang (2010) Purification of H5N1 and H1N1 VLP Based Vaccines, presented at IBC Single-Use Conference, May 1, La Jolla.
13. S. Chatterjee (2011) Role of Models in the Quality by Design (QbD) Paradigm: Regulatory Perspective, presented at the AAPS Annual Meeting, October 23, Washington Convention Center; Washington, DC.
14. H.J. van Walsum and M.C. Thompson (1997) Simulated moving bed in the production of lysine, *J. Biotechnol.*, 59, 127–132.
15. R. de la Fuente, M. Pennings and M. Bisschops (2007) Continuous Multicolumn Ion Exchange for Vitamin C Production, poster presented at EFB Workshop on Downstream Processing, May 10–11, Delft, the Netherlands.
16. W.K. Lewis and W.G. Whitman (1926) Principles of gas absorption, *Ind. Eng. Chem.*, 16 (12) 1215–1220.
17. M.A.T. Bisschops, J.A.M.M. Pennings and J.A. Tijsterman (2010) Device for chromatographic separations, US Patent 7,846,335.
18. D.M. Ruthven (1984) *Principles of Adsorption and Adsorption Processes*, John Wiley & Sons, Inc., New York.
19. G. Carta and A. Jungbauer (2010) *Protein Chromatography*, Wiley VCH Verlag, Weinheim.
20. G. van Zee, M. Bisschops and J. de Graauw (1997) Internal Sherwood Numbers for diffusion in sorption processes, *Chem. Eng. Prog.*, 36 (3) 201–208.
21. M.T. Tyn and T.W. Gusek (1990) Prediction of diffusion coefficients of proteins, *Biotechnol. Bioeng.*, 35 (4) 327–338.
22. J.T. McCue, G. Kemp, D. Low and I. Quiñones-Garcia (2003) Evaluation of protein-A chromatography media, *J. Chromatogr. A*, 989, 139–153.
23. R. Hahn, P. Bauerhansl, K. Shiahara, C. Wizniewski, A. Tscheliessnig and A. Jungbauer (2005) Comparison of protein A affinity sorbents II. Mass transfer properties, *J. Chromatogr. A*, 1093, 89–110.
24. W.-D. Chen, X.-Y. Dong and Y. Sun (2002) Analysis of diffusion models for protein adsorption to porous anion-exchange adsorbent, *J. Chromatogr. A*, 962, 29–40.

16

Chemometrics Applications in Process Chromatography

Anurag S. Rathore and Sumit K. Singh

Department of Chemical Engineering, Indian Institute of Technology, New Delhi, India

16.1 Introduction

Chromatographic separations are integral to the downstream processing of biotech therapeutics. Sources of variability that typically affect process chromatography performance include pumping systems, temperature gradients, buffer composition, feed material quality, stability of stationary phases, and detection systems [1, 2]. In addition, challenges could include the need for baseline/background corrections [3], data alignment or normalization [4, 5] for analytes exhibiting sample-to-sample variation, and quantification of overlapping peaks in cases of components eluting together (co-elution) [6]. Successfully meeting these challenges requires use of advanced statistical tools that can handle these intricacies and extract the relevant information.

Chemometrics/multivariate data analysis (MVDA) has gradually emerged as an approach that is capable of handling complex datasets and identifying parameters that contribute toward the variability in the process/system through the use of advanced mathematical and statistical algorithms [7]. In addition to analyzing chromatographic data [1, 2, 6, 8, 9], MVDA has been instrumental in facilitating a variety of applications in biopharmaceutical industry including:

- NIR spectral data analysis for antibiotic production processes [10]
- Multivariate statistical process monitoring during processing of pharmaceutical granules [11]
- Assessment of seed inoculum quality in a manufacturing process [12]
- Prediction of product attributes [13, 14]
- Quantitative assessment of process comparability [13, 15]

- Root cause analysis [16]
- Raw material characterization [17]

MVDA uses projection techniques such as principal component analysis (PCA) and partial least squares (PLS) to extract useful information from datasets [18]. The primary focus is typically to obtain a model that captures the maximum variation and underlying trends, if any are present, in the data. Such a model is, therefore, an accurate representative of the data to be analyzed. However, before subjecting the dataset to any of the dimensionality reduction algorithms, it is important to preprocess and transform the data into suitable form for analysis in a way that all important assumptions are adequately fulfilled [5, 14].

In this chapter, we present a review of chemometrics applications in process chromatography. The various data preprocessing methods and modeling approaches are briefly discussed. Further, two case studies illustrating the utility of chemometrics in analyzing process chromatographic data are summarized. Finally, a description of common pitfalls and recommendations for chemometric users are presented.

16.2 Data Types

An important requirement of data to be analyzed by any multivariate statistical approach is that the data should be arranged in a $(m \times n)$ matrix array format, where n is the number of dimension in the dataset and m is the number of data points in the matrix array [19]. Once the data is organized in the desired format, a relationship is sought between the row (observations) and column (variables) of the matrix. Another important necessity is with regard to the number of variables included in the data analysis. The typical recommended minimum number of variables is five [19]. If the variables are less than this number, the information in the data can be analyzed by plotting histograms and calculating averages, variance and/or covariance. Table 16.1 classifies the different data types based on the number of variables included in the analysis.

Table 16.1 Classification of data types based on the number of variables. For obtaining good quality models, the minimum number of variables included in the study should be five.

Data types	Number of variables
Univariate	1
Bivariate	2
Low variate	≤5
Multivariate	≥6

Details adapted from Ref. [19].

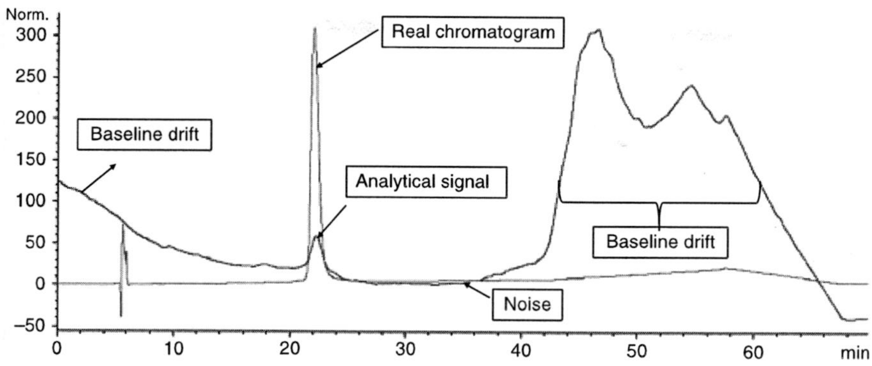

Figure 16.1 Illustration of the basic structure of a typical chromatogram.

16.2.1 Basic Structure of Chromatographic Data

An overall chromatogram is a sum of three different signals as illustrated in Figure 16.1:

- **Analytical signal**: It is the signal that is characteristic of the analyte of interest and is a function of sensitivity of the detector and capability of the chromatograph.
- **Background signal/baseline**: It is the systematic variation in the chromatogram that is not related to the analyte of interest.
- **Noise**: It is the unsystematic/random variation in the chromatographic signal.

An ideal case is when all the three listed components of the chromatogram exist as separate entities so that data analysis can be focused on the analytical signal alone. However, the analysis in such a case is not always so straightforward, especially when the peaks corresponding to the analyte of interest are overlapping. This situation may arise due to a number of reasons including complexity of the samples, drift in baselines, low signal-to-noise ratio, shift in elution times, or combination of any of these factors [6].

As far as the data structure of process chromatography is concerned, the order of the data matrix formed will depend on the number of separation steps (e.g., LC or LC–LC) employed and the detectors used (single/multiple channel detectors). A typical chromatographic data structure hyphenated to a detector device and a possible multivariate modeling approach are illustrated in Figure 16.2.

16.3 Data Preprocessing

Preprocessing of multivariate data comprises of a number of steps. First and foremost, the absolute values of the various parameters in the data are observed, and based on this appropriate preprocessing, methods are chosen to transform

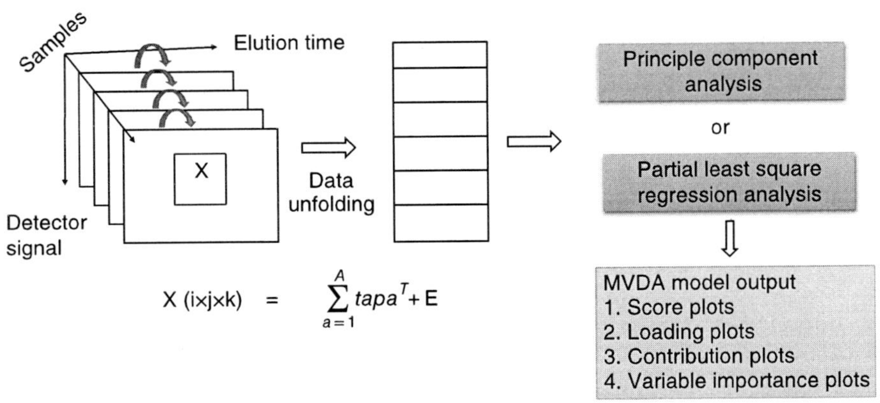

Figure 16.2 An illustration of typical chromatographic data and the possible multivariate data modeling approach. The higher dimensional data matrix is unfolded by preserving the direction of variables. This gives lower dimensional matrix containing data points that are representative of the actual process.

the data in a form, which facilitates data analysis and model interpretation. Common pretreatment techniques that are used in MVDA include scaling, mean centering, transformation [20], trimming, and winsorizing [19].

16.3.1 Scaling

This step involves multiplication of the variables with a scaling vector to modify the numerical values of a scale. However, the choice of the scaling vector varies according to the situation. For example, in cases where variables of different origin and numerical values are encountered, the weight of the scaling vector typically used is reciprocal of the standard deviation of the variables [14]. In other situations, for example, with process data, the scaling vector is defined relative to the tolerable spread in the variables [19]. If scaling is applied in a cautious and careful manner, it enhances the quality of data analysis. However, the relevance of a model based on inappropriately scaled data is severely questionable, and due care should be taken to avoid obtaining spurious results. Techniques used for scaling in MVDA include [14]:

- **Scaling to unit variance (UV)**: UV scaling implies that "long" variables are shrunk and "short" variables are stretched so that all the variables have equal opportunity of influencing data analysis. The scaling weight employed in UV scaling is $1/s_k$, where s_k represents standard deviation of variable "k." UV scaling is employed when no prior information is available about the importance of variables.

- **Pareto scaling**: Pareto scaling method involves assigning each variable a variance numerically equal to its initial standard deviation [21]. The scaling weight employed is $1/\sqrt{s_k}$. Pareto scaling is especially useful when small and medium features in the data are important (e.g., NMR data).

Apart from this, there are data types in which all the variables are expressed in the same units (e.g., spectroscopic data). For such cases, scaling is usually not necessary. However, the data are mean centered (described in Sections 16.3.2–16.3.5).

16.3.2 Mean Centering

Mean centering corresponds to subtraction of mean values of the variables from the data. For majority of data types, mean centering improves the interpretability of the model [18]. In context of chromatographic data, this allows for reducing the rank of the matrix for the case when there is a constant baseline and if the component concentrations add up to a constant sum for each sample [3]. However, there are instances wherein mean centering partially removes interesting information and therefore should be avoided in data pretreatment. As an example, a typical situation when data should not be centered is when the observations for a particular experiment exist as pairs. Doing so would eliminate the information with regard to the magnitude of change taking place in the variable due to treatment. In such situations, the best way to analyze the data is to form a new difference variable that reflects the numerical changes in the original variables before and after treatment [19].

16.3.3 Transformation

As a rule of thumb, high efficiency in data analysis is achieved for normally distributed data [22]. However, in many instances the distribution of data is rather skewed. A very common example is in drug screening where it is common to find many compounds exhibiting low activity and fewer compounds showing high activity. In such a situation, data transformation makes a non-normal distribution more normally distributed. Among different transformations that are used for approaching data normality are logarithmic, negative logarithmic, logit, square root, fourth root inverse, and power transformations [19]. However, logarithmic transformation is the most frequently applied variable transformation method. The important benefits accrued by data transformation include:

- Simplification of the response function by linearizing the nonlinear response factor relationship

- Stabilization of the variance of the residuals
- Makes the distribution of the residuals more normal, thereby eliminating outliers (observations with extreme numerical values)

16.3.4 Trimming and Winsorizing

It has been realized that data homogeneity and absence of any influential outliers are necessary for optimal performance of the various data preprocessing techniques such as scaling and mean centering. However, occurrence of the outlier increases with the increase in number of data points (a case common with multivariate data). The outliers exert their influence while computing statistical parameters like average and standard deviation, which, in turn, are used for scaling and centering of the data. Historically, two different approaches are commonly resorted to overcome this challenge. One is to remove the lower and upper extreme values of variables in the dataset. This approach is referred to as trimming, and it allows the user to better model the systematic structure in the data. The other option that is explored is that of winsorizing. This involves replacing the extreme values of the dataset with certain percentile values from each end (upper and lower extremes) in the training dataset for model building, thereby nullifying the influence of gross outliers [19]. The outliers are then scrutinized in their original form in the prediction set during model validation.

As a general guideline, trimming should be employed when the tails of data distribution does not really add to the information that can be extracted from the data analysis. On the other hand, winsorizing should be done in cases where the main aim is to minimize the skewness in the data while still accommodating the outliers by diminishing their leverage in the forthcoming model.

16.3.5 Data Preprocessing of Chromatographic Data

As mentioned earlier, baseline drift changes in elution times, and low signal-to-noise ratio make the retrieval of the pertinent information from the chromatographic profile a nontrivial task as the effect of all of these process inconsistencies gets reflected in the final data matrix (column scattering) [23]. Typical steps involved in preprocessing of the chromatographic data include retention time adjustment, baseline subtraction, and intensity normalization [1]. This is then followed by data reduction by any of the multivariate modeling approaches (PCA/PLS) depending upon the questions that are to be answered.

Table 16.2 summarizes the various data preprocessing methods in chemometrics analysis that are commonly used in biotech and other related process industries where chemometrics has been applied with a high degree of success and acceptability.

Table 16.2 Summary of the literature on the various preprocessing methods applied on datasets prior to chemometrics analysis.

Application	Multivariate model	Data preprocessing	Reference
Detection of impurity from chromatographic spectra obtained using HPLC-DAD	PCA	Transformation, mean centering, normalization, double centering	23
Determination of the effect of process variables on size exclusion chromatography	PCA	Transformation, mean centering, normalization	5
Quality assessment of herbal medical using chromatographic fingerprint	CASE	Background and retention time shift correction, normalization, trimming, winsorizing	4
Evaluation of *Saccharomyces cerevisiae* metabolic profiles by LC-MS	MCR-ALS, PLS-DA, PCA	Background and retention time shift correction (COW algorithm), normalization	24
Classification of wines based on their origin using chromatographic profiles	PCA, PLS-DA	Normalization, scaling, and mean centering	25
Evaluation of different experimental conditions on the wheat development using LC-MS profiles	ANOVA-PCA, ASCA	Baseline correction, retention time alignment (COW algorithm)	26
Prediction of column integrity and impurity clearance during reuse of chromatography resin	PCA	Scaling and mean centering	2
Assessing comparability of two processes differing in the pooling strategy at chromatographic steps	PCA, PLS-DA	Scaling and mean centering	15

ANOVA, analysis of variance; ASCA, ANOVA simultaneous component analysis; CASE, computer-aided similarity evaluation; COW, correlation optimized warping; PCA, principle component analysis; PLS, partial least square regression; PLS-DA, PLS discriminant analysis. COW is data preprocessing algorithm used for time alignment to eliminate shift-related artifacts from measurements by correcting a sample vector toward a reference.

16.4 Modeling Approaches

It is absolutely essential that the user examines the suitability of the modeling approach with respect to the intended use before proceeding with statistical data analysis. In view of this, we briefly discuss the most commonly used modeling approaches in this section.

The principle function of MVDA is to capture most of the variation in the data and to identify underlying trends. However, MVDA models are unable to include the effects of process parameters that are maintained constant in the dataset (as is typically the case in commercial manufacturing) [13]. For instance, during bioreactor operation in commercial manufacturing, the temperature and pH are kept under tight control as microbial cultures are sensitive to these parameters. But if there is a need to develop a model including these as the parameters, then multivariate analysis of the commercial manufacturing data will not provide the necessary information. The optimal way to gain such information is to perform design of experiment (DOE)-based experimentation and analyze the data from the DOE to create the model [13, 27, 28]. On the other hand, applications where a large amount of manufacturing data comprising of a large number of product quality attributes, process attributes, process parameters, and raw material attributes has to be analyzed, then MVDA is the right approach [14, 16].

16.4.1 Principal Component Analysis

PCA has been defined in a variety of ways by different researchers in the literature. Some examples include:

- Finding lines and planes of closest fit to systems of points in space [29]
- Deriving model that fits the data to the extent possible in the least square sense [30]
- Statistical means of maximizing the variance of the projection coordinates [31]

Stated simply, PCA is an approach involving reduction in dimensionality of data without loss of key information. This is achieved by use of a set of variables called principle components (PCs) and involves data analysis to reveal groups of observation, trends, and outliers. PCA is thus particularly useful for analyzing high dimensional data (generally >3) where commonly used approaches of visual/graphical representation cannot be applied.

PCA involves performing the following steps [14]:

- Arranging the data in matrix array form
- Preprocessing of the data (usually scaling and mean centering)
- Calculation of covariance matrix to measure the spread in the data between different dimensions
- Estimation of eigenvectors (PCs) and eigenvalues
- Inclusion of those PCs that account for most of the variability in the dataset (typically PCs with normalized eigenvalues <2 are considered insignificant and are thus excluded from data analysis)

The model derived upon performing these steps is represented by the equation [14, 19]

$$X = 1 \times X' + \mathbf{TP}' + \mathbf{E} \tag{16.1}$$

where data matrix X is approximated by a matrix product of lower dimension ($\mathbf{TP}'$), called the PCs, plus a matrix of residuals ($\mathbf{E}$). While $\mathbf{T}$ is a matrix of scores that summarizes the X variables, $\mathbf{P}$ is a matrix of loadings showing the influence of the variables, and $\mathbf{E}$ is a matrix of residuals reflecting the deviations between the original values and the projections. The first term in RHS of Equation 16.1 represents variable averages and originates from the preprocessing step. Geometrically, Equation 16.1 corresponds to fitting a line, plane, or hyperplane to the data in the multidimensional space with the variables as axes. This concept is illustrated in Figure 16.3.

Model Output: PCA analysis produces several types of outputs that need to be taken into consideration while drawing conclusions. An overview of key PCA model outputs is summarized in Table 16.3.

16.4.2 Partial Least Squares Regression

PLS is commonly referred to as a projection to latent structures and is another multivariate statistical approach that is particularly useful for generating calibration and/or prediction models for experiments that are otherwise difficult, expensive, and/or time consuming to undertake [34, 35]. The usefulness of PLS is evident from its ability to analyze data with many noisy, collinear, and even missing data (generally up to 50%). PLS models relate two blocks of data formed by input variables/independent variables (X) and output/dependent variables (Y). Chronologically the sequence of operations performed in PLS modeling includes [14] fitting a "PCA-like" model each for X and Y and then aligning the two models so as to predict Y from X using the following set of equations [19]:

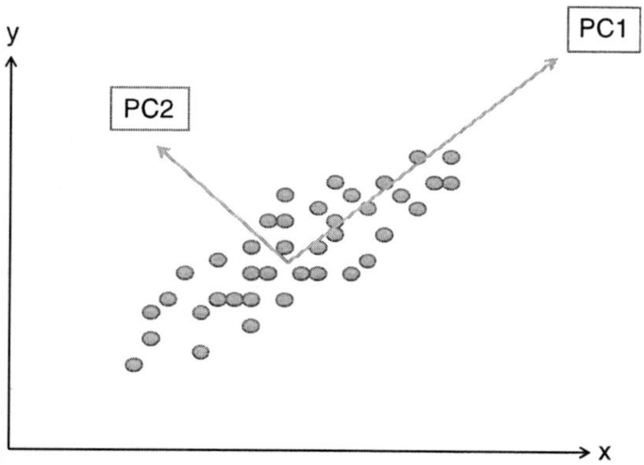

Figure 16.3 A visual representation of PCA showing two orthogonal components accounting for maximum possible variability of the data. *Source:* Adapted from Umetrics.

Table 16.3 Summary and meaning of key PCA model outputs (Umetri [32] and Kirdar et al. [33]).

Model output	Interpretation	Potential usage
Score plot	Score scatter plot of t1 versus t2 is a window in the X space that displays how the X observations are situated with respect to each other. The intended purpose is to pinpoint the presence of outliers, groups, similarities, and other patterns in the data using the score scatter plot. The scores, $t1$ and $t2$, are the variables summarizing the X variable	Scale up, process comparability, fault diagnosis
Loading plot	Loading of an original variable in a principal component is a measure of how much that variable has contributed to that component. Variables with high (positive or negative) loadings contribute significantly to that principle component. Variables that lie in the same quadrant are positively correlated, and those lying in opposite quadrants are deemed to be negatively correlated. Also, variables that form clusters are considered to behave similarly and depend on the same input variables. Loading plot complements the respective score plots	Scale up, process comparability, process characterization
Hotelling's T^2	Diagnostic statistics that measures the distance from the sample to the center of the model. Samples with high Hotelling's T^2 are different from all other samples and can have large influence in the model obtained	Pinpointing of model components exhibiting anomalous/atypical behavior
Variance explained	This value tells how much of the variance included in the dataset is explained by the model/each principal components. It is generally expressed as a number between 0 and 1 or as a percentage.	Measure of degree to which the model is representative of the original data

$$X = 1 * \overline{x'} + TP' + E \tag{16.2}$$

$$Y = 1 * \overline{y'} + UC' + F \tag{16.3}$$

where $1 * \overline{x'}$ and $1 * \overline{y'}$ represent variable averages that originate from data preprocessing operations. While the information related to observations are stored in the score matrices T and U, those related to the variables are stored in the X-loading matrix P' and Y-weight matrix C'. The variation in the data that is left out of modeling is captured in the form of E and F residual matrices [19].

Thus, it is evident that PLS regression is related to the PCA as each data block is decomposed into a low dimension matrix. However, the key difference between the two techniques is that while the former is a maximum *covariance* model of the relationship between *X* and *Y*, the latter is a maximum *variance* least square projection of *X* [36]. In other words, while in PCA the PCs are sequentially extracted with the aim of maximizing the variance captured by each component, the PCs in PLS are extracted in the direction that explains the larger amount of variance in *X* related to the variance in *Y*. This is of particular significance to multivariate calibration models as the model parameters do not change with the addition of new samples/observations to the dataset unless a strong correlation exists between the two data blocks.

Model Output: PLS regression also has similar outputs to PCA in addition to some additional parameters explaining the predictive ability of the model. An overview of key PLS model outputs is presented in Table 16.4.

Table 16.4 Summary and meaning of key PLS model outputs (Umetri [32] and Kirdar et al. [33]).

Model output	Interpretation
Variance explained in Y (R^2)	Indicates how much of the total variance included in the dependent variable (Y) is explained by the model. The parameter provides an indication of how well the model fits the training set
Variance predicted in Y (Q^2)	Indicates how well the model predicts new data. Thus, it provides a realistic assessment of the model's predictive ability when applied to new samples/datasets
Root-mean-square error (RMSE)	A measure of the difference between values predicted by a model and the values actually observed from the environment that is being modeled. The more the error, the poorer the model. RMSE is calculated for calibration (RMSEC), cross validation/internal validation (RMSECV), and external prediction (RMSEP)
Regression coefficient	PLS model parameters are used to reexpress the Y variable as multiple regression models of X variable: $Y = XB$. The values of regression coefficients depend on the scaling of X and Y, and the scaled regression coefficient is indicative of the relative importance of the variables to the model and can be used to assess which are the original variables with stronger relationship with the output variable (Y)
Variable importance in the projection (VIP)	The VIP plot is a reliable means of ranking variables according to their importance. It gives a visual display of the observations made from the score and loading plots and helps to understand the relative importance of each variable included in the analysis. The VIP values reflect the importance of the terms in the model with respect to Y, that is, its correlation to all the responses and with respect to X (the projection). VIP display is the most condensed way of representing PLS model information as it takes into account a weighted summary of all principal components of the model unlike coefficient and loading plots, which considers only the first two principal components

16.4.3 PLS-Discriminant Analysis (PLSDA)

Once the grouping/clusters are identified with PCA/PLS, PLS-discriminant analysis (PLS-DA) can be performed to sharpen the separation between the groups of observations by rotating the latent variables/PCs, such that the maximum separation among the classes are obtained and variables carrying the class-separating information are identified [13, 14].

PLS-DA modeling, in particular, has been used for assessing process comparability at different operational scales. Various steps include:

- Fitting a PLS model between the X variables and an artificial Y matrix of dummy variables (used for describing the class membership of each observation).
- Finding a discriminant plane in which the observations are well separated according to their class membership.
- Q^2 values obtained by PLS-DA modeling are used as a good measure to quantify the degree of dissimilarity between datasets.
- The higher the value of Q^2, the lesser the similarity between the datasets.

It should be mentioned that the Y variable consists of only 0s and 1s, because of which all the observations of the X variable having similar values will occupy a small pocket in the X space, and the method fails when some of the classes suffer from a lack of homogeneity and similarity.

In addition to these modeling approaches, other important MVDA techniques that are used include multivariate curve resolution (MCR) [37], multiblock PCA/PLS [38, 39], and hierarchical PCA/PLS [39, 40].

16.5 Case Studies of Use of Chemometrics in Process Chromatography

Case Study I Improved Process Analytical Technology (PAT) for Protein A Chromatography Using Predictive PCA

Protein A chromatography is widely used for the capture and purification of antibodies and Fc fusion proteins [9]. However, cost of protein A resins is significantly higher than the conventional resins and is therefore the single largest contributor toward the cost of downstream processing [41]. Resin integrity and product yield are the critical attributes of protein A chromatography that are routinely monitored to obtain optimal purification without a loss with respect to process economics [9]. Typically, height equivalent to theoretical plate (HETP) and peak asymmetry are determined after a fixed number of cycles during commercial operations to monitor resin integrity and thus column performance. If either of these criteria is found to exceed preset ranges, the resin is taken out of column and the column is repacked. However, this approach suffers from a major disadvantage that the failure in resin integrity typically lags column underperformance (with respect to product yield and quality). Thus, there is a likelihood of column failure before repacking has been initiated.

In this case study, PCA has been suggested as the appropriate tool for obtaining quantitative determination of column integrity and yield decay from analysis of the chromatograms [9]. The proposed approach thereby enables the operator to take real-time decisions of when to repack the column prior to column failure. The data under consideration was obtained from a scale-down resin lifetime study (96 cycles) using a single resin lot. The chromatographic run consisted of the following process steps performed sequentially: protein loading, two washes for unbound host cell protein removal, elution of the purified protein, stripping, flushing, sanitizing, and finally re-equilibrating the column for storage. Operating conditions were different for different process steps with each step in the chromatographic run being monitored by pH, conductivity, and UV measurements. A typical chromatogram displayed transitions in the measured values of the monitored parameters at each step. Nine transitions (significant in magnitude) were selected for building a predictive PCA model. These pertained to transitions from:

1) Load to wash 1 (detected by UV, conductivity, and pH)
2) Wash 1 to wash 2 (detected by UV)
3) Entire elution profile (detected by UV, conductivity, and pH)
4) Flush to sanitization (detected by conductivity and pH)

As mentioned earlier, it is of utmost necessity to pretreat and reorganize the data in a form suitable for multivariate statistical analysis. In this study, the raw data (the selected transition/profiles) was first aligned both in horizontal and vertical directions. While the vertical alignment was performed by normalizing the data with respect to maximum and minimum values in each step, horizontal alignment was performed by first converting the raw data into CV units and then collecting the data at the same CV intervals so that number of data points remained the same. Subsequently, the aligned data was arranged in $(m \times n)$ matrix form, which was further mean centered so that overall analysis focused on the deviation of each cycle from the mean rather than on the degree of similarity between the cycles.

Development of PCA model for the dataset involved decomposing the data matrix into a series of paired column vectors, with scores "*t*" and loadings "*p*." PCs capturing the maximum variability in the data structure were extracted by the covariance method. The PCs were extracted until the combined variance explained by them exceeded 98%. In this study, the number of PCs satisfying the said criterion was three. In addition, reconstruction of the chromatogram using the PC showed similarity with the original profile thereby confirming the fact that PCs are representative of the dataset being analyzed. PC1 accounted only for the general shape of the profile without capturing the deviations between the cycles. On the other hand, PC2 and PC3 captured most of the deviations, and based on these observations, the data modeling was further carried out using these two components.

Experimental results revealed resin integrity breach occurring at cycles 48 and 54. For multivariate data modeling, the dataset from 1 to 48 was selected for developing PCA analysis for column bed integrity failure detection. Within this, the first 36 cycles were used as training set, while the remaining cycles were used as test set data for cross validation of the model results.

Data analysis provided the following additional information (Figure 16.4):

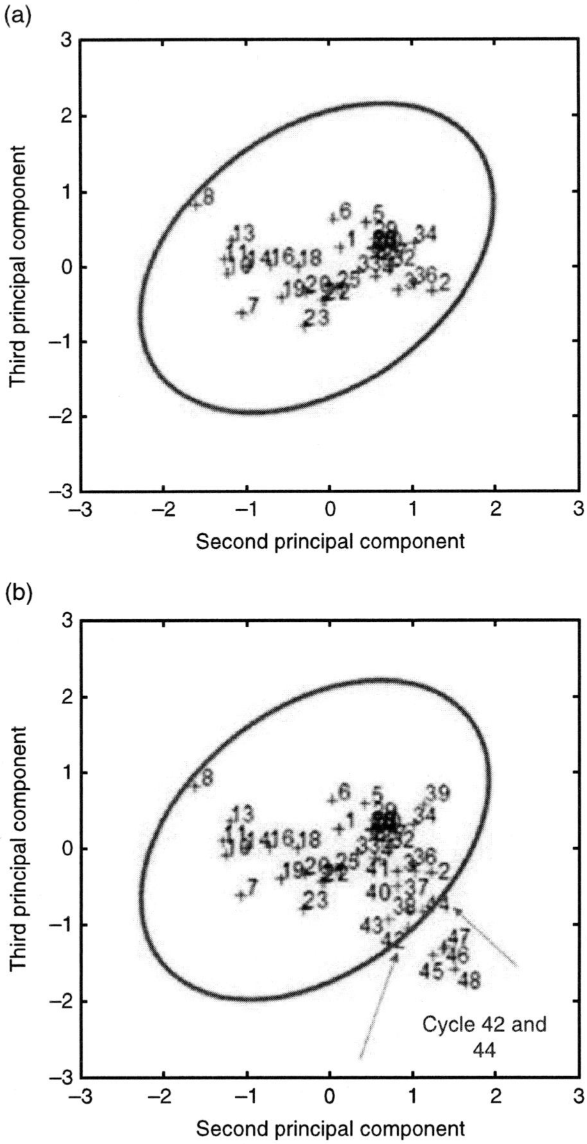

Figure 16.4 Score plots for second and third PCs calculated from UV elution curves: (a) training set cycles 1–36, (b) training set cycles 1–36 plus test set cycles 37–48. Figure adapted from Hou et al. [9] with copyright permission. It is evident that column performance starts deteriorating right from 42nd cycle even though experimentally it was observed only at 48th cycle.

- Cycles 45–48 lied outside the T^2 Hotelling ellipse indicating them as being outliers and thus failing the integrity test.
- Cycles 42–44 lied on the boundary of the ellipse indicating that column performance starts deteriorating from the 42nd cycle even though experimentally it is observed only at the 48th cycle.

This information from the PCA model can be advantageously capitalized for *a priori* determining if a trend toward integrity failure is exhibited and using this knowledge to repack the column before the actual decline in column performance starts. The proposed approach can thus be used as a PAT tool for ensuring smooth process run (particularly at large scale) and achieving consistent product quality.

Case Study II A Tool for Selective Inline Quantification of Co-eluting Proteins in Chromatography Using Spectral Analysis and PLS Regression

This case study presents a chemometrics-based protein quantitation model built using UV spectral data collected during the chromatographic separation of a ternary protein mixture consisting of lysozyme, ribonuclease A, and cytochrome c obtained from different sources [6]. The rationale of selection of the protein mixture for the case study was that the isoelectric point distributions allowed the separation of the various components by cation exchange chromatography at neutral pH.

Traditionally, single wavelength-based UV absorbance measurements (mostly at 280 nm) of the eluted fractions are the routinely adopted procedure for quantifying proteins. However, this method does not allow to differentiate between product and contaminants [42]. This is particularly disadvantageous at a polishing step in bioprocessing where the primary aim is to remove product-related impurities that have nearly identical physicochemical properties to the main product. To overcome this analytical conundrum, at line (using fluorescence spectroscopy) [43], online (using HPLC and UHPLC) [42, 44–48], and rigorous mathematical model-based PAT tools [48, 49] for titer determination have been proposed. However, all these proposed tools involve quantification of a two-component system. Use of chemometrics in combination with the spectral analysis is explored in this case study as an alternative for selective quantification of co-eluting proteins in a chromatographic process [6].

The proposed approach involves integration of software and chromatography hardware to create a specific setup comprising liquid chromatography, diode array detector, and software interface using MATLAB. The proposed approach allows for pooling decisions based on product purity and permits differentiation between product and co-eluting contaminants using multivariate PLS modeling of the mid-UV spectral data (240–300 nm) acquired during the process run. The basis of the approach is that different protein species will normally have different number of aromatic amino acids and thereby exhibit different mid-UV spectra. Thus, PLS was applied to calibrate models that correlate mid-UV absorption spectra with selective concentrations of three model proteins (lysozyme, cytochrome c, and ribonuclease a). The calibrated PLS models, in turn, were successfully applied for selective inline quantification and peak deconvolution of the co-eluting components by spectral analysis of various multicomponent protein mixtures. The software interfacing to the data collection unit of the established setup was performed in such a way that once the criterion has been set for pooling (concentration >0.002 g/L and mass fraction of target protein >0.8) and detected by the MATLAB function that performs calculation of selective protein concentrations based on the mid-UV spectral data acquired by DAD and calibrated PLS model, a digital signal is sent to the pump subunit of the chromatograph to start pooling. The decision for stopping the pooling is also managed in the same manner based on a preset criterion (mass fraction of target protein <0.8).

The results obtained with the presented approach were rigorously validated by comparing with offline analytical chromatography, and a good agreement between the two approaches was established as seen in Figure 16.5. However, a

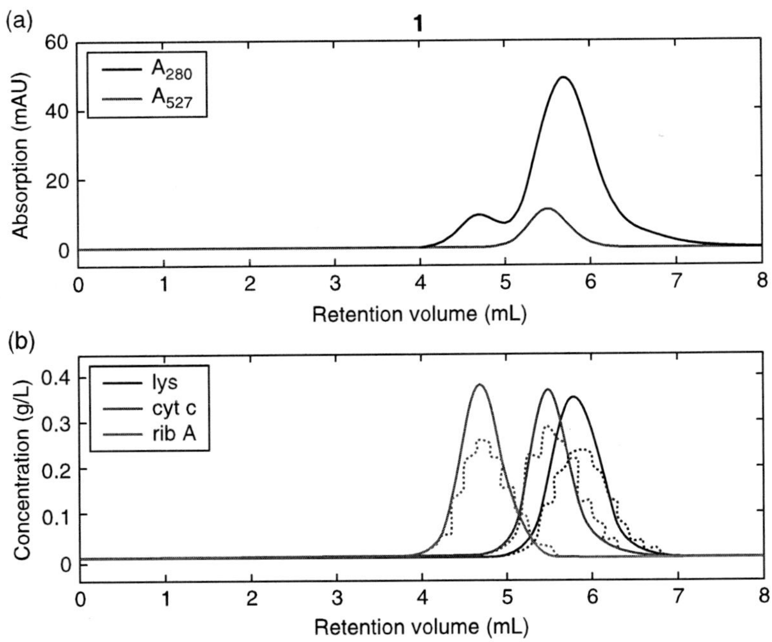

Figure 16.5 Illustration 1: Resulting chromatograms of the inline measurement and off-line analytical chromatography. Illustration 2: Resulting chromatograms of the product purity-based real-time pooling during a chromatographic separation of lys, cyt c, and rib A on SP Sepharose FF. It is observed that the developed model delivers precise results for the retention times and peak flanks for a ternary model protein system. Figure adapted from Brestrich et al. [6] with copyright permission. (*See insert for color representation of the figure.*)

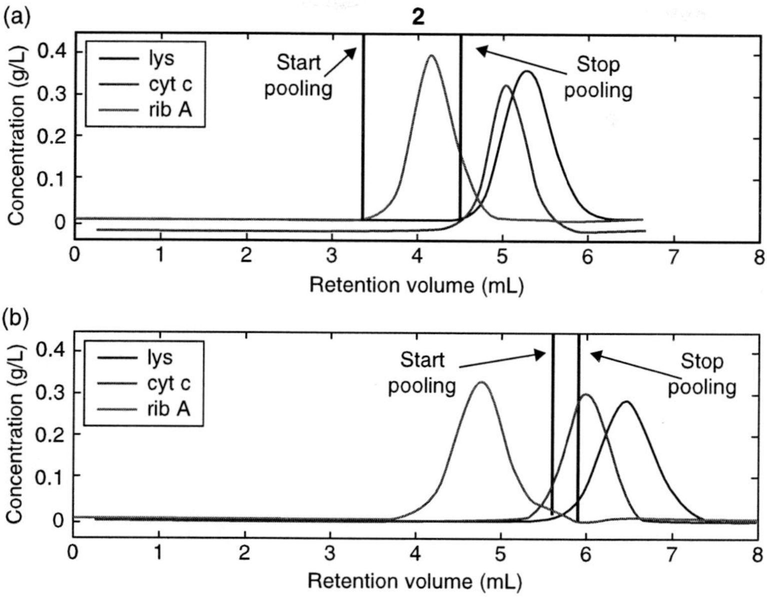

Figure 16.5 (Continued)

major limitation of the proposed method is the requirement of pure protein for PLS model calibration. This may be an issue when performing industrial process development as obtaining pure forms of all species may be difficult or at times not possible. Barring this, the proposed method has the potential of managing batch-to-batch variability by real-time pooling and thereby facilitating consistent pool purity and host cell protein monitoring during biopharmaceutical process/product development.

The obtained results indicate that the developed model delivers precise results for the retention times and peak overlaps for the ternary model protein system under consideration. Also, the tool for selective inline quantification can be successfully applied for making product purity-based, real-time pooling decisions.

16.6 Guidance on Performing MVDA

The most attractive feature of MVDA methods is that they do not require any mechanistic understanding of the system being analyzed. The relationship between and among the variables in the dataset is derived empirically. However, to take advantage of this strength and to prevent drawing of any spurious conclusion, a good understanding of limitations of these techniques is of

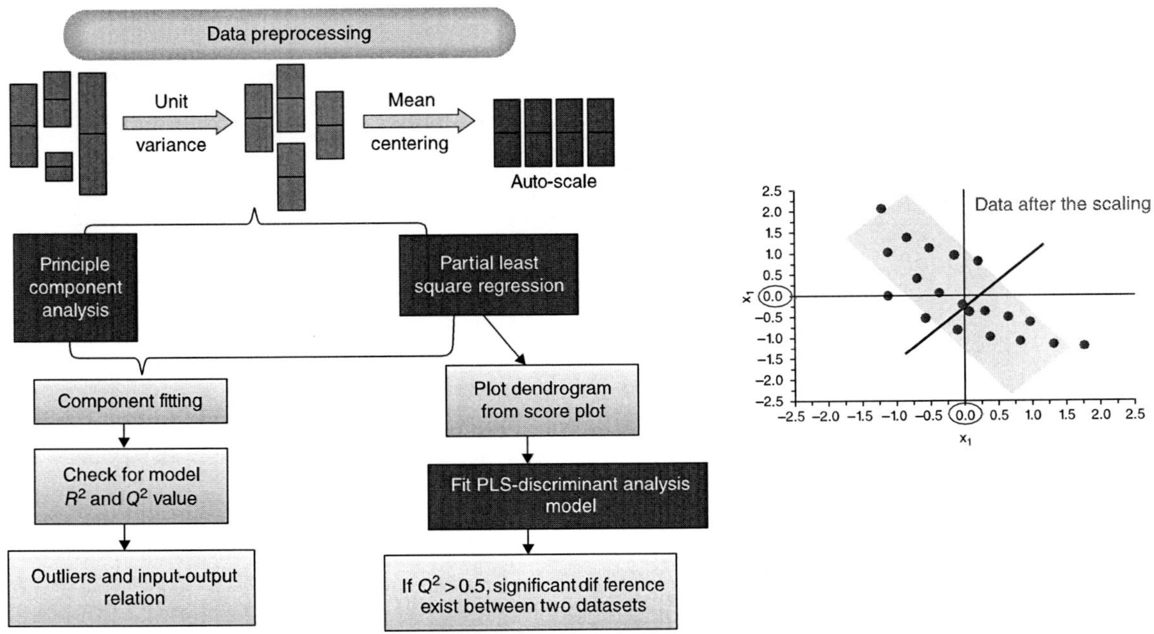

Figure 16.6 Flowchart illustrating the general workflow of multivariate data analysis (MVDA) of bioprocessing data. Adapted from Rathore et al. [14] with copyright permission. (*See insert for color representation of the figure.*)

utmost necessity. A description of the various pitfalls that commonly occur while performing MVDA of bioprocessing data has been recently published [14]. In this section we summarize some of the key points:

- **Amount of data**: It is of prime importance that the data included in the analysis truly represents the variability in the system. It is important to have confidence that the relationships identified are not spurious. However, the minimum amount of data required to develop robust and accurate models is likely to vary from application to application. In general, the larger the number of variables, the better the quality of the model. Also, in the case when data is limited, more effort should be made to unveil all possible information with a versatile and clear procedure.
- **Selection of the best model**: It is necessary that the developed model is not over-fitted as such model results in poor predictive ability for any new data set. To overcome this issue, cross validation of the model is necessary to assess its predictive ability. Cross validation can be done in two stages. One is by internal validation that is performed during model development. The other is external validation that is performed on a new dataset not included in model development.
- **Model interpretation**: The outputs from the developed models should be carefully interpreted, and it should be ensured that the conclusions drawn are not vague. Also, all the characteristics of the developed models (RMSE, R^2, number of principal components/latent variables, variance explained by each PC) should be clearly analyzed and consistency in the obtained results be maintained.

A general workflow for performing MVDA analysis of bioprocessing data is presented in Figure 16.6.

References

1. Amigo, J.M., Skov, T. & Bro, R., 2010. ChroMATHography: solving chromatographic issues with mathematical models and intuitive graphics. *Chemical Reviews*, 110(8), pp.4582–4605.
2. Rathore, A.S. et al., 2012. Chemometrics applications in biotechnology processes: predicting column integrity and impurity clearance during reuse of chromatography resin. *Biotechnology Progress*, 28(5), pp.1308–1314.
3. Andersson, R. & Hämäläinen, M.D., 1994. Simplex focusing of retention times and latent variable projections of chromatographic profiles. *Chemometrics and Intelligent Laboratory Systems*, 22(1), pp.49–61.
4. Gong, F. et al., 2005. Data preprocessing for chromatographic fingerprint of herbal medicine with chemometric approaches. *Analytical Letters*, 38(14), pp.2475–2492.

5. Pate, M.E. et al., 1998. Pre-processing of chromatographic data for principal component analysis. *Bioprocess Engineering*, 19(4), pp.297–305.
6. Brestrich, N. et al., 2014. A tool for selective inline quantification of co-eluting proteins in chromatography using spectral analysis and partial least squares regression. *Biotechnology and Bioengineering*, 111(7), pp.1365–1373.
7. Rathore, A.S. & Mhatre, R., 2011. *Quality by Design for Biopharmaceuticals: Principles and Case Studies*, Hoboken, NJ: John Wiley & Sons.
8. Brestrich, N. et al., 2015. Advances in inline quantification of co-eluting proteins in chromatography: process-data-based model calibration and application towards real-life separation issues. *Biotechnology and Bioengineering*, 112(7), pp.1406–1416
9. Hou, Y. et al., 2011. Improved process analytical technology for protein a chromatography using predictive principal component analysis tools. *Biotechnology and Bioengineering*, 108(1), pp.59–68.
10. Vaidyanathan, S. et al., 2001. Assessment of near-infrared spectral information for rapid monitoring of bioprocess quality. *Biotechnology and Bioengineering*, 74(5), pp.376–388.
11. Undey, C. & Cinar, A., 2002. Statistical monitoring of multistage, multiphase batch processes. *IEEE Control Systems Magazine*, 10 pp.40–52.
12. Cunha, C.C.F. et al., 2002. An assessment of seed quality and its influence on productivity estimation in an industrial antibiotic fermentation. *Biotechnology and Bioengineering*, 78(6), pp.658–669.
13. Rathore, A.S., Mittal, S., Pathak, M. & Mahalingam, V., 2014a. Chemometrics application in biotech processes: assessing comparability across processes and scales. *Journal of Chemical Technology and Biotechnology*, 89(9), pp.1311–1316.
14. Rathore, A.S., Mittal, S., Pathak, M. & Arora, A., 2014b. Guidance for performing multivariate data analysis of bioprocessing data: pitfalls and recommendations. *Biotechnology Progress*, 30(4), pp.967–973.
15. Bhushan, N., Hadpe, S. & Rathore, A.S., 2012. Chemometrics applications in biotech processes: assessing process comparability. *Biotechnology Progress*, 28(1), pp.121–128.
16. Kirdar, A.O., Green, K.D. & Rathore, A.S., 2008. Application of multivariate data analysis for identification and successful resolution of a root cause for a bioprocessing application. *Biotechnology Progress*, 24(3), pp.720–726.
17. Kirdar, A.O. et al., 2010. Application of near-infrared (NIR) spectroscopy for screening of raw materials used in the cell culture medium for the production of a recombinant therapeutic protein. *Biotechnology Progress*, 26(2), pp.527–531.
18. Rathore, A.S., Bhushan, N. & Hadpe, S., 2011. Chemometrics applications in biotech processes: a review. *Biotechnology Progress*, 27(2), pp.307–15.
19. Eriksson, L. et al., 2013. *Multi-and megavariate data analysis basic principles and applications*, Malmö, Sweden: Umetrics Academy.

20. Van den Berg, R.A. et al., 2006. Centering, scaling, and transformations: improving the biological information content of metabolomics data. *BMC Genomics*, 7(1), p.142.
21. Kubinyi, H., 1993. *3D QSAR in drug design: volume 1: theory methods and applications*, Dordrecht: Springer Science & Business Media.
22. Davies, O.L. & Goldsmith, P.L., 1972. *Statistical methods in research and production* (4th edn.), Edinburgh: Oliver and Boyd, p.51.
23. Sánchez, F.C., Lewi, P.J. & Massart, D.L., 1994. Effect of different preprocessing methods for principal component analysis applied to the composition of mixtures: detection of impurities in HPLC—DAD. *Chemometrics and Intelligent Laboratory Systems*, 25(2), pp.157–177.
24. Farrés, M., Piña, B. & Tauler, R., 2015. Chemometric evaluation of Saccharomyces cerevisiae metabolic profiles using LC–MS. *Metabolomics*, 11(1), pp.210–224.
25. Serrano-Lourido, D. et al., 2012. Classification and characterisation of Spanish red wines according to their appellation of origin based on chromatographic profiles and chemometric data analysis. *Food Chemistry*, 135(3), pp.1425–1431.
26. Farrés, M. et al., 2012. Chemometric evaluation of different experimental conditions on wheat (Triticum aestivum L.) development using liquid chromatography mass spectrometry (LC–MS) profiles of benzoxazinone derivatives. *Analytica Chimica Acta*, 731, pp.24–31.
27. Harms, J. et al., 2008. Defining process design space for biotech products: case study of Pichia pastoris fermentation. *Biotechnology Progress*, 24(3), pp.655–662.
28. Kumar, V., Bhalla, A. & Rathore, A.S., 2014. Design of experiments applications in bioprocessing: concepts and approach. *Biotechnology Progress*, 30(1), pp.86–99.
29. Jackson, J.E., 2005. *A user's guide to principal components*, Hoboken: John Wiley & Sons, Inc.
30. Wold, S., Esbensen, K. & Geladi, P., 1987. Principal component analysis. *Chemometrics and Intelligent Laboratory Systems*, 2(1), pp.37–52.
31. Jolliffe, I., 2014. Principal component analysis, Wiley Online Library, DOI: 10.1002/9781118445112.stat06472.
32. Umetri, A.B., 2004. SIMCA-P 10.5, User Guide and Tutorial. Umeå: Umetri.
33. Kirdar AO et al., 2007. Application of multivariate analysis toward biotech processes: case study of a cell-culture unit operation. *Biotechnology Progress*, 23(1), pp.61–67.
34. Geladi, P. & Kowalski, B.R., 1986. Partial least-squares regression: a tutorial. *Analytica Chimica Acta*, 185, pp.1–17.
35. De Jong, S. & Phatak, A., 1997. Partial least squares regression. *Recent Advances in Total Least Squares Techniques and Errors-in-Variables Modeling*, 11(4), pp.311–338.

36. Ferreira, A.P. & Tobyn, M., 2014. Multivariate analysis in the pharmaceutical industry: enabling process understanding and improvement in the PAT and QbD era. *Pharmaceutical Development and Technology*, 20(5), pp.513–527.
37. Tauler, R. & Barceló, D., 1993. Multivariate curve resolution applied to liquid chromatography—diode array detection. *Trends in Analytical Chemistry*, 12(8), pp.319–327.
38. Smilde, A.K., Westerhuis, J.A. & Boque, R., 2000. Multiway multiblock component and covariates regression models. *Journal of Chemometrics*, 14(3), pp.301–331.
39. Westerhuis, J.A., Kourti, T. & MacGregor, J.F., 1998. Analysis of multiblock and hierarchical PCA and PLS models. *Journal of Chemometrics*, 12(5), pp.301–321.
40. Janne, K. et al., 2001. Hierarchical principal component analysis (PCA) and projection to latent structure (PLS) technique on spectroscopic data as a data pretreatment for calibration. *Journal of Chemometrics*, 15(4), pp.203–213.
41. Jiang, C. et al., 2009. A mechanistic study of Protein A chromatography resin lifetime. *Journal of Chromatography A*, 1216(31), pp.5849–5855.
42. Rathore, A.S., Yu, M., et al., 2008b. Case study and application of process analytical technology (PAT) towards bioprocessing: use of on-line high-performance liquid chromatography (HPLC) for making real-time pooling decisions for process chromatography. *Biotechnology and Bioengineering*, 100(2), pp.306–316.
43. Rathore, A.S. et al., 2009. Case study and application of process analytical technology (PAT) towards bioprocessing: use of tryptophan fluorescence as at-line tool for making pooling decisions for process chromatography. *Biotechnology Progress*, 25(5), pp.1433–1439.
44. Fahrner, R.L. & Blank, G.S., 1999. Real-time control of antibody loading during protein A affinity chromatography using an on-line assay. *Journal of Chromatography A*, 849(1), pp.191–196.
45. Fahrner, R.L. et al., 1998. Real-time control of purified product collection during chromatography of recombinant human insulin-like growth factor-I using an on-line assay. *Journal of Chromatography A*, 827(1), pp.37–43.
46. Kaltenbrunner, O. et al., 2012. Risk–benefit evaluation of on-line high-performance liquid chromatography analysis for pooling decisions in large-scale chromatography. *Journal of Chromatography A*, 1241, pp.37–45.
47. Rathore, A.S., Wood, R., et al., 2008a. Case study and application of process analytical technology (PAT) towards bioprocessing: II. Use of ultra-performance liquid chromatography (UPLC) for making real-time pooling decisions for process chromatography. *Biotechnology and Bioengineering*, 101(6), pp.1366–1374.
48. Rathore, A.S. et al., 2010. Large scale demonstration of a process analytical technology application in bioprocessing: use of on-line high performance liquid chromatography for making real time pooling decisions for process chromatography. *Biotechnology Progress*, 26(2), pp.448–457.
49. Mendhe, R. et al., 2015. Comparison of PAT based approaches for making real-time pooling decisions for process chromatography—use of feed forward control. *Journal of Chemical Technology and Biotechnology*, 90(2), pp.341–348.

17

Mid-UV Protein Absorption Spectra and Partial Least Squares Regression as Screening and PAT Tool

Sigrid Hansen[1*], *Nina Brestrich*[2*], *Arne Staby*[3], *and Jürgen Hubbuch*[2]

[1] *Protein Purification Development, Novo Nordisk A/S, Gentofte, Denmark*
[2] *Institute of Engineering in Life Sciences, Section IV: Biomolecular Separation Engineering, Karlsruhe Institute for Technology, Karlsruhe, Germany*
[3] *CMC Project Planning & Management, Novo Nordisk A/S, Bagsværd, Denmark*

17.1 Introduction

Since the promotion by the FDA, quality by design (QbD) and process analytical technology (PAT) have become new buzzwords in the biopharmaceutical industry and have been receiving a lot of attention. The implementation of QbD involves a thorough understanding of both the product and the process by which it is manufactured [1]. This leads to a careful understanding of the relationship between clinical properties of the product and critical quality attributes (CQAs) as well as an investigation of the relationship between CQAs and the process. Parameters in the process influencing CQAs are called critical process parameters (CPPs). By understanding the relationships between CPPs and CQAs, the process can be designed such that the desired product quality is achieved in the presence of process variability. In order to further ensure a consistent product quality, the PAT concept promotes the monitoring and control of both CQAs and CPPs [2]. This helps to manage batch-to-batch variability and to develop a process, in which all variability is managed by the process itself (cf. Chapter 1). In this chapter, a screening tool and PAT tool for chromatography are presented addressing typical issues during process development according to the QbD concept and during large-scale production.

During process development, the examination of relationships between CPPs and CQAs can be performed using high-throughput experimentation

* Contributed equally.

(HTE). The term HTE designates miniaturized experiments performed automated with a high throughput and often parallelized. For column chromatography, HTE was realized through the development of miniaturized packed columns (e.g., RoboColumn, Atoll, Weingarten, Germany) with resin volumes down to 50 μL, which can be operated in parallel by a liquid-handling station. This enables the generation of more data with the often limited amount of material at hand and within the short time frame available for purification process development, so-called high-throughput process development (HTPD). In 2005, Bench et al. [3] published the first review on HTPD. In this review and in a publication by Rege et al. [4], a key problem of the lack of matching analytical throughput was identified. It was described as an analytical bottleneck that creates a situation with an inherent trade-off between analytical information and analysis time. The screening tool presented in this chapter allows for a significant increase in analytical throughput compared to analytical chromatography.

A key goal in biopharmaceutical manufacturing is the control of process variability. The management of process variability by the process itself is also a prerequisite for the implementation of continuous processing in the biopharmaceutical industry. For chromatographic purification, process variability can result in deviating retention volumes or peak shapes from one cycle to another. This complicates the decision on when to start and stop product collection, if the contaminants are co-eluting with the target protein. As pooling decisions are often based on univariate UV absorption measurements that cannot be used to differentiate between the eluting species, process variability usually leads to inconsistent pool purities or yields. To tackle this issue, various PAT tools for chromatographic purification have been published in the past years. One group of PAT tools comprises online applications. At-line fluorescence spectroscopy has been performed to differentiate between correctly folded and misfolded proteins in order to make pooling decisions [5]. Further, at-line HPLC analysis has been used to analyze the product stream in a twin column multicolumn countercurrent solvent gradient purification (MCSGP) process to gain a feedback for a PID controller regulating the peak cutting of the upcoming cycle [6]. However, at-line analysis requires personnel and carries the risk of human error and thus delayed data. Another group of PAT tools is formed by online-(U) HPLC applications, which have been successfully applied to make real-time pooling decisions [7–11]. Depending on the elution flow rate of the preparative separation in question, a problem of this technique is the rather long response time of the analysis and often decreased productivity. An alternative to these applications is the PAT tool presented in this chapter, which allows an in-line quantification of co-eluting proteins. The tool has a response time of <1 s and allows real-time pooling decisions.

Both the screening and the PAT tool are based on partial least square (PLS) regression models, correlating mid-UV protein absorption spectra with selective protein concentrations. The fundamentals of intrinsic protein absorption and PLS as well as their application for selective protein quantification are covered in the first section. In subsequent sections, the similarity of protein absorption spectra and its effect on the precision of PLS models is described. Finally, the screening and PAT tool are discussed in detail, including several case studies.

17.2 Mid-UV Protein Absorption Spectra and Partial Least Squares Regression

Spectroscopic methods for quantitative and qualitative protein analysis are well established and widely applied. A common statistical method to extract qualitative and quantitative information from any kind of spectral data is multivariate data analysis (MVDA). One tool of MVDA, which is often applied to correlate spectra with quantitative physical or (bio-)chemical properties, is PLS regression. In the following text, the application of PLS for correlating protein absorption spectra with selective protein concentrations is described.

17.2.1 Intrinsic Protein Mid-UV Absorption

Protein quantification via intrinsic mid-UV absorption (200–300 nm) has been a standard method for decades. The delocalized electrons of both peptide bonds and the aromatic structures in the amino acid residues are the main originators of the mid-UV absorption of proteins (cf. Figures 17.1 and 17.2). In the lower mid-UV range from 200 to 250 nm, the absorption of the aromatic

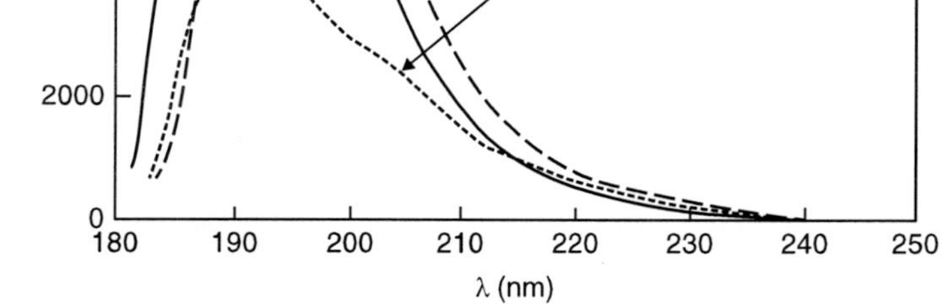

Figure 17.1 Absorbance in the far- and mid-UV range of peptide bonds in different conformations. *Source*: Rosenheck and Doty [12]. Reproduced with permission of Proceedings of the National Academy of Sciences of the United States of America.

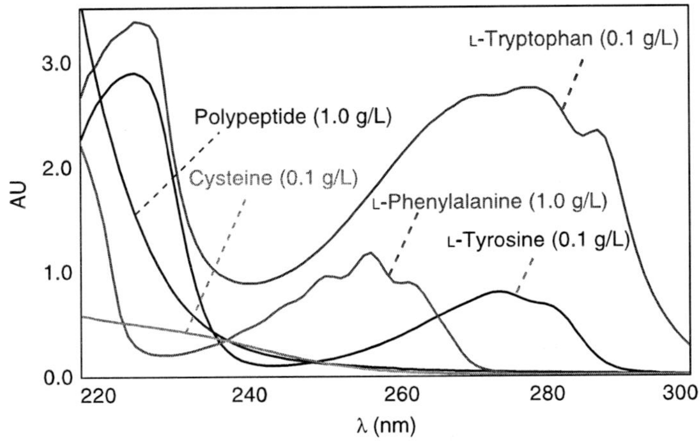

Figure 17.2 Spectra of the components mainly contributing to protein mid-UV absorption spectra. The measurements were performed with a Lambda 35 spectrophotometer (Perkin Elmer, Waltham, USA). *Source*: Adapted from Hansen et al. [13].

residues and the peptide bonds overlap. Also the amino acid residues containing sulfur absorb in this range. In the upper part of the mid-UV range 250–300 nm, only the aromatic residues contribute to absorbance. This range was termed the "simple region" [14] due to the small number of contributors to the intrinsic protein absorbance. The aromatic structures of the three amino acids absorb with different intensities. The absorption intensity of tryptophan is the highest, being approximately 7 times stronger than tyrosine and 70 times stronger that phenylalanine (based on molar absorptivity).

In this section, the term "spectrum" refers to the intrinsic absorption of proteins in the range 250–300 nm. Where nothing else is stated, the proteins were dissolved in 20 mM sodium phosphate buffer and the absorption spectra measured at ambient temperature. In Figure 17.3, the absorption spectra of a wide range of proteins are displayed. With the exception of those proteins containing a heme group, the main difference in the spectra is related to the amount of tryptophan present in the proteins. The two spectra in Figure 17.4 exemplify the difference in protein absorption spectra caused by tryptophan. One spectrum is narrow, representing a protein lacking tryptophan, and the other is wider, representing a spectrum of a protein rich in tryptophan. This correlates with tryptophan being the only residue that absorbs above 295 nm (cf. Figure 17.2).

The direct environment of the chromophores will also influence the spectra. In Figure 17.5 an example of a hypsochromic shift caused by changes in the polarity of the environment of tryptophan is shown. When tryptophan is dissolved in water, the polarity of the environment of the aromatic structure is high compared to the case where the aromatic structure of tryptophan is

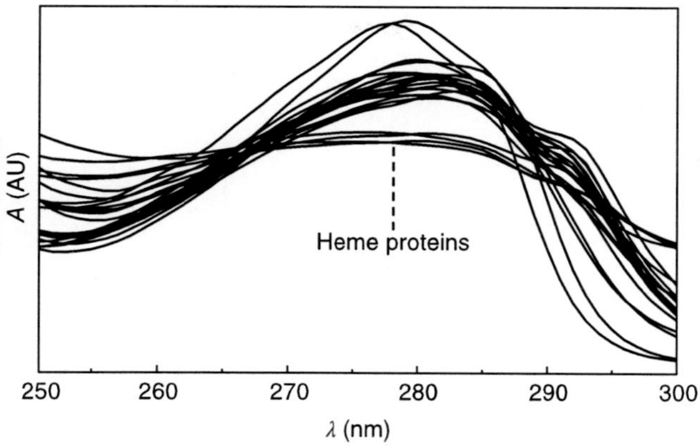

Figure 17.3 UV absorption spectra of 24 proteins. The spectra were recorded in protein solutions of 1.0 g/L. The spectra are normalized to equal total intensity for better visual comparison. The measurements were performed with a Lambda 35 spectrophotometer (Perkin Elmer, Waltham, USA).

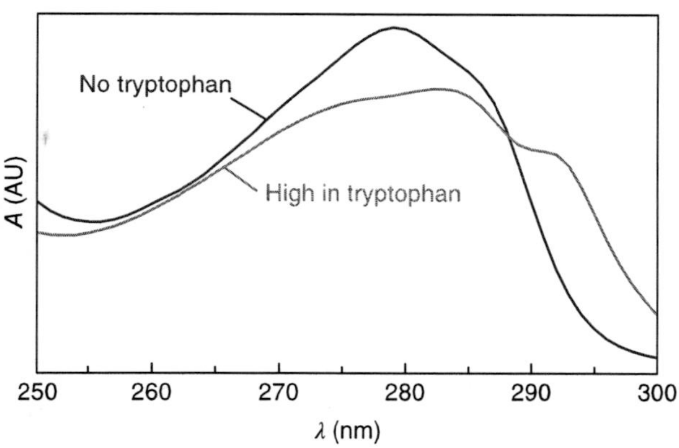

Figure 17.4 UV spectra representative of the extremes within the 24 measured spectra (proteins containing a heme group were omitted). For more details on the protein spectra, see Figure 17.11. The measurements were performed with a Lambda 35 spectrophotometer (Perkin Elmer, Waltham, USA).

present as amino acid residues in a folded protein. Hence, the absorbance of free tryptophan is blueshifted compared to the absorbance of a tryptophan-rich protein. The same effect can also be observed when comparing the spectra of the three aromatic amino acids with the protein spectra (cf. Figures 17.2 and 17.3). The amino acids exhibit a minimum in their spectra in the range 230–240 nm, whereas the minima of the protein absorption spectra are shifted to the range 250–255 nm.

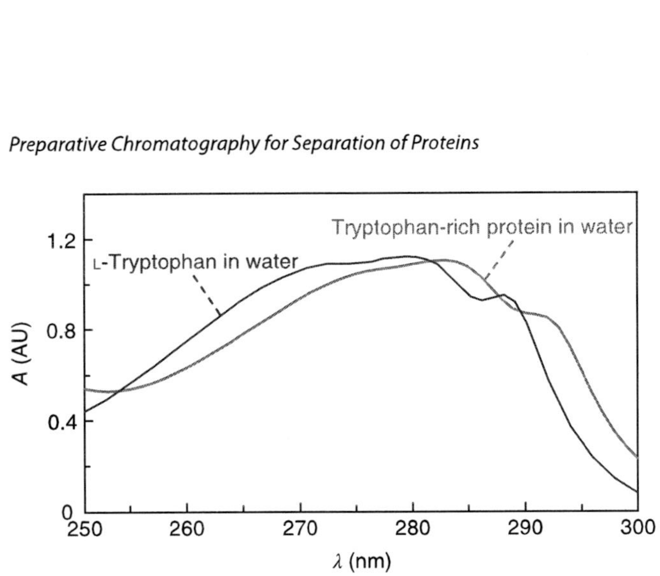

Figure 17.5 UV absorption spectra of L-tryptophan and a tryptophan-rich protein. The spectra have been normalized to equal total intensity. The measurements were performed with a Lambda 35 spectrophotometer (Perkin Elmer, Waltham, USA).

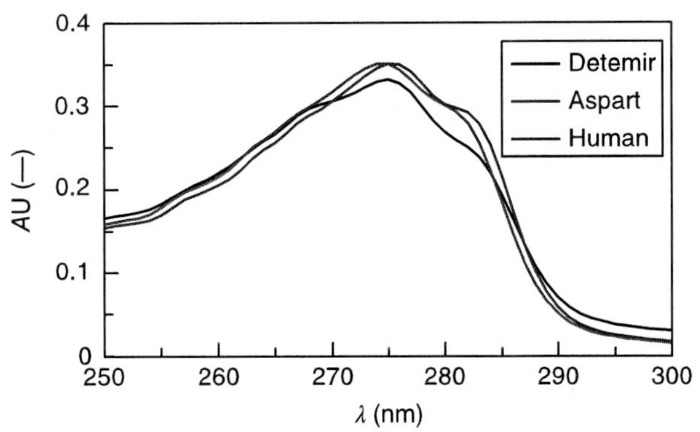

Figure 17.6 UV absorption spectra of insulin detemir, insulin aspart, and human insulin. The spectra have been normalized to equal total intensity. The measurements were performed with a Lambda 35 spectrophotometer (Perkin Elmer, Waltham, USA).

Changes in the environment of some of the chromophores in a protein, for example, due to conformational changes, can be detected in the absorption spectra. As an example, spectra of the human insulin, of the insulin analog aspart, and the insulin derivative detemir are shown in Figure 17.6. Insulin aspart differs from human insulin only in one amino acid; the proline at position B28 has been substituted with an aspartic acid. This introduces a charged residue at a position where a hydrophobic residue was previously positioned. As can be seen, the mid-UV absorption spectrum of insulin aspart is blueshifted compared to the spectrum of human insulin, as the change in structure has changed the environment of at least some of the chromophores in the insulin aspart molecule. Insulin detemir is a truncated version of human insulin (threonine in position B30 has been omitted) with a fatty acid (myristic acid) attached to B29. This change is also detectable in the mid-UV spectra, and the spectrum of insulin detemir is also different from that of human insulin. The detectability of these small molecular changes in the mid-UV spectrum points to the feasibility of using mid-UV spectroscopy for detection of protein product-related species in protein-based pharmaceuticals. Examples will be presented in Section 17.6.3.

17.2.2 Partial Least Squares Regression (PLS)

If a high degree of the mid-UV absorption data recorded from a protein mixture is caused by the intrinsic absorbance of the proteins, it is possible to correlate the spectral data with selective protein concentrations [13, 15]. This requires a multivariate regression to correlate the matrix of protein absorption spectra **X** (factors) with the matrix of selective protein concentrations **Y** (response) as displayed in Figure 17.7. Several multivariate regression methods exist such as multilinear regression (MLR), principal component regression (PCR), or PLS regression. If the data contains collinearities (which is the case for protein absorption spectra), PCR or PLS should be the methods of choice as they only apply relevant and are noncollinear information of **X** for the regression [16]. In some cases, PLS has been shown to be more efficient than PCR in extracting the most relevant information of **X** for the correlation with **Y** [17]. Due to the high quality of the regression models and the simplicity of its implementation, PLS has become the most popular among the multivariate regression methods [18].

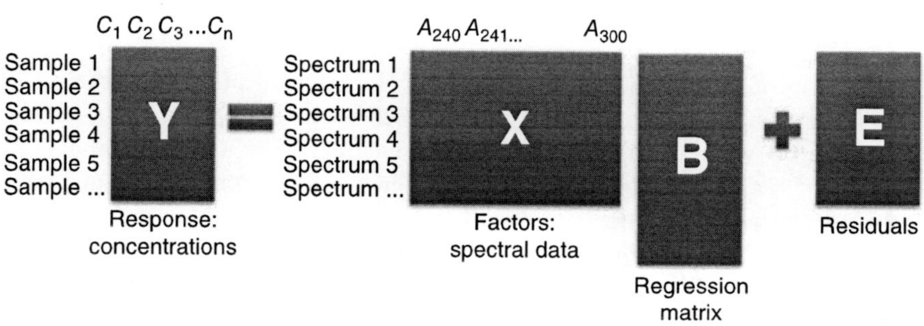

Figure 17.7 Partial least square regression between factor matrix **X** (mid-UV absorption spectra) and response matrix **Y** (selective protein concentrations). The matrix **B** contains the regression coefficients and the matrix **E** the residuals (information not included in the regression).

In biotechnology, numerous applications of PLS and spectral data for selective quantification of different species in multicomponent mixtures have been reported during the past decade. In upstream processing, PLS has been widely used for regression modeling between fluorescence, near-infrared, or mid-infrared spectra and biomass [19–23], substrate [22–26], product [20, 25, 26], and metabolite concentrations [22, 23]. Applications of spectral data and PLS for quantification in multicomponent mixtures in downstream processing may also be found [13, 15, 27–31].

In order to correlate only relevant information (variance) of the data matrix **X** with **Y**, the data is reduced when performing PLS. This data reduction is achieved by summarizing variables containing similar information (wavelengths in the case of spectra) in so-called latent variables (LVs). While the first LV contains the most information (variance of the original data), the information content decreases for higher LVs. The higher LVs usually contain detector noise. Consequently, the variance in the data can be explained by only a few LVs, leaving out higher LVs. This means that the number of LVs applied in PLS is a measure of the extent of data reduction. The result of PLS is a regression matrix **B** containing the regression coefficients and the residual matrix **E** containing the information not applied in the regression (cf. Figure 17.7). Usually, the matrix **B** is referred to as a PLS model and the generation of **B** as model calibration. In addition to the PLS model calibration, validation of the model and a careful application for prediction of samples with unknown composition are crucial. These three steps will be explained later for protein mid-UV absorption spectra. More detailed information about PLS and its practical application may be found elsewhere [32].

17.2.3 Application of PLS and Mid-UV Protein Absorption Spectra for Selective Protein Quantification

As displayed in Figure 17.7, the selective protein concentrations and the mid-UV absorption spectra of several samples are required for the multivariate regression between **X** and **Y**. To perform PLS, samples are usually arranged in the rows of **X** and **Y**, while the selective protein concentrations and absorption at each wavelength form the columns of **X** and **Y**. The calculation of **B** and **E** in dependence on the number of LVs used in the PLS model may easily be performed using suitable software. However, there are several crucial questions that have to be answered: Which samples should be selected for the PLS model calibration? What is the reasonable number of LVs for the PLS model (PLS model validation)? What should be considered when applying the calibrated and validated PLS model as screening or PAT tool? These questions will be answered later.

17.2.3.1 PLS Model Calibration

The samples applied in the PLS model calibration should be representative and cover the data space as completely and evenly as possible. For chromatographic steps, the data space consists of mixing ratios and concentration levels of the proteins to be separated. If the pure protein components are available, this data space can be covered by generating mixtures from pure protein stock solutions according to a design of experiments (DoE) setup. Figure 17.8 displays the procedure for a three-component system. The three components are first mixed according to a DoE setup (here, four-level D-optimal onion design), and the corresponding mid-UV absorption spectra are subsequently measured to obtain **X**. The protein concentration matrix **Y** is known from the DoE setup. However for true biopharmaceutical

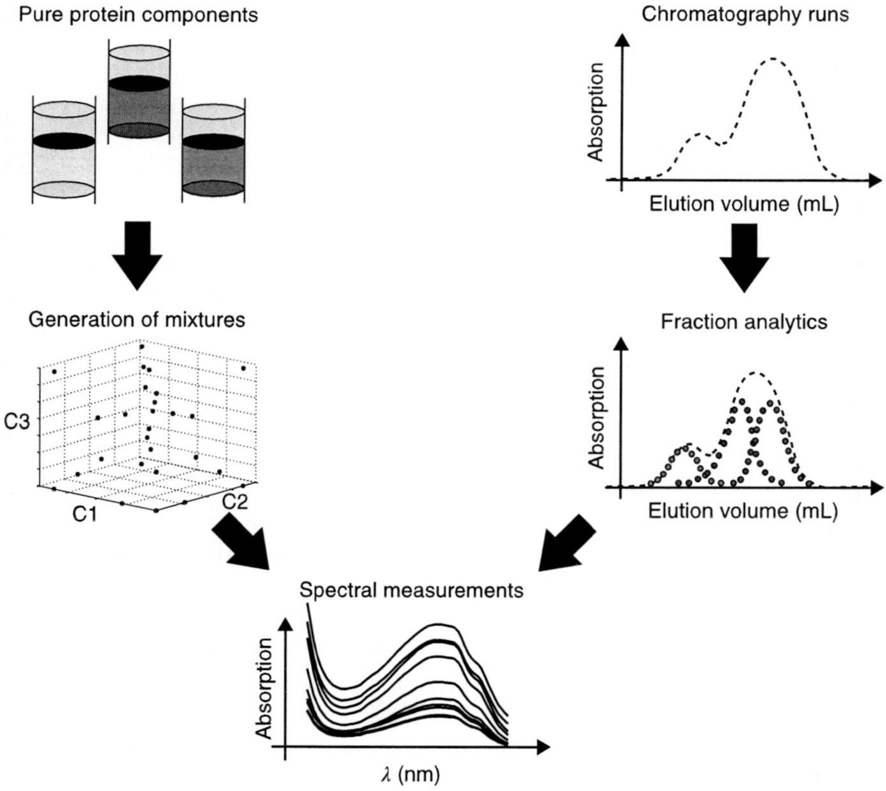

Figure 17.8 PLS model calibration from pure protein components (left side) or process data (right side): Pure components can be applied to generate mixtures according to a DoE setup, while fraction analytics of chromatography runs has to be performed to obtain **Y**, if no pure components are available. Finally, the mid-UV absorption spectra of the mixtures or fractions are measured to obtain **X**.

applications, pure components are often not available as this would require an elaborate isolation of the target protein and all relevant contaminants. This is usually impractical with respect to the time and material limitations in pharmaceutical process development, and thus process data (data from chromatography runs) may be applied for the PLS model calibration. Model calibration could, for example, be performed using data of runs from process robustness studies, which cover the expected data space of the final process. The concentration matrix **Y** can then be obtained by fraction analytics, and the spectral matrix **X** can be obtained by recording the absorption spectra of the corresponding fractions. Finally, **X** and **Y** can be correlated using PLS. To examine the optimal number of LVs during the model validation, various PLS models with different number of LVs should be calibrated.

17.2.3.2 PLS Model Validation

The higher the number of LVs used in a PLS model calibration, the more precise the established model for the prediction of protein concentrations in the calibration samples. Thus, the root mean square error of the calibration (RMSEC) decreases with the increasing number of LVs. However, the use of too many LVs can result in overfitting. Detector noise is fitted into the model, and the error increases for prediction of samples with different compositions than the calibration samples. To determine an optimal number of LVs, calibrated PLS models with different number of LVs can be applied to predict the protein concentrations of several validation samples (external validation). These samples should exhibit a different composition to that of the calibration samples and ideally be situated in between the calibration samples in the data space. If the model calibration and validation are performed using mixtures from pure protein components, this can be easily realized using a DoE setup. The root mean square error for prediction of protein concentrations in the validation samples (RMSEV) can then be calculated and plotted as a function of the number of LVs applied in the model calibration. The minimum corresponds to the optimal number of LVs. An alternative to an external validation sample set is cross validation, where all samples are applied for both calibration and validation. This approach might be useful, if it is not possible to identify or generate representative validation samples. To perform cross validation, some samples are excluded during model calibration, and the prediction error for these samples is calculated subsequently. For every number of LVs, this procedure is performed until each sample was excluded. As for the external validation, the optimal number of LVs corresponds to the minimum in a plot of the root mean square error for the prediction with respect to the number of LVs applied in the model. In addition to determination of the optimal number of LVs, residuals for prediction of both the calibration and validation samples should be examined for normal distribution.

17.2.3.3 Prediction of Unknown Samples

After calibration and validation, the PLS model can be applied for protein quantification in samples with unknown composition. When applying the model as a screening or PAT tool, concentrations and mixing ratios of the proteins should be within the calibrated data space.

17.3 Spectral Similarity and Prediction Precision

A PLS model based on spectral data and protein concentrations can be applied both as analytical tool in combination with HTE and as in-line process control during large-scale production. The value of this approach will depend on the number of other factors influencing the spectra and to what levels they can be controlled (e.g., solvent ionic strength, impurities not included in calibration, etc.). Further, the value will depend on how pronounced the spectral differences are compared to measurement noise and external influences (e.g., temperature). In general, it is not expected that the protein absorption spectra will change significantly under common mild conditions of standard protein purification processes. An overview of mid-UV protein absorption spectra and the influence of spectral similarity on the prediction precision of PLS models will be given in Section 17.3.1.

17.3.1 Overview of Protein Spectra

Spectra of a wide range of proteins are shown in Figure 17.9. In order to facilitate a visual comparison, the spectra are normalized to equal total intensity. In Figure 17.9a, all 24 spectra are shown and colored according to similarity. Here, two groups of similar spectra are directly identifiable among the 24 spectra. One group consists of spectra of proteins containing a heme group (spectra colored red, see also Figure 17.9b) and spectra of proteins without tryptophan (spectra colored green, see also Figure 17.9c). In the Figure 17.9d–f, three groups of spectra that are not immediately identifiable among the 24 spectra are shown. Table 17.1 presents the number of aromatic residues in each protein except for those containing a heme group. By comparing the protein groups with their number of aromatic residues, it was realized that the ratio of tryptophan to tyrosine correlates with the similarity of the spectra: Figure 17.9d, low tryptophan/tyrosine ratio; Figure 17.9e, medium tryptophan/tyrosine ratio; and Figure 17.9 f, high tryptophan/tyrosine ratio. This is in accordance with the absorption intensity of tryptophan and tyrosine being higher compared to the absorption intensity of phenylalanine (cf. Figure 17.2).

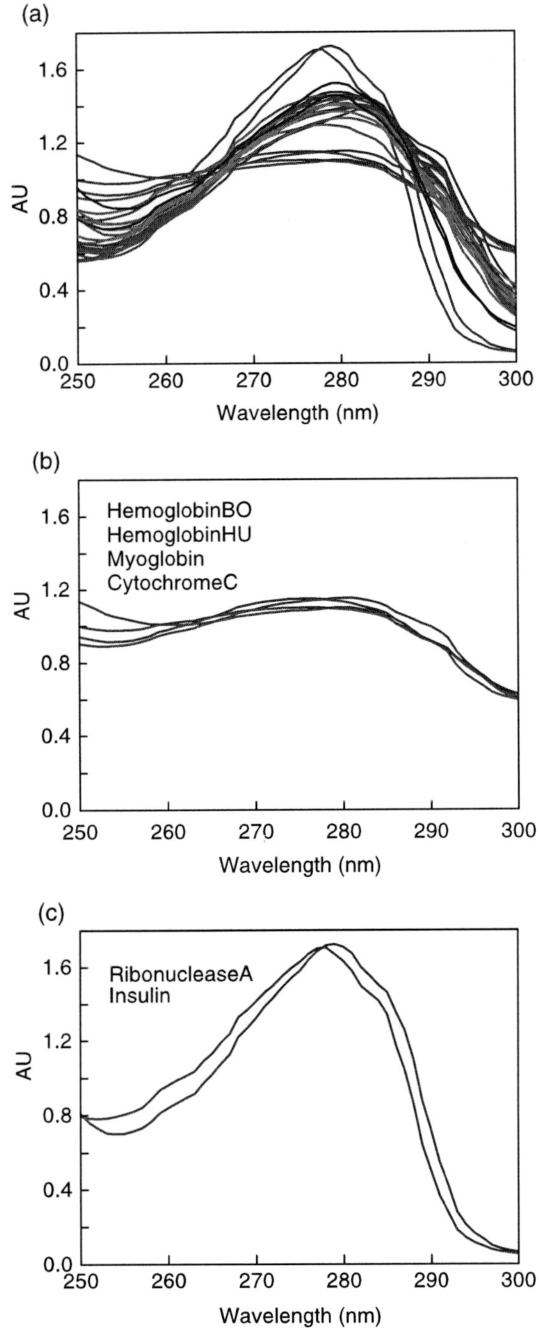

Figure 17.9 Spectra of 24 proteins measured in microtiter plates with a multimode microplate reader (Infinite 200, TECAN, Germany). All spectra are normalized to equal total intensity to facilitate comparison. Spectra are colored according to their similarity. (a) Spectra for all proteins. (b) Heme proteins. (c) Spectra of proteins without tryptophan. (d–f) Spectra of proteins with increasing ratio of tryptophan to tyrosine. *Source*: Hansen et al. [13]. Reproduced with permission of Wiley. (*See insert for color representation of the figure.*)

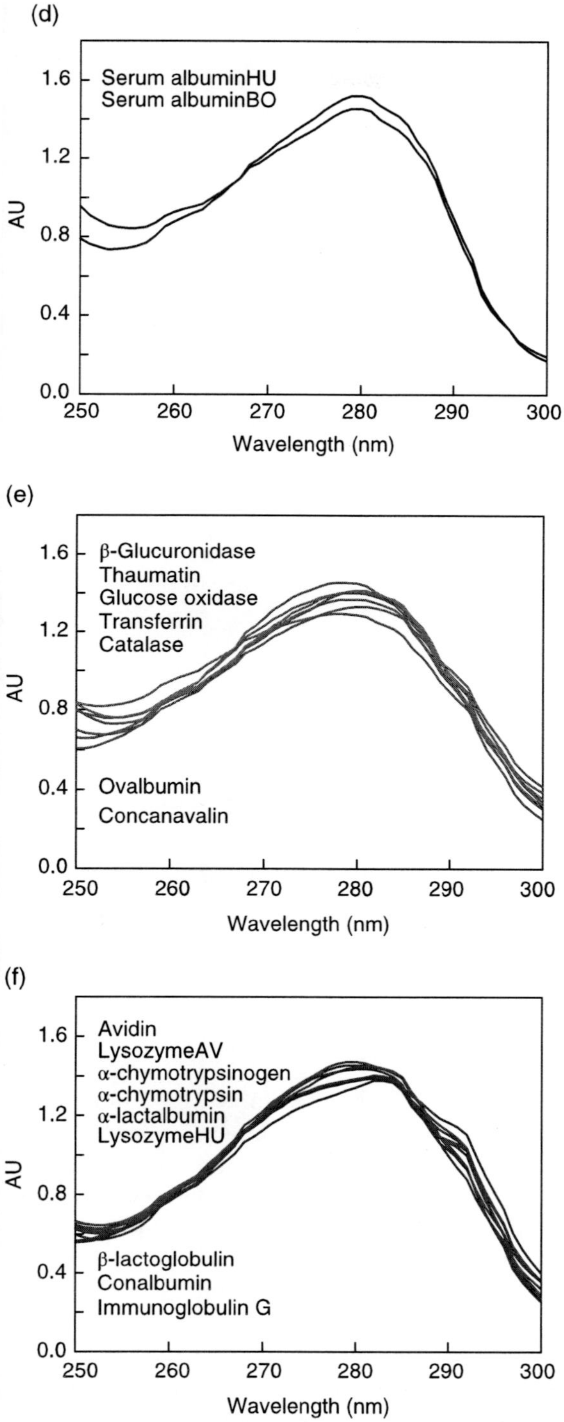

Figure 17.9 (Continued)

Table 17.1 Number of tryptophan, tyrosine, and phenylalanine residues in proteins without a heme group.

Protein	trp/tyr ratio	trp residues	tyr residues	phe residues	Total aa
Avidin	4.00	4	1	7	128
LysozymeAV	2.00	6	3	3	129
α-Chymotrypsinogen	2.00	8	4	7	245
α-Chymotrypsin	2.00	8	4	7	243
α-Lactalbumin	1.00	4	4	4	123
LysozymeHU	0.83	5	6	2	130
β-Lactoglobulin	0.50	2	4	4	162
Conalbumin	0.48	10	21	26	686
β-Glucuronidase	0.47	17	36	30	652
Thaumatin	0.38	3	8	11	207
Glucose oxidase	0.37	10	27	18	583
Transferring	0.35	8	23	28	685
Catalase	0.30	6	20	31	526
Ovalbumin	0.30	3	10	20	385
Concanavalin	0.26	5	19	13	299
Serum albuminBO	0.10	2	20	27	583
Serum albuminHU	0.06	1	18	31	585
Ribonuclease A	0.00	0	6	3	124
Insulin	0.00	0	4	3	51

The ratio of tryptophan to tyrosine was determined for each protein (IgG) is not included as no sequence was available). The proteins are listed in descending order according to their trp/tyr ratio.

17.3.2 Spectral Similarity and Prediction Precision

Differences in protein absorption spectra can be used for selective quantification. Precision of the method will mainly depend on the similarity of the absorption spectra. Figure 17.10 presents three examples of selective quantification based on absorption spectra. PLS was used to calibrate multivariate models based on spectral data. Calibration of the PLS models was based on 16 samples (six-level D-optimal onion design) when including two proteins and 29 samples (four-level D-optimal onion design) when including three proteins. The number of LVs applied in the regression was identical to the number of proteins included. For proteins with highly different spectra, the method delivers very precise results as can be seen for ribonuclease A and cytochrome c (cf. Figure 17.10a, d). Also for proteins with very similar spectra, for example, myoglobin and hemoglobin HU, the selective quantification is very precise (cf. Figure 17.10b, e). In the

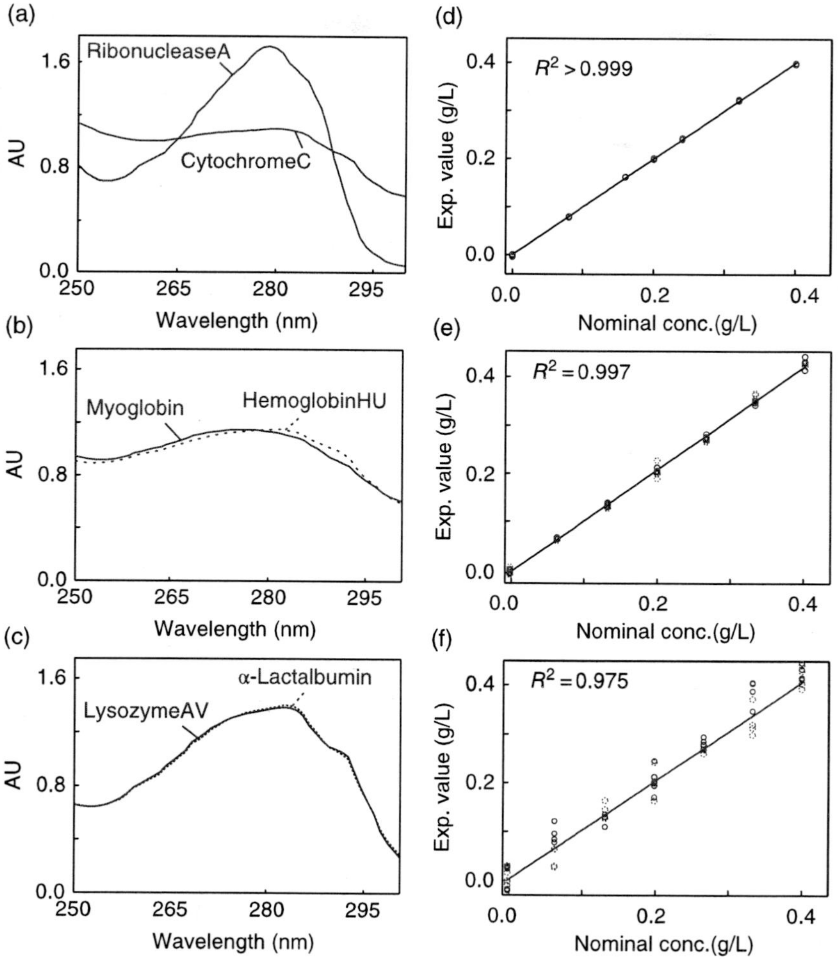

Figure 17.10 Protein spectra and calibration validation of three binary protein combinations. (a–c) Absorption spectra measured in microtiter plates with a multimode microplate reader (Infinite 200, TECAN, Germany). Spectra have been normalized to equal total intensity. (d–f) Visual display of validation results. *Source*: Hansen et al. [13]. Reproduced with permission of Wiley.

case of α-lactalbumin and lysozyme AV, the absorption spectra are similar to a point where the differences are barely recognizable and quantification is accordingly less precise (cf. Figure 17.10c, f). It must be kept in mind that all absorption spectra were recorded in microtiter plates, which is less precise than measurements in cuvettes. To illustrate this, spectral data of two monoclonal IgGs, with spectra of even higher similarity, were obtained in a traditional photometer system using cuvettes as sample containers. In Figure 17.11a, absorption spectra of the two monoclonal IgGs are shown. The spectral difference is ~0.5% of the

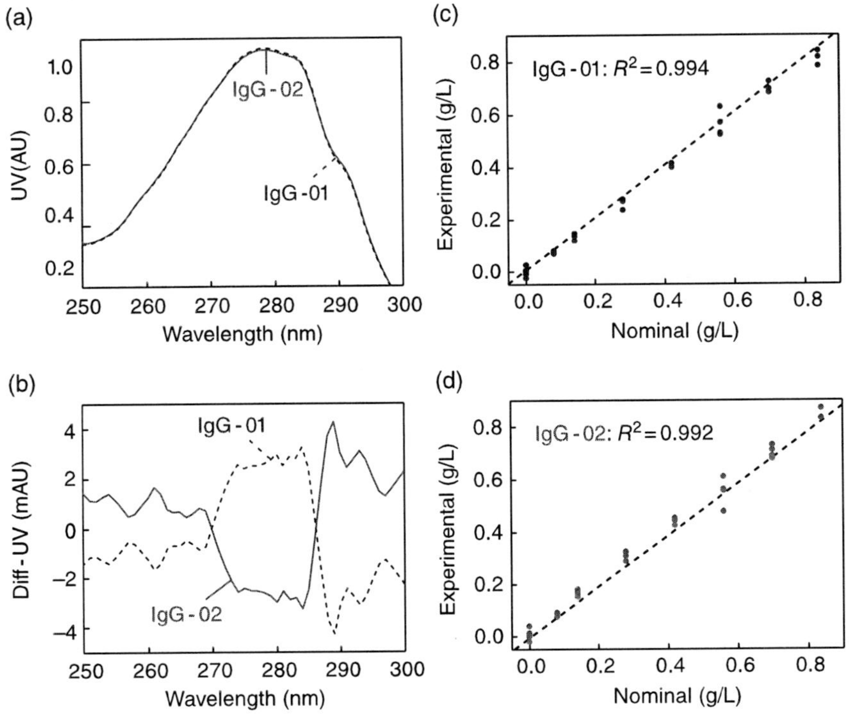

Figure 17.11 Validation result of a calibration of two monoclonal IgGs. (a–b) Absorption and absorption difference spectra of the two IgGs. (c–d) Validation results and resulting coefficient of determination. *Source*: Hansen et al. [13]. Reproduced with permission of Wiley.

average absorption intensity. As this is hardly visible to the naked eye, difference spectra are shown to visualize the spectral distinctions (Figure 17.11b). By changing the sample container from microtiter plate to cuvette, a significant increase in precision compared to measurements in microtiter plates was achieved. The coefficient of determination increased from 0.944 (IgG-01) and 0.932 (IgG-02) to 0.994 (IgG-01) and 0.992 (IgG-02).

17.4 Application as a Screening Tool: Analytics for High-Throughput Experiments

The application of mid-UV absorption spectra in combination with PLS for selective protein quantification in multicomponent mixtures can increase the analytical throughput of samples generated by HTE significantly. As a proof of concept, a resin screen in HTE mode is presented in Figure 17.12. The spectral-based quantification generated high quality data, which were not influenced by

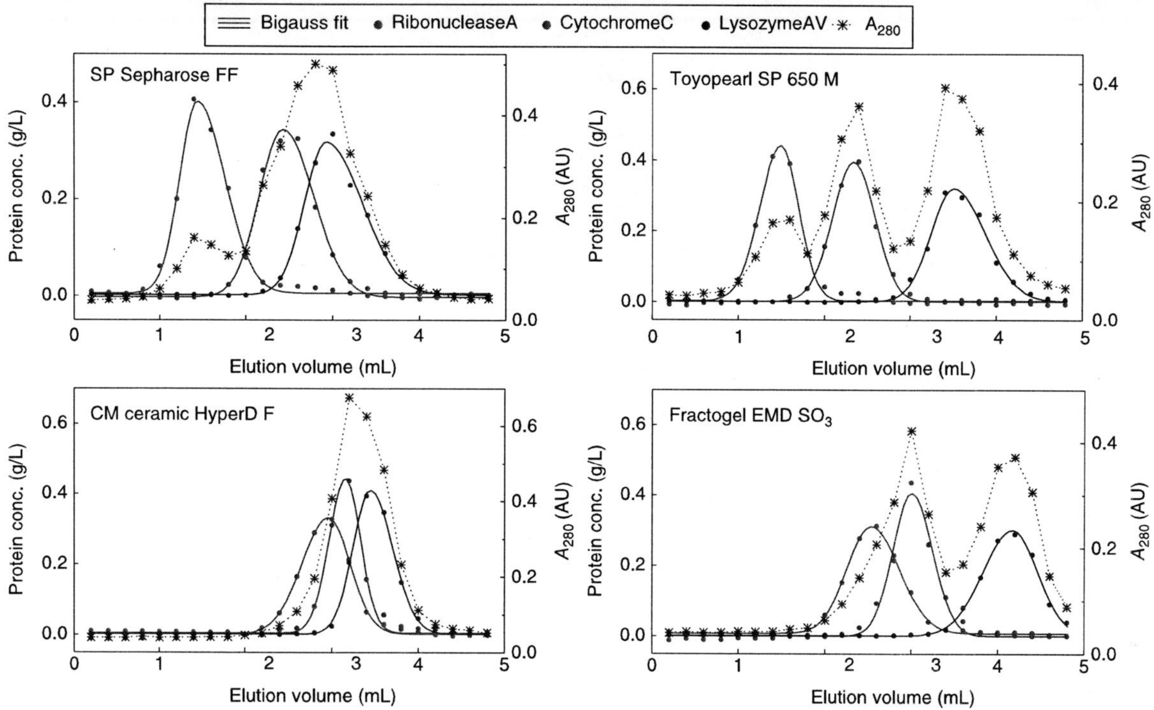

Figure 17.12 Result of a HTE resin screen for separation of three proteins with 200 μL packed columns (Atoll, Germany). The separation was performed using a sodium chloride gradient (0–500 mM over 24 CV) and the eluate was collected in fractions of 200 μL. UV absorption spectra of each fraction were measured using an infinite plate reader (Tecan), and protein concentrations were calculated based on a previously calibrated PLS model. Peak fitting was performed to calculate resolution that was used for method validation purposes employing an alternative analytical method (analytical chromatography). *Source*: Hansen et al. [15]. Reproduced with permission of Wiley.

co-elution or increasing sodium chloride concentrations in the samples (i.e., collected fractions). This was verified by reference analytics based on analytical chromatography. The chromatographic resolution determined for each resin based on the HTE screen deviated by less than 2% when compared to the results based on the reference analytics. The mass balances were 83–109% when applying spectral-based quantification and 86–104% when applying analytical chromatography-based quantification. The mass balances do indicate that the applied analytical chromatography is slightly more precise. However, the high throughput of the spectral-based quantification method (>3 samples per min) makes it very advantageous for certain application despite the small decrease in precision.

In the resin screening example, absorption spectra of the applied proteins (cytochrome c, ribonuclease A, and lysozyme AV) were of low similarity (cf. Figure 17.10a, c). However, also for proteins with higher spectral similarity, the spectral-based selective quantification method can deliver sufficient precision. The result of a HTE chromatography experiment with lysozyme HU and lysozyme AV is shown in Figure 17.13. In Figure 17.13a, b, spectra of lysozyme HU and lysozyme AV and the precision of the calibrated PLS model used for calculation of the concentrations are shown. In Figure 17.13c–j, the four HTE chromatograms show the A_{280} signal of four separate runs: without protein, with both proteins, with lysozyme HU only, and with lysozyme AV only. In the following text, the respective HTE chromatograms based on protein concentrations calculated based on spectral data are shown (Figure 17.13e–h). Except for small differences in peak height, the elution profiles are identical regardless of whether one or both proteins were included in the run. The mass balances were between 102 and 84%, which is also highly acceptable for HTE chromatography when using univariate absorption-based methods or analytical chromatography [33–35]. The analytical speed of spectral-based selective protein quantification combined with the low requirement for material due to the miniaturized scale makes the combination with HTE very well suited for large screens including many different resins, modifiers, and salts. Further, mechanistic modeling can be performed based on the generated data and used for process optimization and process scale-up [36, 37].

17.5 Application as a PAT Tool: Selective In-line Quantification and Real-Time Pooling

Batch-to-batch variability in combination with the lack of suitable real-time analytics for co-eluting proteins can complicate product collection in preparative chromatography. Application of protein absorption spectra in combination with the PLS modeling for in-line process control might overcome this,

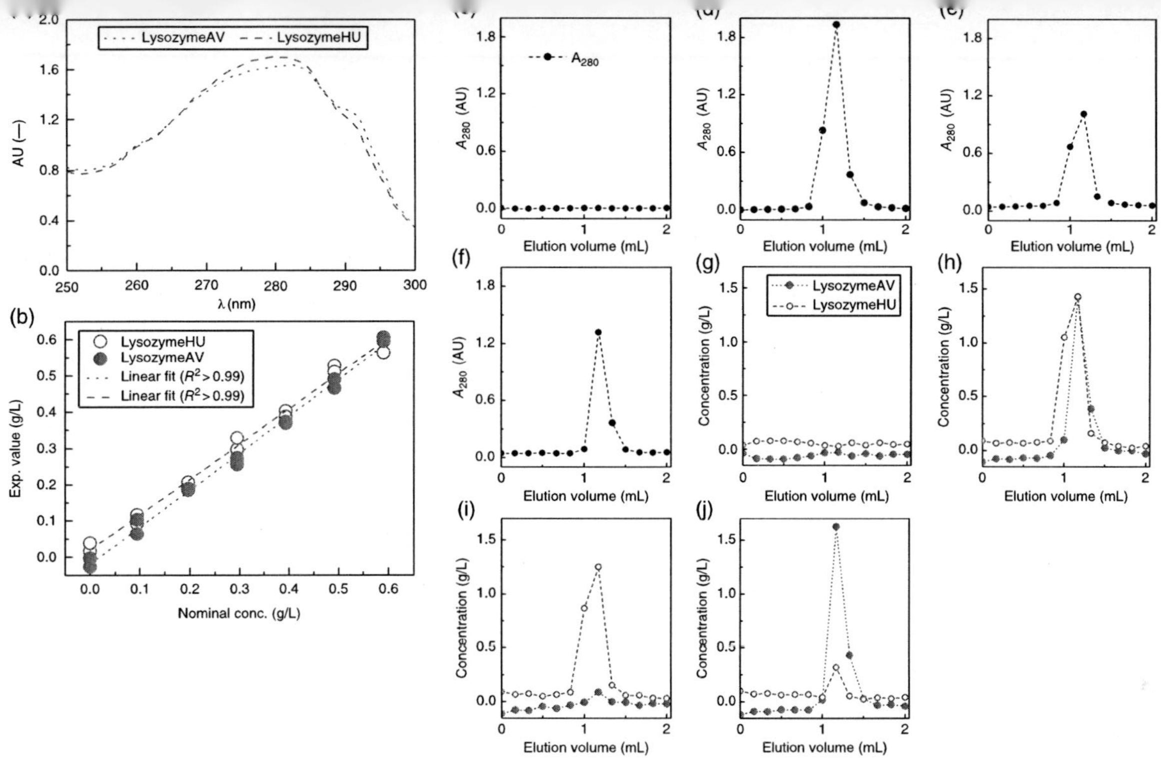

Figure 17.13 UV absorption spectra of human breast milk lysozyme (lysozyme HU) and hen egg white lysozyme (lysozyme AV). The spectra have been normalized to equal total intensity. (a) UV absorption spectra of lysozyme HU and lysozyme AV. (b) Result of selective concentration determination in validation samples containing lysozyme HU and lysozyme AV. (c–j) HTE chromatography with 200 μL columns prepacked with Macro-Prep ceramic hydroxyapatite type 1, 80 μm (Atoll, Germany). Load: no protein (C/G), 1.0 g/L resin of each protein (D/H), 1.0 g/L resin of lysozyme HU (E/I), 1.0 g/L resin of of lysozyme AV (F/J). Load buffer: 1 mM phosphate buffer pH 6.8. Elution: linear step gradient from 1 to 300 mM phosphate at pH 6.8 over 18 CV. The eluate was collected in fractions of 1 CV. *Source*: Hansen et al. [15]. Reproduced with permission of Wiley

and the method was recently set up as PAT tool for this purpose [38, 39]. The setup as well as its application for selective in-line protein quantification and real-time pooling is described in the following text.

17.5.1 PAT Tool Setup

Application of mid-UV absorption spectra and PLS as a PAT tool requires a fast spectral measurement of the column flow stream. If a diode array detector (DAD) is applied for this purpose, in-line monitoring is possible. The setup displayed in Figure 17.14 was applied. The flow stream of an ÄKTApurifier 10 (GE Healthcare, Chalfont St. Giles, UK) and an ultimate 3000 DAD (Thermo Fisher Scientific, Waltham, MA, USA) were connected so that spectral measurements of the column flow stream were possible. The data acquisition (DAQ) of the DAD was triggered and stopped by sending digital signals from the ÄKTA to a software tool chain via a USB-6008 DAQ device (National Instruments, Austin, TX, USA). The software tool chain then started or stopped the DAD DAQ. The software tool chain consisted of MATLAB (DAQ and Spreadsheet Link EX toolbox) (MathWorks, Natick, MA, USA) as well as an Excel Visual Basic for Applications (VBA) macro (Microsoft, Redmond, WA, USA) that was using the application programmable interface of the Chromeleon Software Developer Kit (Thermo Fisher Scientific, Waltham, MA, USA).

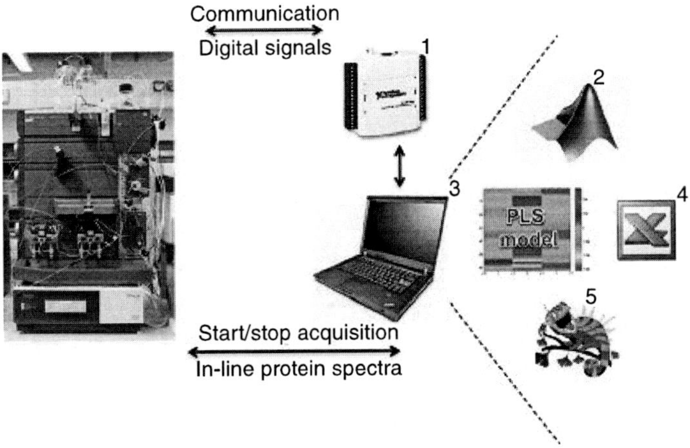

Figure 17.14 PAT tool consisting of a Dionex Ultimate 3000 DAD, an ÄKTA purifier 10, a DAQ device, and a software tool chain. The software tool chain can communicate with the DAD and the ÄKTA. Communication with the ÄKTA is handled via the DAQ device using digital signals. The software tool chain consists of MATLAB and an Excel Visual Basic for Application Macro that is using classes of the Chromeleon Software Developer Kit. Copyright by 1: National Instruments, 2 MathWorks, 3 Lenovo, 4 Microsoft, and 5 Thermo Fisher Scientific. *Source*: Hansen et al. [13]. Reproduced with permission of Wiley.

While MATLAB was handling the communication with the DAQ device, the communication with the DAD was managed via the VBA macro. An interface between MATLAB and Excel was realized using the Spreadsheet Link EX toolbox of MATLAB. Besides triggering and stopping the DAD DAQ, the software tool chain retrieved in-line absorption spectra from the DAD and determined the selective protein concentrations from the absorption spectra. Data retrieval was managed via the VBA macro, while calculation of the selective protein concentrations from the absorption spectra was handled by MATLAB. Therefore, a calibrated PLS model was integrated into the MATLAB script. Using this setup, an in-line peak deconvolution was achieved. The reaction on the calculated peak deconvolution was communicated back to the ÄKTA via the DAQ device using MATLAB.

17.5.2 Selective In-line Protein Quantification

To demonstrate the functionality of the methodology as a PAT tool, the setup was applied for a selective in-line quantification of a three-component model protein system. The system consisted of lysozyme, cytochrome c, and ribonuclease A (all purchased from Sigma-Aldrich, St. Louis, MO), which were co-eluting after their chromatographic separation on an SP Sepharose FF column (GE Healthcare). Elution was performed using a linear salt gradient from 0 to 500 mM sodium chloride in 5 column volumes (CVs). Figure 17.15a displays the resulting standard chromatogram obtained without peak deconvolution. The absorption at 280 and 527 nm (specific for cytochrome c) is plotted with respect to the retention volume. Figure 17.15b displays the resulting chromatogram from the in-line peak deconvolution for comparison. Good agreement was observed between the retention volume of the first peak in the A_{280} signals and the ribonuclease A concentration profile from the in-line peak deconvolution. The retention volume of cytochrome c from the in-line peak deconvolution corresponded to the peak maximum of the A_{527} signal, which is specific for cytochrome c. To further evaluate the functionality of the PAT tool, the column flow stream was collected in fractions. Fractions were analyzed for their selective protein concentrations using analytical cation exchange chromatography (CEX). The results of the fraction analytics are displayed in Figure 17.15b as dashed stair functions. An excellent agreement between the results of the in-line peak deconvolution and the off-line analytical chromatography was observed. This clearly demonstrates the functionality of the PAT tool and its potential for real-time pooling decisions.

17.5.3 Real-Time Pooling Decisions

In-line peak deconvolution allows for real-time pooling decisions. To further demonstrate this application in the model protein study, ribonuclease A was

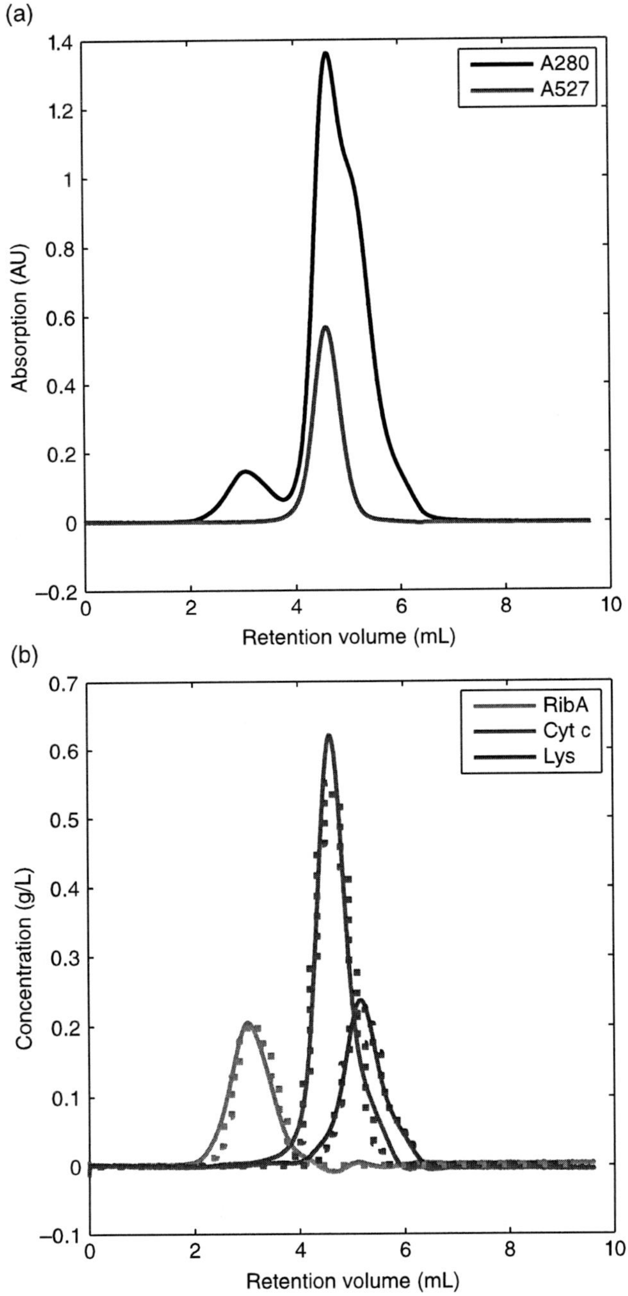

Figure 17.15 Resulting chromatograms of the in-line peak deconvolution and off-line analytical chromatography. (a) Absorption at 280 and 527 nm plotted against the retention volume. Absorption was measured using a path length of 0.4 mm and then recalculated for a path length of 10 mm. (b) Predicted protein concentrations by the PLS model (solid lines) and determined protein concentrations in collected fractions using off-line analytical chromatography (dashed lines) plotted against the retention volume. *Source*: Brestrich et al. [39]. Reproduced with permission of Wiley.

declared as target protein in the first experiment and separated from the two "contaminants" cytochrome c and lysozyme on the SP Sepharose FF column. Elution was performed using a 3 CV sodium chloride gradient from 0 to 500 mM (however, with different loadings and protein-to-protein content than in the previous section). In the second experiment, cytochrome c was the target protein to be separated from the other two proteins using a 5 CV salt gradient. In both experiments, the pooling criteria were defined as follows: As soon as the target protein concentration was above 0.002 g/L and the mass fraction of the target protein was above 0.8, the pooling was started. The stop pooling criterion was a mass fraction of the target below 0.8. Figure 17.16 displays the resulting chromatograms of both experiments. The black, vertical lines visualize the pooling criteria that were automatically detected by Matlab. In addition to real-time pooling, the PAT tool was furthermore applied for a rapid pool purity and yield determination. The mass balances gained by the selective in-line quantification and the pool borders were used to calculate the product yield and purity of the collected pools. Calculated yields were 88.8% for ribonuclease A and 36.3% for cytochrome c. The determination of the pool purity resulted in 96.9% for the ribonuclease A pool and in 83.5% for the cytochrome c pool.

17.6 Case Studies

While the functionality of the screening and PAT tool was tested with model proteins, three case studies were performed to demonstrate the value for true biopharmaceutical applications. For the first and the second case study, process data was applied for PLS model calibration. All PLS models were calibrated using several chromatography runs exhibiting "batch-to-batch" variability. Collected fractions of the runs contained different protein concentrations and compositions, spanning the calibration space of the PLS model. To validate the models, they were subsequently applied to predict a run not included in the calibration. The first case study was performed using the PAT tool displayed in Figure 17.14. Consequently, absorption spectra from the DAD were averaged over the size of collected fractions and were then correlated with the results of fraction analytics using PLS. As the resolution of collected fractions was high, the PLS model could afterward be applied for in-line peak deconvolution. The second case study was however performed with off-line absorption spectra from collected fractions as it was performed at CSL Behring, Broadmeadows. This means that the calibrated model can only be applied for at-line monitoring as it is specific for the applied detector. However, the study could be easily transferred to the PAT tool described in Figure 17.14 to enable in-line monitoring. In the third case study, it was assessed whether spectral analysis and PLS modeling can be applied for the quantification of deamidation

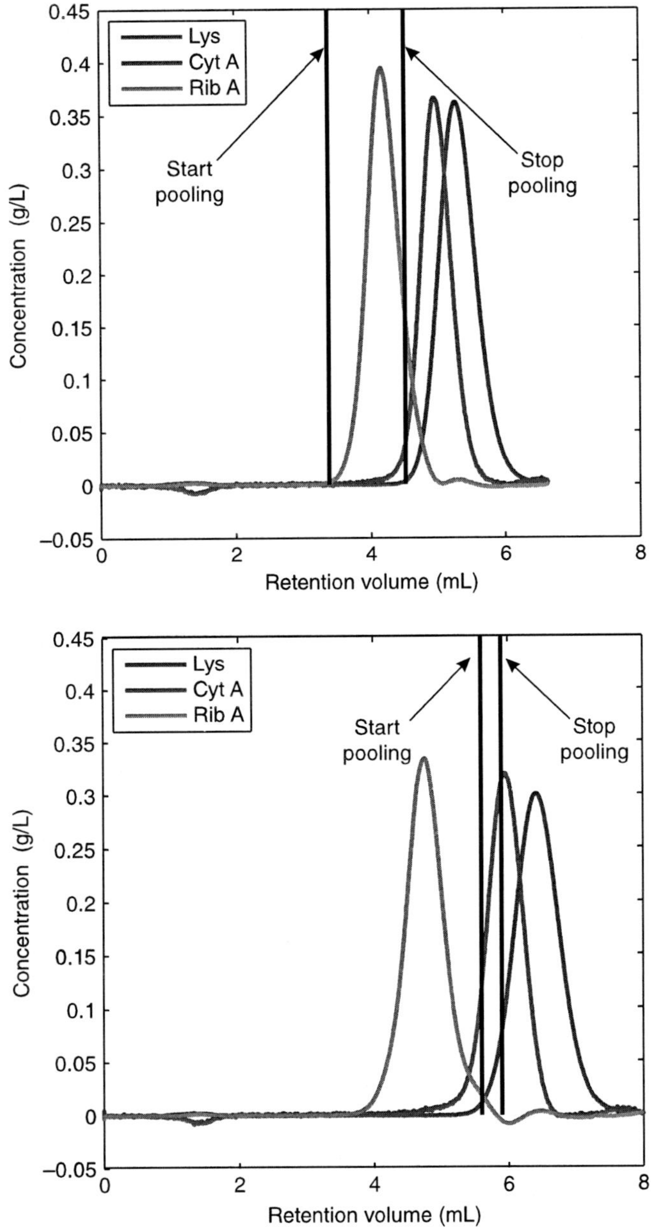

Figure 17.16 Resulting chromatograms of real-time pooling during a chromatographic separation of lysozyme, cytochrome c, and ribonuclease A. The black, vertical lines in the chromatograms visualize the pooling criteria that were automatically detected by MATLAB and forwarded to the control software of the ÄKTA. (a) Ribonuclease A was declared as target protein and separated with a gradient length of three CVs. (b) Cytochrome c was declared as target protein and separated with a gradient length of five CVs. *Source*: Brestrich et al. [38]. Reproduced with permission of Wiley.

in insulin. Model calibration and validation in this case study was performed by generating mixtures from purified deamidated and native insulin according to a DoE setup.

17.6.1 mAb Monomer, Aggregates, and Fragments

The protein A capture and low pH virus inactivation steps of monoclonal antibody (mAb) platform process are usually followed by two polishing chromatography steps aiming to reduce remaining host cell proteins, aggregates (high molecular weight (HMW) mAbs), fragments (low molecular weight (LMW) mAbs), DNA, and leached protein A. One polishing step is typically CEX, which is performed to reduce mAb HMWs and LMWs. The resolution of these two product-related contaminants and the mAb is often poor due to similar physicochemical properties. The goal of this case study was calibration of a PLS model for the deconvolution of the elution peaks in a CEX step. Application of the peak deconvolution method is possible due to spectral differences of mAb monomers, aggregates, and fragments. Absorption spectra of mAb fragments differ from the monomer due to a different number and ratio of aromatic amino acid residues. One effect for spectral differences between mAb monomer and aggregates in molecular structure might be conformational changes, leading to a different microenvironment around the aromatic amino acid residues [40–42]. More recently, Thakkar et al. [43] found that even protein–protein interactions in highly concentrated solutions can lead to changes in the microenvironment around the aromatic amino acid residues and thus to changes in UV absorbance. For larger aggregates, however, scattering effects might be overruling. Scattering of incident light on solute protein multimers prevents the light from reaching the detector and results in an artificial increment of the absorbance values [44]. This scattering artifact is corrected in most applications, but is desired in this study. Figure 17.17a displays the change in the mid-UV spectrum due to increasing HMW level in the elution peak. Absorption beyond 320 nm indicates scattering effects. Consequently, a lot of information for the peak deconvolution was included in wavelengths beyond 300 nm, and protein absorption spectra between 240 and 410 nm were thus applied for the PLS model calibration.

To calibrate the PLS model, pulse injections of pH-adjusted protein A pool were performed. The pool contained 11.0% aggregates and 5.8% fragments. Batch-to-batch variability was imitated by different gradients from 0 to 700 mM sodium chloride on a POROS 50HS (Life Technologies, Carlsbad, CA, USA) column. Variable gradient lengths lead to different concentrations and variable resolution of the species. The average absorption spectra of collected fractions and results from off-line fraction analytics of a 10 CV and a 30 CV gradient run were applied to calibrate the PLS model, while a 20 CV gradient run was used to validate the model. Off-line fraction analytics were performed using analytical

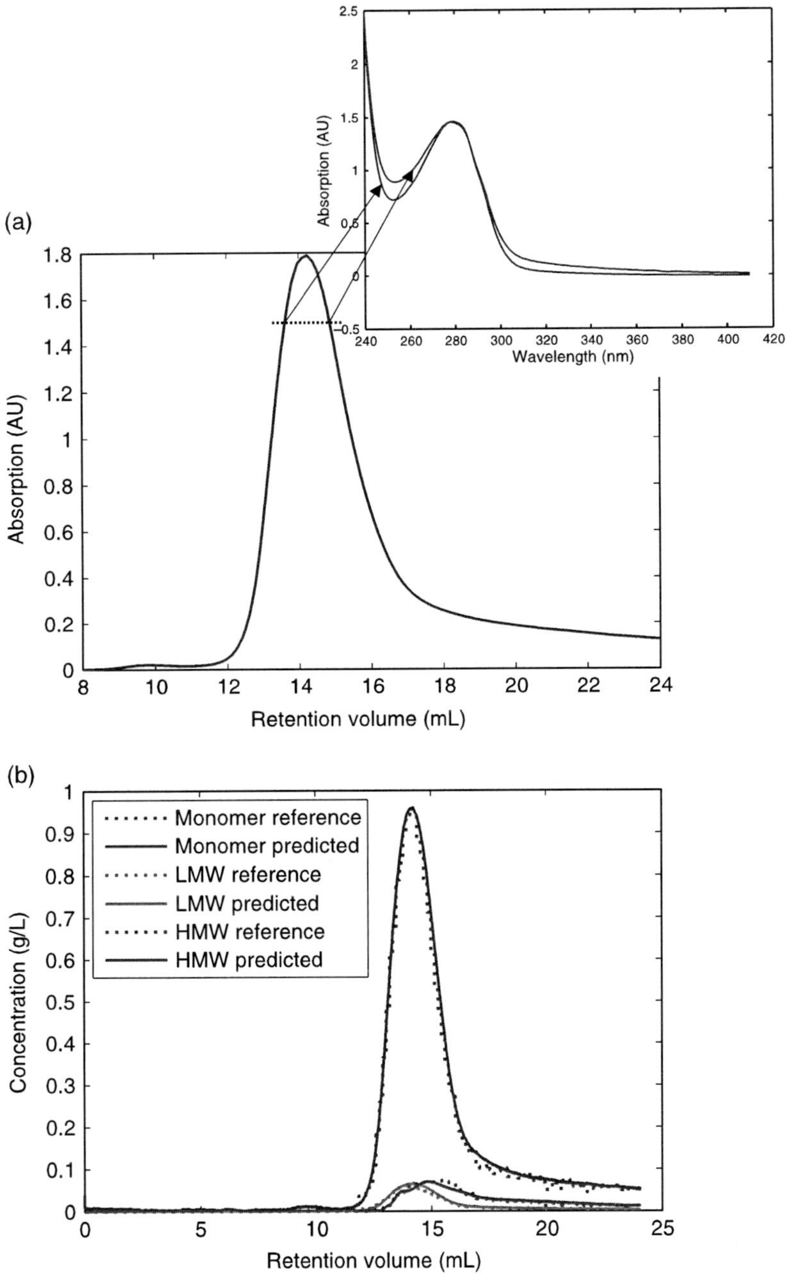

Figure 17.17 Validation run of the case study with mAb monomer, HMW, and LMW. (a) Change in the mid-UV absorption spectrum due to increasing aggregate level. Absorption beyond 320 nm indicates scattering effects. Absorption was measured using a path length of 0.4 mm and then recalculated for a path length of 10 mm. (b) In-line peak deconvolution for the 20 CV gradient validation run. (c) PLS model prediction for LMW and reference analytics. (d) PLS model prediction for HMW and reference analytics. *Source*: Brestrich et al. [39]. Reproduced with permission of Wiley. (*See insert for color representation of the figure.*)

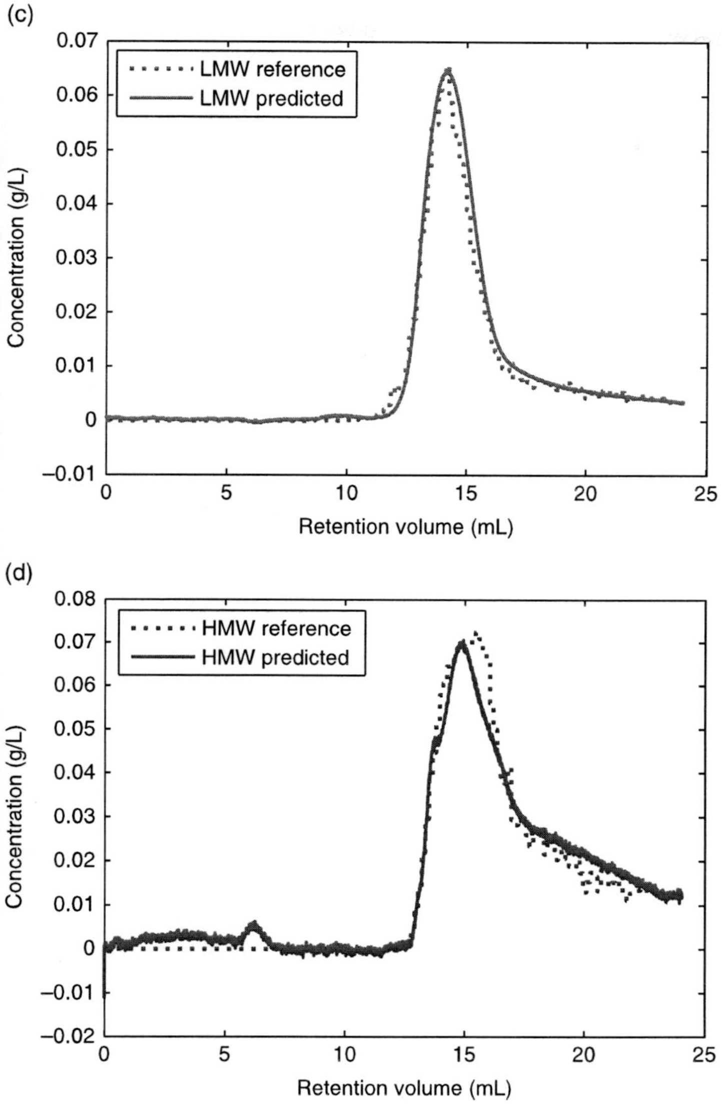

Figure 17.17 (Continued)

size exclusion chromatography for the monomer, HMW and LMW ratios, while the total mAb concentration of each fraction was determined using a NanoDrop 2000 (Thermo Fisher Scientific). The calibrated PLS model was applied for an in-line peak deconvolution of the validation run, which is displayed in Figure 17.17b and compared with the off-line fraction analytics (dashed stair functions). Figure 17.17c and d display the PLS model prediction and off-line reference analytics for HMW and LMW separately for a better evaluation of model precision. A good agreement between prediction of the PLS model and reference analytics was observed for all monitored species, demonstrating the applicability of the PAT tool for in-line quantification in this case study. The slightly scattered PLS prediction of HMW can most likely be ascribed to a certain noise in the off-line fraction analytics applied for the model calibration.

17.6.2 Serum Proteins

In the blood plasma fraction process of CSL Behring, a DEAE Sepharose FF (GE Healthcare) chromatography step is used to purify albumin, IgG, and transferrin from Cohn fractionation supernatant I. The major components of supernatant I are albumin (~73%), IgG (~19%), and trf (~5%). Other minor components are IgA and IgM. Supernatant I is loaded onto the column at pH 5.2, allowing albumin (p*I* 4.8) to bind to the resin, while IgG (p*I* 4.35–9.95) passes through the column unretained [45]. Other proteins in Cohn supernatant I such as transferrin (p*I* 5.0–6.0), IgA (p*I* 4.0–7.1), and IgM (p*I* 4.0–9.1) are partially retained and therefore present in both the IgG- and the albumin-rich fraction [45, 46]. In order to collect both transferrin and IgG, the flow-through of the DEAE Sepharose FF column is collected in two fractions. The resolution between the two proteins is strongly dependent on the sample loading, on the protein concentration in supernatant I, as well as on the conductivity of supernatant I and of the equilibration buffer [47]. In this case study, the main goal was to calibrate a PLS model aiding the quantification of the main co-eluting species IgG and transferrin. Such a model could enable real-time pooling of these two target proteins at a later stage. IgM and IgA are separated at a later point of the process. Although their quantification by spectral measurement and PLS modeling was not the main purpose of this study, IgM and IgA were examined as well.

The standard process at CSL exhibits batch-to-batch variability, which usually results in slightly variable retention times and concentrations of the co-eluting proteins. Therefore, the absorption spectra of collected fractions of two runs were correlated with the results of a corresponding reference analysis using PLS. The reference analysis for serum protein quantification was performed by nephelometry using an Immage 800 Immunochemistry System by Beckman Coulter (Pasadena, California, USA). The PLS model was subsequently applied for peak deconvolution of a validation run, which was not included in the PLS model calibration. Results of the peak deconvolution are displayed in Figure 17.18 and compared with the reference analysis. A good agreement between PLS model prediction and reference analytics of the higher concentrated target proteins IgG and transferrin could be observed (cf. Figure 17.18a).

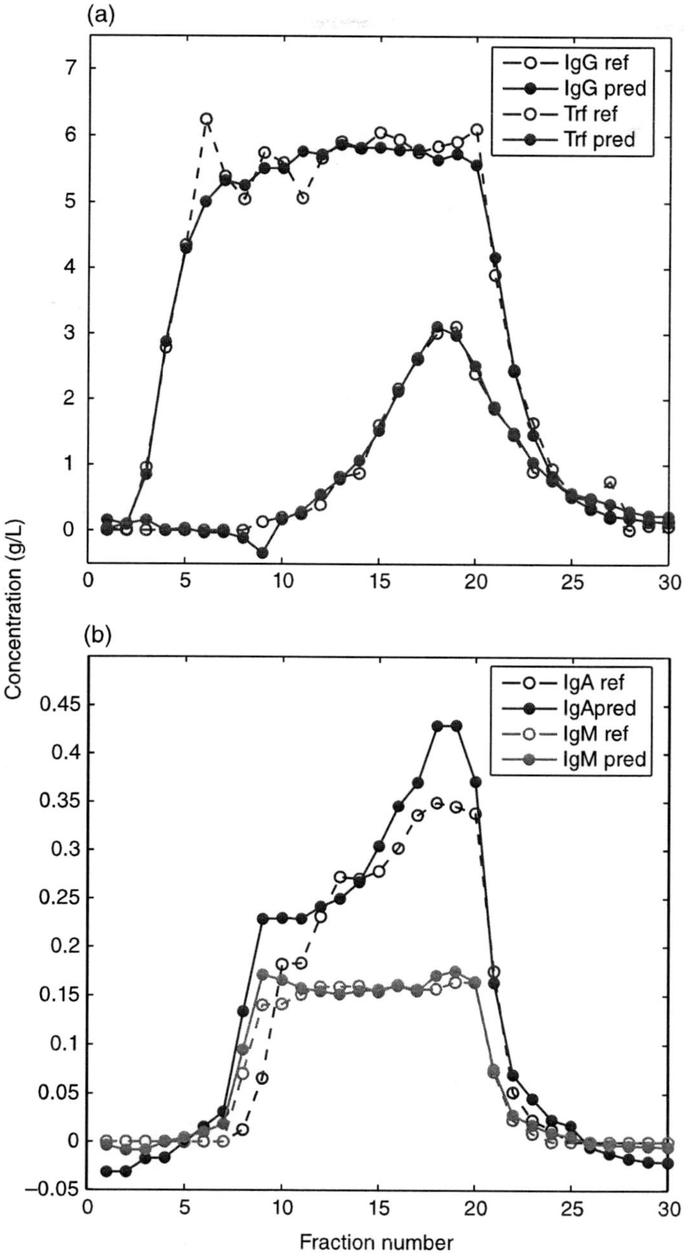

Figure 17.18 Case study with serum proteins. (a) PLS model prediction and corresponding results of the reference analytics for IgG and transferrin. (b) PLS model prediction and corresponding results of the reference analytics for IgA and IgM. *Source*: Brestrich et al. [39]. Reproduced with permission of Wiley.

Moreover, a good prediction of the elution profile of the lower concentrated IgM could be achieved (cf. Figure 17.18b). Differences in the total concentrations were found solely for IgA, indicating a lower precision of the model for this protein (cf. Figure 17.18b). However, the overall trend of the elution profile could be predicted. As the main goal of the study was the accurate prediction of IgG and transferrin, a better PLS model prediction for IgA was not pursued. However, if the exact determination of IgA was important for the process, more calibration runs would be needed in order to calibrate a model that can predict IgA more precisely.

17.6.3 Selective Quantification of Deamidated Insulin Aspart

Nonenzymatic deamidation is a common challenge during production and storage of biopharmaceuticals. The amino acid most prone to deamidation is asparagine, and the rate of deamidation is mainly related to primary structure, pH, and temperature [48]. As with all other product-related impurities, deamidated product variants must be tightly monitored and controlled and if possible avoided. To assess whether spectral analysis can be employed for the quantification of deamidation in insulin, the spectra of enriched deamidated insulin aspart and native insulin aspart were compared. The mid-UV absorption spectra of enriched deamidated insulin aspart (97% aspB3 and isoaspB3) and native insulin aspart are shown in Figure 17.19. There are detectable differences between the spectra of deamidated and native insulin aspart. This is quite remarkable and can possibly be ascribed to a structural change caused by the charge shift introduced by the deamidation. It was then tested whether the observed spectral differences were sufficiently pronounced and robust for

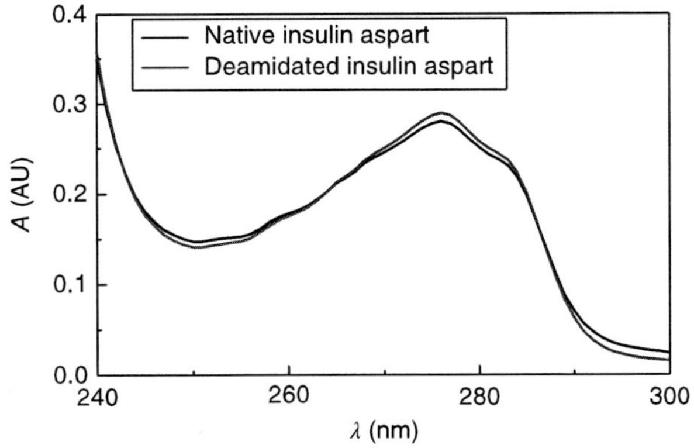

Figure 17.19 Mid-UV absorption spectra of native insulin aspart and highly enriched deamidated insulin aspart (97% aspB3 and isoaspB3). The measurements were performed with a Lambda 35 spectrophotometer (Perkin Elmer, Waltham, USA).

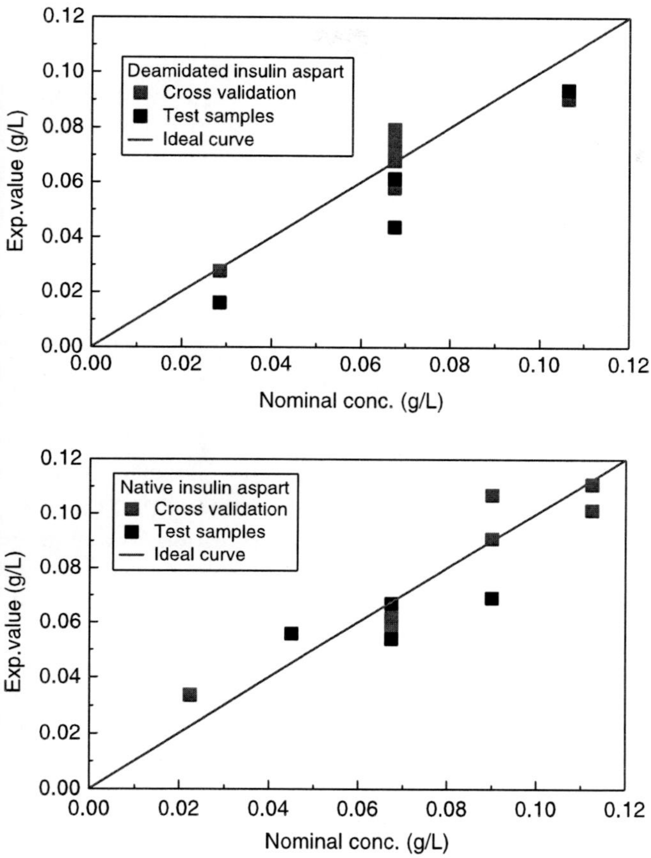

Figure 17.20 Validation result of a calibration of highly enriched deamidated insulin aspart and native insulin aspart.

selective quantification purposes. In Figure 17.20 the validation result of a calibration based on eight calibration samples and four independent test samples are shown. The validation is comprised of both independent test samples and cross validation. The data shows that a correlation between spectral data and selective concentrations of deamidated and native insulin aspart can be built; however, a higher precision is desirable.

Based on the observed differences in the absorption spectra of deamidated and native insulin aspart, it was anticipated that a correlation of high precision should be obtainable (cf. Figure 17.11). A this point however some optimization is required to obtain a precise calibration for deamidated insulin aspart and native insulin aspart before a final implementation of the methodology may be accomplished as it was the case for the two monoclonal IgGs.

17.7 Conclusion and Outlook

Differences in mid-UV protein absorption spectra due to varying ratios of aromatic residues or structural influences on the environment of the chromophores can be utilized for selective protein quantification. A suitable multivariate regression technique to correlate mid-UV absorption spectra with selective protein concentrations is PLS. The precision of calibrated PLS models depends on the spectral similarity of the proteins to be distinguished. However, even proteins with spectral differences hardly visible to the naked eye might be quantified selectively using a corresponding PLS model. In this case, the spectral measurement has to be exact so that detector noise does not overrule the spectral difference.

Mid-UV protein absorption spectra and PLS can be applied as a screening tool to increase the analytical throughput. The application as a screening tool was successfully demonstrated in several HTE chromatography experiments. Besides the application as a screening tool, a hardware and software setup where mid-UV protein absorption spectra and PLS are applied as PAT tool was established. This PAT tool allows a selective in-line quantification of co-eluting proteins in chromatography as well as real-time pooling of a target protein. The value of the presented screening and PAT tool for true biopharmaceutical applications was successfully demonstrated in three case studies. Consequently, the application of mid-UV protein absorption spectra and PLS for selective protein quantification poses a valuable approach for high-throughput analytics in biopharmaceutical process development and for in-line analysis in production. By applying the presented screening tool in future process development, the analytical bottleneck posed by the large amount of samples generated in HTE might be overcome. This again might allow a more efficient analysis of the relationships between CPPs and CQAs. By applying the presented PAT tool for in-line quantification and real-time pooling at future large-scale production, batch-to-batch variability might be more efficiently managed. This should also enable consistent pool purities and thus product quality, ultimately leading to real-time release of intermediates and/or final API. Future work will focus on the applicability of the presented PAT tool for preparative chromatography with high load densities.

References

1. ICH. 2008. Pharmaceutical development. Harmonised Tripartite Guideline: Q8(R1).
2. FDA. 2004. Guidance for Industry. PAT—A framework for innovative pharmaceutical development, manufacturing, and quality assurance.
3. Bensch M, Schulze Wierling P, von Lieres E, Hubbuch J. 2005. High throughput screening of chromatographic phases for rapid process development. *Chem Eng Technol* 28:1274–1284.

4. Rege K, Pepsin M, Falcon B, Steele L, Heng M. 2006. High-throughput process development for recombinant protein purification. *Biotechnol Bioeng* 93:618–30.
5. Rathore AS, Li X, Bartkowski W, Sharma A, Lu Y. 2009. Case study and application of process analytical technology (PAT) towards bioprocessing: Use of tryptophan fluorescence as at-line tool for making pooling decisions for process chromatography. *Biotechnol Prog* 25: 433–1439.
6. Krättli M, Steinebach F, Morbidelli M. 2013. Online control of the twin-column countercurrent solvent gradient process for biochromatography. *J Chromatogr A* 1293:51–59.
7. Fahrner RL, Lester PM, Blank GS, Reifsnyder DH. 1998. Real-time control of purified product collection during chromatography of recombinant human insulin-like growth factor-I using an on-line assay. *J Chromatogr A* 827: 37–43.
8. Kaltenbrunner O, Lu Y, Sharma A, Lawson K, Tressel T. 2012. Risk-benefit evaluation of on-line high-performance liquid chromatography analysis for pooling decisions in large-scale chromatography. *J Chromatogr A* 1241: 37–45.
9. Rathore AS, Parr L, Dermawan S, Lawson K, Lu Y. 2010. Large scale demonstration of a process analytical technology application in bioprocessing: Use of on-line high performance liquid chromatography for making real time pooling decisions for process chromatography. *Biotechnol Prog* 26: 448–457.
10. Rathore AS, Wood R, Sharma A, Dermawan S. 2008a. Case study and application of process analytical technology (PAT) towards bioprocessing II: Use of ultra-performance liquid chromatography (UPLC) for making real-time pooling decisions for process chromatography. *Biotechnol Bioeng* 101:1366–1374.
11. Rathore AS, Yu M, Yeboah S, Sharma A. 2008b. Case study and application of process analytical technology (PAT) towards bioprocessing: Use of on-line high-performance liquid chromatography (HPLC) for making real-time pooling decisions for process chromatography. *Biotechnol Bioeng* 100: 306–316.
12. Rosenheck K, Doty P. 1961. The far ultraviolet absorption spectra of polypeptide and protein solutions and their dependence on conformation. *Proc Natl Acad Sci USA* 47:1775–85.
13. Hansen SK, Jamali B, Hubbuch J. 2013. Selective high throughput protein quantification based on UV absorption spectra. *Biotechnol Bioeng* 110:448–460.
14. Wetlaufer D. 1963. Ultraviolet spectra of proteins and amino acids. *Adv Protein Chem* 17:303–390.
15. Hansen SK, Skibsted E, Staby A, Hubbuch J. 2011. A label-free methodology for selective protein quantification by means of absorption measurements. *Biotechnol Bioeng* 108:2661–2669.

16. Næs T, Mevik BH. 2001. Understanding the collinearity problem in regression and discriminant analysis. *J Chemom* 15:413–426.
17. Martens H, Næs T. 1989. *Multivariate Calibration*. New York: Wiley.
18. Lavine BK. 2000. Chemometrics. *Anal Chem* 72:91–97.
19. Arnold SA, Gaensakoo R, Harvey LM, McNeil B. 2002. Use of at-line and in-situ near-infrared spectroscopy to monitor biomass in an industrial fed-batch *Escherichia coli* process. *Biotechnol Bioeng* 80:405–413.
20. Boehl D. 2003. Chemometric modelling with two-dimensional fluorescence data for *Claviceps purpurea* bioprocess characterization. *J Biotechnol* 105:179–188.
21. Haack MB, Eliasson A, Olsson L. 2004. On-line cell mass monitoring of *Saccharomyces cerevisiae* cultivations by multi-wavelength fluorescence. *J Biotechnol* 114:199–208.
22. Navrátil M, Norberg A, Lembrén L, Mandenius CF. 2005. On-line multi-analyzer monitoring of biomass, glucose and acetate for growth rate control of a *Vibrio cholerae* fed-batch cultivation. *J Biotechnol* 115:67–79.
23. Tamburini E, Vaccari G, Tosi S, Trilli A. 2003. Near-infrared spectroscopy: A tool for monitoring submerged fermentation processes using an immersion optical-fiber probe. *Appl Spectrosc* 57:132–138.
24. Jung B, Lee S, Yang InH, Good T, Cote GL. 2002. Automated on-line noninvasive optical glucose monitoring in a cell culture system. *Appl Spectrosc* 56:51–57.
25. Kara S, Anton F, Solle D, Neumann M, Hitzmann B, Scheper T, Liese A. 2010. Fluorescence spectroscopy as a novel method for on-line analysis of biocatalytic CC bond formations. *J Mol Catal B: Enzym* 66:124–129.
26. Mazarevica G, Diewok J, Baena JR, Rosenberg E, Lendl B. 2004. On-line fermentation monitoring by mid-infrared spectroscopy. *Appl Spectrosc* 58:804–810.
27. Capito F, Skudas R, Kolmar H, Hunzinger C. 2013a. Mid-infrared spectroscopy-based antibody aggregate quantification in cell culture fluids. *Biotechnol J* 8:912–917.
28. Capito F, Skudas R, Kolmar H, Stanislawski B. 2013b. Host cell protein quantification by Fourier transform mid infrared spectroscopy (FT-MIR). *Biotechnol Bioeng* 110:252–259.
29. Capito F, Skudasb R, Kolmar H, Hunzinger C. 2015. At-line mid infrared spectroscopy for monitoring downstream processing unit operations. *Process Biochem* 50(6): 997–1005.
30. Elshereef R, Budman H, Moresoli C, Legge RL. 2010. Monitoring the fractionation of a whey protein isolate during dead-end membrane filtration using fluorescence and chemometric methods. *Biotechnol Prog* 26:168–178.
31. Yeung KSY, Hoare M, Thornhill NF, Williams T, Vaghjiani JD. 2000. Near-infrared spectroscopy for bioprocess monitoring and control. *Biotechnol Bioeng* 63:684–693.

32. Eriksson L, Johansson E, Kettaneh-Wold N, Trygg J, Wikström C, Wold S. 2006. *Multi- and megavariate data analysis.* Umeå: Umetrics Academy. 95 p.
33. Kelley BD, Switzer M, Bastek P, Kramarczyk JF, Molnar K, Yu T, Coffman J. 2008. High-throughput screening of chromatographic separations: IV. Ion-exchange. *Biotechnol Bioeng* 100:950–963.
34. Kramarczyk JF, Kelley BD, Coffman JL. 2008. High-throughput screening of chromatographic separations: II. Hydrophobic interaction. *Biotechnol Bioeng* 100:707–720.
35. Treier K, Hansen S, Richter C, Diederich P, Hubbuch J, Lester P. 2012. High-throughput methods for miniaturization and automation of monoclonal antibody purification processes. *Biotechnol Prog* 28:723–32.
36. Baumann P, Huuk T, Hahn T, Osberghaus A, Hubbuch J. 2016. Deconvolution of high-throughput multi-component isotherms using multivariate data analysis of protein spectra. *Eng Life Sci,* 16, 194–201.
37. Osberghaus A, Drechsel K, Hansen S, Hepbildikler S, Nath S, Haindl M, von Lieres E, Hubbuch J. 2012. Model-integrated process development demonstrated on the optimization of a robotic cation exchange step. *Chem Eng Sci* 76:129–139.
38. Brestrich N, Briskot T, Osberghaus A, Hubbuch J. 2014. A tool for selective inline quantification of co-eluting proteins in chromatography using spectral analysis and partial least squares regression. *Biotechnol Bioeng* 111:1365–1373.
39. Brestrich N, Sanden A, Kraft A, McCann K, Bertolini J, Hubbuch J. 2015. Advances inline quantification of co-eluting proteins in chromatography: Process-data-based model calibration and application towards real-life separation issues. *Biotechnol Bioeng,* 112(7):1406–1416.
40. Donovan JW. 1969. Chemistry and metabolism of macromolecules: Changes in ultraviolet absorption produced by alteration of protein conformation. *J Biol Chem* 244:1961–1967.
41. Kueltzo LA, Ersoy B, Ralston JP, Middaugh CR. 2003. Derivative absorbance spectroscopy and protein phase diagrams as tools for comprehensive protein characterization: A bGCSF case study. *J Pharm Sci* 92:1805–1820.
42. Ragone R, Colonna G, Balestrieri C, Servillo L, Irace G. 1984. Determination of tyrosine exposure in proteins by second-derivative spectroscopy. *Biochemistry* 23:1871–1875.
43. Thakkar SV, Allegre KM, Joshi SB, Volkin DB, Middaugh CR. 2012. An application of ultraviolet spectroscopy to study interactions in proteins solutions at high concentrations. *J Pharm Sci* 101:3051–3061.
44. Jiskoot W, Crommelin D. 2005. *Methods for structural analysis of protein pharmaceuticals.* Arlington: AAPS Press. 9 pp.
45. Prin C, Bene MC, Gobert B, Montagne P, Faure GC. 1995. Isoelectric restriction of human immunoglobulin isotypes. *Biochim Biophys Acta* 1243:298–290.

46. Morgan EH. 1981. Transferrin, biochemistry, physiology and clinical significance. *Mol Aspects Med* 4:1–123.
47. McCann KB, Hughes B, Wu J, Bertolini J, Gomme PT. 2005. Purification of transferrin from Cohn supernatant I using ion-exchange chromatography. *Biotechnol Appl Biochem* 42:211–217.
48. Lindner H, Helliger W. 2001. Age-dependent deamidation of asparagine residues in proteins. *Exp Gerontol* 36:1551–1563.

18

Recent Progress Toward More Sustainable Biomanufacturing

Practical Considerations for Use in the Downstream Processing of Protein Products

Milton T. W. Hearn

Victorian Centre for Sustainable Chemical Manufacturing, Monash University, Melbourne, Australia

18.1 Introduction

Over the past two decades, there has been much debate and many policy initiatives by governments and industry associations in response to the emerging mega trends in resource utilization, to address public needs and expectations, and in recognition that developments in productivity, product performance, and economic value have to be matched with reduction in the environmental footprint of manufacturing processes and practices. These circumstances as they apply to sustainable development in product manufacture have been motivated in large measure by the Brundtland Report, also known as *Our Common Future* [1], commissioned in 1982 by the United Nations General Assembly, which cogently enunciated in 1987 two key concepts, namely, that of societal needs and the limitations imposed by the state of technology. According to the Brundtland Report, sustainable development is intended to "*meet the needs of the present without compromising the ability of future generations to meet their own needs.*"

Although some aspects of the Brundtland Report and its definitions have been criticized due to their ambiguity, openness to interpretation, or lack of detailed analysis on specific areas of industrial production where emphasis should be placed, they have provided a core set of principles that enable the meaning of sustainable development to now be quantitatively and objectively evaluated from the perspective of the environmental and economic impact of mega trends on chemical and bioproduct manufacturing practices from the scale of bulk chemicals and polymers, to commodity materials such as concrete, to technical progress in downstream processing of biopharmaceuticals.

The Brundtland Report strongly influenced the agenda and the derived action plans such as the United Nations (UN) Millennium Development Goals; the Earth Summit in Rio de Janeiro, Brazil, in 1992; the third UN Conference on Environment and Development in Johannesburg, South Africa, in 2002; the UN Copenhagen Climate Change Conference in 2009; and subsequently numerous other UN-sponsored and intergovernmental commissions related to sustainable developments. Importantly, these action plans have also captured the attention of many national and multinational company associations and industrial alliances involved with the manufacture of chemicals, medicines, food products, water supply reagents, and energy-related materials. Generally, for the chemical and biotechnology industries, these action plans have resonated with the thinking and motivations of many industrial peak bodies as evident from various initiatives, for example, the GSC-7 JACI—Hitotsubashi Declaration [2] in July 2015—and thus are gaining acceptance with regard to the nexus between more benign manufacturing practices, economic viability, and resource utilization.

Since 1987, sustainable development has been defined in many ways and interpreted in many contexts, but, overall, the developments over the past 30 years have provided a framework that represents the most prevalent basis to globally move forward the economic, social, and environmental well-being of society for today and tomorrow. Sustainable biomanufacturing represents a subset of these initiatives, and downstream processing forms a component of this sustainable biomanufacturing subset. In terms of mega trends, at least six drivers can be identified for their known or potential impact on the future development and practice of downstream processing—emerging markets, SET advances, resource scarcity, the greater need for food and water security, advances in disease treatments, health and well-being, and the obligations and roles of responsible industries (Figure 18.1).

All of these drivers are interdependent and essential. Their implications and consequences need to be simultaneously taken into account in a comprehensive manner if downstream processing of bioproducts is, in the future, to be equipped to capture many of the significant targeted developments that can be translated to more sustainable manufacturing practices. Previous definitions of sustainable development have their basis in systems thinking that connects mass, space, time, and energy into a holistic framework [3]. The quantitative elucidation of this framework enables the impact of inputs and outputs to be better understood and the origins of complex multi-factorial phenomena to be revealed.

Often, in providing these solutions, disruptive, enabling scientific, engineering, and technological breakthroughs have to be made. Importantly, the creation of these solutions, leading to the adoption and deployment of new manufacturing innovations, requires integration of process imperatives that frequently have their basis in fields distant to science and engineering *per se*, such as the need for economic growth, the goals set by governmental/political agendas and

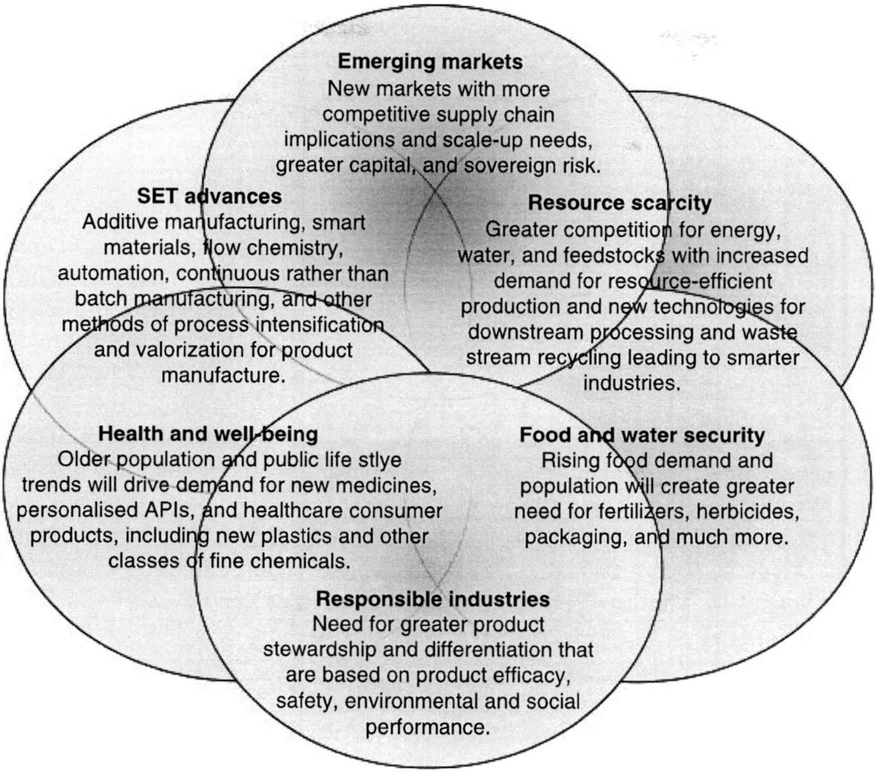

Figure 18.1 An integrated set of mega trend drivers that currently impact on and significantly underpin developments in the sustainable manufacturing of bioproducts. (*See insert for color representation of the figure.*)

regulations, and the environmental protection issues. However, concordance can be realized through the linkages provided from more benign technologies and associated developments created from new scientific and engineering research and its translation to practice.

Thus, according to the OECD and the US Department of Commerce Sustainable Manufacturing Toolkits and Handbooks, sustainable manufacturing is the "*creation of manufactured products that use processes that are non-polluting, conserve energy and natural resources, and are economically sound and safe for employees, communities, and consumers*" and represents a set of circumstances that are "*no longer just nice-to-have, but a business imperative*" [4, 5]. Within the context of downstream processing of protein products, the OECD Toolkit provides a set of 18 internationally applicable key performance indicators that can be employed to measure and improve the environmental performance of manufacturing facilities and their derived bioproducts. Process-specific score cards related to sustainability outcomes derived from the deployment of alternative approaches to downstream processing of feedstocks from fermentation or cell culture can be generated based on generic templates available from the US Department of Commerce [6], the US Department of Energy [7], or other international agencies, for example, the European Union (EU) Environment Directorate-General [8], which have been charged with similar responsibilities.

Successful integration of the technical potential of scientific and engineering advances with the concepts of productivity and profitability, the triple bottom line, as well as the triple top line [9, 10] has important implications to redefine the meaning of product quality and performance made possible through new downstream processing technologies. This outcome will be achieved through adoption of innovative process design criteria that allow industry to produce right from the start of the manufacturing cycle products that enhance the well-being of society, generate economic value, and technical viability with reduced impact on the environment. In this manner, solutions can be found that lead to a comprehensive framework that links the current mega trends associated with societal perspectives and needs with new ways to move beyond cost control in manufacturing to the stage where greater performance is achieved through innovations in technology development and implementation.

In the following text, a set of 12 principles relevant to sustainable manufacturing has been collectively proposed as a means to capture many of these issues, as they apply to all essential stages of downstream processing of bioproducts. Forward-thinking companies and industrial associations are now progressively implementing such principles into more sustainable manufacturing practices as part of their business strategies. The BASF Verbund Strategy [11], the Pfizer Green Chemistry Initiative [12], or the GlaxoSmithKline Fast Life-Cycle Assessment of Synthetic Chemistry (FLASC™) tool [13] are but three examples of many similar strategic developments, where new product opportunities have been garnered from the more comprehensive understanding of the eco-financial benefits derived from such principles, which include the establishment of:

- An encompassing social license to operate
- A strong customer base for the derived products and processes
- Innovation linked to a strong intellectual property base
- A skilled and productive workforce talent
- Supply and value chain integration
- Access to competitive capital
- Access to feedstock security and quality
- Energy and utilities accessed on a cost-competitive basis
- Location of facilities in countries with stable government
- A balanced regulatory environment
- An ability to implement tiered governance/product stewardship
- Nimble and benign adaptive technologies

In order to implement these principles of sustainable manufacturing in the downstream processing space, a number of criteria have to be addressed within the framework of corporate compliance and policies set by the governing environmental protection authorities. Firstly, recognition needs to be given to the fact that product manufacturing is becoming increasingly dependent on new ways to achieve process optimization and valorization. Traditionally, manufacturing processes have tried to balance cost versus performance with sustainability captured through comprehensive cost reduction within predefined boundary conditions of the cost of materials, plant infrastructure, and staff (Figure 18.2a). Arising from the demands of the mega trends, product design and manufacturing optimization must now transition to include the total environmental footprint of the product and process (Figure 18.2b).

This transition to more sustainable manufacturing processes and products implies that production campaigns will in the future be assessed within the wider perspective of integrated supply chains with product stewardship going beyond the traditional approaches of cost reduction and life-cycle analysis. Irrespective of whether the products are made by chemical synthetic or biotechnological processes, this paradigm shift leads to outcomes involving new ways to consider product attributes, including approaches to better define product purity, particularly in terms of the context of "*how pure is pure*," as part of the choice of a specific manufacturing approach and product use. As a consequence, downstream processing campaigns with two different classes of bioproducts often require quite different process analytical technologies (PATs)

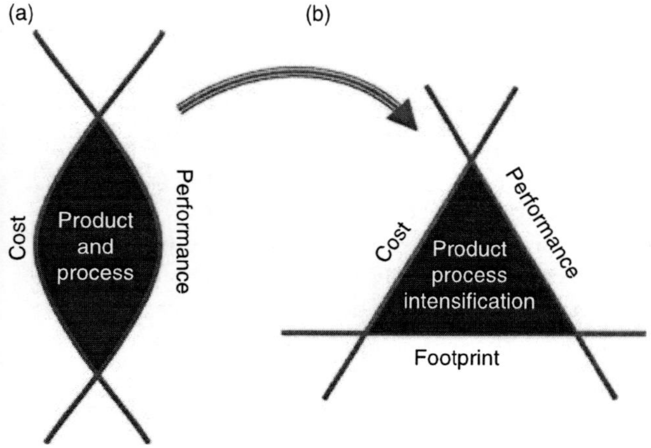

Figure 18.2 Schematic representation of the transition that is occurring from (a) the traditional manufacturing approaches solely based on cost and performance where local optimization of individual unit operations were practiced to (b) the paradigm shift associated with sustainable manufacturing where the overall environmental footprint of the product and global optimization of the overall process are also incorporated. (*See insert for color representation of the figure.*)

to validate product purity, efficacy, and safety and yet may still be superficially based on similar chromatographic, membrane, or other fractionation steps. To illustrate this issue, intravenous immunoglobulin preparations produced by fractionation of human plasma are widely used in the clinical therapy of a broad range of neurological, hematological, immunological, nephrological, rheumatological, and dermatological disorders, and yet the products are at totally different levels of molecular purity and structural diversity compared to the levels required, for example, for a recombinant insulin used in the treatment of Type 2 diabetes [14, 15]. This does not mean that the fitness for use of intravenous immunoglobulin preparations are in any way impaired but simply that these different products necessitate access to and implementation of different integrated process chains of operation and methods of product analysis that are tailored to the product's molecular characteristics, safety and efficacy profiles, and intended use.

Ways to implement this transition toward more sustainable manufacturing practices are not fully understood by many industry sectors, and the biotechnological industries involved in the manufacture of industrial enzymes, biopharmaceuticals, or other protein products are no exception. To provide the level of quantitative information that is needed by these industries, creation of new types of contract manufacturing organizations or alternatively access to "*in-house*" facility service centers and specialist multidisciplinary teams are required. Development of such facilities are becoming important imperatives behind manufacturing process initiatives with their practical and robust outcomes essential to better understand the broader implications of specific manufacture processes, such as a preferred chromatographic unit operation. The robustness of these approaches is essential if this information is to gain full acceptance by industry. When this occurs through integration of full life-cycle analysis with full system analysis, then this information can be used with greater confidence to guide the implementation of the new technological approaches and simultaneously de-risk the deployment of the new concepts and technologies. The availability of creditable and accredited facilities to carry out this experimentation and assessments will become increasingly important in the future, since quantification of the impact of all of the parameters that contribute to an environmental footprint is complex and needs a full understanding of product efficacy and quality, energy, water, and resource balances and their costs, whether reuse, recycle, or "one-pot" processes are viable options and whether the product satisfies the criterion of "safe and benign by design."

This chapter examines recent progress toward the incorporation of these concepts into approaches that are increasingly being employed for the more sustainable manufacturing of protein-based products, with emphasis of the downstream aspects of the recovery and purification of value-added protein products derived from biotechnological procedures. Lessons gained from the use of similar approaches developed within the chemical, chemical pharmaceutical, and food ingredient industries will be examined in terms of their applicability to the downstream processing of protein products derived from genetic engineering, cell culture, and associated biotechnology strategies.

18.2 The Impact of Individualized Unit Operations versus Integrated Platform Technologies on Sustainable Manufacturing

The uptake of sustainable manufacturing technologies for the productions of chemicals and biologicals by industry is an interplay between what is made possible through scientific advances, increasingly now generated through innovated developments in green chemistry and engineering, and what needs are being created by the changing social, economic, and political forces. There is thus tension between natural and man-made forces in terms of sustainable manufacturing issues, which are often identified as *market pull* versus *technology push*. Increasingly, developments in green chemistry and engineering in the chemical manufacturing sector, which represent major and essential components of the larger and more encompassing aspects of sustainable manufacturing *per se*, are seen as being equally relevant to advanced biomanufacturing. For example, the paradigm changes in sustainable manufacturing as they apply to biotechnology equally involve the same consideration of the potential for climate changes and resource depletion in response to changes in IPAT *I*mpact = *P*opulation × *A*ffluence × *T*echnology. Despite the fact that the relationships between population, affluence, and technology are interdependent and complex and differ significantly between highly industrialized, prosperous countries, and countries where advanced manufacturing economies have yet to be established, IPAT considerations ultimately drive chemical, bioproduct, and material development and manufacturing.

It is widely perceived that downstream processing procedures for many bioproducts, including biopharmaceuticals, frequently represent major and costly technical bottlenecks that typically involve a complex set of multiple unit operations [16, 17]. Considerable developments have occurred during the past two decades in terms of upstream methodologies, for example, greatly increased expression levels of the target product or removal of interfering proteins or other secondary metabolites by genetic engineering of the host cells used in fermentation or cell culture. However, these outcomes have not been fully matched by downstream developments, leading to increased costs. For example, the downstream processing requirements in the production of many biopharmaceuticals generated by fermentation or cell culture processes can alone represent between 50 and 85% of the total manufacturing costs and generate the largest amount of waste [18–20]. Because historically, most processes have been treated as a unique set of modular tasks based on a set of individualized unit operations, traditional practice has been to optimize each task rather than the process as a whole. Individualized unit operation approaches do not maximize the knowledge transfer or minimize the process risk during scale-up, while guidance for application to the next product remains empirical. For these reasons, more integrated platform technology alternatives have gained some favor, particularly for the downstream processing of monoclonal antibodies (mAb).

Further development of these trends would act as important drivers to gain greater sustainability in the manufacturing procedures with the potential to reduce the environmental impact compared to current processes. As apparent from the representative examples and data shown in Table 18.1, significant reduction in the environmental footprint could be realized through process intensification that leads to reduction in the use of water and other chemicals (buffers and salts) and other attendant consumables such as membranes, filters, disposable plasticwares, and even chromatographic resins. Consumption of the order of several hundred thousands of kilograms of water per kilogram of the final recombinant therapeutic protein product has been reported [21, 22] for some processes. Additional process requirements also increase the level of water usage such as water (and other chemicals) for biowaste stream treatment, equipment and facility cleaning, and water for cooling and heating. Ways to capture the benefits in manufacturing processes through waste mitigation

Table 18.1 Representative order-of-magnitude estimates of the process water and chemicals and consumable materials used in manufacture of recombinant proteins.

	Microbially derived proteins		**mAbs from mammalian cell culture**	
	Highly optimized, large scale	Typical multistage "composite" process	Optimized large scale	Highly intensified, large scale
Water usage	~1000–2000	15 000	4500	1500
Salts + buffers	1–10	400	300	100
Consumables (solid wastes)	1–10	20–35	4–15	2–10
Organic solvents	0 (for some industrial enzymes) 5–10 (for specialty proteins)	100 (alcohols, may involve some hazardous solvents)	8 (alcohols)	8 (alcohols)

Source: Adapted from Ho [21], Royal Society of Chemistry.
All values expressed in kilogram of reagent per kilogram recombinant protein.

from the concepts of "reduce," "reuse," and "recycle" (3Rs) are thus important. Nevertheless in these cases, holistic understandings derived from the application of rational design principles based on systems thinking, incorporating global rather than local individual step optima, have not yet been well developed. This limitation has constrained the ability to maximize efficient supply- and process-chain integrations and restricted rational process intensification and valorization. The use of platform technologies coupled to process intensification protocols, in principle, goes well down the path to overcome many of these constraints. These benefits become even more reinforced if continuous rather than batch processes are utilized [23–25]. In this manner, better integration is achieved for both the material supply chains associated with the acquisition and use of the reagents, materials, and other consumables required for the overall process, as well as the individual bioprocess supply chain steps that form the overall process *per se*, allowing more sustainable manufacturing procedures to be better implemented and products of high quality, efficacy, and safety to be obtained.

Within the context of either an individualized single-unit operation or generic platform technologies, the response has been toward the development of closed loop systems for materials usage, where extended product lifetimes, functionality, and efficacy have been linked to reduced supply chain material consumption and the use of materials that have been *"designed for disassembly"* (the so-called "D4days"). Such approaches are now forming the basis for continuous manufacturing [26] with the employment of materials and technologies that are not so specialized that make them impossible or uneconomic to put into practice at larger scales or to put them back to their original condition for reuse/recycling. Adoption of the concepts of the 3Rs as they apply to waste hierarchies—recycle, reclaim, and remanufacture—as captured, for example, in the EU Directive 2008/98/EC [27] and attention to implications drawn from the so-called Factor 4 resource usage concepts [28], or the subsequently adapted so-called Factor 10 concepts [29], can provide beneficial guidance in circumstances where the intention is to undertake an overall dematerialization of the product life cycles and overall reduction of the carbonization of those stages of the processes that form the key components of a specific biomanufacturing approach.

To achieve these outcomes, serious consideration must be given to process selections that (i) minimize environmental impact through attention to feedstock choices, with substitution where possible of renewables for nonrenewable reagents and materials, (ii) permit a system approach to the production and analysis, (iii) enable enhanced automation, miniaturization, and parallelism, (iv) utilize process separation and purification systems that have been designed from the onset to act in an integrated manner to provide improvements in mass intensification, and (v) implement cost-effective technologies that differentiate product attributes from other contaminants present in the complex feedstock mixture. In these contexts, the need to discover, deploy, and enable disruptive innovations, such as new platform technologies based, *inter alia*, on new modalities of molecular recognition, becomes obvious. Although individualized single-unit operations in a sequence of batch modes have largely been the mainstay of product manufacture in the chemical and biotechnology industries, to more productively achieve the aforementioned outcomes, attention should now be focused on methods for continuous manufacturing based on the incorporation of platform technologies. For example, platform technologies involving scalable multipurpose equipment, a common set of technologies, and a rational hierarchy of work tasks for product separations have the potential to (i) lower risk, (ii) provide greater insight and knowledge on the impact of process variables, (iii) increase speed within the manufacturing cycle, (iv) lower costs, and (v) provide an easier transfer of knowledge from discovery to development and then to production departments within a specific company or to contract CMOs. Similar considerations apply to the adoption of process analytical techniques (PATs) which allow systematic real-time or near-real-time analyses of product process streams that are based on multipurpose instrumentation for multiple analyses that are nondestructive for the remaining portions of the sample.

The respective benefits and differences between individualized unit operations and platform technologies can thus be objectively and quantitatively differentiated in terms of how they measure up against a set of key manufacturing goals that lead to:

- Reduction in the environmental impact of a product
- Reduction in material's usage to achieve a specific level of product and process performance
- Reduction in the use of auxiliary materials for the manufacture of the product
- Reduction in the use of nonrenewable resources
- Reduction in the waste and emissions generated during the manufacture
- Reduction in the energy used to manufacture the product
- Reduction in the overall risk and hazard from the manufacturing approach
- Reduction in product toxicity or adverse biological side effects
- Reduction in the life-cycle costs to validate the product(s)

The aforementioned issues can be illustrated with the example of hydrophobic interaction chromatography (HIC), one of the key chromatographic adsorption techniques in protein purification. Historically, the choice of a HIC resin and the elution conditions have largely been based on empirical procedures derived from trial and error and experience. This can mean that a HIC resin of lower performance may end up being used in a purification process, leading to the generation of an excessive amount of waste. The increased use of design of experiment (DoE) and quality by design (QbD) methods, linked to high-throughput procedures (HTPs) for resin screening [30, 31], have to some extent allowed the selection of a specific HIC resin and associated loading, wash and elution conditions to be systematized in the context of its use in a process based on several individualized single-unit operation steps. However, to establish a more sustainable process, even more sophisticated approaches that guide the proper choice of the HIC adsorbent, the salt system, and other operational conditions, as an integral component of the overall manufacturing process, are still needed. These new approaches can deal with separation tasks of increasing complexity and scale and allow more efficient and predictable options to be explored with regard to the reuse and recyclability of some or many of the so-called consumables that are integral to the overall process. The solution, of course, is to take advantage of the more rational systems design methodologies to move from the local optima implicit to a single-unit operations associated, for example, with a single type of HIC resin to a global optimum associated with a fully integrated process.

Attainment of improved overall process efficiencies in a downstream campaign often involves the full integration of two or more process steps, with the order of the sequential separation procedures scheduled according to robust process design criteria. The objective here is to seek an overall global optimum rather than a local optima, which may at first pass appear counter to established practice. Traditionally, each step in a downstream process involving a sequence of single-unit operation is optimized individually in order to achieve step productivity rather than an overall process productivity. If attainment of the latter criterion is the goal, then a lower yield at an earlier stage in the process may be necessary to ensure a higher purity of the product is achieved subsequently. This can be realized, for example, with chromatographic systems based on packed or fluidized beds [32, 33] through optimization of the appropriate choice of column configuration, chromatographic resin, column loading, flow rate, and product yield, leading to the identification of operational conditions whereby a production rate [34], as shown in Figure 18.3 for the fractionation of a human serum albumin feedstock, can be selected to better suit the subsequent stage of the downstream processing of the product.

With products generated by fermentation or cell culture, these efficiency dividends include not just the enhanced removal of host cell proteins and other contaminants but also the attainment of overall optimal process economics and product yield concomitantly with reduction of the environmental footprint. Case studies associated with mAb purification are particularly useful exemplars in this regard. Such case studies with mAb of different isotypes have generally involved the sequential use of either an affinity (Protein A), cation exchange (CEX), HIC, mixed mode (MM), hydroxyapatite (HT), or fluorapatite (FT) chromatographic step and have been captured in numerous scientific publications and diverse patent applications, for example [35–40]. Typically, many of these former approaches have been based on individualized single-unit operation practices, whereby local rather than global optima for the overall

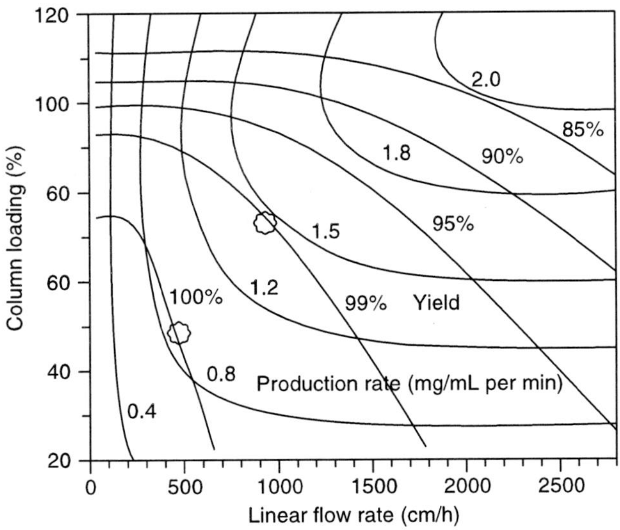

Figure 18.3 Operating regions related to the product yield and processing rate for the purification of human serum albumin with a DEAE Trisacryl M ion exchange chromatographic column. The open circles are the maximum production rates (0.87 and 1.46 mg/mL per min) for the 100 and 99% product yield curves derived from the SAM approach. *Source*: Mao and Hearn [34]. Reproduced with permission of Wiley.

purification and recovery of the mAb have been targeted. Due to their operational implications, these local optima approaches do not fully take into account considerations on how to best utilize consumables employed in the process, such as plasticwares, columns, or membrane materials, the characteristics and dimensions of the associated process infrastructure, or the energy consumption to achieve the best overall outcome from the perspective of yield, purity, cost, performance, and environmental impact.

For consumables and materials to be utilized in a more sustainable manner, options for their operation in a recycling mode are needed, including system regeneration and reuse. Equally, the same considerations relate to the choice and use of chemicals and other materials employed in the process, for example, water, salts, solvents, etc., and whether they or other potentially reusable "waste" stream components can, or alternatively should, be recycled based on an enhanced understanding of their molecular properties, toxicity, or risk. Generally, the modality of in-process recycle is usually very difficult to carry out during batch operations, and for this reason post-processing or out-of-process recycle are more commonly undertaken. To achieve these outcomes, technologies leading to more sustainable manufacturing approaches must employ (i) process options that allow movement towards zero-waste and zero-impact manufacturing procedures as measured against well-designed process sustainability metrics and indices [41–43], (ii) greater levels of recycling and reuse of material supply chains, (iii) product design methods to reduce lifetime impact and cost, and (iv) the application of multidisciplinary research that supports the analytical systems needed to monitor product purity, function, mass recovery, and fitness for use at the same time that contaminants are monitored for their structure, abundance, toxicity, and potential to interact with, bind to, and impede the desired functional attributes of the protein product.

To support and obtain the best performance from single-unit operation or alternatively platform technologies intended for large-scale process applications, efficient, fast analytical detection procedures, such as multiwavelength second derivative UV-Vis spectroscopy with diode array detection, fluorescence or chemiluminescence measurements, near-infrared (NIR) spectroscopy, or direct injection n-dimensional mass spectrometry (MS^n), preferably linked also to capillary LC × LC, HP-CE, or HP-CEC methods, are usually required. The use of these analytical procedures as PATs is primarily intended to address the issues of product quality. Moreover, these methodologies permit early recognition of operational vagaries falling outside the required process and product specifications, thus reducing the need for product rejection or reprocessing and enabling better utilization of the facility's plant capacity. Additional positive environmental consequences can also ensue from reduction in energy and material consumption and appropriate guidance gained to implement continuous processing. Acquisition of knowledge on the selective, whole-of-cycle attributes of these process analytical systems and their versatility in terms of function and detection capabilities across a large variety of molecular classes, including information on the characteristics and molecular features of the target molecules and contaminants present in the sample, is important. Without a strong base of supporting analytical data, attainment of excellent atom and mass efficiency with optimization of resource utilization associated with low mass intensification factors will not usually occur, and robustness in the scaling-up of the manufacturing process to a more sustainable level will not be achieved.

18.3 Implications of Recycling and Reuse in Downstream Processing of Protein Products Generated by Biotechnological Processes: General Considerations

The opportunities offered through the adoption of overall process design criteria, including rational recyclability regimes for packed chromatographic column or membrane skid reuse, have the potential in the mid to longer terms to significantly contribute to a reduction in plant investment expense, associated facility infrastructure costs, as well as overall material costs once their implications have been ascertained and measured against regulatory requirements. The ability to incorporate recycling approaches within a strictly regulated framework as a driver to achieve greater material sustainability at the downstream processing stages associated with the manufacture of a specific protein product hinges on a large variety of products and process-specific attributes, including *inter alia* the protein concentration in the feed, the structural and biological stability of the protein, the level of purity required, the cost boundary conditions for production, the process scale in terms of mass and volume, and the energy consumption involved.

Moreover, the relevance of scaled-down approaches with "miniaturized" model technologies to assist in the prediction of performance for the validated upscaled industrial process has also to be factored into these considerations. These approaches have found wide advocacy, particularly with chromatographic procedures that exploit either batch (bath) or dynamic binding experimental methods for data acquisition [44–47]. Because the objective is to achieve a global productivity optimum, the outcomes and impacts must be determined not just at the level of product quality *per se* but how the use of other components, raw materials, and the process system overall can benefit from the system integration and process intensification. At one level this can be achieved through the use of process valorization tools, computational modeling, and well-designed experiments. At another practical level, systematic examination of the compositions, mass and volumes of the waste stream can be employed to derive a ranking in terms of risk or regulatory requirements that have to be managed for specific key components that contribute to the waste if recycle options are under consideration. With this knowledge, closed loop systems can be put in place with each stage benefiting due to the improvements derived from the process intensification.

Such transitions from the traditional end-of-pipe, cradle-to-grave technologies with their significant generation of primary and secondary waste streams [48] can be replaced by cradle-to-cradle technologies based on integrated process intensification by taking advantage of a strategy that leads to a dramatic reduction in the size of the facility footprint yet still achieves a given production objective [49, 50]. Such strategies result in improved use of raw materials and greater reduction and recycling of waste streams. For practical reasons, for bulk chemical products, such as essential amino acids, aliphatic acids, and other low molecular weight secondary metabolites, derived from biotechnological processes, the reuse of recycled streams, particular water, is a relatively common practice. Further up the value chain, such practices are less common, primarily due to the higher levels of regulatory stringency. However, if the transition to a lower cost continuous manufacturing regime is to occur with improvements in the sustainability index (SI) realized, it is obvious that recycling of waste streams derived from the downstream processing methods must occur.

Because of its chemical and physical properties, water, being the largest chemical fraction by mass and volume in a biotechnological process, is a very relevant target for recovery and reuse, even within the constraints of downstream processing of protein products. With the upward trend in the cost of water as a raw material at a purity grade appropriate for good manufacturing practice (GMP) processes, deployment of recycled water becomes increasingly attractive provided regulatory requirements are satisfied based on rigorous quality control. In fact, well-defined regulations exist internationally for the reuse of water based on stringent quality standards. A schematic illustration of the type of water deployment that can occur within a typical mAb production facility is shown in Figure 18.4 with multiple exit nodes for the recovery of waste streams.

Recent economic modeling studies have suggested [51] that the use of a recycling facility to recover water at a quality grade suitable for injection (water for injection (WFI)), based on ultrafiltration and reverse osmosis (RO) membranes, rather than sending the waste water to a disposal facility, would provide a return on investment (ROI) of less than 3 years for a typical monoclonal production facility, where ca. 14 000 kg water is consumed per kg of mAb product, with the accumulated waste water amounting to ca. 3400 m^3 per year without clean-in-place (CIP) sanitization. Analogous favorable ROIs would appear feasible for the recovery of injection grade water for other recombinant proteins, for example, human insulin or alternatively organic

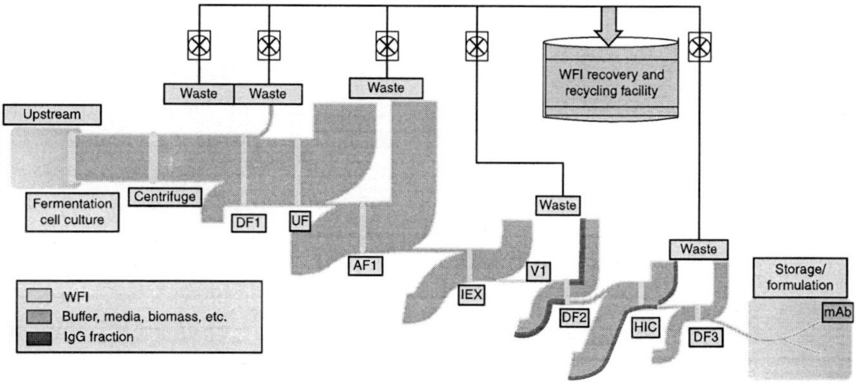

Figure 18.4 Schematic illustration of the water deployment within a typical monoclonal antibody production facility and the different exit ports where waste water recovery could be undertaken singly or in aggregation. *Source*: Grote et al. [51]. Reproduced with permission of Wiley.

solvents in a manner similar to that employed for extraction of low molecular weight secondary metabolites generated by fermentation procedures.

Conceptually, such recovery and recycle facilities represent opportunities for turnkey applications based on a common set of operational activities that encompass advanced design principles, tools, and materials science approaches. In so doing, these approaches are central to the practical realization of the 5 P's of productivity enhancement that lead to greater sustainability, namely,

- Practice, design, and process robustness
- Process intensification
- Precedence and metrics
- Progressive automation
- Pricing and public pressure

and provide a conduit to garner insights into the overall process efficiencies as well as the robustness of different models for process valorization as they relate to the environmental footprint associated with the manufacture of the protein product. In this regard, the deployment of single-use equipment represents a special case, which goes well beyond current life-cycle thinking. Partly, this circumstance is due to the fact that the current generations of plastics and polymeric materials suitable for single-use applications in equipment or as consumables in biomanufacturing have very long half-lives for biodegradation. Various published studies have reported [52–55] that in terms of cost and energy inputs, the expenditure to acquire products using single-use processing technologies based on plastic/engineered polymeric consumables are considerably lower than the corresponding stainless steel (tanks, reactors, etc.) systems. However, the environmental waste burden, particularly with regard to disposable plastic bags, increases significantly while additional issues of corporate product stewardship also emerge. As a consequence, much more detailed life-cycle assessments are required [56] based, for example, on the ReCiPe midpoint and end point impact assessment methodologies [57], which address a range of environmental concerns at the midpoint level and then aggregate the chosen midpoints into a set of end point damage categories to differentiate single use from traditionally used (stainless steel) systems. Preliminary findings, however, suggest that single-use process technologies, at least for mAb production, can have less impact on the environment than traditional process technologies, particularly with regard to the need to use large quantities of steam, process water, and WFI. However, the fundamental challenges set by the 3R criteria have not been fully solved with single-use systems, since disposal of plasticwares remains a longer term environmental impact consideration.

18.4 Metrics and Valorization Methods to Assess Process Sustainability

In undertaking process valorization as part of the overall assessment of the extent of sustainability of a manufacturing campaign, the following aspects are often taken into consideration:

- The material properties of all of the components used in the process, including the nature of the mass intensification indices, defined as $MI_{n = 1, 2, 3...}$
- The scalability and sustainability of footprint indices
- The energy requirements
- The number of unit operations per product chain
- The total assessment of the process costs, including the distributions between Type I and Type V investments
- The potential profitability target and the desired range of productivity gains

The underlying assumption inherent to these considerations, and their use as the *raison d'etre* for process intensification leading to more sustainable manufacturing procedures, is that a set of process capability indices, *Cp*'s, can be derived either by *ab initio*, by *ad valorem*, or by *ex fida bona* statistical techniques that adequately capture the ability of the parameter settings of a process to be in a state of statistical control and the manner an organization goes about controlling them, including (i) the intent, capability, and corporate *modus operandi* behind a specific production pathway, (ii) the nature of the process supply chains, and (iii) the properties of the associated material supply chains set in alignment with the business strategy of the manufacturer to achieve a process performance that falls within a preset upper and lower specification limit. Importantly, knowledge of the process *Cp*'s permits assessment of the robustness of the overall process control. When process control cannot be realized, product and process specifications will not be achieved, and this will lead to the need for either reprocessing or, more commonly, the problem of product disposal with additional waste production.

Process capability indices are thus very useful metrics to assess the statistical control of a downstream processing task, since they link the output of a well-controlled process to predetermined specification limits. For convenience, process capability indices are usually represented as ratios of the product specification "width" to the spread in-process values. Tight specifications can be set to ensure that the process control statistically falls, for example, within the boundary conditions set by six sigma, that is, one occurrence of a noncompliant product/process defect in a million or for biopharmaceutical production, since even more stringent parameter control setting are frequently needed and even nine sigma can be used as a set point. In-process or on-column protein refolding [58, 59], in-process enzymatic activation of zymogens, in-process or on-column removal of polypeptide tags from recombinant proteins [60–64], as well as the purification of potent cascade self-cleaving or auto-proteolytic biopharmaceuticals, such as the activated recombinant factor VII (rFVIIa) [65, 66] or the chimeric recombinant factor IX-Fc fusion protein (rFIXFc) [67] and the downstream processing of recombinant factor VIII (rFVIII) [68], are representative examples where these higher stringency requirements need to be applied.

In industrial biotechnology as well as in the fine and bulk chemical industries, considerable understanding of process pathways exists, often in association with extensive analytical data from control monitoring systems. A well-controlled process will have almost all of the measured data falling inside the defined specification limits. With real-time analytical monitoring of processes now becoming available, stringent controls of processes in the range of six to nine sigma are becoming technically feasible, although typically most industries are still operating within the regime window of three to four sigma. From the perspective of contaminant load in biopharmaceutical protein products, where greater molecular purity, safety, and efficacy are demanded, if operational performance only reach a limit of three or four sigma, then this would mean that the processes would not be under tight control, with the consequence that reduced throughput and lower productivity will occur. If the product was intended to be used as an active pharmaceutical ingredient (API) and if only a level of three or four sigma was achieved, then almost certainly this would mean that the batch was rejected because of purity and quality concerns. Moreover, such processes will consume more raw material and energy per unit of the final product as well as generate considerably more waste, that is, will be less sustainable. Clearly, when $Cp > 1.0$, then most if not all of the process analytical measurements fall within the process specifications. The general form of process capability indices, *Cp*'s, can be defined (for a process involving a potential operational environment of six standard deviations, 6σ), as

$$\text{Process capability index} = Cp = \frac{\text{USL} - \text{LSL}}{6\sigma} \tag{18.1}$$

where USL is the upper specification limit and LSL is the lower specification limit at 6σ. The practical value of knowing the value of the process capability indices, *Cp*'s, is considerably enhanced when linked to life-cycle inventory and assessment methodologies that allow the cumulative environmental impacts of a given process or product to be estimated across its entire life cycle. In combination, these tools thus enable more comprehensive views of the technical, environmental, and economic trade-offs that have to be made when considering alternative process or product options [69–72]. Such approaches permit the impact on overall sustainability of changes in process design and controllability to be considered in a more robust and informed manner than usually undertaken by traditional methods of process assessments. Table 18.2 summarizes exemplars of three different stages of process controllability in terms of their relevance to a SI.

According to the current knowledge of process intensification and the associated valorization procedures as they apply to downstream processing of protein products, process metrics that address sustainable manufacturing requirements can generally be split into two categories. One involves the assembly of extensive sets of indicators to evaluate a target protein product based on the historical experience and depth of process know-how within the field, information that is usually dependent on in-house proprietary data of the specific industry/company, while the other involves the *de novo* development of composite indicators that are employed to replicate a broad set of manufacturing issues through *in silico* modeling and simulation. In the latter case, mathematical formalisms can be derived *inter alia* from Monte Carlo procedures and methods of statistical multivariate analysis (MVA), allowing metric development to be more readily linked to reliable collections of data and life-cycle assessments without requiring commitment of extensive time or resources. The advantages of Monte Carlo simulation techniques are that input probability distributions can be employed to determine the output probability distributions. In this manner, outcome options can be identified at or above critical threshold values defined by the process capability indices, *Cp*'s.

Table 18.2 Three representative stages of process control and their impact on process sustainability.

Consumption of raw materials	State of process	Sustainability index
Process gets out of control	Lacks robustness and operational efficiencies	Process results in the product falling outside specifications, excessive amounts of waste
Process achieves controlled but local optimum for raw material processing with suboptimal efficiency	Near or at local optimum for conversion of raw materials to product occurs for some but not all process stages; full productivity not attained	Steady-state condition achieved but yield versus environmental footprint not optimized
Process generates no unused raw materials, with all by-products recyclable	Process achieves closed loop efficiency; all product falls within specifications; mass, space, energy, and time parameters achieve global optima for the process	Process cycle becomes sustainable and aligned with critical process and product specifications

Ultimately, these criteria can be summarized into frameworks that specify the whole-of-cycle stages of a successful manufacturing process and correlated with the appropriately assembled production data. Importantly, various production planning and production distribution methodologies or decision support models for the prediction of the performance of different stages of a downstream processing campaign as part of the manufacturing of a protein product, such as individual chromatographic stages, have already been developed. Moreover, literature related to both process intensification and process valorization also exists for various manufacturing sectors of the food, chemical, and industrial biotechnology fields [73–78]. Developments in these allied fields of activity are all relevant to downstream processing of proteins and other types of bioproducts. A key challenge with the application of such simulation and modeling procedures is finding the few key metrics that are (i) useful, (ii) feasible to independent measure, and (iii) robust and that have the greatest likely impact on any given process. In this context, the application of the so-called 80 : 20 rule may be beneficial (M.T.W. Hearn, unpublished findings). In general, this challenge represents a boundary condition problem, which can be rigorously tested and potentially solved through the use of several recently developed mathematical formalisms and predictive models as discussed subsequently.

The significance of these approaches is that they enable both qualitative and quantitative outcomes to be derived from mass intensification, energy intensification, or resource intensification assessments and associated metrics [79] and when used in valorization models linked to life-cycle considerations provide systems-wide, holistic views of a process. In this manner, process material flows, plant layouts, or instrumentation configurations do not have to be solely viewed as a set of isolated unit operations carried out across a finite time and discrete space. Rather, the effect of any variance in one particular unit operation can be related to its effect on other unit operations through a network of predictive cause-and-effect relationships that comprise the overall manufacturing process. Importantly, the intent of these modeling approaches is to simplify or guide the decision processes related to the *modus operandi* use for the conversion of raw materials to higher value products and to assist with the overall cradle-to-cradle gated manufacturing outcomes through optimizing the choice of production scheduling tools.

By employing a systems-wide approach, multivariate views of the cradle-to-cradle product requirements emerge that are implicit to the chosen manufacturing system. This strategy permits a more quantitative evaluation of trade-offs that have to be made across the product supply chain in terms of critical process parameters, leading to a better environmental profile, improved process economics, and overall enhancement of the safety of the process. In particular, knowledge of (i) the process input and output masses (volumes), (ii) the total mass (volume) of the raw materials, (iii) the capacity of the manufacturing plant to process all or only a fraction of the total mass (volume) of the raw material, (iv) the compositional equivalence or balance of the input material with the molecular features of the final products, (v) the relationship between revenue and production, marketing, transport, and material purchase costs, and (vi) the product streaming ratios associated with multistep, multiproduct processes. Typically, these aspects form part of the product's business plan, the standard operating procedures' manual, and the approach employed for data assembly along the process supply chain. Emphasis can then be placed on several general subcategories of the process technology and considered *inter alia* from the perspective of the impact of (i) material properties, (ii) scalability, (iii) energy requirement, (iv) costs, and (v) the minimum number of unit operations required to achieve process development. The quantitative significance of each of these categories is captured through knowledge of the different product and process attributes, as set out in Sections (4a) to (4d), respectively:

(4a) *Impact of the Physical Form and Chemical Properties of the Raw Materials and the Product*: The physical form and chemical properties of the feedstock and other raw materials required to generate the final product as well as the protein product itself are directly linked to, and impact upon, the capital investment required and the technical stages needed to deliver the process. Such requirements may range from the level of special high security containment facilities and dedicated fermentation/cell culture suites, all the way through to temperature-controlled buffer preparation and storage facilities. The molecular nature of the raw materials and the protein product in terms of their solubility, potential to precipitate, ability to be preferentially separated between two phases or immiscible liquids and so on all impact on the process attributes and the requirements for energy and water for heating or cooling, the need for recovery or recycling/reprocessing of waste streams if applicable, the cleaning of equipment, or the disposal of the generated waste.

Although the molecular composition and concentrations of the final protein product and many of the other species present in the feedstock or other raw materials may be known or alternatively can be determined by fast analytical methods at different stages of the process chain, it cannot be assumed that all relevant data to achieve improved sustainability have been acquired, nor does it ensure that full control over batch-to-batch variability will be obtained. This situation is often more challenging in the earlier stages of process and materials supply chains, since information on the precise chemical composition and concentrations of all the components in the raw materials are usually not readily available in advance. For example, variations in the composition of cell culture media at the upstream stages of production of a recombinant protein can affect the product yield/titre and the function of the specific protein expressed by the host cell in culture. Routine MVA of raw materials and in-process feed streams based on experimental data acquired with diode array UV absorption spectroscopy [80, 81] or NIR spectroscopy [82] can allow more robust use of ingredients such as soybean, casein, or tissue peptones/hydrolysates at the fermentation/cell culture stage of production as well as buffer additives required for the downstream processing tasks, while the application of other PAT procedures is essential if better control of process performance at the downstream stages are to be realized.

(4b) *Impact of Mass*: In general terms, mass can be divided into several key types—the (anticipated) mass of the protein product and solid compounds and materials, liquids and gases present in the feedstocks, and raw materials. The percentage amount(s) of each type of mass that is converted, consumed, or required to generate the protein product during downstream processing can be represented by a use-fraction (X_{use}) and a productivity-fraction (X_{prod}), both of which are measures of the scale and process logistics. Thus, the compositional mass balance during downstream processing of a crude fermentation or cell culture feed stream can be represented by the summation of the elemental compositions of all of the individual input materials and compounds used in the production of the final protein product P at a facility site location L, with the use of a process i, selected from a set of processes I, according to a predetermined productivity specification ($Prod_{\text{defined}}$), an achieved target productivity specification ($Prod_{\text{actual}}$), and production rates for the overall mass utilization of the feedstock and protein product generation achieved with process i, given by the terms ($ProdRate_{(f,i,r)}$) and the ($ProdRate_{(P,\, i,\, r)}$), respectively The downstream processing system is fully optimized in terms of efficiency when the ratio of $Prod_{\text{actual}}$ to $Prod_{\text{defined}}$ (i.e., $Prod_{\text{actual}}/Prod_{\text{defined}}$) equals 1.0, but typically in the case of the production of recombinant proteins, this ratio falls into a range ≤ 0.001. The mass conversion of raw feedstock to product and the ratio ($Prod_{\text{actual}}/Prod_{\text{defined}}$) in the manufacturing space of chemical products is usually represented by the so-called E factor [83, 84]. In terms of mass, this summated value must equal the total compositional content of the individual materials and compounds, which end up in the resulting protein product(s), any by-product(s), and the waste formed. Although conceptually these mass dependencies can be considered to fall into three or more physical classes and states, for example, solids, liquids, gases, or combinations of each other, the formalisms shown as Equations 18.2 and 18.3 represent a general case for protein products generated by downstream processing procedures and can be expressed in the following manner:

$$\sum_{P\varepsilon P'\backslash E} \text{Mass}_{(t,P)}\, x X_{\text{use}(t,P,P',i,L)} = \sum_{P\varepsilon P'\backslash E} \text{Mass}_{(t,P)}\, x X_{\text{prod}(t,P,P',i,L)} + W_{\text{mass}(t,P',i.L)}, P'\varepsilon P \setminus B, \forall P, \forall i, \forall t \qquad (18.2)$$

where the left-hand side of Equation 18.2 represents the mass conversion of a crude cell culture/ fermentation broth of known weight (derived from volume times density) and containing a known weight of the product with a conversion rate of $X_{use(t,P,P',L)}$ to a protein product, P, and a related product P', ... P'_n (which may be a deamidated, posttranslationally modified, or enzymatically truncated product) but nevertheless are intended to be included in the product pool as, for example, occurs with the manufacture of the naturally occurring and recombinant blood coagulation proteins, factor VII, factor VIII, and von Willebrand factor [65–68, 85, 86], to the by-product(s), b, b', ..., b_n (which may, for example, be a glycoform of a specific glycosylated protein destined for another end use) and to waste, $W_{mass(t,P',\iota P)}$, in a time interval, t, at a specific manufacturing site, L, using a particular process, i. Similar formalisms can be used to represent the cases of other biological products that exist as liquids or states that include both solids and liquids at room temperature as defined by the ICH Q6B Specifications for CMC regulatory compliance [87]. Linkage of the overall mass of all components used in a production run to a raw material conversion rate for product generation provides both an avenue to assess productivity *per se* and a score card against which financial, compliance/regulatory, and environmental requirements can be assessed.

The overall raw material/feedstock consumption in the generation of the protein product(s) can be given by

$$C_{\text{raw material}} = \sum_{i\varepsilon I}\left[\sum_{f\,\varepsilon\,F}\frac{X_{\text{use}(t,f,f,i,L)}}{\text{ProdRate}_{(f,i,r\text{cap})}}\right] + \sum_{P\varepsilon H\cup E}\frac{X_{\text{prod}(t,P,P',i,L)}}{\text{ProdRate}_{(P,i,r\text{cap})}} \qquad (18.3)$$

where b is the mass of a specific protein by-product, for example, a deamidated or aggregated species present at a specific unit operation stage of the process i, chosen from a set of alternative processes I; B is the mass sum of all protein by-products generated from all unit operations of the specific process i with $B \geq b$; D is the mass sum of all specific protein products (or alternatively all by-products or precursor products streamed at each unit operation stage) at the commencement of the process; d is the mass of a specific product (or alternatively a by-product or precursor product streamed at each different unit operation stage) at the commencement of the process with $d \leq D$; i is a specific process or unit operation employed to generate the product (often equated in a SOP as a recipe); f is the specific amount (mass) of feedstock raw material required for the processing of a single batch at the specific location, L; F is the total amount (mass) of feedstock raw material required for the processing of all batches that constitute the manufacturing campaign; r_{Cap} is the specific resource capacity available at the specific location, L; $R_{\text{Cap,overall}}$ is the sum of all specific resource capacities needed to deliver in the time period t the protein product in the required quantities and quality within the desired specifications; t is a specific period of time; and t_{tot} is the sum of all specific periods of time required to deliver the protein product during the course of the manufacturing campaign.

Obviously, the maximum $C_{rawmaterial}$ value that can be processed at a specific manufacturing facility will depend on the overall resource capacity, $R_{Cap,overall}$, of the production facility. Although scheduling procedures with regard to maintenance, cleaning, and refurbishments will clearly strive to maximize the operational utilization and efficiency of the facility, $C_{rawmaterial}$ will always be less than $R_{Cap,overall}$. Moreover, as apparent from Equation 18.2, the quantitative magnitude of the second term $Wsolid_{(t,P',i.L)}, P' \varepsilon P \setminus B, \forall P', \forall i, \forall t$, which captures the mass of all by-products and all waste generated from the process has a significant bearing on the overall process efficiency in terms of the mass utilization in a specific process. In downstream processing, as well as other areas of biotechnological production, these terms to a large extent account for the observed low values of the ratio ($Prod_{actual}/Prod_{defined}$) (or the corresponding very high E values). Moreover, as the process becomes technically more complex, the value of the ratio ($Prod_{actual}/Prod_{defined}$) generally falls further, which drives up cost and increases waste generation in one or more parts of the process life cycle. A further factor that contributes to these process inefficiencies is that downstream facilities are currently rarely designed solely for a single, fit for purpose usage, and this can lead to greater challenges to maintain a high degree of control over the mass and energy utilization for any unit operation.

As apparent from practice, a major contributor to these poor $Prod_{actual}/Prod_{defined}$ or E metrics in many downstream processing activities in terms of process performance is waste water. For example, current downstream processing methods for the purification of mAb with targeting annual throughputs of 250–5000 kg mAb and overall recoveries of ca. 75–80% of the mAb from the stage of cell culture to final API product typically involve at least three chromatographic column purification steps as part of the process chain. A typical mAb production scenario, involving the consumption of about 10 kg of low molecular weight consumables (salts, buffers, etc.) per kilogram of mAb produced, can consume up to 7500 kg of water per kg mAb produced. This level of consumption can result in the WPI costs greatly exceeding the costs of other consumable raw materials and reagents. Similarly, for the purification of industrial enzymes and other recombinant proteins when no recycling is carried out, the waste disposal costs can represent more than 10% of the total manufacturing costs. With innovative reuse, the reduced cost of other goods, such as chromatographic resins and membranes, with recycling rates of ca. 100 cycles per kg product can drop the manufacturing costs two to sixfold [88–91].

In this context, the concept of process mass intensity (PMI) forms an integral component of a sustainability model development and is particularly relevant for simulations and process valorizations. Although water *per se* usually does not constitute a significant environmental impact or risk, the use of highly purified water in some sectors of the biotechnology industry generate significant life-cycle impacts due to the nature of the chemicals and equipment used to purify the water. Moreover, the cost and energy inputs are becoming increasingly significant to generate water at a grade often needed for biomanufacturing, for example, WFI grade. Consequently, water use must be included in the intensification modeling. PMI is related to the mass productivity and as such defines how much of the input mass is integrated into a final, saleable product and ultimately represents a measure of profitability. The specific form of PMI can be represented by Equation 18.4, as follows:

$$\text{Process mass intensity (PMI)} = \frac{\text{total mass involved in the overall manufacturing process (kg)}}{\text{mass of product obtained (kg)}} \quad (18.4)$$

While for multistage integrated processes, involving further in-stream processing, such as the on-column removal of a purification tag from a recombinant protein or alternatively an *in situ* activation of an enzyme zymogen during the downstream processing can be represented by the generalized form of the PMI relationship given as follows:

$$\begin{array}{l}\text{Generalised Process}\\ \text{Mass Intensity (GPMI)}\end{array} = \frac{\sum_{n=1}^{j} \text{total mass involvedin the multistage manufacturing process (kg)}}{\sum_{n=1}^{j} \text{mass of all products obtained from the integrated processes (kg)}} \quad (18.5)$$

Linked to the PMI is the process mass efficiency (PME), which is often presented as either a ratio or alternatively a percentage of the total mass ratios, and given by Equation 18.5, as well as the historical *E* factor, given by Equation 18.6:

$$\text{Process mass efficiency (PME)} = \frac{\text{mass of product obtained (kg)}}{\text{total mass involved in the overall manufacturing process (kg)}} = \left(\frac{1}{\text{PMI}}\right) \quad (18.6)$$

$$E\text{ factor} = \frac{\text{total mass of waste (kg)}}{\text{mass of product (kg)}} = \frac{\text{total mass used in process or process step (kg)}}{\text{mass of product (kg)}} = \text{PMI} - 1 \quad (18.7)$$

In an analogous manner, the impact of the water intensity (WTI) and waste intensity (WSI) (as well as the other mass-linked intensities for salts, solvents, or other consumables) can be formally represented through the same forms of the relationships shown as Equations. 18.8 and 18.9, respectively, as follows:

$$\text{Waste intensity} = \frac{\text{total water input}}{\text{total mass input}} \tag{18.8}$$

$$\text{Waste intensity} = \frac{\text{total waste produced}}{\text{total mass input}} \tag{18.9}$$

The importance of these dependences and their implications on how to reduce the impact of waste in a manufacturing campaign cannot be overemphasized. First, for many downstream processing tasks, purity is certainly a key, but not necessarily a sole driver. In some instances, a higher purity gained at the stage of one-unit operation may limit the productivity of the overall process and add to the mass and energy intensity without necessarily adding value. The PMI is a measure of productivity and thus directly relates to product value adding and sustainability rather than just cost management. Second, recent studies [92–94] with low molecular weight APIs and other chemical compounds have documented that metrics based on PMI, and associated mass and energy intensity inputs, lead to important correlations between PMI expressed in terms of kilogram of raw material processed per kilograms of active product and global warming potential factors based on mass (Figure 18.5). Similar correlations can be anticipated for protein products produced by recombinant DNA methods and associated fermentation or cell culture procedures. Third, low PMI values, often below 0.001, identified with traditional process technologies can be used as a yardstick against which future scientific and engineering improvements can be measured with higher PMI values equating within lower environmental impacts. In addition, the gains in productivity achieved through recycling using low energy technologies, such as membrane-based procedures, can also be better quantified.

(4c) *Impact of Cost*: In determining the economic sustainability of a particular protein production process, the cost of the materials is but one of the considerations. Other costs, classified according to whether they fit into Type I to Type V categories [95] according to total cost assessment (TCA) methods have to be taken into account. Determination of the magnitude and distribution of these costs is of fundamental importance in environmental cost accounting [96]. Type I costs are generally considered to arise as capital investment and operational costs that include the direct cost of labor, raw materials, equipment and plant, and internally managed waste disposal/treatment. Often the approach followed for capital investment can be based on factorial analysis

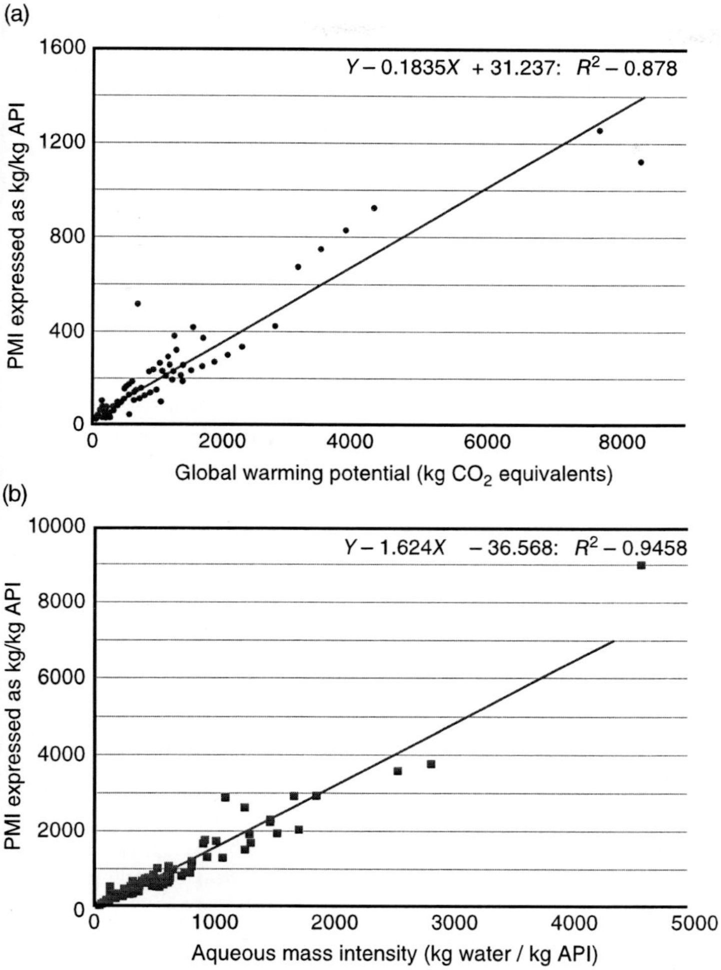

Figure 18.5 Correlation between the PMI expressed in terms of kilogram raw materials per kilogram API and (a) the Global Warming Potential/Carbon Footprint and (b) aqueous mass intensity for APIs in a development portfolio. *Source*: Jimenez-Gonzalez et al., [92]. Adapted with permission of American Chemical Society.

utilizing the so-called "Lang" factors, which typically fall into the range of 3.3–8.1 for biomanufacturing [97–99].

For biomanufacturing, particularly the production of biopharmaceuticals, Type II costs can reach considerable levels, since they include *inter alia* the cost of compliance and regulatory approvals, monitoring, and reporting. Contingency costs (Type III) usually capture the legal cost of operation and costs due to facility incidents and accidents or mandated clean-up liabilities.

Besides enabling the obvious additivity of material, plant/equipment, labor, and resource Type 1 costs, application of TCA methods provide a strategic view of trends targeted toward more sustainable downstream processing practices that can directly affect the long-term impact of Type I to Type III costs of a process. Besides these costs, which can be rigorously evaluated from life-cycle and systems-based perspectives, additional costs need to be formally incorporated into models of sustainability, for example, relationship and branding costs and other intangible internal expenses (Type IV costs), while costs currently associated with external activities but involving indeterminate pricing of commodity waste stream that are currently subsidized to some extent through public or governmental policies, for example, cost of air, water, CO_2, or other intangible costs associated with externally released by-products/ waste, are typically included as Type V costs. Table 18.3 summarizes some of the considerations that drive the nexus between economic and environmental thinking in terms of TCA methods as the decisions for Type I to Type V expenditure are made.

Table 18.3 Several considerations related to the nexus between environmental and economic thinking.

	Thinking environmentally	Thinking economically
Atom economy	Minimal by-product formation; less solid/liquid waste	More from less—incorporate total value of materials
Water/solvent reduction	Less water/solvent waste treatment needed	Higher throughput, less energy
Reagent optimization	Plan for low mass stoichiometries, recyclable reagents that minimize usage	Target higher efficiency and higher selectivities; greater productivity
Convergency	Reduced environmental burden due to increased process efficiency	Higher efficiency—fewer unit operation stages, greater use of platform technologies
Energy reduction	Reduced environmental burden from power generation, transport, and energy use	Reduced energy consumption reflects increased greater efficiency, shorter process cycles, milder conditions
In situ analysis	Reduced possibility for an adverse process incident, exposure, or release to the environment	Real-time data acquisition increases throughput and process efficiency, fewer reworks
Safety	Nonhazardous materials reduce risk of exposure, release, fires, explosions, or biological contamination	Worker safety and reduced time/money spent on special control measures

Recent trends also suggest that the practitioners of downstream processing are progressively favoring a reduction in the scale of traditional batch unit operations and moving more toward continuous processes on the basis of cost. Crystallization, phase extraction, and new modalities of membrane technologies will certainly play an increasing role here since these technologies offer certain advantages over conventional chromatographic methods, particularly at large scale. Innovations in process technology will be needed, however, to allow cycle numbers to be increased to a level that in-process recycling will become much more economic and attractive to use. The size, type, and number of unit operation place demands on the overall process mass and energy efficiencies, which all impact on the overall performance and the ability to obtain the desired product end point [100]. In particular, the potential impact of unit operation size, operational change options, and key process tasks, for example, the efficiency of buffer mixing on the production rate, formation of by-products, or level of control over exo- or endothermic events, requires additional inputs to maintain a high degree of control over mass and energy transfer in specific unit operations and to reduce the impact on the overall cost trajectories.

(4d) *Impact of Energy Requirements*: It is well recognized that energy use in downstream processing of protein products is highly dependent on process conditions. In fact, bioproduct separation and purification is one of the single largest consumers of energy and material when compared to many other manufacturing processes. Various process modeling softwares, for example, SuperPro Designer (Intelligen Inc., Scotch Plains, NJ, USA), Aspen Batch Plus (Aspen Technology Inc., Cambridge, MA, USA), and other types of flow sheet simulation methods, can be employed [101–106] to incorporate energy consumption for facilities of different capacities and productivities using chromatographic, membrane, aqueous two-phase extraction, or crystallization technologies for the downstream purification of proteins, but typically public domain information on process energy consumption within an industrial setting is sparse. Generally, most companies measure and report energy usage by site rather than by process. From the perspective of life-cycle energy requirements, processing energy is but one component, with other levels of energy consumption that are needed to produce raw materials, recycle materials, and waste treatment as well as to run the facilities infrastructure also contributing factors [107]. Nevertheless, the impact of energy requirements for raw material conversion to products, recycling materials if necessary, waste treatment, and cleaning of plants can be assessed from the perspective of the additivity of the relevant metrics for multistep linear or multistep convergent production models, whereby the process energy intensity (PEI), the product life-cycle energy intensity (PLCEI), the waste treatment (including cleaning) energy intensity (WTEI), the water recovery/recycling energy intensity (WREI), the water energy intensity ratio (WEIR), and the waste energy intensity ratio (WAEIR) can be quantitatively evaluated from the following expressions:

$$\text{Process energy intensity}\left(\text{PEI}\right) = \frac{\text{total process energy}\left(\text{MJ}\right)}{\text{kilogram of final product}}$$
$$= \frac{\Sigma\left(\text{process, material manufacturing, treatment, resource load}\right)\left(\text{MJ}\right)}{\text{kilogram of final product}} \tag{18.10}$$

$$\text{Product life cycle energy intensity}\left(\text{PLCEI}\right) = \frac{\text{product life cycle energy requirements}\left(\text{MJ}\right)}{\text{kilogram of final product}} \tag{18.11}$$

$$\text{Waste treatment energy intensity}\left(\text{WTEI}\right) = \frac{\text{waste treatment energy requirements}\left(\text{MJ}\right)}{\text{kilogram of final product}} \tag{18.12}$$

$$\text{Water recover energy intensity}\left(\text{WREI}\right) = \frac{\text{water recovery energy requirements}\left(\text{MJ}\right)}{\text{kilogram of final product}} \tag{18.13}$$

$$\text{Water energy intensity ratio}\left(\text{WEIR}\right) = \frac{\text{total energy for water use and recovery requirements}\left(\text{MJ}\right)}{\text{total energy input}\left(\text{MJ}\right)} \tag{18.14}$$

$$\text{Waste energy intensity ratio}\left(\text{WAEIR}\right) = \frac{\text{total energy for waste treatment and disposal requirements}\left(\text{MJ}\right)}{\text{total energy input}\left(\text{MJ}\right)} \tag{18.15}$$

All of the aforementioned energy intensities, which can be considered as a set of “in-process” and “out-of-process” energy modules, are affected by a large number of factors [108], including whether increases in the cycle times and decreases in process mass intensities have arisen through the choice of less efficient technologies. A further factor that impacts on these energy intensities is cleaning and maintenance. Various approaches are employed for cleaning and maintenance, although two scenarios have gained popularity, for example,

CIP protocols versus breakdown and rebuild protocols. CIP protocols with chromatographic resins are widely used. Often, here, the impact on energy use associated with a CIP protocol is assumed to be mainly linked to the impact of cleaning chemicals (water, detergents, sodium hydroxide, solvents, etc.) on the PMI. However, other impacts leading to high energy consumption per unit time arise due to changes in the cycle time as well as throughput, in the context of handling, delivery, and disposal of these CIP compounds. The disposal of waste water through incineration also represents a particularly high energy demand since evaporative disposal consumes between 90 and 95% of the energy used. Careful consideration of the energy intensities associated with solid waste disposal can have a significant benefit in reducing energy consumption and ensuring that the trade-offs between single-use disposables and stainless steel equipment can be better captured [109].

(4e) *Impact of Scalability, Throughput, Cycle Time, and Life-Cycle Analysis*: The impact of process scalability, throughput, and cycle time on product quality, reproducibility, and yield become key issues if the protein's properties, such as propensity for low stability and tendency to aggregate, are associated with longer cycle and handling conditions. Such adverse incidents lead to greater waste generation and the requirement for larger consumption of raw materials and energy. Critical quality attributes of the protein product thus need to be understood, if the limits of process scalability and the impact of changes in throughput and other operational conditions are determined. Certainly, for continuous processing, it is essential to maintain high throughputs in order to achieve performance and economic viability, so with continuous downstream processing throughput is generally close to maximum capacity of the process step. These conditions can lead to highly concentrated protein solutions with the potential for unexpected aggregation and self-assembly phenomena, including the formation of protein nanoparticles and fibrils, upon storage or chromatographic processing [110, 111].

The impact of key process parameters on the throughput (often defined as the average saleable production output per a given time unit) and the cycle time (often defined as the time taken to manufacture one batch of the product(s)) thus have to be optimized from the perspective *inter alia* of raw material availability, feedstock conversion, process time, number of unit operations, and product yield and quality. Robust downstream processes are not greatly affected by minor variations in the rates of feedstock dilution, concentration, process temperatures, or mixing. However, adequate availability and facility capacity must exist for equipment, warehousing, and labor in keeping with the product intensification regime while seeking to reduce the facility footprint, again strong arguments in favor of movement toward continuous manufacturing approaches to achieve greater sustainability. As upscaling proceeds and technical economies of scale become the objectives [112], the predicted Type I capital cost of upscaled equipment and plant as technical units has historically been based on the so-called six-tenths rule [113], which can be expressed mathematically as:

$$X_U = X_S \times \left(\frac{C_U}{C_S} \right)^{0.6} \tag{18.16}$$

where X_U and X_S are the total costs of larger and smaller set of technical units, respectively, and C_U and C_S are the corresponding capacities of these technical units. Other exponential scaling factors in the range of 0.36–1.0 have been proposed [114] for computer-controlled fermenters and for ultrafiltration rigs. In addition, the productivity, P_{rod}, of a batch process determines the overall time required to achieve the certain mass of product at the quality level defined by the process capability index, Cp, according to the dependency

$$\text{Time of manufacturing} = T_m = \frac{N_{Batch}}{P_{rod}} + \sum_{n=1}^{n=j} T_{misc} \tag{18.17}$$

where N_{Batch} is the number of batches, P_{rod} is the overall process productivity, and the term ΣT_{misc} represents the sum of the times committed for all other activities and procedures, such as plant cleaning, maintenance, andso on shut down. In order to measure the environmental impact of scalable processes and the impact of optimized throughput and cycle time, a variety of process productivity metrics are required and derived from methodologies that also capture life-cycle assessments and inventories. Since the late 1990s, such methodologies have been actively pursued [26, 27, 43, 48, 93, 115, 116] by various sectors of the chemical and pharmaceutical manufacturing industries. Given the need for significant resources and the labor-intensive nature of traditional life-cycle analysis procedures based on ISO 14040:2006 methodologies, the concepts inherent to the generalized form of the PMI relationships and the other attendant operational intensity relationships have provided a way to streamline at the initial stages of upscaling the estimation of the environmental footprints from data acquired in laboratory and pilot scale studies. However, to transform curiosity-driven research discoveries and proof of principle pilot scale endeavors into industrial-scale practice, supply chain life-cycle inputs and metrics are needed with regard to the nature of the materials used within each step in the process and their respective amounts used, in addition to the batch size, expected product mass, expected product yield and product purity, and importantly metrics associated with the process energy.

To garner these inputs, acquisition of the appropriate life-cycle analysis data is required for each stage of the process and protein product, from the pre-manufacturing (including supply chain outsourced raw materials and process infrastructure), manufacturing, product delivery, formulation, use, and, where relevant, information related to the planned obsolescence of the product or process, that is, at what stage of development are the second-generation options. Aspects of the life-cycle implications of downstream process technologies can also be captured and streamlined at the stages of process design, construction, and implementation. At these stages, the possibility to achieve recycling and D4D options can be inherently built into the process technology, while design metrics can be employed to select a suitable process from a suite of options that has operationally built in environmentally responsible features, that is, minimization of the consumption of energy, avoidance, or minimization of the generation of solid, liquid, or gaseous wastes and reduction in the use of hazardous or toxic materials, and so on.

The combined impacts of mass, cost, scale, throughput, cycle time, and process energy thus collectively represented part of the essential information necessary to harness the potential of practical, sustainable strategies that can lead to resolution of the current challenging bottlenecks in the downstream processing of proteins and other bioproducts. To allow sustainability decisions to be rationally embedded into the design and deployment of improved downstream processing technologies for the recovery and purification of proteins and other bioproducts, N-dimensional matrix assessment approaches represent attractive tools, whereby one dimension can be used to represent the life-cycle stage and the other dimensions to represent cost, scale, or environmental concern, and so on. A suitable starting point is to incorporate life-cycle assessments and life-cycle inventories with process operational criteria and metrics that are linked to the intended product usage at defined levels of structure, function, and purity as well as production profitability. Put another way, the steps to improve process and product sustainability with more benign downstream processing procedures can, by analogy to other areas of manufacturing eco-efficiency [117], involve:

- Optimization of material intensities (as predicted by Eqs. 18.2–18.9)
- Reduction in energy intensity (as predicted by Eqs. 18.10–18.15)
- Maximal deployment where possible of renewable resources
- Reduction in waste generation (as predicted by Eq. 18.2)
- Minimization of the release of hazardous/toxic substances
- Increased ability to recycle water and other "waste" streams
- Increased product efficacy/safety with reduction of EHS and OHS incidents
- Increased social license acceptability and service value

Since the impact of a number of these parameters on the sustainability outcomes associated with a downstream activity may not be known with numerical precision, a number of qualitative and semiquantitative matrix-based approaches can be employed [118] as alternatives, with each element of the matrix assigned, for example, an integer rating from 0 (corresponding to the highest impact and representing a very negative evaluation option) to 10 (corresponding to the lowest impact and representing a very positive evaluation option). Following evaluation of their corresponding matrix element values of all of the relevant process parameters in terms of their respective impact, scope, and system boundary conditions, for example, their process capability indices, *Cp*'s, and parameter intensities as given by Equations 18.1–18.17, a *SI* can be computed and given a product or process ranking. If it is assumed that linear additivities across all parameters apply, from the sum of the derived matrix element values, (M_{ij}), the SI value can be given by

$$\mathrm{SI} = \sum_{i \to j} M_{i,j} \tag{18.18}$$

In the more common circumstances where linear additivity does not prevail, then exponential power function terms (A, B, C,...) can be employed to describe the weighting of each matrix element values in terms of their respective environmental impact, such that

$$\mathrm{SI} = \Sigma[M_i^A + M_j^B + M_k^C + \cdots] \tag{18.19}$$

To assess the relevance of individual power function terms, A, B, C,..., combinations of sensitivity analysis, dominance analysis, and Monte Carlo analysis procedures can be employed with inputs and outputs derived from smaller-scale (laboratory scale) experience as indicators. As noted previously, for downstream processing of protein products, a tendency has historically existed to not take a systems-wide or holistic view of a process.

When a systems-wide view is adopted and when a SI is qualitatively or semi-quantitatively computed, several process scenarios may in practice emerge as options. For example, the downstream procedures for the manufacture of a protein product might involve (a) a relatively simple procedure from a single feedstock stream, typical of the production of relatively low value protein product, (b) a multistep, but linear, production strategy more suited to produce only a single highly purified protein product from a single feedstock stream, (c) a multistep but convergent production strategy intended to produce only a single target protein product but from several feedstock streams and using another protein, such as a protease, as a surrogate reagent required for use in production of the target protein, or (d) a multistep production strategy intended to produce several protein products from several convergent feedstock streams. Other downstream strategies of increasing multistage and branching complexity can be contemplated and in fact have already found application within various sectors of the chemical manufacturing sectors.

Two simplified examples can be used to illustrate the scenarios arising under (a) and (c) earlier as part of this approach to acquire process-specific SI. Figure 18.6a shows a common pathway for a multistep but linear production protocol, while Figure 18.6b illustrates a more complex integrated process tree whereby two analogous purification methods are concomitantly carried out,

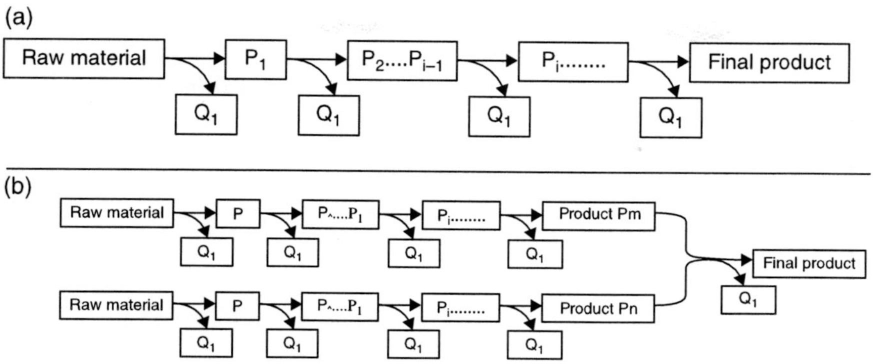

Figure 18.6 Schematic layout for a multistep linear downstream protocol (a) and an integrated multistage downstream protocol (b). Both scenarios accommodate the cascade effect of changes in one particular unit operation on (subsequent) unit operations. The symbols P, P_1, P_2, ...P_{i-1}, P_i, P_m, etc., represent the protein product(s) at each individual unit operation stage, while Q_1 represents the by-products and waste generated. *Source*: M.T.W. Hearn and R.I. Boysen, unpublished findings.

resulting in the generation of two protein products, P_m and P_n, with P_n required as an in-process reagent, for example, a proteolytic enzyme, to generate the final protein product. This latter case study could, for example, involve the generation of a recombinant enzyme, such as a recombinant diaminopeptidase, required to cleave a histidine-rich peptide tag from the desired protein target to obtain the final purified product [119].

Application of either the multistep linear protocol or the more complex integrated two protein multistage protocol goes partly to the requirements of a holistic model of process intensification and valorization. Because of such holistic, systems-wide views of the process, the collection of information of more than a single metric is required, and this implies that multivariate methods of analysis are needed. Trade-off between the various metrics can thus be expected to find a rise in practice, for example, a change in one critical process parameter during a downstream processing campaign with a specific recombinant protein may lead to a better environmental profile but simultaneously results in a reduction in yield. For these reasons, process metrics must be continuously reviewed and adapted to align with their application context, as discussed earlier with regard to the purity/impurity profile for intravenous immunoglobulin preparations vis-à-vis recombinant insulin preparations. Based on the approaches described earlier, a software package called *Icarus*, compatible with Microsoft Excel formats, has recently been evaluated (M.T.W. Hearn and R.I. Boysen, unpublished findings) as a valorization tool for the calculation of SI values and the ranking of process alternatives as feasible production options for the manufacture of chemical and biological products. The procedure follows the qualitative and semiquantitative matrix-based methods described previously with the mass values of all components present in the feedstock stream and other parameters included for each stage of the process, either as a specific unit operation, such as an ion exchange step, conversion of a precursor to a more advanced product, and so on, or alternatively a modification to a unit operation condition, such as a change in ionic strength, pH shift, or addition of an organic solvent. The imported and exported data associated with each input and output for the unit operations of multistep downstream processing procedures can be either captured as downloadable screen shots of the respective data entry pages or linked via ASCII files to suitable graphic software for presentation in a variety of graphical formats, as spider graphs, three-dimensional contour plots, and so on, for process chains involving up to 30 discrete process manipulations. One such example of this methodology to determine global optima, generated by an integrated multistage process involving two biomolecular targets of relevance to the industrial food processing industry and the predicted overall profitability is shown in Figure 18.7. When a substantial

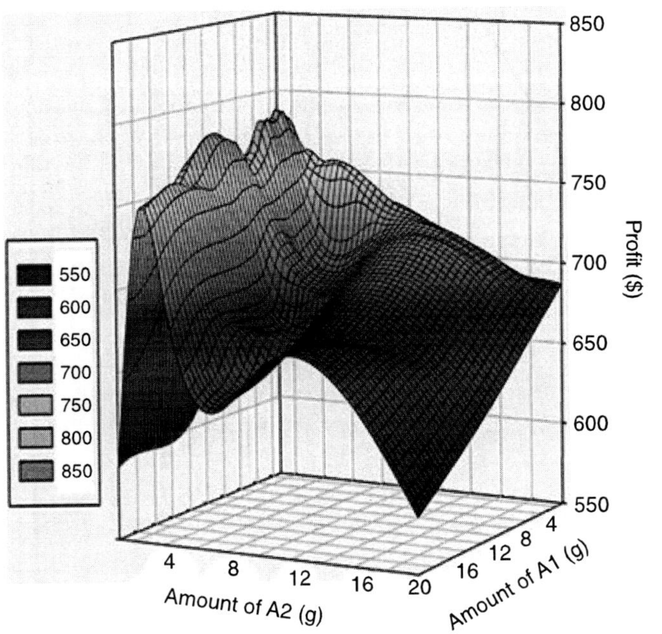

Figure 18.7 Graphical presentation showing the global optima generated for an integrated multistage process involving two related products and the predicted profitability as a measure of the SI derived from the associated data inputs and outputs, which detail the respective normalized mass intensities, production rate, process times and energy costs, and productivity outcomes. *Source*: M.T.W. Hearn and R.I. Boysen, unpublished findings. (*See insert for color representation of the figure.*)

body of experimental data is available, such graphical presentations are useful visual tools to summarize the impact of changes in the PMI and the other process-specific intensities (defined by Eqs. 18.2–18.10) in response to changes in operational conditions and allow different process scenarios and parameters to be compared and contrasted, thus providing a measure of the overall sustainability of different process options.

18.5 Conclusions and Perspectives

As apparent from the preceding discussion, significant progress has been made in recent years to achieve more sustainable biomanufacturing, including areas associated with the downstream processing of protein products. Holistic, whole-of-process approaches based on the determination of global rather than local optima for the chosen process are increasingly finding application. The use of systems thinking increasingly is displacing empirical approaches, with the consequence that greater productivity becomes feasible options, with less waste generated, and overall enhancement of process performance and improved life-cycle assessment profiles realized.

The lessons of the past decade are thus carriers of good news for the biotechnology industry, since through the innovations that are emerging, more sustainable downstream processing technologies are being deployed, former logjams in feedstock processing overcome, and greater process profitability accrued. Products and processes generated by biotechnological procedures are often considered by default to be "green" and highly sustainable when compared to products made by chemical routes. On mass intensity arguments, this is often not the case. The paradigm change in bioprocessing that is occurring will continue to benefit from new, innovative approaches to process integration and intensification due to the increasing availability of robust and relevant process data and comprehensive modeling methods. Greater interest in the selective adoption of the concepts implicit to the 3Rs will undoubtedly continue due to their environmental and economic benefits. The prospect that practitioners of downstream processing technologies will progressively discover ways to be able to replicate many of the processes of molecular separation, sorting, and recycling as achieved by Nature is an exciting one and represents a field well worth the effort along the path toward more sustainable biomanufacturing.

Acknowledgment

The support of the Australian Research Council is gratefully acknowledged.

References

1. Brundtland Report Our Common Future (1987) United Nations General Assembly Resolution 42/187, Oxford University Press, UK.
2. GSC-7 JACI—Hitotsubashi Declaration, Tokyo, July 8, 2015; http://www.jaci.or.jp (accessed on September 19, 2016).
3. Kiss, A.A., Grievink, J. and Rito-Palomares, M. (2015) A systems engineering perspective on process integration in industrial biotechnology. *J. Chem. Technol. Biotechnol.*, 90: 349–355.
4. OECD (2008) About sustainable manufacturing and the toolkit. http://www.oecd.org/innovation/green/toolkit/aboutsustainablemanufacturingandthe toolkit.html (accessed on September 19, 2016).
5. U.S. Department of Commerce (2010) Sustainable business. http://www.trade.gov/competitiveness/sustainablemanufacturing/index.asp (accessed on October 11, 2016).
6. U.S. Department of Commerce (2014) 2014 strategic sustainability performance plan and climate adaptation plan. http://www.osec.doc.gov/ofeq/OSEEP/Annual_Rpts_Scrcrds.html (accessed on October 11, 2016).
7. U.S. Department of Energy 2014 Strategic Sustainability Performance Plan; http://energy.gov/sites/prod/files/2014/10/f18/doe_sspp_2014.pdf (accessed on September 19, 2016).
8. European Commission (2013), The Sustainable Consumption and Production and Sustainable Industrial Policy (SCP/SIP) Action Plan; http://ec.europa.eu/environment/eussd/escp_en.htm (accessed on September 19, 2016).
9. Savitz, A. (2013) *The Triple Bottom Line: How Today's Best-Run Companies Are Achieving Economic, Social and Environmental Success*, Josey-Bass/Wiley, New York, pp 1–352.
10. McDonough, W. and Braungart, M. (2002) Design for the Triple Top Line: New Tools for Sustainable Commerce, http://www.globalcommunity.org/business/Design%20for%20the%20Triple%20Top%20Line.pdf (accessed on October 11, 2016).
11. BASF Verbund Investor relations: our strategic principles. https://www.basf.com/en/company/investor-relations/basf-at-a-glance/strategy/our-strategic-principles.html (accessed on October 11, 2016).
12. Pfizer Green Chemistry Initiative Greener processes. http://www.pfizer.com/responsibility/protecting_environment/greener_processes (accessed on October 11, 2016).
13. Curzons, A.D., Jiménez-González, C., Duncan, A.L., Constable, D.J.C. and Cunningham, V.L. (2007) Fast life cycle assessment of synthetic chemistry (FLASC^{TM}) tool. *Int. J. Life Cycle Assess.*, 12: 272–280.
14. Jolles, S., Sewell, W.A.C. and Misbah, S.A. (2005) Clinical uses of intravenous immunoglobulin. *Clin. Exp. Immunol.*, 142: 1–11.

15. Swinnen, S.G., Hoekstra, J.B. and Devries, J.H. (2009) Insulin therapy for type 2 Diabetes. *Diabetes Care*, 32Supplement 2: S253–S259.
16. Jagschies, G. (2008) Where is biopharmaceutical manufacturing heading?. *BioPharm Int.*: October 72–82.
17. DePalma, A. (2010) Levelling downstream process bottlenecks. *Genetic Eng. Biotechnol. News*, 30: 9–10.
18. Lye, G., Hubbuch, J., Schroeder, T. and Willimann, E. (2009) Shrinking the costs of bioprocess development. *BioProcess Int.*, 7: October Supplement.
19. Gronemeyer, P., Ditz, R. and Strube, J. (2014) Trends in upstream and downstream process development for antibody manufacturing. *Bioengineering*, 1: 188–212.
20. Rathore, A.S., Latham, P., Kaltenbrunner, O., Curling, J. and Levine, H. (2004) Costing issues in the production of biopharmaceuticals. *BioPharma Int.*, http://www.lathambiopharm.com/pdf/ (accessed on September 19, 2016).
21. Ho, S.A., McLaughlin, J.M., Cue, B.W. and Dunn, P.J. (2010) Environmental considerations in biologics manufacturing. *Green Chem.*, 12: 755–766.
22. Environmental sustainability: reducing water use. http://www.gene.com/good/sustainability/water (accessed on October 11, 2016).
23. Jakoby, R., Heim, M., Pernenkil, L., Sabad, A. and Harutunian, S. (2015) *Advanced Biopharmaceutical Manufacturing: An Evolution Underway*, Deloitte Development LLC, London, UK, pp. 1–16.
24. Xenopoulos, A. (2015) A new, integrated, continuous purification process template for monoclonal antibodies: process modeling and cost of goods studies. *J. Biotechnol.*, 213: 42–53.
25. Walther, J., Godawata, R., Hwanga, C., Abeb, Y., Sinclair, A. and Konstantinov, K. (2015) The business impact of an integrated continuous biomanufacturing platform for recombinant protein production. *J. Biotechnol.*, 213: 3–12.
26. Godawat, R., Konstantinov, K., Rohani, M. and Warikoo, V. (2015) End-to-end integrated fully continuous production of recombinant monoclonal antibodies. *J. Biotechnol.*, 213: 13–19.
27. European Commission, Directorate-General for the Environment, Joint Research Centre (2010) *Life Cycle Thinking and Assessment for Waste Management*, Luxembourg: EUR-OP.
28. von Weizsäcker, E., Lovins, A.B. and Lovins, L.H. (1999) *Factor Four: Doubling wealth, halving resource use—A Report to the Club of Rome.* Earthscan, London, UK, pp. 1–224.
29. Schmidt-Bleek, F. (2008). Factor 10: the future of stuff. *Sustainability: Science, Practice, & Policy*, 4: 1–4.
30. Wiendahl, M., Wierling, P.S., Nielsen, J., Christensen, D.F., Krarup, J., Staby, A. and Hubbuch, J. (2008) High throughput screening for the design and optimization of chromatographic processes—miniaturization, automation and parallelization of breakthrough and elution studies. *Chem. Eng. Technol.*, 31: 893–903.

31. Petzold, M., Coghlan, C.J. and Hearn, M.T.W. (2014) Studies with an immobilized metal affinity chromatography cassette system involving binuclear triazacyclononane-derived ligands: automation of batch adsorption measurements with tagged recombinant proteins. *J. Chromatogr. A*, 1351: 61–69.
32. Mao, Q.M., Prince, I.G. and Hearn, M.T.W. (1995) Impact of operating parameters in large-scale chromatography of proteins. *J. Chromatogr. A*, 691: 273–283.
33. Finette, G.M.S., Mao, Q.M. and Hearn, M.T.W. (1996) Studies on the expansion characteristics of fluidized beds with silica-based adsorbents used in protein purification. *J. Chromatogr. A*, 743: 57–73.
34. Mao, Q.M. and Hearn, M.T.W. (1996) Optimization of affinity and ion-exchange chromatographic processes for the purification of proteins. *Biotechnol. Bioeng.*, 52: 204–222.
35. Wang, C., Coppola, G. and Chumsae, C. (2015) Protein Purification Using Displacement Chromatography. US 9,067,990 B2.
36. Hickman, R.K., Huang, Q., Weed, C.L., Ennis, S.T., Perilli-Palmer, B. and Wan, M. (2010) Isolation and Purification of Antibodies Using Protein A Affinity Chromatography, US 9,018,361 B2.
37. Gagnon, P.S. (2009) Purification of monoclonal antibodies by mixed-mode chromatography, in *Process Scale Purification of Antibodies*, U. Gottschalk, ed., John Wiley and Sons, New York, pp. 125–144.
38. Black, D.J., Barford, J.P., Harbour, C. and Fletcher, A. (1997) The effect of different purification schemes on the activity of a monoclonal antibody, in *New Developments and New Applications in Animal Cell Technology*, Merten, O-W., Perrin, P. and Griffiths, B. eds., Kluwer Academic Publishers, New York, pp. 433–435.
39. Gagnon, P., Cartier, P.G., Maikner, J.J., Eksteen, R. and Kraus, M. (1993) A systematic approach to the purification of monoclonal antibodies. *LC-GC*, 11: 26–34.
40. Follman, D.K. and Fahrner, R.L. (2004) Factorial screening of antibody purification processes using three chromatography steps without protein A. *J. Chromatogr. A*, 1024: 79–85.
41. Wrisberg, N., Haes, H.A.U.d., Bilitewski, B., Bringezu, S., BroRasmusen, F., Cliff, R., Elder, P., Elkins, P., Frischknecht, R. and Triebswetter, U. (2002) Analytical tools, in *Analytical Tools for Environmental Design and Management in a Systems Perspective*, Wrisberg, N., Haes, H.A.U.d., Triebswetter, U. Eder, P. and Clift, R. eds., Kluwer Academic Publishers, Dordrecht, the Netherlands, pp. 45–73.
42. Brunner, N. and Starkl, M. (2004) Decision aid systems for evaluating sustainability: a critical survey. *Environ. Impact Assess. Rev.*, 24, 441–469.
43. Nfor, B., Ahamed, T., van Dedem, G., van der Wielen, L., van de Sandt, E. and Eppnik M (2008) Design strategies for integrated protein purification

processes: challenges, progress and outlook. *J. Chem. Technol. Biotechnol.*, 83: 124–132.

44. Mao Q.M., Stockmann R., Prince I.G. and Hearn M.T.W. (1993) Modelling of protein adsorption with non-porous and porous particles in finite baths. *J. Chromatogr. A*, 646: 67–80.
45. Mao, Q.M., Johnston, A., Prince, I.G. and Hearn, M.T.W. (1991) Predicting the performance of non-porous particles in affinity chromatography of proteins. *J. Chromatogr. A*, 548: 147–163.
46. Johnston, A., Mao, Q.M. and Hearn, M.T.W. (1991) Analysis of operating parameters affecting the breakthrough curves in fixed-bed chromatography of proteins using several mathematical models. *J. Chromatogr. A*, 548: 127–145.
47. Mao, Q.M., Prince, I.G. and Hearn, M.T.W. (1993) Optimization of operating parameters for protein purification with chromatographic columns. *J. Chromatogr. A*, 646: 81–89.
48. Ghorpade, A. (2003) New trends in pharmaceutical manufacturing waste treatment. *Swiss Pharma*, 11: 31–35.
49. Ramshaw, C. (1995) "The Incentive for Process Intensification," Proceedings of First International Conference Process Intensification for Chemical Industry. BHR Group, London, 18: 1–12.
50. Stankiewicz, A.I. and Moulijn, J.A. (2000) Process intensification: transforming chemical engineering, process design trends. *Chem. Eng. Prog.* (January), 22–34.
51. Grote, F., Ditz, R. and Strube, J. (2012) Downstream of downstream processing: development of recycling strategies for biopharmaceutical processes. *J. Chem. Technol. Biotechnol.*, 87: 481–497.
52. Eibl, R. and Eibl D. (2011) *Waste Generation, Treatment Options and Environmental Impact of Single-Use Systems in Single-Use Technology in Biopharmaceutical Manufacture*, John Wiley & Sons, Inc, Hoboken, NJ pp. 1–388.
53. Sinclair, A., Leveen, L., Monge, M., Lim, J. and Cox, S. (2008) The environmental impact of disposable technologies. *BioPharm Int.*, 11:1–11.
54. Wells, B. (2007) Guide to disposal of single use bioprocess systems. *BioProcess Int.*, 11: 22–28.
55. Ray, S. and Hirai, M. (2012) Process validation in membrane chromatography, in *Process Validation in Manufacture of Biopharmaceuticasls*, Rathone, A.S. and Sofer, G. eds., CRC Press: Boca Raton, FL, USA, pp. 386–426.
56. Klutz, S., Magnus, J., Lobedann, M., Schwan, P., Maiser, B., Niklas, J., Temming, M. and Schembecker, G. (2015) Developing the biofacility of the future based on continuous processing and single-use technology. *J. Biotechnol.*, 213: 120–130.

57. Goedkoop M.J., Heijungs R., Huijbregts M., De Schryver A., Struijs J. and Van Zelm R (2008) ReCiPe. A Life Cycle Impact Assessment Method which Comprises Harmonised Category Indicators at the Midpoint and the Endpoint Level. http://www.lcia-recipe.net/ (accessed on September 19, 2016).
58. Oganesyan, N., Kim S.H. and Kim, R. (2004) On-column Chemical Refolding of Proteins. *Pharmagenomics* (September) 22–26.
59. Li, M., Su, Z-G. and Janson, J-C. (2004) *In vitro* protein refolding by chromatographic procedures. *Protein Expr. Purif.*, 33: 1–10.
60. Hearn, M.T.W. and Acosta, D. (2001) Applications of novel affinity cassette methods: use of peptide fusion handles for the purification of recombinant proteins. *J. Mol. Recognit.*, 14: 323–369.
61. Kuo W.H. and Chase, H.A. (2010) Process intensification for the removal of poly-histidine fusion tags from recombinant proteins by an exopeptidase. *Biotechnol. Prog.*, 26:142–149.
62. Wood, D.W. (2014) New trends and affinity tag designs for recombinant protein purification. *Curr. Opin. Struct. Biol.*, 26: 54–61.
63. Young C.L., Britton Z.T. and Robinson A.S. (2012) Recombinant protein expression and purification: a comprehensive review of affinity tags and microbial applications. *Biotechnol. J.*, 7: 620–634.
64. Mooney, J.T., Fredericks D.P., Christensen, T. and Hearn, M.T.W. (2014) Removal of cleavage slow points from affinity tags used in the IMAC purification of recombinant proteins. *Biotechnol. J.*, 9: 1023–1032.
65. Jurlander, B., Thim, L., Klausen, N.K., Persson, E., Kjalke, M., Rexen, P., Jørgensen, T.B., Østergaard, P.B., Erhardtsen, E. and Bjørn, S.E. (2001) Recombinant activated factor VII (rFVIIa): characterization, manufacturing, and clinical development. *Semin. Thromb. Hemost.*, 27: 373–384.
66. Sejergaard, L. and Krarup, J. (2009) Method of Controlling a Polypeptide Modification Reaction. WO 2009144318 A1.
67. McCue, J. Osborne, D., Dumont, J., Peters, R., Me I, B., Pierce, G. F., Kobayashi, K. and Euwart, D. (2014) Validation of the manufacturing process used to produce long-acting recombinant factor IX Fc fusion protein. *Haemophilia*, 20: e327–e335.
68. Casademunt, E., Martinelle, K., Jernberg, M., Winge, S., Tiemeyer, M., Biesert, L., Knaub, S., Walter, O. and Schroeder, C (2012) The first recombinant human coagulation factor VIII of human origin: human cell line and manufacturing characteristics. *Eur. J. Haematol.*, 89: 165–176.
69. Wenzel, H., Hauschild, M. and Alting, L. (1997) *Methodology, Tools and Case Studies in Product Development. Environmental Assessment of Products*, Chapman and Hall: New York, 1, pp 1–543.
70. Klöpffer, W. and Grahl, B. (2014) *Life Cycle Assessment (LCA): A Guide to Best Practice*, Wiley-VCH: Weinheim, Germany, pp. 1–457.

71. Pietrzykowski, M., Flanagan, W., Pizzi, V., Brown, A., Sinclair, A. and Monge, M. (2013) An environmental life cycle assessment comparison of single-use and conventional process technology for the production of monoclonal antibodies. *J. Clean Prod.*, 41: 150–162.
72. Sonnemann, G., Castells, F. and Schuhmacher, M. (2004) *Integrated Life-Cycle and Risk Assessment for Industrial Processes*, CRC Press: Boca Raton, FL, USA, pp. 1–362.
73. Roupas, P. (2008) A predictive model of manufacturing processes. *Int Dairy J.*, 18: 741–753.
74. Mula, J., Peidro, I.D., Diaz-Madroneno, M. and Vicena, E. (2010) Mathematical programming models for supply chain production. *Eur. J. Oper. Res.*, 204: 377–390.
75. Ahumada, O. and Villalobos, J.R. (2009) Application of planning models in agri-food supply chains. *Eur. J. Oper. Res.*, 96: 1–20.
76. Banaszewska, A., Cruijssen, F., van der Vorst, J.G.A., Claassen, G.D.H. and Kampman, L.J. (2013) A comprehensive dairy valorisation model. *Dairy Sci.*, 96, 761–779.
77. Chhatre, S., Thillaivinayagalingam, P., Francis, R., Titchener-Hooker, N.J., Newcombe, A.R. and Moore E.K. (2007) Decision-support software for the industrial-scale chromatographic purification of antibodies, *Biotechnol. Prog.*, 23: 888–894.
78. Pleissner, D. and Lin, C.S.K. (2013) Valorisation of food waste in biotechnological processes. *Sust. Chem. Processes*, 1: 21.
79. Marteel, A.E., Davies, J.A., Olson, W.W. and Abraham, M.A. (2003) Green chemistry and engineering: drivers, metrics, and reduction to practice. *Ann. Rev. Environ. Resources*, 28, 401–428.
80. Hubbuch, J., Hansen, S., Skibsted, E., Krarup, J.C., Soerensen, H.H. and Staby, A. (2012) Selective Protein Quantification by Means of Multivariate Evaluation of UV Absorption Spectra, Deutsche Patentanmeldung 10 2010 047 427.4.
81. Hansen, S.K., Skibsted, E., Staby, A. and Hubbuch, J. (2011) A label-free methodology for selective protein quantification by means of absorption measurements. *Biotechnol. Bioeng.*, 108: 2661–2669.
82. Tamburini, E., Marchetti M.G. and Pedrini P. (2014) Monitoring key parameters in bioprocesses using near-infrared technology. *Sensors*, 14, 18,941–18,959.
83. Dicks, A. and Hent, A. (2015) *Green Chemistry Metrics: A Guide to Determining and Evaluating Process Greenness*, Springer Science: Heidelberg, Germany, pp. 1–120.
84. Sheldon, R. (2010) Introduction to green chemistry, organic synthesis and pharmaceuticals, in *Green Chemistry in the Pharmaceutical Industry*, Dunn, P.J., Wells, A. and Williams M.T. eds., Wiley-VCH: Weinheim, Germany, pp. 1–20.

85. Mannucci, P.M., Kempton, C., Millar, C, *et al.* (2013) Pharmacokinetics and safety of a novel recombinant human von Willebrand factor manufactured with a plasma-free method: a prospective clinical trial. *Blood*, 122: 648–657.
86. Winge, S., Dadaian, M., Johansson, E. and Fuchs, B. (2015) A Process for Manufacturing Factor viii Having an Improved Ratio of FVIII:c/FVIII:ag, WO 2015107222 A1.
87. Geigert, J. (2011) *The Challenge of CMC Regulatory Compliance for Biopharmaceuticals and Other Biologics*, Springer: New York, pp. 1–341.
88. Vaňková, K., Antošová, M. and Polakovič, M. (2005) Design and economics of industrial production of fructosyltransferase. *Chem. Pap.*, 59. 441–448.
89. Ernst, S., Garro, O.A., Winkler, S., Venkataraman, G., Langer, R., Cooney, C.L. and Sasisekharan, R. (1997). Process simulation for recombinant protein production: cost estimation and sensitivity analysis for heparinase I expressed in *Escherichia coli*. *Biotechnol. Bioeng.*, 53: 575–582.
90. Farid, S.S. (2007) Process economics of industrial monoclonal antibody manufacture. *J. Chromatogr. B*, 848: 8–18.
91. Werner, R.G. (2004) Economic aspects of commercial manufacture of biopharmaceuticals. *J. Biotechnol.* 113: 171–182.
92. Jimenez-Gonzalez, C., Ponder, C. S., Broxterman, Q.B. and Manley, J.B. (2011) Using the right green yardstick: why process mass intensity is used in the pharmaceutical industry to drive more sustainable processes. *Org. Process. Res. Dev.* 15: 912–917.
93. Cespi, D., Beach, E.S., Swarr, T.E., Passarini, F., Vassura, I., Dunn, P.J. and Anastas, P.T. (2015) Life cycle inventory improvement in the pharmaceutical sector: assessment of the sustainability combining PMI and LCA tools. *Green Chem.*, 17, 3390–3400.
94. McElroy, C.R., Constantinou, A., Jones, L.C., Summerton, L. and Clark, J.H. (2015) Towards a holistic approach to metrics for the 21st century pharmaceutical industry. *Green Chem.*, 17, 3111–3121.
95. Kennedy, M.L. (1997) *Total Cost Assessment for Environmental Engineers and Managers*, John Wiley & Sons, Inc: New York, USA, pp 1–280.
96. Schaltegger, S., Bennett, M. and Burritt, R. eds. (2005) *Sustainability Accounting and Reporting*, Springer Kluwer Publ: Heidelberg, Germany, pp 1–716.
97. Novais, J.L. Titchener-Hooker, N.J. and Hoare, M. (2001) Economic comparison between conventional and disposables-based technology for the production of biopharmaceuticals. *Biotechnol. Bioeng.*, 75: 143–153.
98. Hamers, M.N. (1993) Multiuse biopharmaceutical manufacturing. *Bio/technology*, 11: 561–570.
99. Petrides, D. (2003) Bioprocess design and economics, in *Bioseparations Science and Engineering*, Harrison R.G. ed., Oxford University Press: New York.
100. Heinzle, E., Biwer, A.P. and Cooney, C.L. *Development of Sustainable Bioprocesses: Modeling and Assessment*, John Wiley & Sons: West Sussex, UK, pp. 1–295.

101. Grandison, A.S. (1996) *Separation Processes in the Food and Biotechnology Industries: Principles and Applications*, Woodhead Publishing: Cambridge, UK, pp 1–315.
102. Ahmad, M.M., Hauanb, S. and Przybycien, T.M. (2010) Flowsheet simulation of aqueous two-phase extraction systems for protein purification. *J. Chem. Technol. Biotechnol.*, 85: 1575–1587.
103. Hekmat, D. (2015) Large-scale crystallization of proteins for purification and formulation. *Bioprocess Biosyst. Eng.*, 38: 1209–1231.
104. Heinzle, E. (1999) Ecological and economical evaluation of the production process in early developmental phases. *Initiativen zum Umweltschutz*, 14: 103–113.
105. Barrett, W.M. and van Baten, J (2012) Evaluating process sustainability using flowsheet monitoring. *J. Chem. Eng. Technol.*, 35: 1405–1411.
106. Jimenez-Gonzalez, C., and Woodley, J.M. (2010) Bioprocesses: modeling needs for process evaluation and sustainability assessment Comp. *Chem. Eng.*, 34, 1009–1017.
107. Capparella, J. (2013) Energy benchmarking in the pharmaceutical industry. *Pharm. Eng.*, October, 1–6.
108. Jiménez-González, C. and Overcash, M.R. (2000) Energy sub-modules applied in life-cycle inventory of processes. *Clean Products Processes*, 2: 57–66.
109. Junker, B. (2010) Minimizing the environmental footprint of bioprocesses: part 1: Introduction and evaluation of solid-waste disposal. *BioProcess Int.* (September) 62–71.
110. Gray, J.J. (2004) The interaction of proteins with solid surfaces. *Curr. Opin. Struct. Biol.*, 14: 110–115.
111. Nicoud, L., Owczarz M., Arosio, P. and Morbidelli, M. (2015) A multiscale view of therapeutic protein aggregation: a colloid science perspective. *Biotechnol. J.*, 10: 367–378.
112. Kelley, B. (2007) Very large scale monoclonal antibody purification: the case conventional unit operations. *Biotechnol. Prog.*, 23: 995–1008.
113. Sinnott, R.K. (1993) Costing and project evaluation, in *Coulson and Richardson's Chemical Engineering (Chemical Engineering Design)*, Coulson, J.M. and Richardson, J.F., eds. Pergamon Press: Oxford, pp. 209–244.
114. Remer, D.S. and Idrovo, J.H. (1993) Process equipment cost, biotechnology and pharmaceutical, in *Encyclopedia of Chemical Processing and Design*, 43, McKetta, J., ed., Marcel Dekker Inc., New York: pp. 294–306.
115. Wernet, G., Conradt, S., Isenring, H., Jimenez-Gonzalez, C. and Hungerbuhler, K. (2010) Life cycle assessment of fine chemical production: a case study of Pharmaceutical synthesis. *Int. J. Life Cycle Assess.*, 15: 294–303.
116. Jiménez-González, C., Ollech, C., Pyrz, W., Hughes, D., Broxterman, Q. B. and Neil Bhathela, N. (2013) Expanding the boundaries: developing a streamlined tool for eco-footprinting of pharmaceuticals. *Org. Process Res. Dev.*, 17: 239–246.

117. DeSimone, L. and Popoff, F. (2000) *Eco-efficiency: the business link to sustainable development*, MIT Press: Cambridge, USA. pp. 1–280.

118. Shuaib, M., Seevers, D., Zhang, X., Badurdeen, F., Rouch, K.E. and Jawahir, I.S. (2014) Product sustainability index (ProdSI) a metrics-based framework to evaluate the total life cycle sustainability of manufactured products. *J Ind. Ecol.*, 18, 491–507.

119. Mooney, J.T., Fredericks, D.P., Zhang, C., Christensen, T., Jespergaard, C., Schiødt C.B. and Hearn, M.T.W. (2013) Purification of a recombinant human growth hormone by an integrated IMAC procedure. *Protein Expr. Purif.*, 94: 85–94.

Index